D0498644

ESTUARINE VARIABILITY

Proceedings of the

EIGHTH BIENNIAL INTERNATIONAL ESTUARINE RESEARCH CONFERENCE

University of New Hampshire, Durham

July 28–August 2, 1985

Produced by

The Estuarine Research Federation
with cooperation from
The Society of Wetland Scientists

Sponsored by

Baltimore Gas and Electric Company
Georgia Power Company
UNESCO Marine Science Division
U.S. Environmental Protection Agency
Minerals Management Service
and
National Oceanic and Atmospheric Administration
Estuarine Programs Office
Ocean Assessments Division

ESTUARINE VARIABILITY

Edited by

Douglas A. Wolfe

Ocean Assessments Division
National Oceanic and Atmospheric Administration
Rockville, Maryland

1986

ACADEMIC PRESS, INC.
Harcourt Brace Jovanovich, Publishers
Orlando San Diego New York Austin
Boston London Sydney Tokyo Toronto

ACADEMIC PRESS RAPID MANUSCRIPT REPRODUCTION

ACADEMIC PRESS, INC.
Orlando, Florida 32887

United Kingdom Edition published by
ACADEMIC PRESS INC. (LONDON) LTD.
24–28 Oval Road, London NW1 7DX

Library of Congress Cataloging in Publication Data

International Estuarine Research Conference (8th :
1985 : University of New Hampshire)
Estuarine variability.

Proceedings of the eighth biennial International
Estuarine Research Conference, University of New
Hampshire, Durham, July 28-August 2, 1985; produced
by the Estuarine Research Federation with cooperation
from the Society of Wetland Scientists; sponsored by
Baltimore Gas and Electric Company ... et al.

Includes index.

1. Estuarine ecology—Congresses. 2. Estuaries—
Congresses. I. Wolfe, Douglas A. II. Estuarine
Research Federation. III. Society of Wetland
Scientists (U.S.) IV. Baltimore Gas and Electric
Company. V. Title.
QH541.5.E8I56 1985 574.5'26365 86-47766
ISBN 0–12–761890–2 (alk. paper)

PRINTED IN THE UNITED STATES OF AMERICA

86 87 88 89 9 8 7 6 5 4 3 2 1

CONTENTS

TEMPORAL VARIABILITY IN ESTUARIES

Charles B. Officer, Convenor

PROCESS VARIABILITY IN ESTUARIES

Alejandro Yáñez-Arancibia and John W. Day, Convenors

SPATIAL VARIABILITY IN ESTUARIES

W. C. Boicourt, Convenor

FOREWORD

This is the seventh volume of Estuarine Research Federation biennial conference proceedings, the central themes of which have mirrored trends in the field of estuarine research. The initial two-volume effort treated general estuarine research, with emphasis on biology, geology, and engineering; successive proceedings dealt with estuarine processes, interactions, perspectives, comparisons, and filtration. The current volume directs attention to variability in estuaries. In fact, elements of all these themes have been incorporated in each of the conferences, reflecting the scope of membership interests which embrace natural sciences as well as social and political factors.

The 28 invited papers contained herein treat the current central theme in terms of geographic, temporal, processual, and spatial variability along with modeling. These and other aspects of estuarine variability were represented in numerous concurrent sections of the conference as well. Other treatments of variable biological and physical processes, including long-term data sets, that were given in more than 350 oral and poster presentations will be published separately.

This conference of 519 registrants, unlike those that preceded it, was held by ERF in collaboration with a sister society, the Society of Wetlands Scientists. Their input, along with sessions on environmental health, mangrove systems, politics and coastal sciences, economics of a bay system, scientists and educators, and program administrators' reviews of federal activities in estuaries, served to broaden the program content and to link theoretical to applied activities.

From this mix came perhaps the most well-rounded and mature meeting of the series, a sign that study of estuaries, inherently many-faceted, has further progressed from descriptive science to a comprehension of function, and is beginning to produce enough understanding of natural estuarine systems that human activities may be accommodated to them.

Austin B. Williams
President
Estuarine Research Federation
1983–1985

PREFACE

This book is organized into five sections that correspond to the thematic sessions convened at the Eighth Biennial International Estuarine Research Conference: geographic variability, temporal variability, process variability, spatial variability, and modeling of estuarine variability. The papers in the first four sections draw from examples around the globe to illustrate the sources and nature of variability exhibited by estuarine systems. The section on modeling describes various approaches that have been taken to capture important aspects of spatial and temporal variability in models that can guide estuarine management decisions and focus further research. In the introductory chapter, the editor and the symposium program chairman have collaborated to provide an overview perspective on estuarine variability and to establish a context for the more detailed observations and conclusions presented by the other chapter authors. The individual chapters of this volume tend to emphasize the *differences* that appear at various times within estuaries, or that exist among estuaries at any given time. The overview shows how these local aspects of estuarine variability are, to a large extent, a natural manifestation of processes and interactions that transcend the scales of time and space normally associated with estuarine observations.

The papers presented in each thematic session of the symposium were invited by the convenor to illustrate particular aspects of current research related to estuarine variability. The resultant collection of papers, while clearly not comprehensive in scope, thus addresses a very broad range of topics that should prove to be of significant value to future students in this very active field of scientific research. The papers include broad syntheses of current knowledge and information, punctuated and complemented by detailed accounts of new original research.

The 31 papers submitted for publication in this proceedings volume were subjected to rigorous scientific review, and those that are included here have in many cases undergone substantial revision from the form initially submitted. I wish to thank the reviewers who took time from their already busy schedules to contribute to the technical quality of this volume. The timely publication schedule was achieved through the cooperative efforts of many individuals, including the authors themselves. Special thanks go to Zenobia Neugebauer, of NOAA's Ocean Assessments Division for expert secretarial support and to Angela Ferri, Angie Zile, and Jill Anderson, of Lithocomp Inc., Bethesda, Maryland, for accomplishing the typesetting and page composition of the papers on a very tight production schedule. The financial and technical support of the Ocean Assessments Division, National Oceanic and Atmospheric Administration, must be recognized for providing me with the opportunity to participate in this project, and many of the resources needed to complete it. Finally, I am grateful to Nancy Wolfe for her usually gracious toleration of the many overtime hours consumed by this effort.

Douglas A. Wolfe

OVERVIEW

ESTUARINE VARIABILITY: AN OVERVIEW

Douglas A. Wolfe

Ocean Assessments Division
National Oceanic and Atmospheric Administration
Rockville, Maryland

and

Björn Kjerfve

Belle W. Baruch Institute
for Marine Biology and Coastal Research
Department of Geology and Marine Science Program
Columbia, South Carolina

Abstract: Estuaries are inherently variable in time and space. In this paper we discuss and interpret that variability in terms of broad principles of system function. Variability encompasses the diversity of systems in space and their changeability in time. Changes occur on a continuum of spatial and temporal scales, in which defined systems at one scalar level are embedded successively within other, larger scales. Estuarine systems in various stages of development are affected by regional climatological, geological, hydrological, and oceanographic patterns. Over geologic time, estuarine systems undergo major evolutionary transitions due to sedimentation and coastal sea level changes. On shorter time scales, the character of estuarine systems may be substantially altered by tectonic events or by major climatologic events such as El Niño episodes, droughts, or hurricanes. Such events may cause an estuarine system to shift to a new stable point about which homeostasis occurs until the next disrupting episode. These principles are illustrated with numerous examples drawn from current research on estuaries.

Introduction

Estuaries occur at the interfaces between the earth's land masses and oceans from the Arctic to the south polar seas. All present-day estuaries are geologically young, having been formed when sea level reached its present level approximately 5,000 years ago. Estuaries represent a geologically ephemeral transition zone, in which many features of geomorphology, water circulation, biogeochemistry and ecology are varied and diverse, leading to description and classification of different categories of estuarine systems with unique combinations of characteristics (Hedgpeth 1957; Casper 1967; Pritchard 1967). These various types of estuarine systems all undergo fluctuations in response to exogenous inputs of materials or energy that may themselves vary over wide ranges of magnitude, and exhibit different periodicities over a broad range of time scales. Estuaries are thus diverse in space and changeable in time. *Estuarine Variability,* as discussed here, encompasses both this spatial diversity and temporal changeability. These attributes, as we shall see, are inseparably intertwined.

Geographic Variability

On a geographic scale, variability of estuaries arises from differences in coastal geomorphology, drainage basin and hydrologic characteristics, and climatology. These differences are exemplified well by the contrast between the estuaries entering the Yellow Sea from Korea and the more mature, sediment-laden estuaries of the Yangtze and Yellow rivers on China's coastal plain (Schubel *et al.* 1986). The larger river discharges and smaller tidal ranges of the Chinese estuaries contribute to much lower filtering efficiencies for incoming sediments than are observed in the Korean estuaries. Kjerfve (1986) examined a large suite of lagoonal estuaries, and showed that the character and magnitude of oceanic exchange through the lagoon entrance (relative to the freshwater discharge) are the principal determinants for lagoon characteristics. Historical differences of estuarine exchange and runoff may produce major differences in erosion and sedimentation patterns, and account for variations in coastal habitat characteristics, even among adjacent estuaries (Kelley *et al.* 1986). The geomorphological and hydrological variations among estuaries are probably a major source of ecological differences. For example, Deegan *et al.* (1986) characterized all the estuarine systems around the mainland periphery of the Gulf of Mexico in terms of their primary hydrographic and topological features, and identified through correlative analysis several possible sources of variation in biological productivity.

To a large extent, the important exchange characteristics of contemporary estuarine systems reflect the present height of sea-level in relation to watershed characteristics and morphology of coastal basins. Over geological time, sea-level has varied considerably, in response to cycles of glaciation and interglaciation. These cycles drastically alter the estuarine character, or "estuarinity" of any particular basin. For example the Gulf of Maine (Campbell 1986) exhibits mixing, flushing, and nutrient cycling characteristics similar to those of an estuary, as a result of the constraining effects of the submarine topography on coastal shelf circulation. Some 15,000 years before present, however, when sea level was much lower the basin probably exhibited coastal lagoonal characteristics (Campbell 1986; Kjerfve 1986). The rate of coastal sea level change depends on tectonic factors and on climatological factors that may be influenced by human activities. Examples are the CO_2 balance (Hansen *et al.* 1985) that is changing due to the combustion of fossil fuels and the clearing of forests (human factors); vulcanism and seismicity (plate tectonics); and subsidence (due to coastal land/water use or simply compaction of sediments). These factors exhibit considerable variability, such that along uplifting coastlines apparent sea level is falling (e.g., -13 mm y^{-1} at Juneau, Alaska), whereas in Louisiana, where subsidence occurs, apparent sea level is rising at a rate of 8 mm y^{-1} (Stevenson *et al.* 1986). In parts of Scandinavia apparent sea level is falling at a rate of > 10 mm y^{-1} as a result of isostatic rebound (Stacey 1969).

Geographic variations among estuaries also result from differences among coastal basins in their orientation with respect to global or regional circulation patterns, both in the atmosphere and the ocean. The attitude of a lagoonal system with respect to prevailing wind directions (Kjerfve 1986) or of an estuary with respect to exposure to oceanic wave energy (Kelley *et al.* 1986) may drastically influence the geomorphologic (and consequently the ecologic) characteristics of the system. We may conclude, therefore, that estuarine variability among geographic areas is driven by phenomena on a global scale, and results in a variety of estuarine systems with differing morphological characteristics. Forcing from global- to meso-scale climatology and oceanic circulation, coupled with latitudinal variations of solar insolation and tidal forcing, superimposes upon these geographical differences a high degree of temporal and spatial variability in estuarine processes.

Temporal and Spatial Variability of Estuaries

Estuaries are inherently variable. Temporally, they respond to a combination of forcing functions in a variable fashion over a broad spectrum of frequencies. To compute and consider time-averages or instantaneous "snap-shots" of a state of an estuary function can be highly misleading. Rather, it becomes necessary to consider time series and estimate the degree of variability. Figure 1 illustrates, for example, the variations of water level in North Inlet estuary (South Carolina) over a year, corresponding only to meteorological and seasonal forcing. If the effects of surface waves, tides, multiannual variations and glacial-interglacial sea level variations etc. were included, however, the picture would be much more complicated. Variations on these different temporal scales all contribute to important system characteristics.

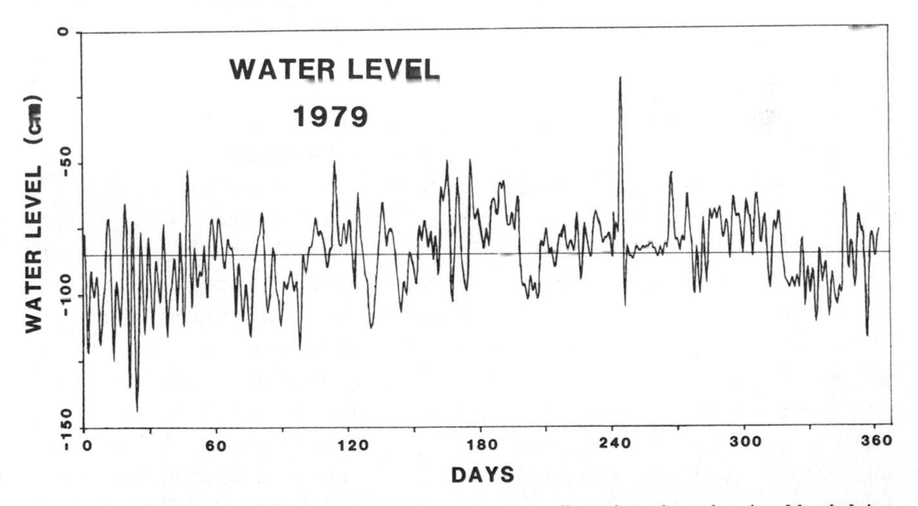

Figure 1. Variations in meteorologically and seasonally induced sea level in North Inlet, South Carolina, during 1979. Hourly measurements have been subjected to a 48-hour low-pass filter to remove wave and tide variations.

Temporal variations interact, however with spatial variations, which are also important determinants of system processes and structure. For example, the time-averaged (net) velocity distribution in a cross-section of the North Inlet estuary, South Carolina (Fig. 2) results directly from circulation processes, dominated by residual tidal circulation (Kjerfve in press). Consideration of the temporal variability of the net velocity gives a much different distribution (Fig. 2B), however, and more complete understanding of the synergism between temporal and spatial variations can be obtained.

Within any single estuary, interannual variability probably has its origins primarily in large-scale climatic or oceanographic forcing. The importance of climate as a determinant of stability in ecosystems is well substantiated. Climate may exhibit short-term variations (5-10 years) as a result of major volcanic eruptions which introduce atmospheric turbidity and depress annual mean surface temperatures (Watt 1969). Variability in seasonal climate regimes are related to rainfall patterns which in turn, successively affect runoff and hydroperiod, the variability in nutrient cycles (Peterson *et al.* 1986) and planktonic populations (Tyler 1986). The timing and duration of seasonal rainfall and runoff are critical for germination and establishment of marsh plants, and successional changes initiated during extreme years may persist over several years (Zedler and Beare 1986). Interannual variability in nutrient inputs and distributions may not be paralleled by subsequent changes in plankton populations due to the influence of other factors, such as the importance of nutrient-exchange with bottom sediments in shallow systems, light limitation in turbid systems, or predation (Ustach *et al.* 1986). The timing, intensity, and duration of wind regimes probably affect the interannual variability of fish larval transport into estuarine nursery areas from coastal or shelf waters (Pietrafesa *et al.* 1986; Nelson *et al.* 1977), and wind direction also influences the spatial distribution of larvae among different estuarine subsystems (Pietrafesa *et al.* 1986). Similarly, phytoplankton distributions may be markedly influenced by seasonal and interannual rainfall patterns, as observed by Tyler (1986) in the Chesapeake Bay. Related climatic effects have also been observed in estuarine benthic populations and communities (Dugan and Livingston 1982; Holland 1985; Nichols 1985).

Episodic events, such as sustained strong onshore or offshore winds, extreme rainfall, hurricanes, or drought can represent major perturbations to the stability of a particular estuarine system. Such events may initiate, through unusually low recruitment processes or high mortalities, successional changes that lead to significantly different ecosystems. This "new" system may persist until it is again perturbed by another unusual event. The importance of such episodic events in determining interannual variability requires that long-time records (multiple decades) are necessary to understand these processes (Peterson *et al.* 1986).

Depth and bathymetric features strongly affect estuarine circulation and mixing (Uncles *et al.* 1986; Simpson 1986; Dyer and New 1986; Powell and Cloern 1986). These features contribute to the formation of fronts that may recur repeatedly in the same location on similar tidal stages (Simpson 1986) and

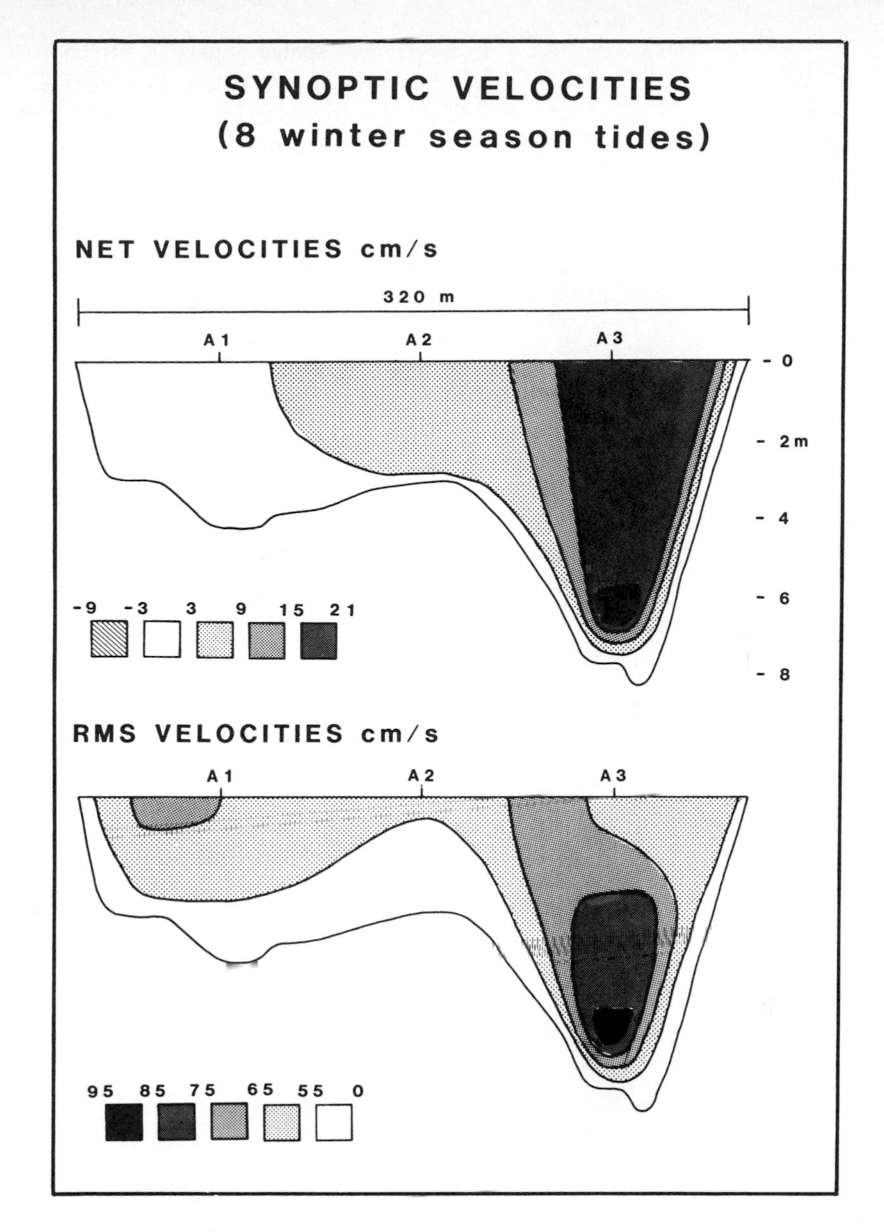

Figure 2. *A) Variability of current velocities in a cross-section of the North Inlet estuary, South Carolina. Distribution of tidally averaged (net) current velocity, showing strong assymetry in the longitudinal circulation. Data were collected every 1.5 hours over eight tidal cycles. B) Root mean square deviations from the net velocities showing distribution of significant differences in current magnitudes across the section.*

may be associated with persistent patterns of phytoplankton distribution (Powell and Cloern 1986). On more restricted spatial/temporal scales, topographic steps or obstructions may induce periods or regions of intensive mixing that can be entrained within larger convection cells and moved about an estuary (Dyer and New 1986). This intermittency in estuarine mixing induces substantial variability in related transport processes on short spatial and temporal scales (in-estuary, less than one tidal cycle). On longer time scales energy variations associated with neap-spring tidal cycles can affect sediment suspension and redeposition rates (Kraeuter and Wetzel 1986) and modify general circulation and estuarine mixing (Granat and Richards 1986; Schroeder and Wiseman 1986).

The seasonal cycle of primary productivity, dependent on both temperature and light, is directly coupled to related cycles of decomposition and nutrient turnover. Thus the seasonal cycles of chlorophyll-a in the water column (Tyler 1986; Ustach *et al.* 1986) or of salt-marsh productivity (Randerson 1986; Wiegert 1986) may reflect not only the nutrient cycles in the water column (Ustach *et al.* 1986; Facco *et al.* 1986), but also the processes of nutrient exchange and metabolism in estuarine sediments (Twilley 1986; Kraeuter and Wetzel 1986; Rizzo and Wetzel 1986). The absolute numerical scale and short-term periodicity, as well as the amplitude of the seasonal productivity fluctuation may also be influenced by other factors, such as availability of nutrient inputs from agricultural activities (Ustach *et al.* 1986) or industrial pollution inputs (Facco *et al.* 1986). These same productivity cycles are of course also influenced directly by seasonal cycles of grazing and predation among the holoplanktonic and meroplanktonic components of the zooplankton. Williams *et al.* (1986) have hypothesized that selective grazing by zooplankton induces diurnal variability in the species composition, total numbers, and depth distribution of phytoplankton observed in the Patuxent River Estuary.

These multiple sources of variability and interaction in estuarine systems pose special problems for design of sampling programs. Rizzo and Wetzel (1986) demonstrated that annual mean rate estimates (e.g., of respiration in estuarine sediments) obtained from random monthly measurements were similar to those obtained from detailed hourly measurement programs. The seasonal trends suggested by these two sampling approaches, however, were radically different. Livingston (In Press) has also examined the effects of extrapolating means from data sets with various sampling frequencies. The main conclusion of these studies is that sampling densities must be carefully fitted to the intrinsic variability of the observational data and to the periodicity of the cycles of interest. Though seemingly obvious, this conclusion is often overlooked in estuarine sampling programs (i.e., with arbitrarily selected monthly or quarterly sampling periods).

Modeling Estuarine Variability

Several types of models have been applied to estuarine studies. These include conceptual models, statistical and empirical models, hydrodynamic circulation-dispersion models, scaled physical models, and ecosystem models.

Conceptual models are usually descriptive, such as that used to describe the relative estuarinity, over time, of the Gulf of Maine (Campbell 1986), or the "salinity gap" model used by Zedler and Beare (1986) to explain salt marsh successional control processes. Statistical and empirical models have received broad application in estuarine science to demonstrate relationships among different characteristics and to evaluate the significance of such correlations. Some examples include analyses of correlations among estuarine characteristics related to fisheries productivity (Deegan *et al.* 1986), and the testing of extent of change or difference before and after experimental manipulation or other (natural or human-caused) perturbation (Ustach *et al.* 1986). Hydrodynamic processes in estuaries can be simulated by scaled physical models, e.g., of the Chesapeake Bay (Granat and Richards 1986; Richards and Granat 1986), or by mathematical models that simulate physical circulation and dispersion processes in an estuary (Blumberg 1977; Klein and Galt 1986).

The scaled physical modeling approach, used by Granat and Richards (1986) to simulate circulation in Chesapeake Bay, relies on actual physical simulation of the estuarine basin's morphology and bathymetry, its riverine inflows, and tidal flushing characteristics. The circulation effects of varying any of these features can then be determined by direct measurement in the physical model. Verification is achieved by comparing synoptic patterns in the scaled system model with those in the actual estuary and checking for similarities of events or processes in localized situations. Spatial variability is estimated by direct measure (logistically much easier in the model than in the prototype), and temporal variability is generated by the nature of the tidal and riverine input conditions selected by the modelers for testing.

Conventional hydrodynamic modeling (Blumberg 1977) relies on solution of well-known physical deterministic equations, and the use of observational data on estuarine morphology and bathymetry, riverine inputs and tidal forcing. The circulation or dispersion characteristics of the estuary are determined from the simulation computations with variable input conditions. Verification is achieved by comparing the synoptic patterns predicted by the model with those observed in the estuary, and by confirming detailed features in selected regions. Spatial variability is estimated directly from the model results, and temporal variability generated by different tidal or riverine input conditions can be estimated from sequential runs of the model. The finite-element hydrodynamic modeling technique described by Klein and Galt (1986) provides estimates of spatial and temporal distribution of estuarine salinity based on freshwater and oceanic inputs to the system from land runoff and exchange with adjacent water bodies. Initial applications of this finite-element approach demonstrated its general reliability at predicting observed salinity distributions and its ease of adaptability to other boundary conditions (i.e., different estuaries).

Ecosystem models are used to simulate the flow of energy and materials among different abiotic and biotic compartments of the system. These models generally assume kinetic relationships among the compartments (e.g., Campbell 1986) and are thus fundamentally different from hydrodynamical modeling approaches. Ecosystem models have been applied frequently to nutrient-phytoplankton-zooplankton-consumer systems in open water (Campbell 1986) and to energy flow in salt-marsh ecosystems (Wiegert 1986; Randerson 1986). Each of these simulation modeling approaches incorporates hypotheses that explain how the modelled system works. Manipulation of the model represents an exercise of the hypotheses to generate predictions about that system. The validity of these predictions can be tested by further observations or experiments, thus lending support to the credibility of the model's hypothetical basis. The predictions generated by the models are highly useful to support decisions regarding modification or management of the modeled system, or regarding priorities for further research into the functioning of that system. The various modeling approaches treat variability differently, depending on the objectives of the management problem and of the modeler.

Ecosystem energetics models (e.g., Wiegert 1986; Randerson 1986) are designed primarily to examine the temporal changes (and associated structural implications) that may ensue—given different hypothetical values for selected processes or structural relationships in the salt marsh ecosystem. In general the salt marsh is assumed to be spatially homogeneous, and the nature of spatial variability is deduced by adjusting the system variables to fit specific subsystems of interest. These models have proven valuable in estimating both the timing and the quantities of material exported from the marsh system to the surrounding estuary. The general ecosystem modeling approach can be extended to include a number of cells, each representing a different spatial ecosystem element that can receive and/or provide energy and material to adjacent cells (Costanza *et al.* 1986). This spatially-articulated approach addresses both spatial and temporal aspects of variability, but for large and diverse systems, may be very difficult to calibrate and verify due to the absence of cell-specific data for important model variables. Nonetheless, preliminary application of this complex modeling approach, using a 3,000-cell formulation, to the extensive and diverse freshwater-estuarine-marsh system of southern Louisiana (Costanza *et al.* 1986) has shown promise for describing the spatial character of habitat changes over a 20-year period of changing land and water usage.

The foregoing simulation modeling approaches can incorporate the temporal variability of important driving functions or processes, and generate either by physical or mathematical simulation a predicted spatial-temporal pattern for the dependent system variables of interest to the modeler. These and other related simulation techniques are probably the only practical and effective means by which the relative importance of the multiple sources of variability in estuaries and other complex natural systems can be examined. Such models then, can provide an effective mechanism for focusing estuarine research on those processes that appear to be most important in determining the spatial and temporal variability of estuarine characteristics.

Sources of Variability in Estuaries

Variability results from motion and change. The spatial diversity and temporal variability within and among estuaries reflect changes that have occurred and are occurring simultaneously over a continuous, wide range of different spatial and temporal scales. In this section, we discuss a broader context for the multitude of processes that lead to estuarine variability on a global scale, and illustrate how some of these processes are related to broader, unifying concepts.

Human perceptions of variability are undoubtedly influenced strongly both by the time span of one's attention to a particular process and by the breadth of the spatial horizon within which one studies that process. We tend to orient our observations around easily discernible (definable) aggregations of matter and/or process, and to be irritated or "surprised" (Holling 1985) when we encounter either unexpected discontinuities within the boundary of a defined system or unexpected continuities between different defined systems. A growing body of literature, however, addresses the broad inter-relationships of temporal and spatial changeability in natural systems (cf. Holling 1985; Pimm 1984; Allen and Starr 1982; Paine and Levin 1981; Levin 1978; May 1977).

Four concepts warrant brief mention in the context of understanding variability or predicting change in estuaries or other natural systems:

(1) Hierarchy (Allen and Starr 1982). Systems contain embedded subsystems *ad infinitum*. The time-space scales of processes that are pertinent to system function vary for different levels in the hierarchy, and events that ensue during the natural course at one level may be viewed as exogenous at another, lower level. For example, large-scale changes in sea level (whether they result from local differences in subsidence, long-period cyclical global temperature fluctuations, from episodic seismic or volcanic events associated with plate tectonic processes, or even from episodic showers of large meteors), can drastically affect the integrity of a salt-marsh system (Stevenson *et al.* 1986) or the character of an entire estuary (Campbell 1986).

(2) Stability and Resiliency (Holling 1973, 1985). Ecosystems exhibit stability—i.e., they tend to function at or near an equilibrium condition, with internal self-regulating homeostatic controls that prevent radical departure from the equilibrium state. More than one stability domain may exist—i.e., the system may fluctuate with different amplitudes or frequencies about different equilibrium values when system components or fluxes are modified either by external forces or through natural evolutionary processes. The tendency of a system to retain its structure and behavior about a particular stable point is termed resiliency (Holling 1985).

Stable oscillations of estuarine variability occur over a broad spectrum of scales including the diurnal oscillations of selected phytoplankton in surface waters (Williams *et al.* 1986), daily and tidal fluctuations in oxygen metabolism in estuarine sediments (Rizzo and Wetzel 1986), fluctuations of currents and circulation processes with periodicities ranging from daily to neap-spring cycles (Simpson 1986; Wilson *et al.* 1986; Uncles *et al.* 1986), and seasonal fluctuations of nutrients and biomass (Facco *et al.* 1986; Campbell 1986; Kraeuter and Wetzel 1986; Ustach *et al.* 1986; Tyler 1986). Over a certain range of inputs and forcing functions, the system retains its stability. For example, modest increases in the inputs and ambient levels of nutrients resulting from watershed modification did not seem to shift either the stable point or degree of variability for estuarine phytoplankton production (Ustach *et al.* 1986). Similarly (based on physical model studies), modest deepening of channel entrances to lower Chesapeake Bay was predicted not to have major effects on distributions or fluctuations of salinity (Richards and Granat 1986).

(3) Dispersion of spatial patterns (Levin 1978). A change introduced at one location in a system may be relocated through diffusion or advection to other parts of the system, thus affecting spatial patchiness at any time. For example, Dyer and New (1986) described the formation and movement of mixing zones within an estuary as a function of current intensity with respect to obstructions and/or topographic features on the bottom.. Tyler (1986) showed how phytoplankton transport and distribution depends on estuarine circulation and thus (ultimately) on rainfall and climate. Pietrafesa *et al.* (1986) demonstrated that movement of larval fish depends on wind events of proper direction and intensity, leading to subsequent spatial variability of juvenile distribution, and (by extension) to variable year-class strength in the adult population. On a larger spatial-temporal scale, the seasonal timing and movement of atmospheric pressure systems was seen to influence local regimes of circulation and nutrient flux (Peterson *et al.* 1986). The spatial diversity introduced by such diffusion or advection processes promotes, and in turn is heightened by, the asynchronous evolution (or succession) of system structure in different parts of the estuary.

(4) Succession (Odum 1969; Hollings 1985). Ecosystems undergo an orderly process of development leading toward increased stability, biomass, and structured interaction among components (Odum 1969). However, this same process may also lead to decreased resiliency (Hollings 1985), such that the system is more susceptible to impact or change from exogenous events that may lead to establishment of new stability domains with different spatial-temporal behavior and variabilities. For example, the Pacific salmon fishery has experienced enhancement and greater predictability through effective protection and supplementation of spawning stocks. The resultant increase in investment and fishing effort, however, has become more dependent on hatchery stocks that may be more vulnerable to collapse (Larkin 1979). Stocks of clupeoid fishes, such as Pacific sardines and Atlantic herring, appear to undergo large cyclic changes in abundance on a period of 50-100 y (Steele and Henderson 1984). During periods of peak abundance, these populations are probably more vulnerable to sudden change as a result of shifts in food supply or predation (exploitation).

We may view succession, then, as any ordered process of change and development that a system exhibits during establishment and maintenance of a stability domain. Periods of stability are interrupted by discontinuities introduced by events exogenous to the particular subsystem of focus; but these events in fact may be regular features of some larger system in the hierarchy. For example, the familiar seasonal succession of phytoplankton-zooplankton and larval fish production is interrupted by the annual temperature/stratification event; and the process of marsh succession and development may be interrupted and steered toward new stability domains by intermittent extremes in climatology (Zedler and Beare 1986). Similarly, the stabilizing process of accretion in a *Spartina* marsh may be reversed and interrupted by local changes in circulation leading to erosion, or by large-scale submergence (Stevenson *et al.* 1986).

We have seen that variability is the result of movement and change over time. Movements occur at each level in a hierarchy of nested systems, and at any level those movements may generate stable oscillations among components of the system and its subsystems. In estuaries, we are growing increasingly familiar and comfortable with such oscillations related to earth orbit around the sun, lunar orbit around the earth, rotation of the earth, and feedback loops between nutrients and plankton, or predators and prey organisms. However, hints of much longer-period oscillations still tweak our imaginations, e.g., the 50-yr cycles of abundance for some fish populations (Steele and Henderson 1984).

At each level in the hierarchy, systems exhibit structure that undergoes ordered change (succession) on its own scale at rates commensurate with its position in the hierarchal arrangement. These changing structural patterns create a spatial diversity (or variability) of regional and local systems with differing histories and asynchronous processes. We know that variability in the structure and function of estuaries is influenced by the past (and ongoing) flows of magma and plate tectonics that influence geologic structure (Kelley *et al.* 1986) and sea level changes (Stevenson *et al.* 1986; Kjerfve 1986; Campbell 1986); by global water cycles and resultant erosionary-depositional processes (Kelley *et al.* 1986; Schubel *et al.* 1986); by successional phases in relatively long-lived marsh communities (Stevenson *et al.* 1986; Zedler and Beare 1986); and by daily changes in availability of food organisms (Williams *et al.* 1986) or nutrients (Rizzo and Wetzel 1986; Kraeuter and Wetzel 1986).

Once initiated, these processes of succession and the accompanying phases of structural development are also subject to dispersion in space, further contributing to spatial variability. The scales of such movement include diffusion of nutrients or oxygen across a sediment-water interface and into the water column (Rizzo and Wetzel 1986; Kraeuter and Wetzel 1986), shifting of mixing zones within an estuary (Dyer and New 1986; Simpson 1986), and (on a much larger scale) submergence/uplift and continental drift. The orderly development of stable structure at any system level is disrupted when its resiliency is exceeded as a result of changing conditions either within the subsystem or at a higher system level. Humans generally describe such events at a higher system level as "exogenous" (Hollings 1985) or as "externalities" (Hardin 1985). Such "externalities" might include (in approximate order of increasing system scale): 1) the appearance of newly photosynthesized substrates within a heterotrophic microbial community, 2) the sunset, 3) the full moon, 4) onset of autumn or winter, 5) episodic storm-flushing of fine sediments accumulated over years-to-decades in an estuary , 6) an El Nino event, or 7) a period of volcanic eruptions and earthquakes. Hardin (1985) suggested that the concepts of "externalities" and "side effects" are rhetorical ploys of the human intelligent order, designed to prevent distraction from one's intended accounting of reality. In dealing with this variability, scientists and managers clearly must remain cognizant of the potential interactions at many different scales in the embedded system hierarchy. To overlook the 50-year or 100-year storm event could easily lead to surprises of catastrophic proportion.

Holling (1985) has incorporated these "externalities" into a logical continuum of system structure and function (Fig. 3). His example is drawn from classical forest succession (Odum 1969) with concepts of pioneer species, stages of climax and senescence, and disruption by fire. Analogies with marine and estuarine systems are readily drawn, however, at all levels of system hierarchy (Steele 1985). Figure 3 emphasizes the principle that the present dominant human perspectives of stability and orderly succession may represent only phases in a broader continuum that also includes structural discontinuities and periods of disorganization. Elaboration of this continuum both upwards and downwards into the hierarchy of systems and spatial-temporal scales leads us

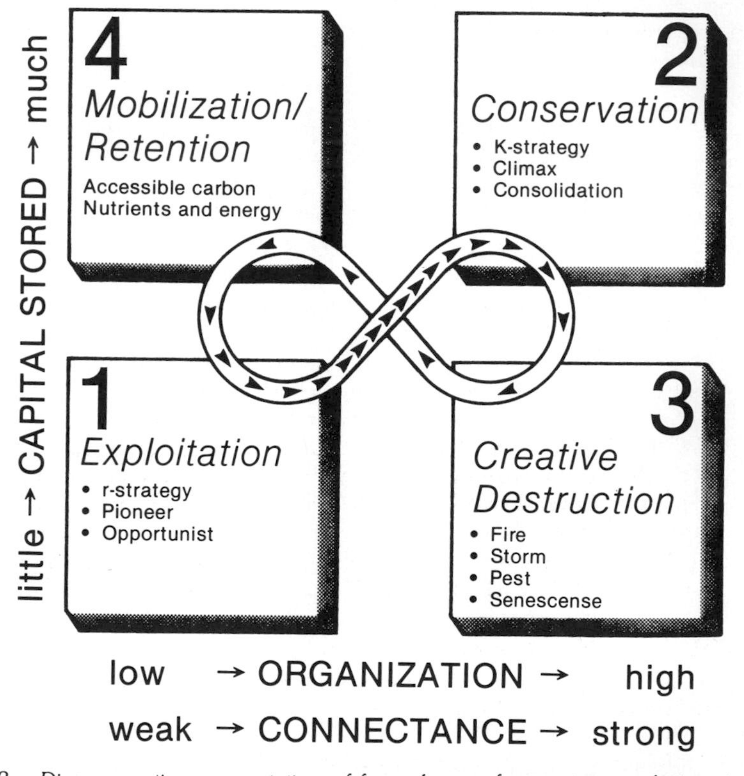

Figure 3. Diagrammatic representation of four phases of ecosystem evolution, in relation to the amount of stored energy or capital (x-axis) and the degree of connectedness (y-axis). The arrowheads trace the sequence of the ecosystem cycle, and the intervals between arrowheads reflect the relative rates of change (short interval represents slow change; long interval represents rapid change). From Holling (1985). Used with permission of the publisher.

eventually and naturally to a concept of evolutionary change, with homeostatic mechanisms operating at all levels in the hierarchy.

Estuaries, like other systems, are evolving. Based on an analysis of the geographic distribution and diversity of major taxonomic categories of fauna over time, Simpson (1969) concluded: "If indeed the earth's ecosystems are tending toward long-range stabilization or static equilibrium, three billion years has been too short a time to reach that condition." As transition zones between land and sea, estuaries are particularly susceptible to cyclical oscillations, whether on a daily period (e.g., tidal cycles) or an epochal one (e.g., glaciation-interglaciation). System reponses to cataclysmic episodic events are superimposed on these cyclical patterns, and variability is manifest on every applicable spatial and temporal scale of the ongoing process. Analyses of estuarine variability over long time scales and large geographic scales offer considerable insight into the processes that influence the characteristics of today's estuarine systems. Effective use and environmental management of these evolving transition areas in the future depend, however, on how well we understand the intrinsic system variability on shorter-scales of time and space.

References Cited

Allen, T. F. H. and T. B. Starr. 1982. *Hierarchy Perspectives for Ecological Complexity*. University of Chicago Press. Chicago. 310 pp.

Blumberg, A. F. 1977. Numerical tidal model of the Chesapeake Bay. *J. Hydraulics Div., ASCE* 103(HY1):1-10.

Campbell, D. E. 1986. Process variability in the Gulf of Maine—A macroestuarine environment. pp. 261-275. *In:* D.A. Wolfe (ed.), *Estuarine Variability*. Academic Press, New York.

Caspers, H. 1967. Estuaries: analysis of definitions and biological considerations. pp. 68. *In:* G. W. Lauff (ed.), *Estuaries*. Publ. 83. Am. Assoc. Advancement Sci., Washington, D.C.

Costanza, R., F. H. Sklar and J. W. Day, Jr. 1986. Modeling spatial and temporal succession in the Atchafalaya/Terrebonne Marsh/Estuarine complex in South Louisiana. pp. 387-404. *In:* D. A. Wolfe (ed.), *Estuarine Variability*. Academic Press, New York.

Deegan, L. A., J. W. Day, Jr., J. G. Gosselink, A. Yáñez-Arancibia, G. S. Chávez and P. Sánchez-Gil. 1986. Relationships among physical characteristics, vegetation distribution and fisheries yield in Gulf of Mexico estuaries. pp. 83-100. *In:* D. A. Wolfe (ed.), *Estuarine Variability*. Academic Press, New York.

Dugan, P. J. and R. J. Livingston. 1982. Long-term variation of macroinvertebrate assemblages in Apalachee Bay, Florida. *Estuar. Coastal Shelf Sci.* 14:391-403.

Dyer, K. R. and A. L. New. 1986. Intermittency in estuarine mixing. pp. 321-339. *In:* D. A. Wolfe (ed.), *Estuarine Variability*. Academic Press, New York.

Facco, S., D. Degobbis, A. Sfriso and A. A. Orio. 1986. Space and time variability of nutrients in the Venice Lagoon. pp. 307-318. *In:* D. A. Wolfe (ed.), *Estuarine Variability*. Academic Press, New York.

Granat, M. A. and D. R. Richards. 1986. Chesapeake Bay Physical Model investigations of salinity response to Neap-Spring Tidal Dynamics: A descriptive examination. pp. 447-462. *In:* D. A. Wolfe (ed.), *Estuarine Variability*. Academic Press, New York.

Hansen, J., G. Russell, A. Lacis, I. Fung and D. Rind. 1985. Climate response times: dependence on climate sensitivity and ocean mixing. *Science* 229:857-859.

Hardin, G. 1985. Human ecology: the subversive, conservative science. *In: Science as a Way of Knowing. II-Human Ecology*. Amer. Zool. 25:467-476.

Hedgpeth, J. W. 1957. Estuaries and lagoons. II. Biological aspects. pp. 673-693. *In:* J. W. Hedgpeth (ed.), *Treatise on Marine Ecology and Paleoecology I.* Memoir No. 67, Geological Society of America, Boulder, Colorado.

Holland, A. F. 1985. Long-term variation of macrobenthos in a mesohaline region of the Chesapeake Bay. *Estuaries* 8(2A):84-92.

Holling, C. S. 1973. Resilience and stability of ecological systems. *Ann. Rev. Ecol. Syst.* 4:1-23.

Holling, C. S. 1985. Resilience of ecosystems: local surprise and global change. pp. 228-269. *In:* T. F. Malone and J. G. Roederer (eds.), *Global Change:* The Proceedings of a Symposium sponsored by the International Council of Scientific Unions (ICSU) during its 20th General Assembly in Ottawa, Canada on September 25, 1985. Cambridge University Press.

Kelley, J. T., A. R. Kelley, D. F. Belknap and R. C. Shipp. 1986. Variability in the evolution of two adjacent bedrock-framed estuaries in Maine. pp. 21-42. *In:* D. A. Wolfe (ed.), *Estuarine Variability.* Academic Press, New York.

Kjerfve, B. 1986. Comparative oceanography of coastal lagoons. pp. 63-81. *In:* D. A. Wolfe (ed.), *Estuarine Variability.* Academic Press, New York.

Kjerfve, B. In Press. Circulation and salt flux in a well-mixed estuary. *In:* J. van de Kreeke (ed.), Physics of Shallow Estuaries and Bays. Springer Verlag, Berlin.

Klein, C. J. III and J. A. Galt. 1986. A screening model framework for estuarine assessment. pp. 483-501. *In:* D. A. Wolfe (ed.), *Estuarine Variability.* Academic Press, New York.

Kraeuter, J. N. and R. L. Wetzel. 1986. Surface sediment stabilization-destabilization and suspended sediment cycles on an intertidal mudflat. pp. 203-223. *In:* D. A. Wolfe (ed.), *Estuarine Variability.* Academic Press, New York.

Larkin, P. A. 1979. Maybe you can't get there from here: history of research in relation to management of Pacific salmon. *J. Fish. Res. Bd. Canada* 36:98-106.

Levin, S. A. 1978. Pattern formation in ecological communities. pp. 433-470. *In:* J. H. Steele (ed.), *Spatial Pattern in Plankton Communities.*

Livingston, R. J. In press. Field sampling in estuaries: the relationship of scale to variability. *Estuaries.*

May, R. M. 1977. Thresholds and breakpoints in ecosystems with a multiplicity of stable states. *Nature* 269:471-477.

Nelson, W. R., M. C. Ingham and W. E. Schaaf. 1977. Larval transport and year-class strength of Atlantic menhaden, *Brevoortia tyrannus.* U.S. Natl. Mar. Fish. Serv. *Fish. Bull.* 75(1):23-41.

Nichols, F. H. 1985. Abundance fluctuations among benthic invertebrates in two Pacific estuaries. *Estuaries* 8(2A):136-144.

Odum, E. P. 1969. The strategy of ecosystem development. *Science* 164:262-270.

Peterson, D., D. R. Cayan and J. F. Festa. 1986. Interannual variability in biogeochemistry of partially mixed estuaries: dissolved silicate cycles in northern San Francisco Bay. pp. 123-138. *In:* D. A. Wolfe (ed.), *Estuarine Variability.* Academic Press, New York.

Paine, R. T. and S. A. Levin. 1981. Intertidal sandscapes: disturbance and the dynamics of pattern. *Ecol. Monogr.* 51:145-178.

Pietrafesa, L. J., G. S. Janowitz, J. M. Miller, E. B. Noble, S. W. Ross and S. P. Epperly. 1986. Abiotic factors influencing the spatial and temporal variability of juvenile fish in Pamlico Sound, North Carolina. pp. 341-353. *In:* D. A. Wolfe (ed.), *Estuarine Variability.* Academic Press, New York.

Pimm, S. L. 1984. The complexity and stability of ecosystems. *Nature* 307:321-326.

Powell, T. M., J. E. Cloern and R. A. Walters. 1986. Phytoplankton spatial distribution in South San Francisco Bay: mesoscale and small-scale variability. pp. 369-383. *In:* D. A. Wolfe (ed.), *Estuarine Variability.* Academic Press, New York.

Pritchard. D. W. 1967. What is an estuary: physical viewpoint. pp. 3-5. *In:* G. H. Lauff (ed.), *Estuaries.* Publ 83. Am. Assoc. Advancement Sci., Washington, D.C.

Randerson, P. F. 1986. A model of carbon flow in the *Spartina Anglica* Marshes of the Severn Estuary, U.K. pp. 427-446. *In:* D. A. Wolfe (ed.), *Estuarine Variability*. Academic Press, New York.

Richards, D. R. and M. A. Granat. 1986. Salinity redistributions in deepened estuaries. pp. 463-482. *In:* D. A. Wolfe (ed.), *Estuarine Variability*. Academic Press, New York.

Rizzo, W. M. and R. L. Wetzel. 1986. Temporal variability in oxygen metabolism of an estuarine shoal sediment. pp. 227-239. *In:* D. A. Wolfe (ed.), *Estuarine Variability*. Academic Press, New York.

Schroeder, W. A. and W. J. Wiseman, Jr. 1986. Low-frequency shelf-estuarine exchange processes in Mobile Bay and other estuarine systems on the northern Gulf of Mexico. pp. 355-367. *In:* D. A. Wolfe (ed.), *Estuarine Variability*. Academic Press, New York.

Schubel, J. R., H-T Shen and M-J Park. 1986. Comparative analysis of estuaries bordering the Yellow Sea. pp. 43-62. *In:* D. A. Wolfe (ed.), *Estuarine Variability*. Academic Press, New York.

Simpson, G. G. 1969. The first three billion years of community evolution. pp. 162-177. *In: Diversity and Stability of Ecological Systems*. Brookhaven Symp. Biol., No. 22. Brookhaven National Laboratories, Upton, New York.

Simpson, J. H. and W. R. Turrell. 1986. Convergent fronts in the circulation of tidal estuaries. pp. 139-152. *In:* D. A. Wolfe (ed.), *Estuarine Variability*. Academic Press, New York.

Stacey, F. D. 1969. *Physics of the Earth*, Wiley, New York, 324 pp.

Steele, J. H. 1985. A comparison of terrestrial and marine ecological systems. *Nature* 313:355-358.

Steele, J. H. and E. W. Henderson. 1984. Modeling long-term fluctuations in fish stocks. *Science* 224:985-987.

Stevenson, J. C., L. G. Ward and M. S. Kearney. 1986. Vertical accretion in marshes with varying rates of sea level rise. pp. 241-258. *In:* D. A. Wolfe (ed.), *Estuarine Variability*. Academic Press, New York.

Tyler, M. A. 1986. Flow-induced variation in transport and deposition pathways in the Chesapeake Bay: the effect on phytoplankton dominance and anoxia. pp. 161-175. *In:* D. A. Wolfe (ed.), *Estuarine Variability*. Academic Press, New York.

Twilley, R. R. and W. M. Kemp. 1986. The relation of denitrification potentials to selected physical and chemical factors in sediments of Chesapeake Bay. pp. 277-293. *In:* D. A. Wolfe (ed.), *Estuarine Variability*. Academic Press, New York.

Uncles, R. J., M. B. Jordan and A. H. Taylor. 1986. Temporal variability of elevations, currents and salinity in a well-mixed estuary. pp. 103-122. *In:* D. A. Wolfe (ed.), *Estuarine Variability*. Academic Press, New York.

Ustach, J. F., W. W. Kirby-Smith and R. T. Barber. 1986. Effect of watershed modification on a small coastal plain estuary. pp. 177-192. *In:* D. A. Wolfe (ed.), *Estuarine Variability*. Academic Press, New York.

Watt, K. E. F. 1969. A comparative study on the meaning of stability in five biological systems: insect and furbearer populations, influeza, Thai hemorrhagic fever, and plague. pp. 142-149. *In: Diversity and Stability in Ecological Systems*. Brookhaven Symp. Biol., No. 22. Brookhaven National Laboratories, Upton, New York.

Wiegert, R. G. 1986. Modeling spatial and temporal variability in a salt marsh: sensitivity to rates of primary production, tidal migration and microbial degradation. pp. 405-426. *In:* D. A. Wolfe (ed.), *Estuarine Variability*. Academic Press, New York.

Williams, J., J. Foerster and F. Skove. 1986. Diurnal variation of surface phytoplankton in the Patuxent River. pp. 193-202. *In:* D. A. Wolfe (ed.), *Estuarine Variability*. Academic Press, New York.

Wilson, R. E., M. E. C. Vieira and J. R. Schubel. 1986. Tidal rectification in the Peconic Bays Estuary. pp. 153-160. *In:* D. A. Wolfe (ed.), *Estuarine Variability*. Academic Press, New York.

Zedler, J. B. and P. A. Beare. 1986. Temporal variability of salt marsh vegetation: the role of low-salinity gaps and environmental stress. pp. 295-306. *In:* D. A. Wolfe (ed.), *Estuarine Variability*. Academic Press, New York.

ESTUARINE GEOGRAPHIC VARIABILITY

Scott W. Nixon, Convenor

VARIABILITY IN THE EVOLUTION OF TWO ADJACENT BEDROCK-FRAMED ESTUARIES IN MAINE

Joseph T. Kelley
Maine Geological Survey
Augusta, Maine

and

Alice R. Kelley, Daniel F. Belknap, and R. Craig Shipp
Department of Geological Sciences
University of Maine
Orono, Maine

Abstract: Casco and Saco Bays are large, adjacent, bedrock-framed embayments in southern Maine. Bottom samples, underwater vibracores, bridge borings, and 500 km of high resolution seismic reflection profiles were used to compare the Holocene evolution of the bays. Each has experienced glaciation, marine inundation, subaerial emergence, as well as contemporary drowning in the past 13,000 years, but they nonetheless differ profoundly in bedrock composition, shoreline environments, bottom sediment, and bathymetry. Casco Bay is divided by islands of metamorphic rock ridges which parallel the coast. Glacial mud, derived from erosion of the "soft" rocks, filled the bay 12,000 years ago, but was deeply gullied and eroded during later emergence. Recent drowning left ravines partly filled with organic-rich sediment (and gas) and interfluves of bare rock. Bluff erosion of marine clay provides most sediment for today's ubiquitous mudflats. Saco Bay is more exposed to waves than Casco Bay and is bordered by granitic rocks. Sand ultimately derived from glacial erosion of granite, formed barrier spits offshore during emergence. Recent sea level rise has resulted in a smooth sandy seafloor, and sand beaches with salt marshes are the predominant coastal environments.

Introduction

As transitional regions between strictly terrestrial and marine environments, estuaries are ephemeral over geological time. Although considerable documentation exists regarding estuarine variability due to latitude, climate, or fluvial influences, less has been reported on variation in estuaries as a result of their geologic evolution (Roy 1984). Yet, throughout the Quaternary Period (past 2 million years), eustatic sea level changes have repeatedly shifted estuarine environments back and forth across continental shelves and coastal plains. These movements have regularly altered the character of individual estuaries by varying their geometry, river gradient and discharge, and wave climate; which in turn have influenced salinity and circulation as well as biota. Along the U.S. East Coast, the evolution of estuaries has nowhere been more rapid than in northern New England. Coastal Maine has experienced a glaciation, two marine inundations and a withdrawal of the sea within the past 13,000 years. As a result, Maine's estuaries are highly variable in geomorphology depending on local bedrock structure and sediment supply (Kelley 1985a). This paper describes estuaries in barrier spit (Saco) and drowned river (Casco) embayments that are adjacent to one another. The estuarine offshore stratigraphy suggests that the embayments experienced a similar history until the start of the most recent marine transgression about 9,000 years ago. Due to glacially-induced drainage derangement at that time, sand was introduced to Saco Bay resulting in the formation of barrier spit estuaries while muddy environments continued to dominate Casco Bay.

Location and Previous Work

The study area in south central coastal Maine extends 50 km northeast from Biddeford Pool to Bailey Island and 10 km inland from the coast to waterfalls or the termination of tidally influenced shoreline (Fig. 1). The bedrock geology is dominated by granitic plutons bordering and inland from Saco Bay and northeast trending metamorphic rocks surrounding Casco Bay. The seaward boundary of the study area is at the 80 m isobath, although most discussion will focus on the region shallower than 40 m.

South central Maine is dominated by the coastal re-entrants of Saco and Casco bays. Though adjacent, these bays differ in physiography and belong to separate coastal compartments (Kelley and Timson 1983; Kelley 1985a). Saco Bay is a member of the Arcuate Embayments compartment which extends south from Cape Elizabeth into Massachusetts. Embayments within this compartment possess long sandy beaches which protect relatively large salt marshes (Table 1). Small rivers empty into the larger marsh areas while larger rivers, like the Saco, possess steep gradients and valley walls, with a very restricted intertidal area. Sand beaches and salt marshes are the most abundant intertidal environments of Saco Bay (Table 1).

Casco Bay and the embayments extending north to central Maine, constitute the Indented Shoreline compartment (Kelley 1985a). This region is characterized by long bedrock peninsulas separating narrow, elongate estuaries (Fig. 1). The dominant intertidal environments of this compartment include mudflats and salt marshes on the protected landward margins of peninsulas, while bedrock is exposed on the outer headlands (Table 1).

With the exception of the surficial geology of estuarine margins, there has been little systematic geologic research on any of Maine's estuaries (Fefer and Shattig 1980). Farrell (1970, 1972) compared the hydrography of the Saco and Scarboro Rivers and studied the bottom sediment characteristics of Saco Bay. The Saco Estuary is only partly mixed, whereas the smaller Scarboro Estuary is completely mixed by the region's 2.5-m tides. Several meters of muddy peat fill in the Scarboro River mouth, while clean, well sorted sand and occasional outcrops of rock floor the mouth of the Saco River and Saco Bay. The U.S. Army Corps of Engineers collected several vibracores from Saco Bay and determined that, at least where sampled, the surficial sand of Saco Bay represents a thin veneer overlying a thicker section of Pleistocene mud and sand (Williams 1984,

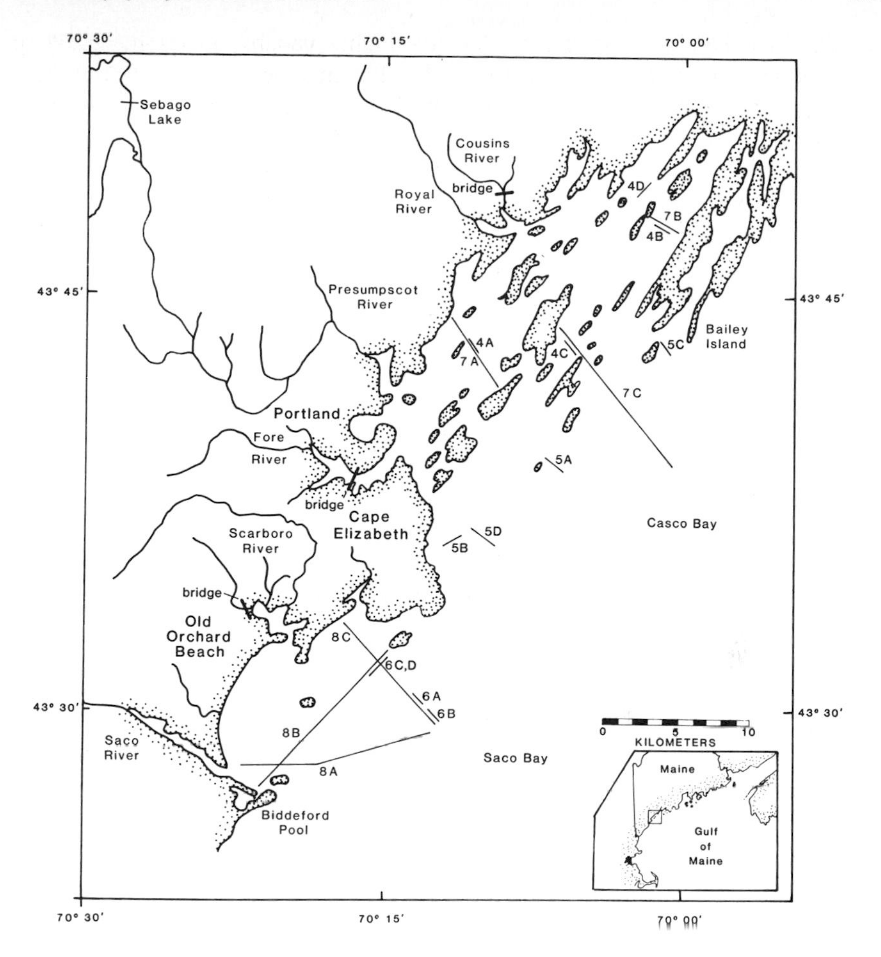

Figure 1. *Location of the study area. Of the 500 km of seismic profile lines, those with figure numbers are shown. The location of bridge crossings is also indicated.*

pers. comm.). Hulmes (1981), in a vibracoring study of Biddeford Pool, also found a relatively thin (<5 m) layer of Holocene littoral sand and mud unconformably overlying Pleistocene sediment.

Despite its size and proximity to Maine's largest city, Casco Bay is essentially unstudied. A few water samples suggest that the bay's estuaries, with their relatively small discharge (Table 1) and 2.75 m tidal range, are well mixed except during periods of high discharge (Larson and Doggett 1979; Fefer and Shattig 1980). Bottom sediment in the bay is generally muddy (Larson *et al.* 1983; Folger *et al.* 1975), though gravel deposits occur near abundant rock outcrops (U.S. Environmental Protection Agency 1983), and were observed during a recent submersible dive (Kelley, unpublished notes).

Table 1. Geological Variation Between Casco and Saco Bays, Maine

Estuary	River Discharge Area (km^2)	River Discharge ($m^3 s^{-1}$)	Estuarine Environmental Area (km^2) Salt Marsh	Mud Flat	Beach	Bedrock	Offshore Bottom	Natural Gas	Mass Movements
Saco R.	3349	77	0.94(36%)	0.25(9%)	1.37(52%)	0.81(3%)			
Scarboro[1]	<50	<5.0	20.2 (78%)	0.67(3%)	4.98(19%)	<.01(<1%)			
Saco Bay							sandy rocky	absent	absent
Fore[2]	72	<5.0	0.78(22%)	2.72(76%)	0.45(1%)	0.03(1%)			
Presumpscot	1549	28.4	1.06(14%)	6.08(82%)	0.05(1%)	0.20(3%)			
Royal	365	7.8	0.25(15%)	1.31(78%)	0.03(2%)	0.09(5%)			
Cousins	<50	<5.0	0.37(17%)	1.78(80%)	0.03(1%)	0.04(2%)			
Casco Bay							muddy rocky	common	common

[1]includes developed beach
[2]does not include filled mud flats

Late Quaternary Geological History of South Central Maine

The Quaternary events of south central Maine can be divided into four time periods:

(1) Before 13,000 years before present (BP). No unconsolidated materials described in south central Maine are older than the Wisconsinan Glaciation (Thompson 1979). On the basis of interbedded till and marine sediments (including shells), the melting ice margin is inferred to have reached the present coast of Maine 20 km south of Saco Bay around 13,800 BP (Smith 1985).

(2) 13,000 BP to 11,500 BP. Because the crust of Maine was depressed by the weight of the ice sheet, glaciomarine sediments were deposited in the wake of the retreating ice. The glaciomarine deposits are highly variable in thickness and texture and are collectively referred to as the Presumpscot Formation (Bloom 1963; Thompson and Borns 1985). Generally the Presumpscot Formation is thicker in valleys than on hills and grades from a coarse-grained (sand and mud layers with dropstones) ice-proximal base, to a finer-grained (silt and clay layers), ice-distal upper section (Bloom 1963; Thompson 1979; Smith 1985). The landward extent of the marine inundation (the "marine limit") is well documented, but the timing is not precisely known (Belknap and Borns 1986).

(3) 11,500 BP to 9,000 BP. Following dissipation of the ice sheet, the crust rebounded and the sea withdrew to a point below present sea level (Bloom 1963; Schnitker 1974). The regression was probably rapid since no prominent shoreline features have been identified. The drainage network that developed on the emergent seafloor was at least partly deranged since most pre-glacial bedrock valleys were filled with sediment. As the new streams began to erode glacial deposits, an increasingly steep gradient permitted sandy deltas to be built into the ocean (Oldale 1985; Belknap *et al.* 1985a). It has been estimated that the sea dropped as low as 50 m (Oldale 1985) to 65 m (Schnitker 1974) below present sea level although no time for the maximum lowstand has yet been obtained.

(4) 9,000 BP to the present. At the time of maximum emergence the rate of land rebound and sea level rise were equal before the present transgression began. Evidence for this includes a barrier island developed on a delta offshore northern Massachusetts (Oldale 1985) and an apparent wave-cut shoreline at many locations offshore Maine (Kelley and Kelley 1985; Shipp 1985). Sea level rose to about 4 m below present sea level at 4,000 BP (Belknap *et al.* 1985b) as determined by dating basal salt marsh peats throughout Maine. It continues to rise, albeit unevenly (Anderson *et al.* 1984), to the present day.

The ideal stratigraphic section that has resulted from the events described above includes the following: a) glacial till unconformably overlying bedrock; b) glaciomarine sediment (Presumpscot Formation) often grading upward from the till and becoming a massive mud near its top; and c) Holocene sand and mud unconformably overlying a gullied weathered Presumpscot Formation surface. While this stratigraphic column is rarely seen completely in outcrop on land, variations of it have been inferred on the basis of seismic reflection profiles elsewhere in Maine (Schnitker 1974; Ostereicher 1965; Knebel and Scanlon 1985; Belknap *et al.* 1985a; Kelley *et al.* 1985).

Methods

During 1983 and 1984 approximately 500 km of high resolution seismic reflection profiles were collected from Saco and Casco Bays (Kelley *et al.* 1985). While the Casco Bay data were collected largely with a Raytheon RTT 1000A 3.5 kHz profiler, the observations in Saco Bay simultaneously utilized an Ocean Research Equipment Geopulse "boomer" system. In addition, 150 bottom samples were collected from Saco Bay on a 1-km^2 grid extending from within the river mouths to the 100-m isobath (Kelley 1985b). Navigation was by Loran C as well as radar and visual sightings off numerous islands and buoys.

Following collection, the bottom samples were analyzed for grain size, organic and carbonate carbon, and heavy minerals by standard methods (Carver 1971). The seismic data were interpreted and digitized at a 20:1 or 50:1 vertical exaggeration.

Stratigraphy of the Embayments

Casco Bay

On the basis of bathymetry (Fig. 2) Casco Bay may be divided into three regions. The inner region includes the estuaries of the Fore, Presumpscot, Royal, and Cousins Rivers. Though each of these estuaries possesses a shallow channel, they are predominantly intertidal mudflats with subordinate salt marshes (Table 1). There are no significant beaches in these estuaries and the marshes fringing the land are typically thin (<1 m thick) and eroding (Kelley, unpublished field notes) except at their most landward locations.

Because shallow water prevented seismic profiling here, bridge borings were used to interpret the stratigraphy (Maine Dept. of Transportation 1948, 1983; Fig. 3). Despite their small discharge (Table 1) the Fore and Cousins Rivers possess the deepest bedrock valleys entering Casco Bay (Fig. 3). Bridge borings met refusal in till at depths greater than 60 m. The till was overlain by up to 40 m of stratified sand and mud of the Presumpscot Formation. This material possessed a dried, weathered surface that was unconformably covered by several meters of Holocene mud. In contrast with the small estuaries, the larger rivers, the Presumpscot and Royal, pass over falls within several kilometers of Casco Bay. Although modern mud is accumulating in their channels today, bedrock commonly exists less than 10 m beneath the surface, and locally has been removed to maintain a 1.5-m navigation channel (U.S. Army Corps of Engineers 1980).

Seaward of the inner bay, seismic reflection profiles were used to interpret the stratigraphy. Interpretation of the seismic data is discussed at length elsewhere (Kelley *et al.* 1985; Belknap *et al.* 1985a; Knebel and Scanlon 1985). While one cannot be certain of the identity of the remotely-sensed reflectors discussed below, intertidal vibracores and subtidal bridge borings lend confidence to the interpretation (Belknap *et al.* 1985a; Knebel and Scanlon 1985). In addition, the four stratigraphic units discussed below outcrop on the margin of the bay and its islands, and may be traced directly from land to beneath the sea.

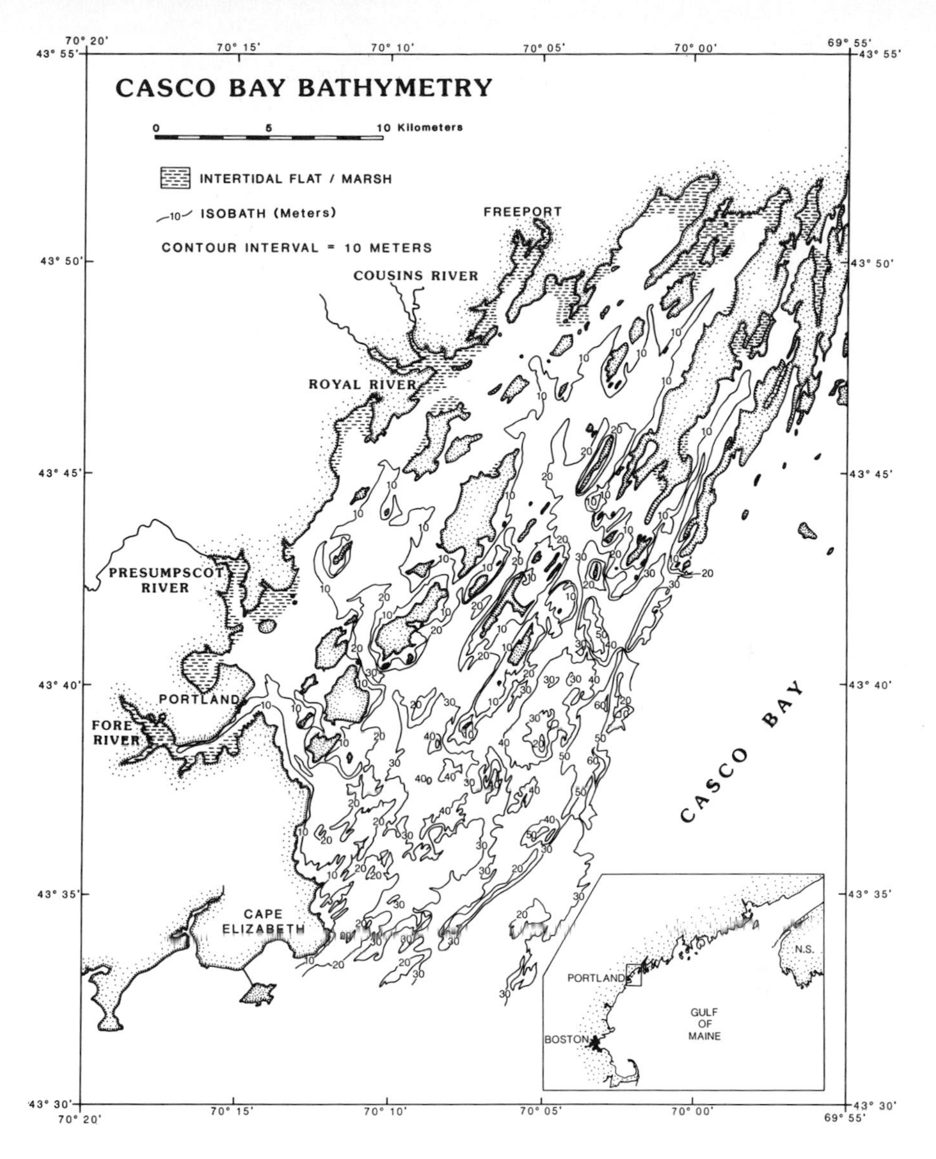

Figure 2. Bathymetry of Casco Bay.

The stratigraphic units recognized are:

(1) Bedrock: The bedrock surface was recognized as a strong, high relief reflector (Figs. 4-6). No internal reflectors were seen within this unit and it could always be traced from the seismic record to an outcrop on land.

(2) Till: The surface of till was also a strong reflector and it was often difficult to pinpoint the contact between its base and the bedrock surface (Fig. 5).

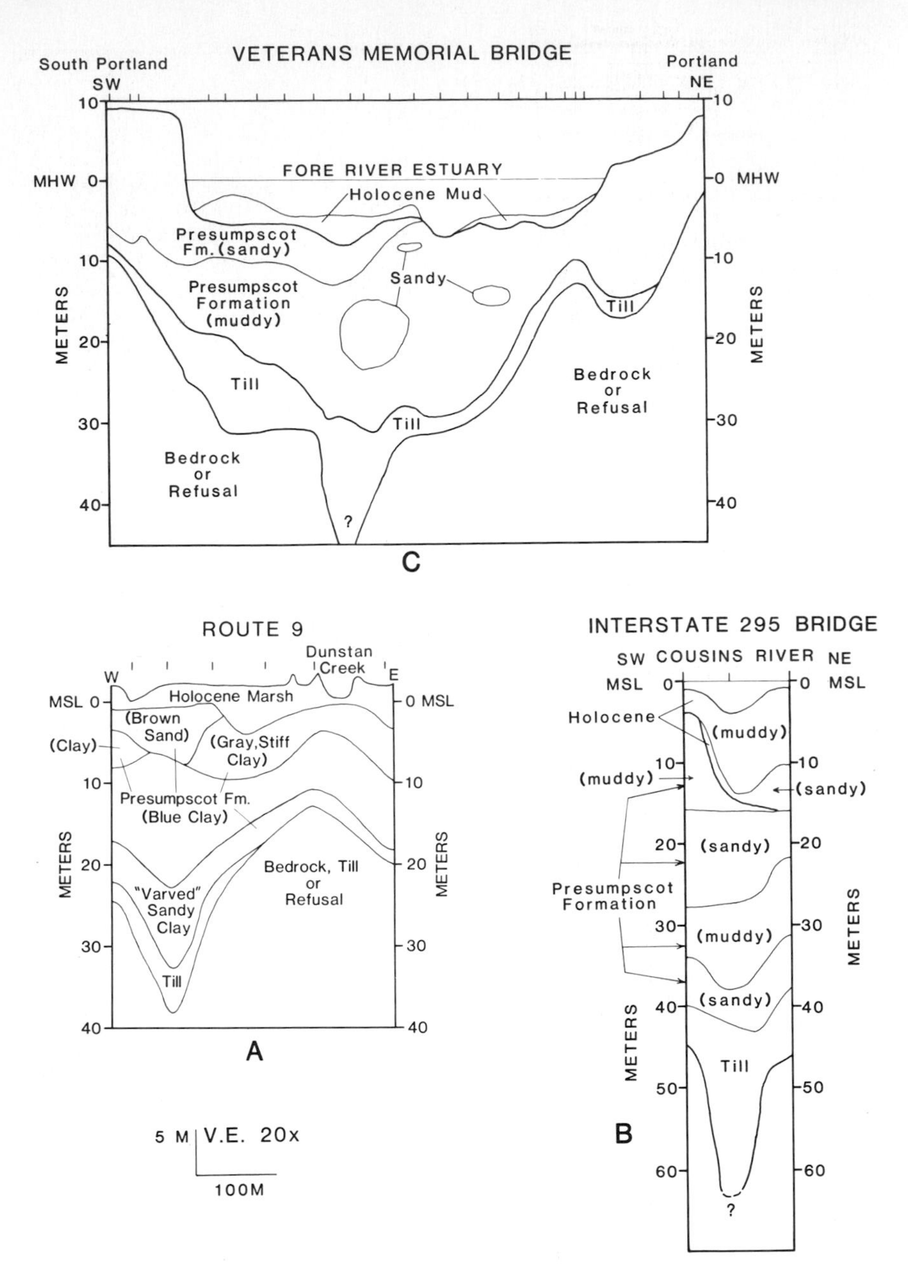

Figure 3. *Stratigraphy of three estuaries within the study area. These small estuaries possess the deepest bedrock valleys in the study area. The estuaries with larger river discharge have bedrock channels with less than 10 m of Holocene sediment. A, B) Redrawn from Maine Department of Transportation borehole logs (1948, 1983). Vertical lines at top of cross section locate boreholes. C) Redrawn from Bloom (1963).*

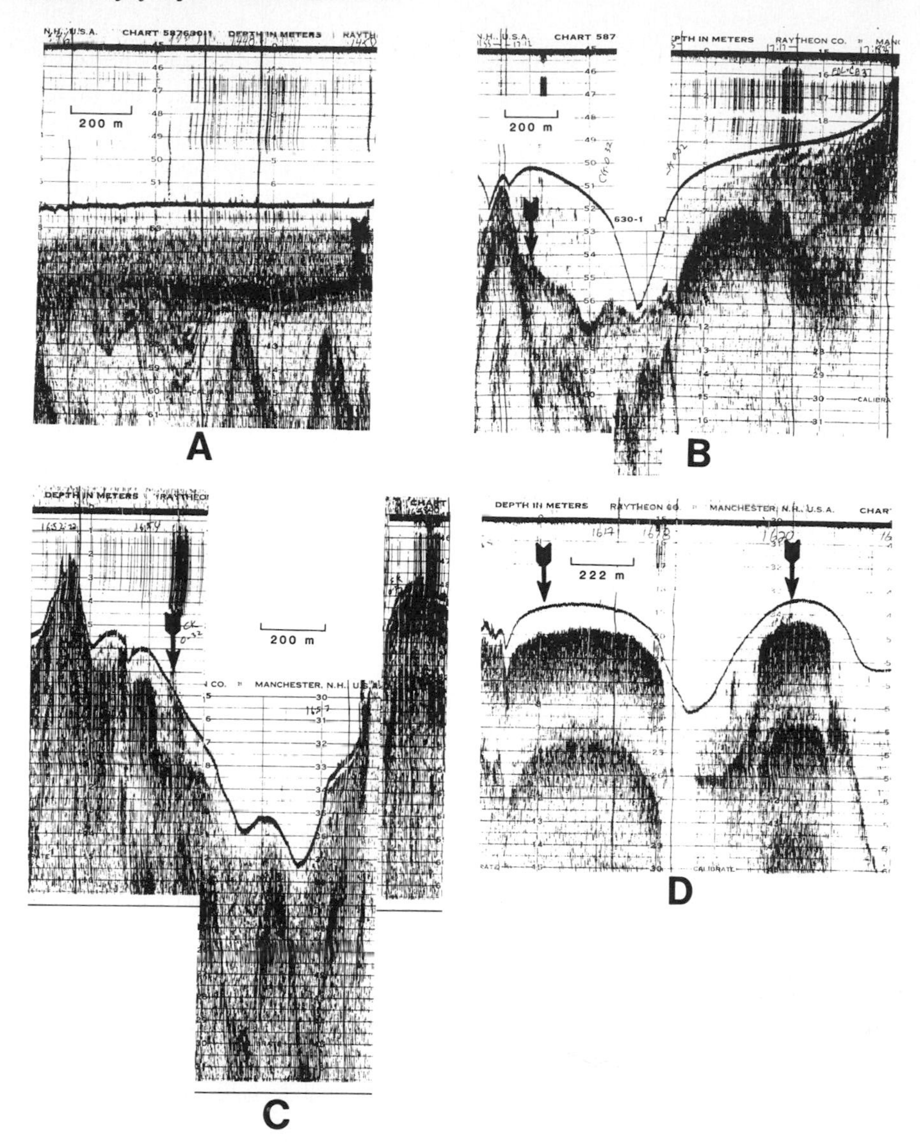

Figure 4. Sections of original seismic record of central Casco Bay. Location shown in Figure 1. A) An arrow indicates the strong reflector at the surface of the Presumpscot Formation. Holocene mud is above, and bedrock is the lowest reflector. B) Acoustically transparent, Holocene mud overlies the reflector representing the unconformity on the Presumpscot Formation surface (arrow). Natural gas blocks the record to the side of the tidal channel. C) Apparent slump block (arrow). Mound at bottom of channel may be slumped debris. D) Gas zones beneath bulges in seafloor (arrows) are similar to those described elsewhere (Schubel and Schiemer 1973). Many slumps were observed in this region and may be related to loss of sediment shear strength caused by natural gas. Horizontal lines are 1-m apart on each record.

Reflectors within the till were short and discontinuous, resulting in a chaotic pattern (Figs. 5, 6).

(3) Presumpscot Formation: The glaciomarine mud was recognized as an acoustically transparent unit easily resolved from underlying till or bedrock. Near its base, reflectors drape the underlying topography while the unit's surface is marked by a strong, flat or occasionally channel-shaped reflector (Figs. 4, 7, 8). This surface is inferred to represent a regressive or transgressive unconformity (Kelley *et al.* 1985).

(4) Holocene: The upper unit in much of Casco and Saco Bays is Holocene sand or mud. Within Casco Bay the acoustically transparent nature of the unit suggests it is mud (Figs. 4, 5). In most of Saco Bay, however, the upper unit is acoustically opaque to the 3.5 kHz seismic system, and bottom samples indicate it is sand (Farrell 1972; Fig. 6). In Casco Bay, within the Holocene, a concave down reflector often obscures channels cut into the Presumpscot Formation. On the basis of comparison with similarly described features elsewhere (Schubel and Schiemer 1973) and limited coring (Schnitker, pers. comm.) this reflector is inferred to be natural gas (Figs. 4, 7).

The central region of Casco Bay possesses a relatively flat, shallow seafloor which is protected from the open sea by a prominent chain of islands and peninsulas (Fig. 2). Channels greater than 10-m deep extend through the chain of islands and dissect the shallow central region. Bottom samples from this area (Larson *et al.* 1983; Folger *et al.* 1975) as well as notations on navigation charts indicate a soft muddy bottom. The transparent nature of the upper section of the seismic record also suggests mud of probable Holocene age (Fig. 4) on the seafloor. The Holocene sediment is about 5-m to 10-m thick in much of this part of the bay, though it thins over bedrock pedestals and near channels (Fig. 7). Beneath this unit, a strong flat reflector extends across much of the central bay (Figs. 4, 7). This reflector truncates the reflectors beneath it and is inferred to be the unconformity at the Presumpscot Formation surface.

In places the Presumpscot Formation appears to have been cut by gullies during the lowstand of sea level and no strong surface reflector is seen. Instead, natural gas obscures the seismic record and seems to fill the channels cut into the Presumpscot Formation (Kelley *et al.* 1985). Natural gas is also frequently associated with bulges or domes on the seafloor and slump deposits (Fig. 4; Kelley *et al.* 1985). Although it is difficult to envision the gas causing seafloor bulges, gas could easily migrate into topographic highs and lead to slumps by reducing sediment shear strength on slopes (Booth *et al.* 1985). In the central region of Casco Bay the Presumpscot Formation is up to 20-m thick and its base usually rests on bedrock. Where till was observed it was a thin deposit on the flank of a bedrock ridge (Fig. 7).

The outer area of Casco Bay possesses a deep irregular bottom (Fig. 2). Seismic profiles show abundant exposed bedrock there with generally thin (<10 m) pockets of sediment confined to low regions between bedrock ridges (Figs. 4, 5, 7). The sediment, where observed from submersible, is coarse grained (Kelley, unpublished notes), and its surface commonly dips in a seaward

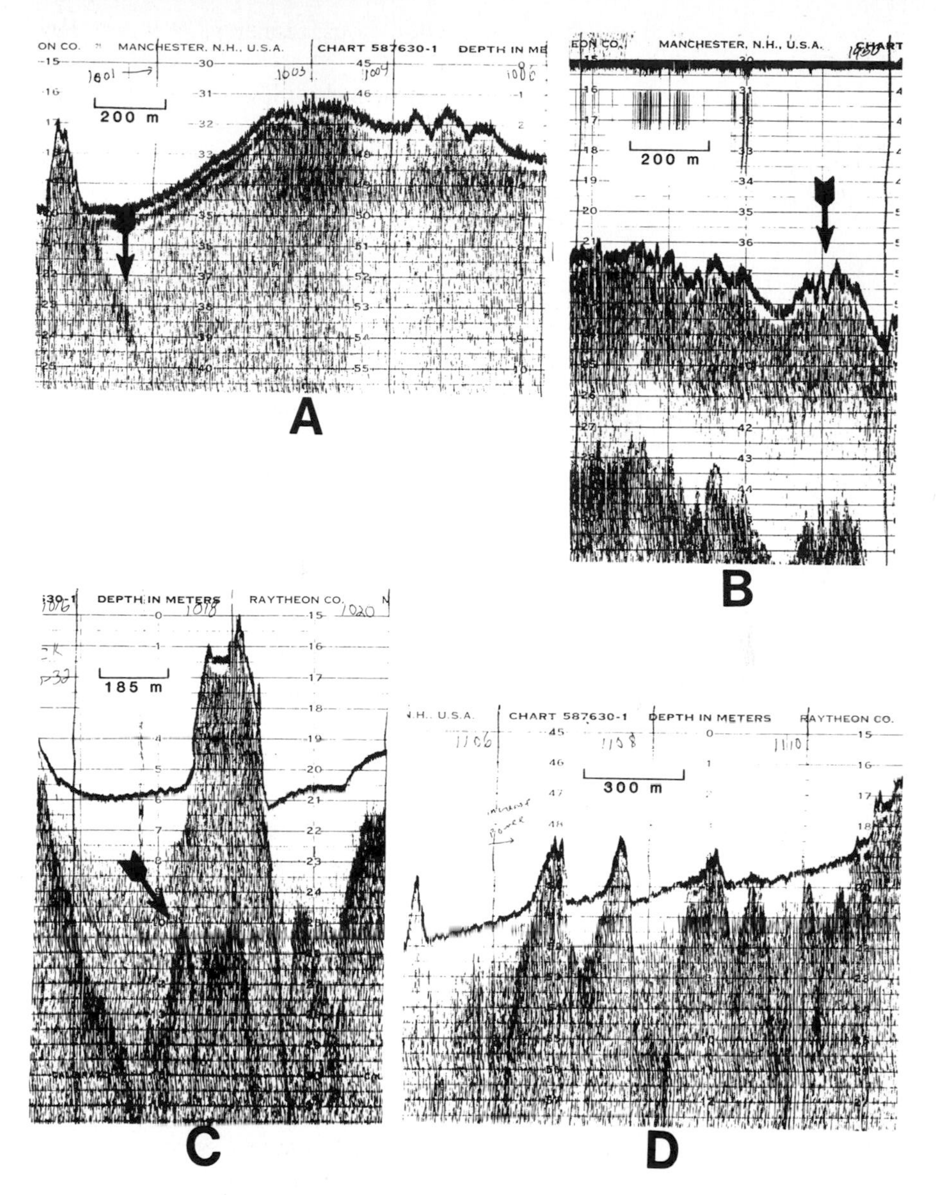

Figure 5. *Sections of original seismic record from outer Casco Bay. Location shown in Figure 1. Horizontal lines on recording paper are 1-m apart. A) Bedrock reflector (arrow) dipping beneath till deposit. B) Typical bedrock reflector (arrow) with no overlying sediment. C) Till overlying bedrock reflector (arrow). Repeated crossings of this feature suggest the till is an extensive, muddy moraine and confirm that the bedrock is real, and not a multiple. D) Thin, seaward dipping ponds of sediment between ridges of bedrock.*

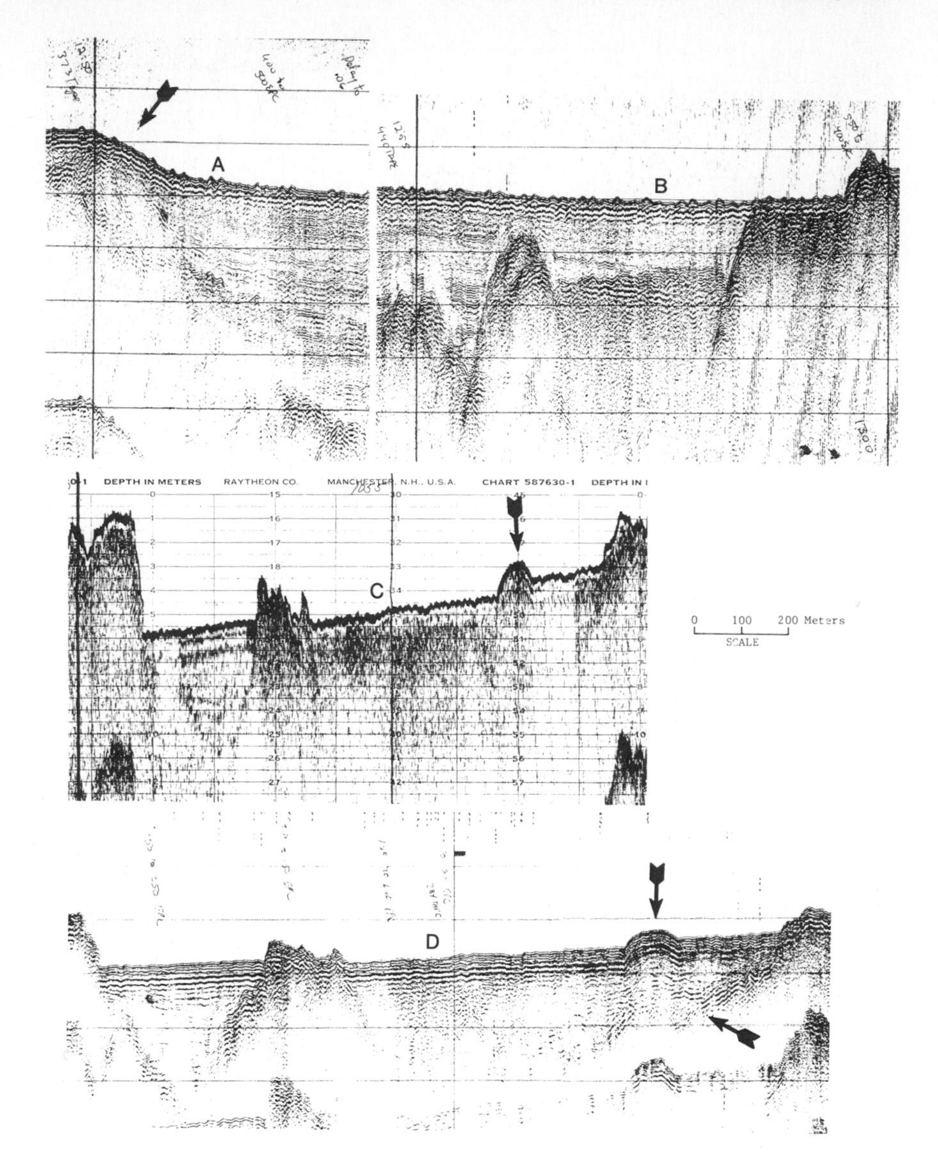

Figure 6. *Sections of original seismic records from Saco Bay. Locations shown in Figure 1. Horizontal lines on A, B, D are 10-m apart and 1-m apart on C. A) Prominent shoreline (arrow) on border of outer basin. Repeated crossings of this feature show that it varies in depth between 60 and 70 m. B) Thick pile of sediment in outer basin. C, D) 3.5 kHz record (C) and O.R.E. "boomer" trace of same area (D). Note the sandy surface reduces the depth of penetration of the 3.5 kHz record while the "boomer" record shows the bedrock basement (arrow). Surface relief may be a drowned spit overlying a till deposit (arrow), or simply a reworked glacial feature.*

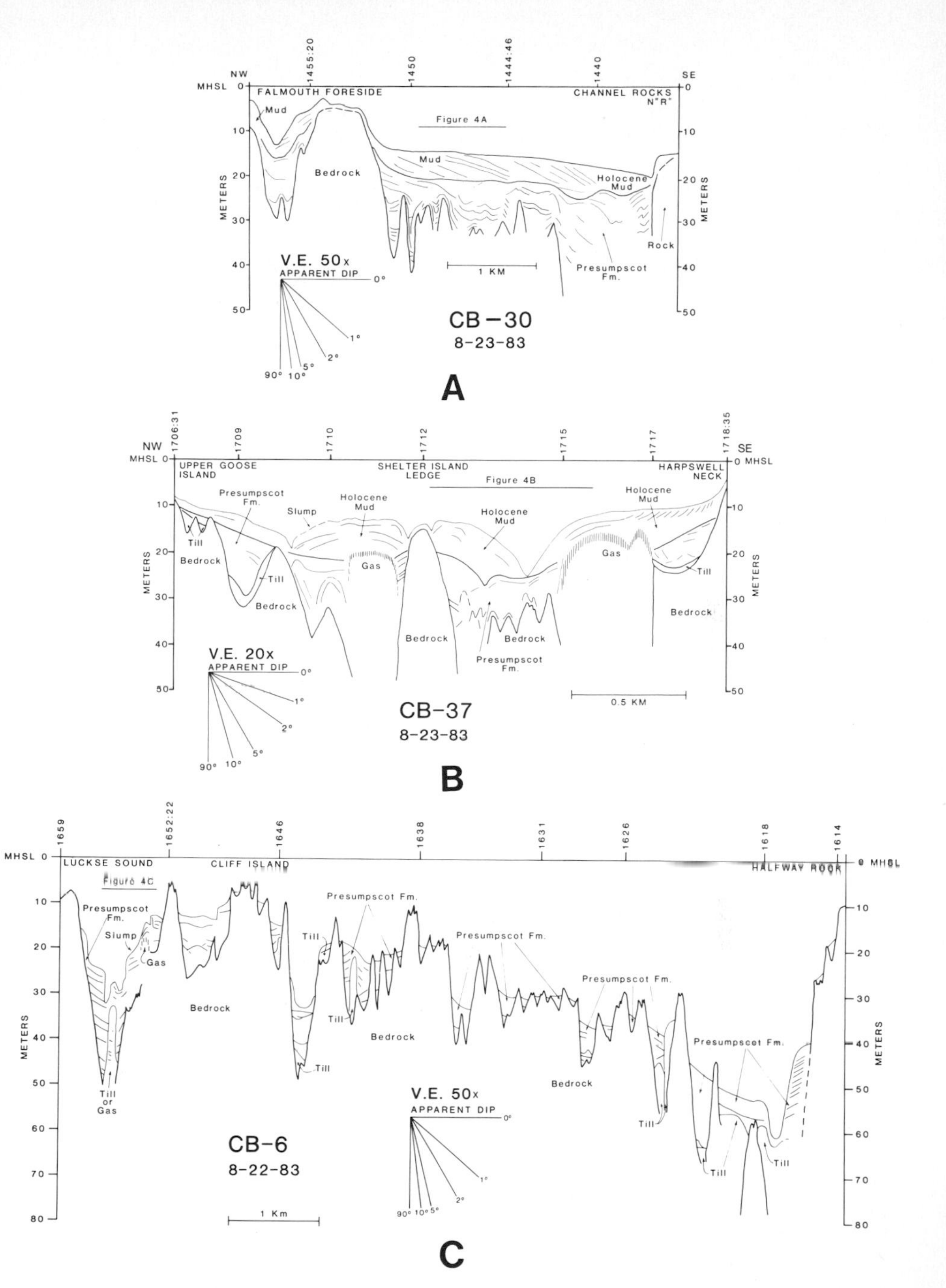

Figure 7. *Interpreted seismic lines digitized to convenient scales. Location of lines shown in Figure 1. Original records for areas indicated A, B, C shown in Figure 4 A, B, C.*

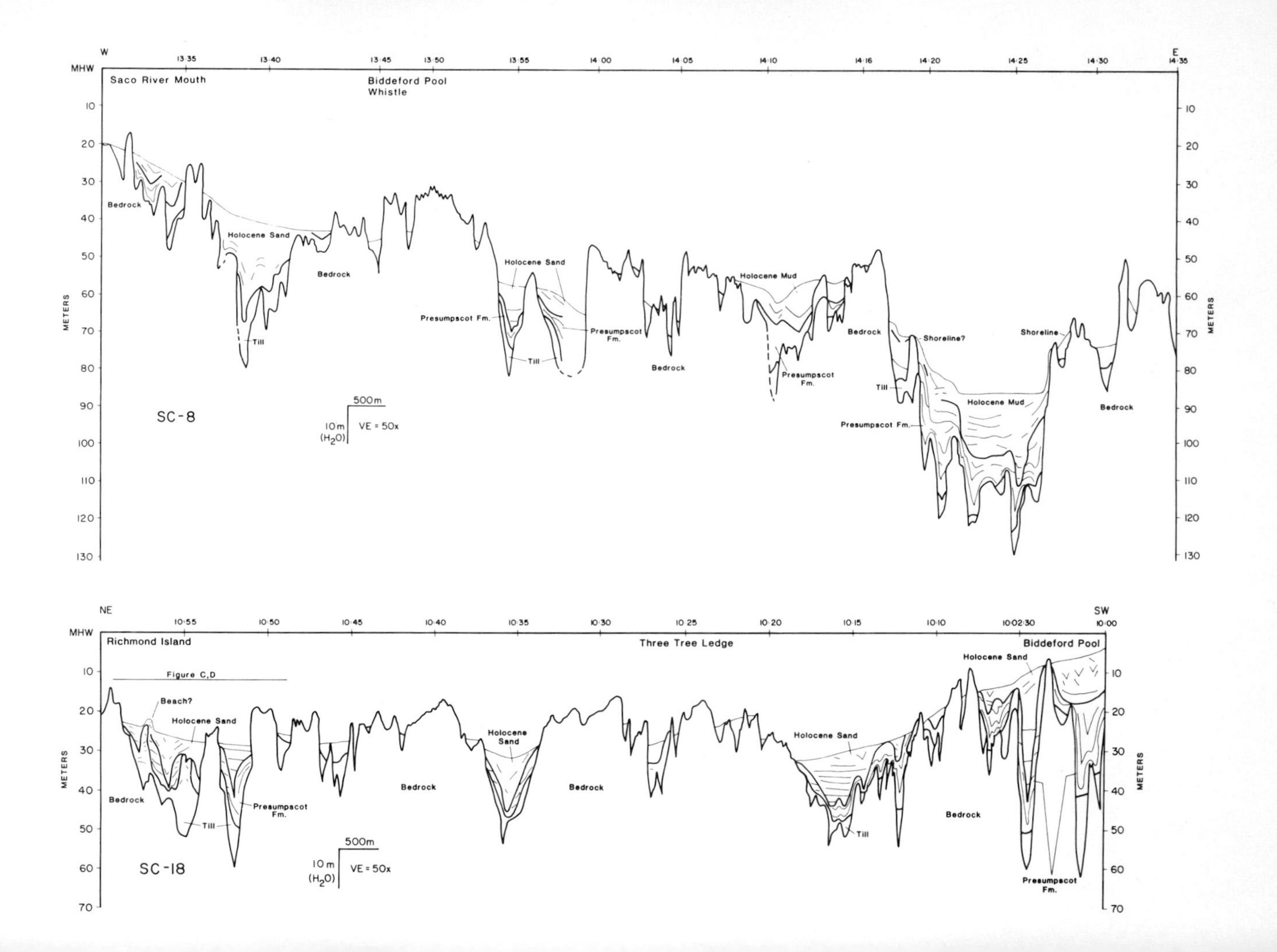

W
E
SW
NE
MHW
METERS
Saco River Mouth
Biddeford Pool Whistle
Bedrock
Holocene Sand
Holocene Mud
Till
Presumpscot Fm.
Shoreline
Shoreline?
500m
VE = 50x
10 m (H2O)
SC-8
Richmond Island
Three Tree Ledge
Biddeford Pool
Figure C,D
Beach?
SC-18

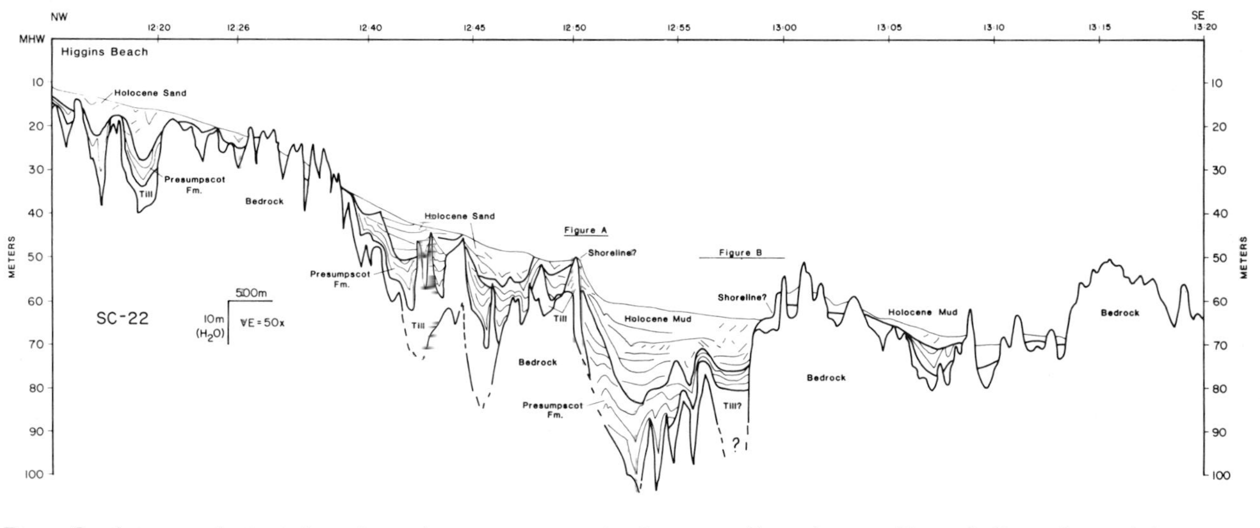

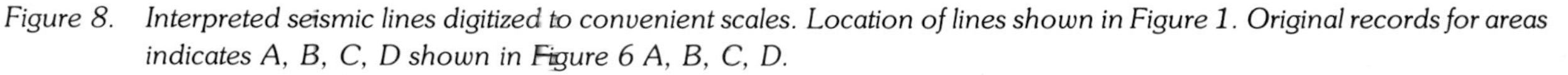

Figure 8. *Interpreted seismic lines digitized to convenient scales. Location of lines shown in Figure 1. Original records for areas indicates A, B, C, D shown in Figure 6 A, B, C, D.*

direction (Figs. 5, 7). Because of its acoustically transparent internal nature, it is inferred to be Presumpscot Formation material which has undergone considerable reworking. These sediments are generally absent over the till deposits of the outer bay, which are presumably armored with a coarse-grained surface. The till forms 10-m high ridges 0.5-km wide which parallel the northeast strike of local bedrock ridges (Fig. 5).

Saco Bay

On the basis of bathymetry Saco Bay may be divided into four regions (Fig. 9). The inner region includes the estuaries and mouths of the Saco and Scarboro Rivers. As with the estuaries of Casco Bay, these estuaries are largely intertidal, although unlike Casco Bay the Saco and Scarboro Rivers are protected by extensive sand beaches (Table 1). Seismic data in the Saco River mouth showed extensive outcrops of bedrock mantled by a few meters of sand (Kelley 1985b). The Scarboro River estuary has up to 60 m of sediment covering its bedrock floor and closely resembles the Fore and Cousins River estuaries (Fig. 2).

Seaward of the extensive sand beach shoreline of Saco Bay is a gently sloping sandy area (Farrell 1972). Seismic profiles here reveal Holocene sand from less than one to greater than 10-m thick (Fig. 8). Cores from this area show the sand to overlie poorly sorted pebbly muds characteristic of the Presumpscot Formation (Williams, pers. comm.). The Presumpscot Formation possesses an unconformity at its surface, as in Casco Bay, but it is less prominent on the seismic records because it is overlain by acoustically similar sand (Fig. 6). As in Casco Bay, the Presumpscot Formation possesses reflectors near its base which mimic the underlying topography. The Presumpscot Formation generally overlies thin till or bedrock and ranges in thickness up to 50 m in bedrock valleys (Fig. 8).

At about 30-m depth, the gently sloping sandy area abruptly terminates against a third region which is of somewhat chaotic bathymetry (Fig. 9). As in outer Casco Bay, this area is largely one of exposed bedrock and small isolated basins of sediment (Figs. 6, 8). The sediment is inferred to be reworked till or Presumpscot Formation material which is gravelly on the surface and possesses up to 50% calcium carbonate in places (Kelley 1985b). Where observed, the surface of all sediment ponds dipped in a seaward direction (Fig. 8).

Four prominent channels extend from the nearshore sandy region, through the area of exposed bedrock, to a nearly flat, muddy outer basin (Fig. 9). Though small in area, the channels contain large volumes of sediment (Fig. 8). In each of these, reflectors of the Presumpscot Formation are truncated by inferred Holocene sediment (Fig. 8). Prominent ridges overlie till adjacent to several channels (Fig. 6) and may represent drowned barrier spits. Bottom samples from the channels show an increasing mud content and a generally decreasing carbonate content in a seaward direction (Kelley 1985b).

The outer basin contains more than 30 m of sediment over an irregular bedrock basement. Till deposits are locally thick and covered by draping reflectors

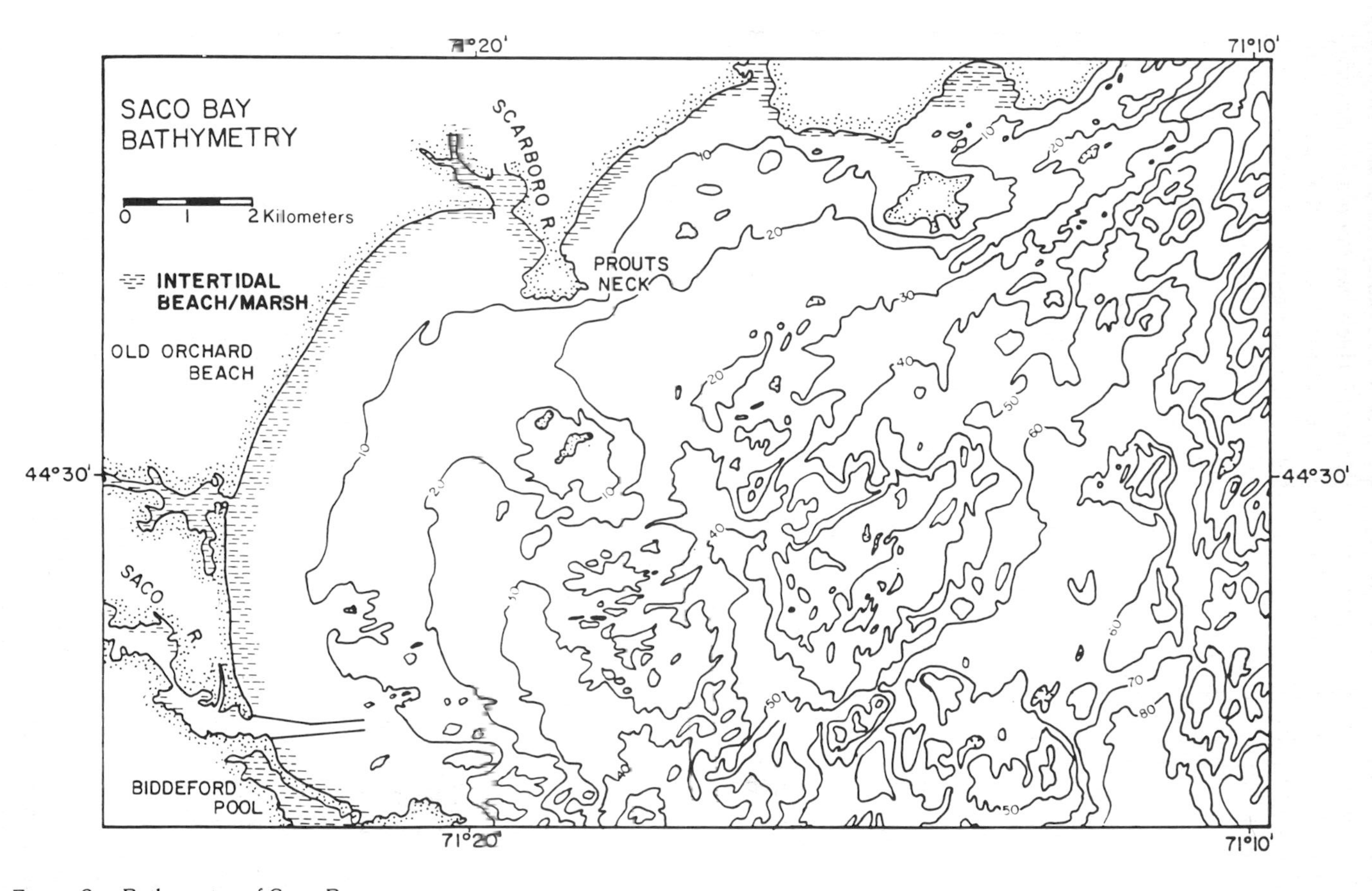

Figure 9. Bathymetry of Saco Bay.

of the Presumpscot Formation. Even at depths of more than 70 m the Presumpscot Formation reflectors are truncated by reflectors of inferred Holocene-age sediment. Along all observed margins of the outer basin, prominent seafloor scarps exist (Figs. 6, 8). Often they appear to have developed on a deposit of till, though recent submersible observations show only mud at the surface (Kelley, unpublished notes). Since these scarps occur within the depth range suggested for the early Holocene lowering of seal level (Schnitker 1974), it is reasonable to infer that they represent shorelines from that time (Kelley and Kelley 1985; Shipp 1985).

Discussion and Conclusions

The estuaries of Casco and Saco Bays are adjacent, yet differ profoundly in their geomorphology and depositional environments. About 96 percent of the inner region of Casco Bay is filled with intertidal mudflats and fringing salt marshes (Table 1) while most of the remaining intertidal portion of the bay is exposed bedrock (Kelley 1985b). About 93 percent of the intertidal estuarine area of Saco Bay is salt marsh fronted by sandy barriers (Table 1) while the remainder of the bay is sand beach or bedrock (Kelley 1985a). These differences in contemporary estuarine environments are also reflected in the geology of the embayments surrounding the estuaries (Fig. 10).

The nearshore area of Saco Bay is a gently sloping sandy area whereas Casco Bay has a generally flat-bottomed muddy central region (Fig. 10). Each of the embayments possesses an outer rocky area with ponds of sediment dipping seaward and cut by channels. In Casco Bay the channels frequently show slump scars on the margins and gas deposits in the subsurface, whereas in Saco Bay they do not. Finally, Saco Bay possesses a basin on its outer margin which is filled with more than 50 m of sediment, much of it apparently Holocene in age. While no equivalent feature exists off Casco Bay, no work has yet been performed in the deeper water off that embayment (Fig. 10).

On the basis of the seismic stratigraphy described above it appears that Casco and Saco Bays experienced similar glacial and late post-glacial histories. They each possess moraine deposits which are aligned with the regional bedrock trend and which may be traced from one bay across land to the next (Fig. 10). Overlying till or bedrock, late glacial marine sediment, the Presumpscot Formation, blankets most of the inner and nearshore region of each bay. The Presumpscot Formation appears identical in each bay and generally fills the deepest bedrock valleys leading to the embayments. This blocking of pre-glacial valleys with glacial sediment is significant because it led to drainage derangement and separate evolutionary paths for the two embayments.

During the sea level regression which followed deposition of the Presumpscot Formation, new streams entering Casco Bay possessed very small drainage basins in relation to the size of the embayment and its filled bedrock valleys. Two of the largest rivers in Maine, the Kennebec and Androscoggin Rivers, are prevented from entering Casco Bay by glacial deposits, and pass within 5 km of the bay before joining and turning to the northeast (Thompson and Borns 1985). The

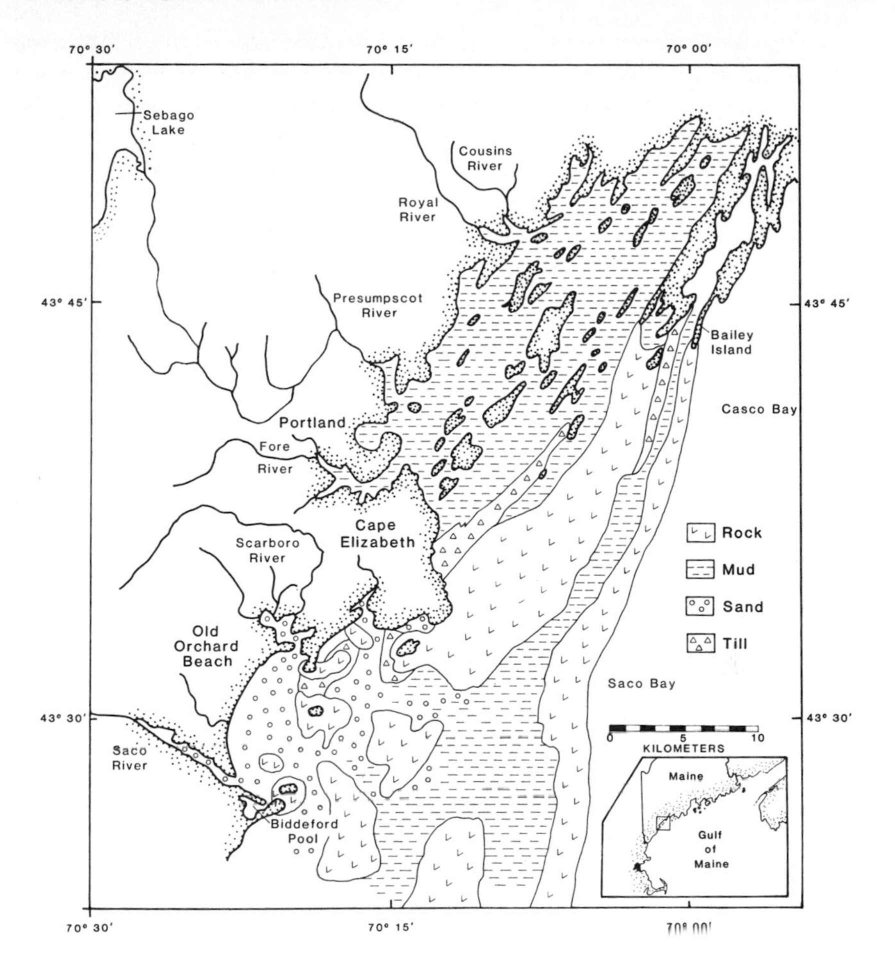

Figure 10. General bottom sediment map of south central, nearshore Maine.

present, deranged streams which enter Casco Bay originate within the marine limit (maximum elevation around 100 m) and so are relatively low gradient streams passing over mostly Presumpscot Formation mud before entering the sea.

The watershed of Saco Bay also experienced drainage derangement and its largest bedrock valley is today occupied by the very small Scarboro River. The Saco River which passes over waterfalls on the southern end of the bay, is a large stream (Table 1), however, which originates in the White Mountains (maximum elevation around 1,900 m). This relatively high gradient stream passes through many sandy glacial deposits, derived from glacial erosion of underlying granites, before entering the ocean (Thompson and Borns 1985).

As sea level reached its lowstand around 9,000 years ago (Belknap and Borns 1986) an increasingly steep gradient permitted the Saco River to erode into muddy Presumpscot Formation sediments in Saco Bay channels (Fig. 8) as well as sandy deposits outside the marine limit (Thompson and Borns 1985). The reworked mud was largely deposited into today's offshore basin (Fig. 8) while the river's sand fraction remained nearshore as the prominent shoreline bordering the outer basin (Figs. 6, 8). As sea level began its most recent rise the Saco Bay channels were filled with sand which unconformably rests on Presumpscot Formation material (Fig 8.) Barrier spits and tombolos were probably common environments adjacent to the channels, although only wave-washed remnants of them exist today (Figs. 6, 8; Farrell 1972). Till and any other material resting on the bay's rocky central area were removed as sea level washed over it and left only seaward-dipping ponds of sediment covered with a gravel armor (Fig. 8). Environments within Saco Bay today are probably similar to those that existed in the recent geological past with the exception of the extensive salt marshes that have formed during the present reduced rate of sea level rise (Belknap and Borns 1986).

During the sea level regression in Casco Bay, rivers also cut into the glaciomarine mud deposits filling bay channels (Figs. 4, 7). The small rivers of Casco Bay originated within the Presumpscot Formation, however, and so did not deliver significant volumes of sand to the sea. Instead mud was introduced, and as sea level rose muddy environments predominated. Muddy estuaries, much like today's, probably evolved up the bedrock-framed channels as sea level rose. Organic matter, resulting from biological productivity in these estuaries, was buried by mud and decomposed into natural gas as described in other estuaries (Schubel and Schiemer 1973). The gas was trapped by impermeable muds and so could not escape as it might have from the more permeable sands of Saco Bay. Gas in the Holocene sediments filling the channels may have reduced the shear strength of the deposits leading to frequent slumps (Figs. 4, 7; Kelley *et al.* 1985). As in Saco Bay, sediment was winnowed from the outer rocky area and left as seaward-dipping ponds between rock ridges (Figs. 5, 7). The estuaries of Casco Bay today are probably very similar to those of the recent past. During the present slow rise in sea level (Belknap and Borns 1986) they have filled with mud derived from local rivers as well as the eroding bluffs of the Presumpscot Formation which rim their margins.

Many questions remain to be answered in this area. When did sea level reach its lowstand, and at what depth? If large quantities of material were reworked, was it transported offshore? Finally, the remotely-sensed reflectors and submarine geomorphic forms (drowned beaches, slumps) need to be cored to establish their existence confidently and possibly to determine their ages.

Acknowledgments

We wish to acknowledge financial support for this work from the Maine Geological Survey, the U.S. Nuclear Regulatory Commission, the U.S. Minerals Management Service, and the NOAA Office of Sea Grant. The seismic equipment was purchased on funds from the National Science Foundation EPSCOR program.

References Cited

Anderson, W. A., J. T. Kelley, W. B. Thompson, H. Borns, D. Sanger, D. Smith, D. Tyler, S. Anderson, A. Bridges, K. Crossen, J. Ladd, B. Andersen, and F. Lee. 1984. Crustal warping in coastal Maine. *Geology* 12:677-680.

Belknap, D. F., R. C. Shipp, and J. T. Kelley. 1985a. Provisional Quaternary seismic stratigraphy of the Sheepscot Estuary, Maine. *Geographie Physique et Quaternaire.* In press.

Belknap, D. F., J. T. Kelley, R. C. Shipp, and H. Borns. 1985b. Holocene sea level rise as measured by dating basal salt marsh peats. *Maine Geological Survey Bulletin no. 25.* Augusta, ME. In press.

Belknap, D. F., and H. Borns. 1986. Late Quaternary sea level changes in Maine. *In:* D. Nummedal, O. Pilkey, and J. Howard, (eds.), *Sea Level Rise and Coastal Evolution.* Society of Economic Paleontologists and Mineralogists Special Publication. In press.

Bloom, A. 1963. Late-Pleistocene fluctuations of sea level and postglacial crustal rebound in coastal Maine. *Amer. J. Sci.* 261:862-879.

Booth, J.S., D. Sangrey, and J. Fugate. 1985. A nomogram for interpreting slope stability of fine-grained deposits in ancient and modern marine environments. *J. Sedim. Petrol.* 55:29-36.

Carver, R. (ed.). 1971. *Procedures In Sedimentary Petrology.* J. Wiley and Sons. New York. 589 pp.

Farrell, S. C. 1970. Sediment distribution and hydrodynamics: Saco and Scarboro Rivers Estuaries, Maine. M.S. thesis, University of Massachusetts, Amherst. 129 pp.

Farrell, S. C. 1972. Present coastal processes, recorded changes, and the post-Pleistocene geologic record of Saco Bay, Maine. Ph.D. thesis, University of Massachusetts, Amherst. 296 pp.

Fefer, S. I., and P. Shattig. 1980. An ecological characterization of coastal Maine. Vol. I (Chapt. 2) and Vol. II (Chapt. 5). U.S. Fish and Wildlife Service Rep. FWS/OBS-80/29. Washington, D.C. Multiple pagination.

Folger, D., C. O'Hara, and J. Robb. 1975. Maps showing bottom sediments on the continental shelf off the northeast U.S.: Cape Ann, MA to Casco Bay, ME. *U.S. Geological Survey Misc. Inv. Ser.* Map I-839. Washington, D.C.

Hulmes, L. J. 1981. Holocene stratigraphy and geomorphology of the Hills Beach and Fletcher Neck tombolo system, Biddeford, Maine. *Northeastern Geol.* 3:197-201.

Kelley, J. T. 1985a. Sedimentary environments along Maine's glaciated, estuarine coastline. *In:* D. FitzGerald, and P. Rosen (eds.), *A Treatise on Glaciated Coasts.* Academic Press, NY. In press.

Kelley, J. T. 1985b. Sedimentary environments and sand resources of the southern Maine inner continental shelf. Maine Geological Survey Open-File Report. Augusta, ME. In press.

Kelley, J. T., and B. Timson. 1983. Environmental inventory and statistical evaluation of the Maine coast. Geol. Soc. Amer. Abs. with Programs. 15:193.

Kelley, J. T., and A. R. Kelley. 1985. The sedimentary framework of Saco Bay, Maine. Geol. Soc. Amer. Abs. with Programs. 17:28.

Kelley, J. T., D. F. Belknap, R. C. Shipp, and S. Miller. 1985. An investigation of neotectonic activity in coastal Maine by seismic reflection methods. *Maine Geological Survey Bulletin 25.* Augusta, ME. In press.

Knebel, H. J., and K. Scanlon. 1985. Sedimentary framework of Penobscot Bay, Maine. *Marine Geology* 65:305-324.

Larson, P. F., and L. Doggett. 1979. *The Salinity and Temperature Distributions of Selected Maine Estuaries.* Maine State Planning Office Publication, Augusta, Maine. 112 pp.

Larson, P. F., A. Johnson, and L. Doggett. 1983. Environmental benchmark studies in Casco Bay-Portland Harbor, Maine. *Tech. Memo.* NMFS-F/NEC-19. National Marine Fisheries Service, Woods Hole, MA. 173 pp.

Maine Department of Transportation. 1948. Soils report for the new Fore River Bridge. Materials and Research Division. Bangor, ME. 7 pp.

Maine Department of Transportation. 1983. Subsurface investigation for I-95 bridge over the Cousins River, Yarmouth, ME. Materials and Research Division Project I-95-4, Bangor, ME. 6 pp.

Oldale, R. N. 1985. A drowned Holocene barrier spit off Cape Ann, Massachusetts. *Geology* 13:375-377.

Ostereicher, C. 1965. Bottom and subbottom investigation of Penobscot Bay, Maine. *U.S. Naval Oceanogr. Off. Tech. Rep.* 173. Washington, D.C. 177 pp.

Roy, P. S. 1984. New South Wales estuaries: Their origin and evolution. pp. 99-121. *In:* B. G. Thom (ed.), *Coastal Geomorphology In Australia.* Academic Press, NY.

Schnitker, D. 1974. Postglacial emergence of the Gulf of Maine. *Geol. Soc. Amer. Bull.* 85:491-494.

Schubel, J., and E. Schiemer. 1973. The cause of acoustically impenetrable or turbid character of Chesapeake Bay sediment. *Mar. Geophys. Res.* 2:61-71.

Shipp, R. C. 1985. Late Quaternary evolution of the Wells embayment, southwestern Maine. Geol. Soc. Amer. Abs. with Programs 17:63.

Smith, G. 1985. Chronology of late Wisconsinan deglaciation of coastal Maine. *Geol. Soc. Amer. Spec. Paper* 197:29-44.

Thompson, W. B. 1979. *Surficial Geology Handbook for Coastal Maine.* Maine Geological Survey Publication, Augusta, Maine. 68 pp.

Thompson, W. B., and H. Borns. 1985. *Surficial Geologic Map of Maine,* Maine Geological Survey, Augusta, Maine.

U.S. Environmental Protection Agency. 1983. *Environmental Impact Statement for the Portland, Maine Dredged Material Disposal Site Designation.* U.S.E.P.A. Criteria and Standards Div., Washington, D.C. pp. 3-1 to 3-19.

U.S. Army Corps of Engineers. 1980. Report of the New England Division, Waltham, MA. pp. 44-45.

COMPARATIVE ANALYSIS OF ESTUARIES BORDERING THE YELLOW SEA

J. R. Schubel

Marine Sciences Research Center
State University of New York at Stony Brook
Stony Brook, New York

Huan-Ting Shen

Institute of Estuarine and Coastal Research
East China Normal University
Shanghai, China

and

Moon-Jin Park

Marine Sciences Research Center
State University of New York at Stony Brook
Stony Brook, New York

Abstract: Like all estuaries throughout the world, estuaries bordering the Yellow Sea were formed by the most recent rise in sea level and are less than 10,000 years old. The rivers and estuaries which enter the Yellow Sea from the west (China) are distinctly different from those entering from the east (Korea). The rivers on the west coast have larger water discharges, larger sediment loads and smaller tidal ranges. Because of these factors, the Chinese estuaries have reached a much more advanced stage of geological evolution—infilling—than the Korean estuaries and have lower filtering efficiencies for the fluvial sediment they receive. The Huang Ho (Yellow River) no longer has an estuary; the Changjiang's (Yangtze) estuary is only 15-20 km long during the wet season and 85-125 km long during the dry season. Chinese rivers make much greater contributions of sediment to the Yellow Sea than do the Korean rivers because of their much larger sediment discharges and, to a lesser extent, because of the lower filtering efficiencies of their estuaries.

Introduction

The Yellow Sea is a semi-enclosed epicontinental shelf sea whose sediments are derived primarily from the rivers and estuaries along its margin (Fig. 1). Each river has a distinctive flow regime and sediment discharge determined primarily by the size and geological characteristics of its drainage basin and by the climate. The distinctive physical and geological characteristics of each estuary determine its filtering efficiency for fluvial sediment, and its contribution of fluvial sediment to the Yellow Sea. Although there are marked differences between estuaries on the two sides of the Yellow Sea, the estuaries have been studied individually and a general comparison of Yellow Sea estuaries is lacking. This paper summarizes previous studies of individual estuaries, and presents a comparative overview of some of the important physical and geological characteristics of estuaries bordering the Yellow Sea. A secondary objective is to stimulate interest in these estuaries. The data are very limited, and there is much to be learned about the geological evolution of estuaries and their biological importance by studying estuaries bordering the Yellow Sea.

Our definition of an estuary is from Pritchard (1967): "An estuary is a semi-enclosed coastal body of water which has a free connection with the open sea and within which sea water is measurably diluted with freshwater derived from land drainage." All the estuaries bordering the Yellow Sea are submerged river valley estuaries. Like all other modern day estuaries, they were formed during the most recent rise in sea level and are less than 10,000 years old.

Once formed, estuaries are modified rapidly by sedimentation. On geological time scales they are ephemeral features having life spans measured in thousands to perhaps a few tens of thousands of years. Because most estuaries are traps for fine-grained sediment, they sometimes are described as being effective filters (Schubel and Carter 1984). Characteristically, the sedimentation rate is highest near the head of an estuary where a delta forms. The delta grows progressively seaward within the estuary to extend the realm of the river and force the intruding sea out of the semi-enclosed coastal basin. Estuaries around the margin of the Yellow Sea vary greatly in their stage of geological evolution.

The filtering efficiency of an estuary is that fraction of the total mass of suspended sediment introduced to the estuary, which is retained by it (Schubel and Carter 1984). The filtering efficiency of estuaries varies with estuarine circulation type and changes as an estuary evolves. An estuary's filtering efficiency is determined primarily by the strength of its gravitational circulation and by the size of the estuary, particularly its length, which is defined as the distance from the mouth of the estuary upstream to the last measurable traces of sea salt. The length of an estuary is a function of river discharge, and can vary by many tens of kilometers with fluctuations in river flow. In all estuaries, except sectionally homogeneous estuaries, the landward limit of sea salt penetration coincides roughly with the "null zone"—the zone in which the net upstream flow of the lower layer dissipates until the net flow is seaward at all depths. Any reduction in the length of an estuary leads to a reduction in its filtering efficiency. This effect is particularly important in highly stratified and partially mixed estuaries. When the estuary is shortened below some threshold length, the entire estuary—the zone of penetration of sea salt—may be expelled from its basin during periods of high riverflow. The estuary's filtering efficiency approaches zero during such periods. Since periods of high riverflow usually are periods of high fluvial suspended sediment discharge, the fraction of this sediment discharged directly to the ocean increases as the estuary is shortened.

Several of the estuaries bordering the Yellow Sea have evolved to a stage where they are expelled, or nearly expelled, from their basins during periods of high river flow. In the following sections, we review some of the most distinctive physical and geological features for the major estuaries bordering the Yellow Sea, and estimate their filtering efficiencies (Fig. 1).

The Changjiang (Yangtze) Estuary

The Changjiang is the longest river in China (6,300 km), ranking third among the world's rivers behind the Nile (6,650 km) and the Amazon (6,437 km). It traverses ten provinces, and discharges into its estuary near Shanghai (Fig. 2). On its way to the sea, hundreds of large and small tributaries join the main stem of the river. The Changjiang's drainage area of 1.8×10^6 km²

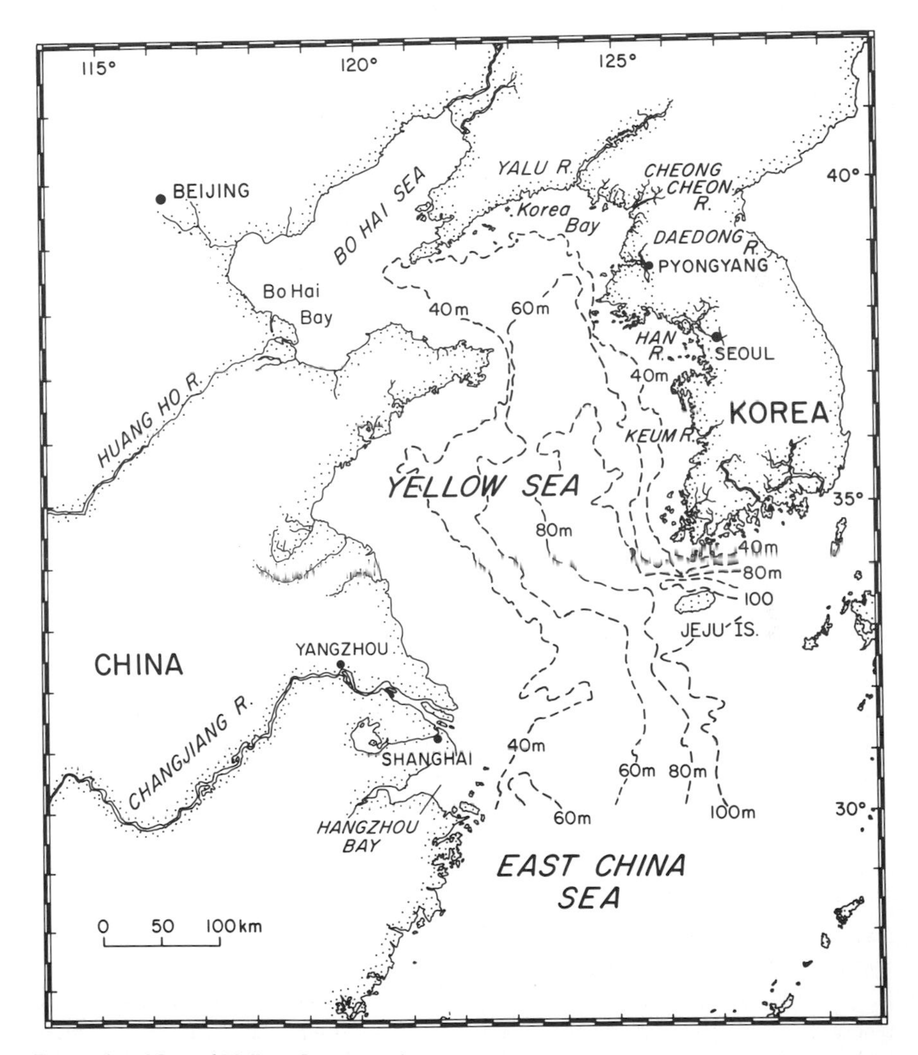

Figure 1. Map of Yellow Sea area showing major estuaries.

accounts for about one-fifth of the total land area of China, and some 400 to 500 million people live within it.

The Changjiang River has large water and sediment discharges. The maximum river discharge at peak flow is 92,600 $m^3 s^{-1}$ and the minimum discharge at low flow is 6,020 $m^3 s^{-1}$. The long-term average river discharge is 29,300 $m^3 s^{-1}$. The average volume discharge is 924 $km^3 y^{-1}$. The flood season extends from May to October and runoff during these months accounts for an average of 71.7% of the total annual volume discharge (Fig. 3).

The annual average suspended sediment concentration is 544 mg l^{-1} in the lower reaches of the Changjiang River and the annual suspended sediment discharge averages 486 $\times$ 10^6 metric tons (Shen *et al.* 1982). The seasonal variation of sediment discharge is more marked than that of runoff (Fig. 3).

Because of the large water and sediment discharges and strong tidal currents, the river channel has been progressively bifurcated over the recent past. Downstream from Xulujing, the Changjiang is divided by Chong Ming Island into North and South Branches (Fig. 4). The South Branch is in turn divided into North and South Channels. The South Channel is further subdivided into North and South passages. Consequently, near its mouth the estuary is divided into four distributaries.

Tides are semi-diurnal and become irregular inside the mouth. The average tidal range near the mouth of South Passage is 2.66 m; the maximum 4.62 m. The tidal range inthe North Branch is larger than that in the South

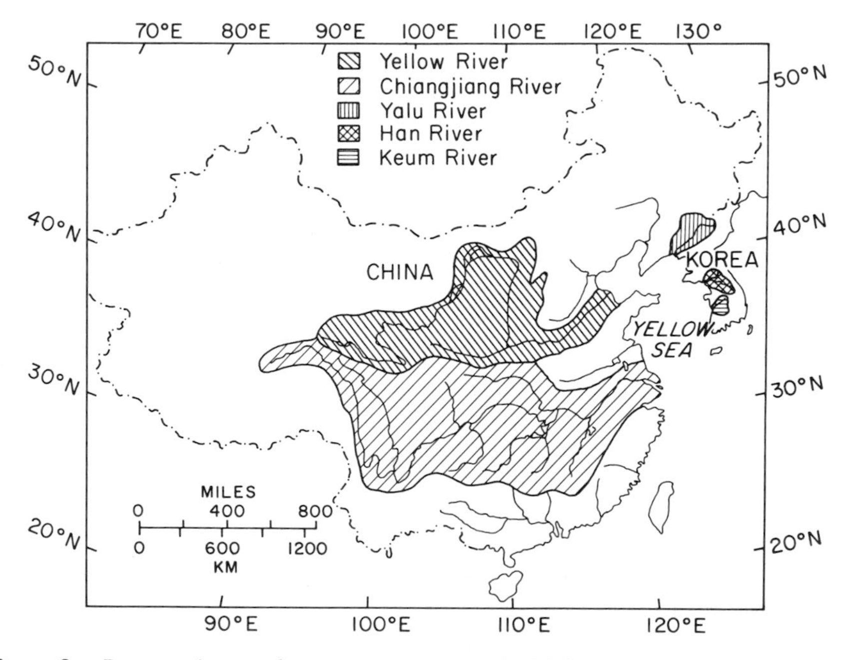

Figure 2. Drainage basins of major rivers entering the Yellow Sea.

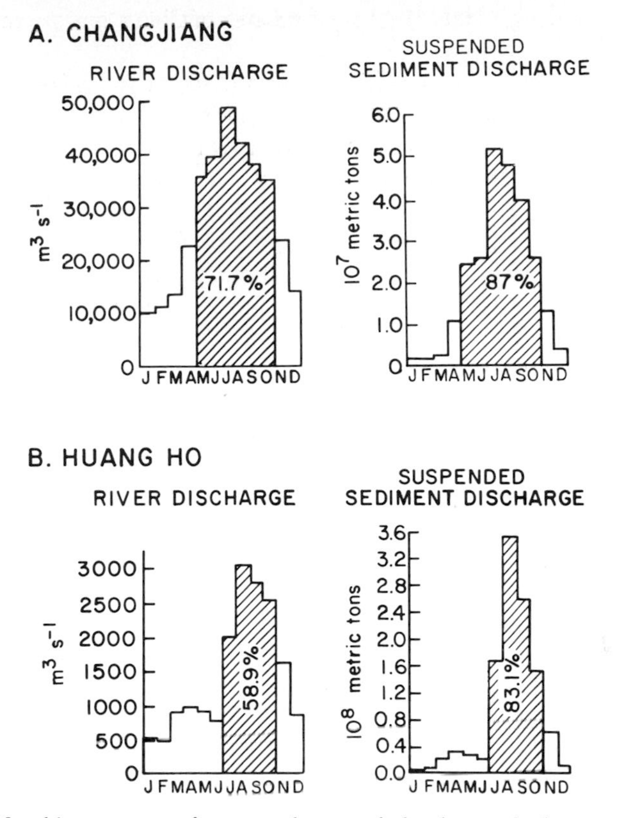

Figure 3. Monthly variations of water and suspended sediment discharges of the Changjiang (Yangtze) and Huang Ho (Yellow) Rivers. The shaded areas indicate the flood season. All units are metric.

Branch and in the reach between Port Lingtian and Port Qinglong is characterized by a tidal bore. The estuary is meso-tidal (Shen and Pan 1979).

The estuarine circulation in the South Branch is partially mixed (Type B) most of the year, and the estuarine circulation in the North Branch is well-mixed (Type C) most of the year. Less than 10% of the total river flow of the Changjiang is discharged presently through the North Branch, although it accounted for 25% of the total before 1915. In the North and South passages of the South Channel, and at times in the lower reaches of the North Channel, the estuarine circulation is typical of partially mixed estuaries with a net seaward flow in the upper layer and net landward flow in the lower layer. A turbidity maximum occurs near the null zone in passages of the South Channel. Locations of the null point during the flood season coincide with the channel sand bars (Shen *et al.* 1982).

About 6,000 to 7,000 years ago, the Changjiang had a large estuary; the upstream limit of salt penetration was located near Zhenjiang and Yangzhou, approximately 230 km from the mouth of the estuary. During the last 2,000

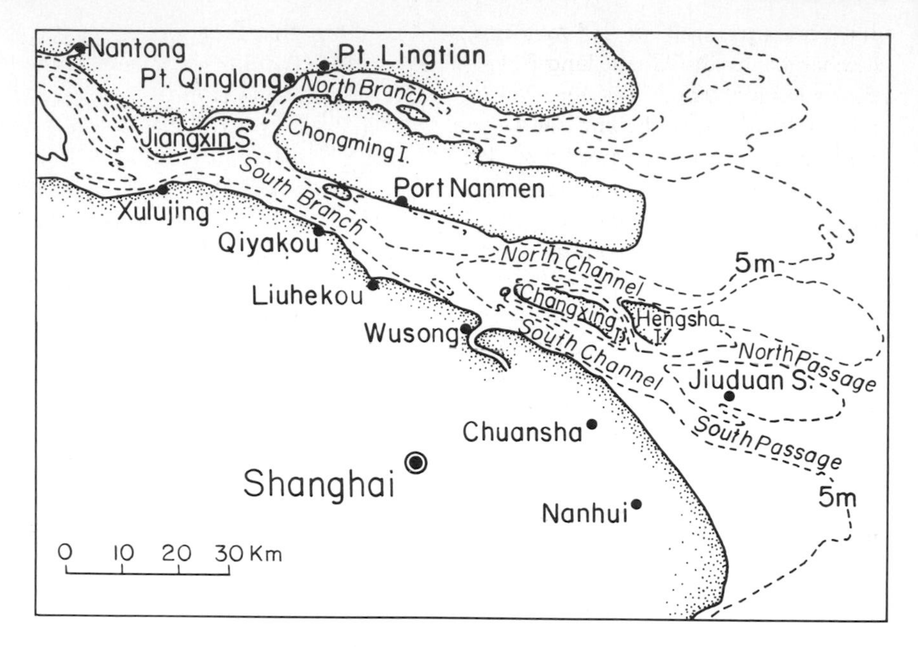

Figure 4. Map of the lower reaches of the Changjiang River and its estuary.

years, shoals (silt and sandbars) have been merged successively into the north bank, and shoals near the south bank have been driven seaward at an average rate of 1 km every 40 years. Recently, this rate has increased to 1 km in every 23 years. These processes have led to a progressive narrowing and shortening of the estuary (Chen *et al.* 1982, 1983) with a consequent decrease in filtering efficiency. Over the past 2,000 years, the mouth has narrowed from 180 km to 90 km. The main channel has shifted toward the south and deepened. As a result of narrowing, prograding, bifurcating, and southward deflection of channels, the estuary has evolved from a typical drowned river valley estuary into a complex, forked deltaic estuary.

In the four channels that connect the Changjiang with the Sea there is an extensive network of bars. The channel sand bars in the North Branch are located well inside the mouth whereas those in the South Passage, in the North Passage, and in the North Channel all are located near the mouths of those distributaries. The minimum equilibrium water depth of the channel sand bars is about 6 m. The aggregate length of channel sand bars that are less than 7-m deep in the South Passage is about 24.5 km; for bars less than 10 m in depth, it is about 60 km. The bars are a natural barrier to maintaining a navigable waterway.

Comparison of historical bathymetric data (1958-1978) suggests that the total volume of sediment deposited near the mouths of South Channel's two Passages and North Channel over this period was 3.11×10^9 m^3, for an average annual depositional rate of 1.56×10^8 m^3. Assuming a sediment density of 1.6 metric tons m^{-3}, the mass of sediment deposited each year would have averaged 2.49 × 10^8 metric tons. Thus more than 50% of the sediment discharged by the Changjiang River probably was deposited near and outside the mouths of South Passage, North Passage and North Channel, forming the wide subaqueous delta and sand-bar system. The submerged delta outside the estuary has an area of more than 10,000 km^2. Over the past two decades the main depositional zone has been the subaqueous delta off the mouths of South Passage and North Branch. The region between 122°30′E and 123°E is an important boundary region; little suspended sediment is transported beyond it.

The suspended sediment that is not deposited on the delta is dispersed mainly in a southeasterly direction; much of it enters Hangzhou Bay. The Changjiang is now a major source—perhaps the major source—of freshwater and sediment to Hangzhou Bay; both entering through its mouth. Suspended sediment discharge of the estuary also changes markedly with tides (Shen *et al.* 1983; Milliman *et al.* 1983).

During periods of high river and sediment discharge (May to October) the Changjiang river has only a short estuary. Riverflow is sufficiently large to restrict the penetration of sea water into the estuarine basin to only 15-20 km, except in the North Branch. During periods of low to moderate river flow (April to November), the flow is still sufficiently large that the estuary is only about 85-125 km long (compared to 230 km some 2000 years ago). The filtering efficiency of the Changjiang estuary for fluvial sediment has been progressively reduced as the estuary has filled and now averages only about 10-15% over the year. This means that the estuary is now by-passing 85 to 90% of the river's sediment discharge to the sea. It also means that the estuary's rate of geological aging—of infilling—has decreased.

The Huang Ho (Yellow) Estuary

The Huang Ho (Yellow River), the second longest river in China, is 5,464-km long with a drainage area of 752,000 km^2. It flows through nine provinces before debouching into Bo Hai Bay (Fig. 2).

The Huang Ho is characterized by moderate water discharge, very large sediment discharge, weak tides, and an advanced stage of geological evolution of its estuary. The long-term mean river flow is 1,400 $m^3\ s^{-1}$. The maximum discharge at peak flow is 10,400 $m^3\ s^{-1}$. Its average annual discharge of water is 44.28 km^3; 1/20 that of the Changjiang River. The maximum yearly river volume discharge is 97.31 km^3; the minimum is 9.15 km^3. The runoff of the Huang Ho has a well established seasonal variation (Fig. 3). Minimum flow in February accounts for only 2.8% of the annual total; maximum flow in August accounts for 17.5% of the total.

The Huang Ho is well known for its concentrations of suspended sediment which are among the highest of the world's rivers (Milliman and Meade 1983). The yearly average suspended sediment concentration near the river's mouth is 25.3 g l^{-1}. The long-term mean annual maximum concentration is 222 g l^{-1}; the long-term minimum is 11.3 g l^{-1}. Among the world's rivers, the Huang Ho ranks second in sediment discharge behind the Ganges-Brahmaputra

(Milliman and Meade 1983). The long-term average annual suspended sediment discharge amounts to 1.1 × 10^9 metric tons. The maximum recorded annual suspended sediment discharge is 2.1 × 10^9 metric tons; the minimum, 0.2 × 10^9 metric tons.

The seasonal variation of sediment disharge is even more dramatic than that of runoff (Fig. 3). Sediment discharge is greatest in August, averaging 31.2% of the annual total. Minimum sediment discharge occurs in January when it accounts for an average of only 0.3% of the annual total.

The tide has little influence on the lower reaches of the Huang Ho River. The limit of tidal current reversal is only 1 to 2 km upstream from the mouth during the dry season. During the wet season there is no reversal of currents within the river. In neither season does sea salt penetrate even as far as 1 km into the semi-enclosed basin. In the offshore area the tide is a standing wave. Tides are semi-diurnal. The tidal range near the middle of the delta averages 0.6 m. The range increases gradually to both sides, but is less than 2 m throughout. The tidal current velocity is inversely related to the tidal range. The maximum current velocity of 1.5 m s^{-1} is found near the center of the delta and velocity decreases to less than 0.5 m s^{-1} to the sides.

The geological evolution of estuaries is characterized by infilling and deltaic progradation. Because of the large sediment discharge of the Huang Ho and the shifting of the lower course of the river, large overlapping deltaic plains have been formed off the mouth. The modern delta has an area of 5,400 km^2.

Most of the river's sediment load is deposited on the shelf. Comparison of bathymetric data collected since 1958 suggests that about 24% of all the sediment discharged has accumulated off the river mouth and on the delta. About 40% of the total was deposited in the zone seaward of the delta front. The growth of sub-aerial deltaic deposits averages about 50 km^2 y^{-1} and the shoreline is moving seaward at about 1.5 km y^{-1} (Pang and Si 1979, 1980). Most of the sediment transported beyond the delta is carried toward the head of Bo Hai Bay.

For all practical purposes, the Huang Ho estuary has been filled and essentially all of the sediment discharged by the river directly enters the sea. The filtering efficiency approaches zero.

Yalu (Aprock) River Estuary

The Yalu River is 800-km long and marks the boundary between China and North Korea (Fig. 2). Its drainage area is 61,000 km^2. The Yalu has a long-term mean discharge of 1,100 m^3 s^{-1}, and its average annual volume discharge is 34.7 km^3. Before reservoirs were built along the river, its discharge was distributed very unevenly throughout the year with runoff in July and August accounting for an average of 48% of the total annual discharge. Maximum discharge in August was 73 times the minimum discharge in March. After the Shui-Feng and Yun-Feng hydroelectric plants were completed in 1942, discharge was decreased during the wet season and increased during the dry season, but there is still a difference in discharge between the two seasons. Discharge in July and August accounts for 29% of the annual discharge (Fig. 5).

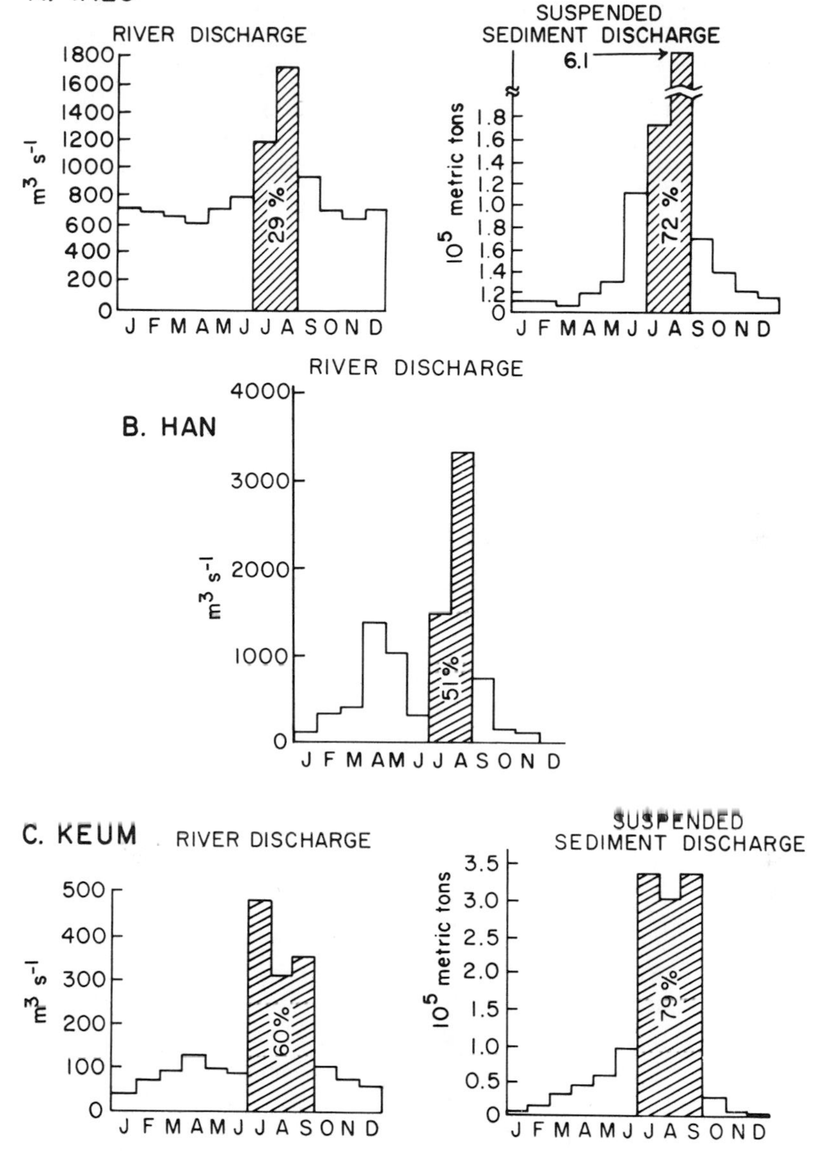

Figure 5. Monthly variations of water and suspended sediment discharges of the Yalu and Keum Rivers and monthly variation of water discharge of the Han River. Suspended sediment data for the Han are inadequate to plot monthly variation. The shaded areas indicate the flood season. All units are metric.

According to statistics for 1957 to 1969 at Huang-Gou, the maximum suspended sediment concentration may reach 455 mg l^{-1}, but the annual mean concentration in the mouth of the river is less than 10 mg l^{-1}. The mean annual suspended sediment discharge of 2 × 10^6 metric tons accounts for 60% of the total fluvial sediment input to the coastal zone of the northern Yellow Sea from the east coast of Liao-Dong Peninsula. The suspended sediment discharge is distributed unevenly over the year with discharge in July and August accounting for 72 to 86% of the annual total.

There is a turbidity maximum between Wen-Zi-Gou and the estuary mouth, a distance of about 25 km. Maximum suspended sediment concentrations are in the upper reaches of the Western Waterway, the segment of the estuary with severe shoaling.

Tides are semi-diurnal with a diurnal inequality. The average tidal range near the mouth of Western Waterway is 4.6 m; the maximum is 6.7 m. Because of the effects of dams on river discharge, tidal influence on estuarine circulation has increased over the past several decades. Tidal limits have moved farther upstream, and the estuarine circulation pattern has shifted toward a tidally-dominated (well-mixed) estuary, thereby decreasing filtering efficiency of the estuary. Dams have further retarded the natural geological aging process of the estuary because of trapping of sediment by the reservoirs. The upstream limit of salt water intrusion is about 25 km from the river mouth.

Three waterways (Western, Middle and Eastern) connect the estuary with the Yellow Sea. Over the past several decades, a significant amount of sediment has been deposited in the upper reaches of the Western Waterway, and the river flow through this branch now is very small. The flood current predominates, and the net water and sediment transports in Western Waterway are landward. Most of the river discharge is through the Middle Waterway. As a result, the ebb current is predominant.

Our estimate is that about half of the Yalu's sediment load is transported out of the estuary; most is carried westward. The Yalu, therefore, has a filtering efficiency for fluvial sediment of about 50%.

Han River Estuary

The Han River, the largest river entering the Yellow Sea from South Korea, passes through Seoul, South Korea's capitol and its largest city (Fig. 6). The Han River is 488-km long, and has a drainage area of 26,000 km^2. Two major tributaries, the North and South Han Rivers, converge about 90 km above the junction with the Imjin River.

The average flow of the Han River is about 800 m^3 s^{-1} and its average annual volume discharge is 25 km^3. The Han has a characteristic seasonal variation of flow (Fig. 5), with maximum flow in July and August when summer monsoon and occasional typhoons carry heavy rainfall. The secondary peak in April and May probably is caused by melting of snow. About 5 to 10% of the total annual river discharge comes from the melting of snow, 20 to 25% from groundwater, and 65 to 75% from rainfall which is concentrated in the summer (Korea

Ocean Research and Development Institute 1978; Korean Industrial Site Development Corporation 1971).

The concentration of suspended sediment varies seasonally from about 3 to 50 mg l^{-1} in the upper reaches of the estuary. The variation is caused primarily by changes in river discharge and tidal currents. In the lower reaches of the estuary where tidal currents are strong, maximum concentrations slightly lag maximum current speeds, and minimum concentrations occur near times of high water slack. About 10 km upstream from the mouth a maximum concentration of suspended sediment of 500 mg l^{-1} throughout the water column occurred at spring tide of flood season and was reduced to 100 mg l^{-1} at neap tide (Oh 1984). Sediment discharge has been reduced greatly by dams. In the Whacheon and Cheongpyeong Dam reservoirs sedimentation rates are 5.5×10^6 m^3 y^{-1} and 1.7×10^6 m^3 y^{-1}, respectively (Ministry of Construction 1974).

Tides are semi-diurnal near the mouth and the tidal range varies from about 3.5 m at neap to 8.0 m at spring at Incheon (Oh 1982). Because of the extreme geomorophological complexity around the mouth, tidal currents show wide variation. According to Oh's (1982) summary of previous works (Lee 1972a,b), flood currents are 100 to 150 cm s^{-1} at spring and 55 to 83 cm s^{-1} at neap tide. Ebb currents are slightly stronger than flood currents and magnitudes are 100 to 155 cm s^{-1} at spring and 55 to 88 cm s^{-1} at neap tide. In narrow straits, currents become even stronger.

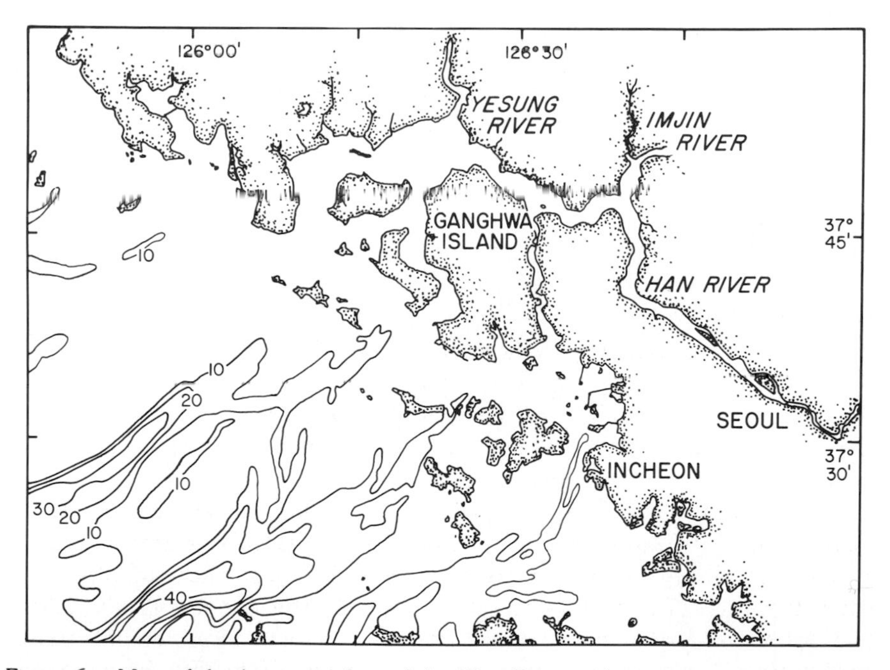

Figure 6. Map of the lower reaches of the Han River and its estuary. Depths are in meters.

The maximum penetration of sea salt has been observed near the Indogyo Bridge, approximately 90 km from the mouth of the estuary. This distance fluctuates, of course, with river discharge and with tidal conditions. When river discharge is between 650 and 700 $m^3 s^{-1}$ (average is 800 $m^3 s^{-1}$), the limit of sea salt penetration is about 60 km above the mouth (Y.A. Park, pers. comm. 1984).

Recent observations (Oh 1984) indicate that the estuary varies between partially mixed (Type B) during periods of high river discharge and neap tides and well-mixed (Type C-D) during periods of low river discharge and spring tides.

Numerous sandbars and sandbanks exist because of the sediment discharge of the river and strong tidal currents near the mouth of the estuary. Vast tidal flats also are formed because of the large tidal range. The source of the sediment which makes up the tidal flats is obscure. It is unlikely that it comes from the present supply of fluvial sediment of the Han or from any other Korean rivers, individually or collectively.

In view of the similarity in physical and geomorphological characteristics of the Han and Keum River estuaries, we estimate that the filtering efficiency of the Han River estuary is greater than 50%.

The Keum River Estuary

The Keum River estuary, one of the typical drowned river valleys of Korea, is located on the southwestern coast of Korea (Fig. 1). The Keum River is 401 km long and has a drainage area of 10,000 km² (Park 1981). The estuary is funnel shaped with its maximum width of about 3 km at the mouth (Fig. 7). Numerous shoals exist inside and outside the estuary and depths are generally less than 10 m.

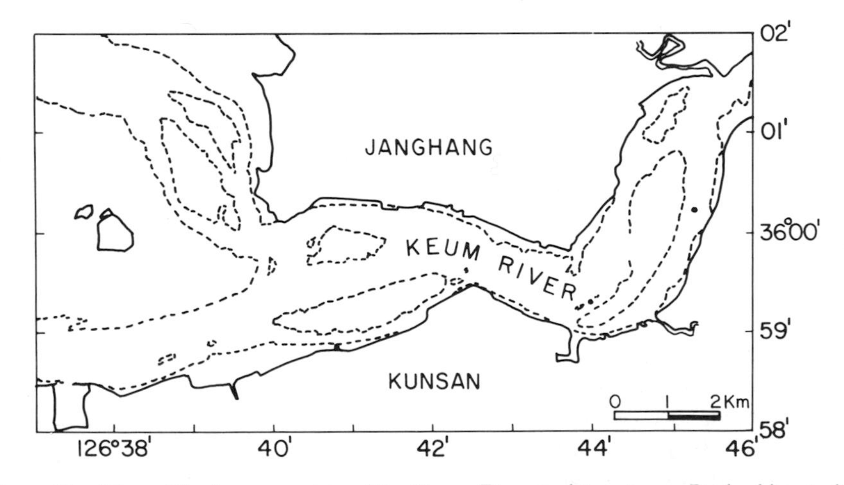

Figure 7. Map of the lower reaches of the Keum River and its estuary. Dashed line indicates the mean low water contour.

The average freshwater discharge is about 200 m^3 s^{-1} and annual river volume discharge is 6.4 km^3. There is a large seasonal variation in river discharge (Fig. 5). During the flood season, July through September, the River discharges about 60% of its total annual discharge (National Institute for Environmental Research 1982).

The concentration of suspended sediment in the estuary varies spatially and on a variety of temporal scales, including semi-tidal (ebb-flood), spring-neap and seasonal (Park 1981; Lee 1984). Although the mean concentration of suspended sediment in the lower river, about 100 km from the mouth, is only about 30 mg l^{-1} in flood season, it varies from 280 mg l^{-1} for the entire water column near the mouth to about 1000 mg l^{-1} in the turbidity maximum 30 km upstream from the mouth at spring tide of the dry season (Lee 1984). Maximum concentrations over 3 g l^{-1} were observed near the bottom of the turbidity maximum that extends over 40 km (Lee 1984). In this same region, however, the concentration during neap tide in the flood season was only 5% of that during spring tide of dry season, apparently because of reduced resuspension by tidal energy. Relocation of the turbidity maximum may also be a factor.

The annual suspended sediment discharge of the Keum River is about 1.3 $\times$ 10^6 metric tons and approximately 79% occurs in the flood season (Fig. 5). The few data we could find on bedload suggest it may be roughly equivalent to the suspended load.

Tides are semi-diurnal with diurnal inequalities. Spring and neap tidal ranges are 5.7 and 2.8 m at the mouth of the estuary. The spring tidal range near the head of estuary, 60 km from the mouth, is about 50% of that at the mouth (Lee 1984).

Generally, flood currents are stronger than ebb currents at all depths (Chung 1981; Lee 1984). Both exceed 2 m s^{-1} near the mouth, but decrease to less than 0.5 m s^{-1} at the head of the estuary on spring tide (Lee 1984).

Because of intense turbulent mixing by strong tidal currents, the vertical variation of salinity between surface and bottom is only about 1‰. In addition there are occasional salinity inversions (Chung 1981; Lee 1984). The maximum vertical salinity gradient exists near the surface rather than near mid-depth as observed in other macrotidal estuaries of small stratification,e.g., the Mersey estuary in the U.K. (Bowden 1963). Even with a small salinity difference of only 1‰ or more between surface and bottom layers, the tidally-averaged net flow may be seaward at the surface layer and landward at the bottom layer as Bowden (1963) has shown.

Salinity variations during the tidal cycle are generally regular showing the maximum at high water slack and minimum at low water slack. On spring tide in the dry season, salinity varies up to 20‰ during the tidal cycle near the mouth, but this variation decreases gradually to less than 1‰ 60 km upstream where salinity fluctuates irregularly and appears to be independent of tidal cycle (Lee 1984). The maximum salinity intrusion is over 60 km at spring tide, but is reduced to about 30 km at neap tide of flood season (Lee 1984). Because there is great variability seasonally in freshwater discharge and in tidal prism on spring-neap cycles, the Keum varies from well-mixed (Type C-D) at spring tide in dry season to partially-mixed (Type B) at neap tide of flood season.

Recent studies on the distribution of clay minerals and trace metals on suspended and bottom sediments near the Keum estuary and adjacent continental shelf (Park 1981; Chough 1981) indicate that suspended sediment discharged from the Keum estuary is dispersed mainly southwestward. The most probable forcing function responsible for this dispersal is the northwesterly monsoon from China in late autumn through early spring. A net southward flux of wave energy along the southwestern coast of Korea occurred for six months from November, 1980 through April, 1981 (Lee and Park 1982), and satellite images also showed a south-westward plume of suspended sediments in winter (Wells *et al.* 1984).

The Keum is still in a relatively young stage of geological evolution. Using the available data (Korea Industrial Site Development Corporation 1979), we estimate the filtering efficiency of the Keum estuary to be about 65%.

A Comparison of Estuaries on the East and West Coasts of the Yellow Sea

There are distinct differences between rivers entering the Yellow Sea along its east and west coasts (Table 1). Relative to rivers on the Korean side,

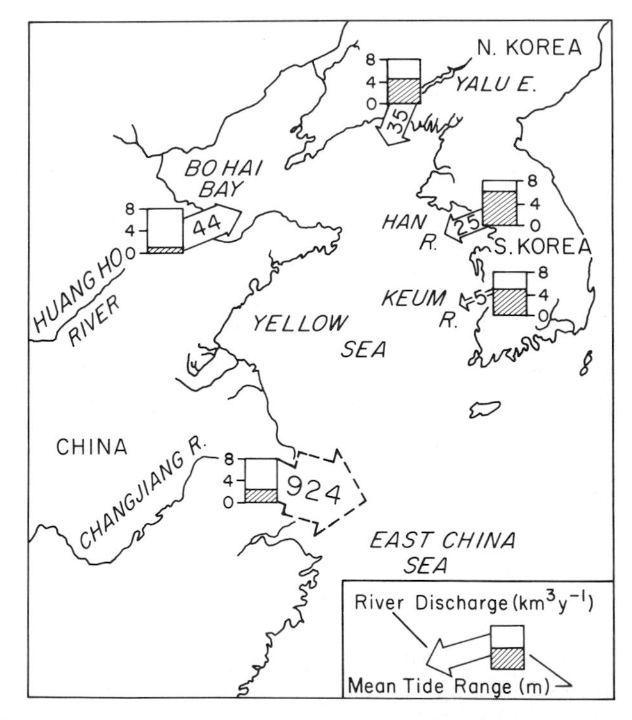

Figure 8. *Annual river volume discharge and mean tidal range at the mouths of major estuaries bordering the Yellow Sea. Note: River discharge for the Changjiang is not to scale of the others.*

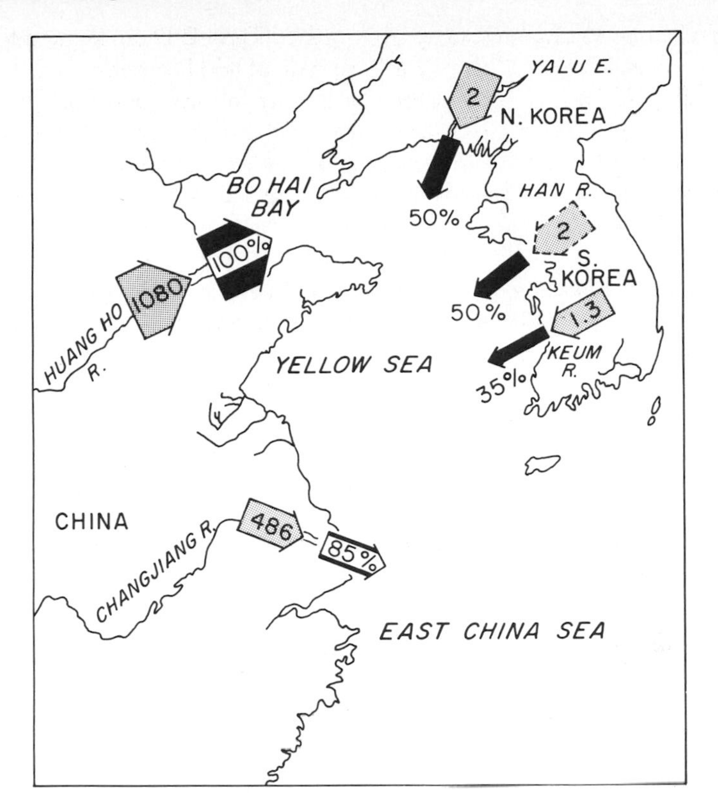

Figure 9. Annual suspended sediment discharges of major rivers bordering the Yellow Sea and the percent of each that is passed through the estuary to the Yellow Sea. Note the difference in scale between Korea and China. Units are in millions of metric tons per year. The arrow for the Han River is dashed to indicate that its sediment discharge is poorly documented.

rivers on the west coast have large water discharges, large sediment loads and small tidal ranges (Figs. 8 and 9). Because of these factors, estuaries on the two sides have different circulation types and are in quite different stages of geological evolution.

Most of the sediment discharged from the Chinese side of the Yellow Sea is fine-grained and is transported as suspended load. On the Korean side, by contrast, slopes of rivers are steeper, tidal currents stronger, sediments are coarser grained and the fraction of total sediment load carried as bed load is greater.

Because of the large sediment discharges of the Huang Ho and Changjiang Rivers on the west coast, their post-Holocene, transgressive drowned river valleys were filled quickly with sediment resulting in the formation of extensive deltas. These two drowned river valley estuaries were transformed into deltaic estuaries. The Huang Ho no longer has an estuary, and the estuary of the

Table 1. Characteristics of Yellow Sea estuaries

Characteristic	Changjiang Estuary	Huang Ho Estuary	Yalu River Estuary	Han River Estuary	Keum Estuary
Location	121°52′E-31°20′N	118°50′E-37°30′N	124°25′E-39°50′N	126°15′E-37°40′N	126°40′E-30°00′N
Length of River (km)	6,300	5,464	800	488	401
Drainage Area (km^2)	1.8×10^6	0.752×10^6	6.1×10^4	2.6×10^4	1.0×10^4
Length of Estuary (salt reach) (km)					
Low River Flow	85	1-2	25 (mean)	90	60 (spring)
High River Flow	South Branch 15 North Branch 75	0			20-30 (neap)
Tide type at mouth	Semi-diurnal	Semi-diurnal Irregular	Semi-diurnal Irregular	Semi-diurnal	Semi-diurnal
Tide range at mouth (m)	2.66 (average) 4.62 (max.)	0.6-2	4.6 (average) 6.7 (max.)	3.5-8.0 (range) 5.7 (mean)	2.8-5.7 (range) 4.3 (mean)
Tidal Current Velocity ($m\ s^{-1}$)	1.02 (mean)* 2.50 (max.)	0.5-1.5	1.3-1.5	1.0-1.5 (spring) 0.5-0.8 (neap)	1.3-1.6 (spring) 1.0-1.3 (neap)

Table 1. (Continued)

Characteristic	Changjiang Estuary	Huang Ho Estuary	Yalu River Estuary	Han River Estuary	Keum Estuary
River discharge ($m^3 s^{-1}$) max	92,600	10,400			
mean	29,300	1,400	1,100	800	200
River discharge ($km^3 y^{-1}$)	924	44.28	34.7	25	6.4
Suspended Sediment at Estuary mouth ($kg m^{-3}$)	1.32 (mean)*	23.5 (mean)	0.1-0.2	0-0.1 (mean)	0.1-0.3
	4.31 (max.)	222 (max.)		1.0 (max.)	
Average grain size of suspended sediment near mouth (mm)	0.0016-0.016	0.02-0.04	n/a	n/a	0.002
Suspended sediment discharge (metric tons y^{-1})	486×10^6	1.1×10^9	2.04×10^6	2×10^6 (est.)	1.3×10^6
Circulation type	South Branch B North Branch C		B-C	B-D	B-D
Filtering efficiency (%)	10-15	~0	<50	>50	65

*mean of the tidal cycle in spring tide during high flow season

Changjiang is small. The Huang Ho and the Changjiang are mature geologically in terms of their stage of infilling, and their low filtering efficiencies for river-borne sediments. The filtering efficiency of the Huang Ho is near zero and that of the Changjiang about 10 to 15%. Most sediment discharged by these rivers is deposited near their river mouths or transported out to sea. The Chinese rivers make greater contributions of sediment to the Yellow Sea than do the smaller Korean rivers which have much smaller sediment loads, less mature estuaries and, as a result, higher filtering efficiencies (Fig. 9).

Rivers on the east coast such as the Han and Keum Rivers have relatively small sediment discharges. Their drowned river valleys are only partially filled and they retain the characteristic outline of drowned river valleys. These estuaries are immature relative to those on the west coast, have a greater potential to accommodate sediment and have relatively high filtering efficiencies for the fluvial sediments they receive. The Korean estuaries contribute far less to sedimentation of the Yellow Sea than do the Chinese rivers. The differential in sediment input on the two sides of the Yellow Sea (Fig. 9) contributes to shallower depths on the western side of the Yellow Sea compared to those on the eastern side (Fig. 1).

The numerous sand bars near the mouth of the Yalu estuary at the apex of the western Korean Bight may indicate a relatively large sediment discharge, but they could be relict. The large tidal range ($\sim$ 7 m) and strong tidal currents (1 to 1.5 m s^{-1}) outside the mouth of the Yalu produce a hydraulic regime which inhibits the formation of a delta. The Yalu River estuary may represent a transition in filtering efficiency and stage of evolution between the estuaries on the two sides of the Yellow Sea.

Acknowledgments

We thank Kenneth Swider and James Liu for assistance in gathering data, Marie Eisel, Marie Gladwish and Vivian Rieger for graphics, and Connie Pawl and Lisa Mayer for typing. Contribution 504 of the Marine Sciences Research Center.

References Cited

Bowden, K. F. 1963. The mixing processes in a tidal estuary. *Internatl. J. Air Water Pollution* 7:343-356.

Chen, Ji-yu, C. X. Yun and H. G. Xu. 1982. The Model of Development of the Changjiang Estuary During the Last 2000 Years. pp. 655-675. *In:* V. S. Kennedy (ed.), *Estuarine Comparisons.* Academic Press, New York.

Chen, Ji-yu, Hui-tang Zhu, Yong-fa Dong and Jie-min Sun. 1983. Development of the Changjiang Estuary and its subaqueous delta. pp. 34-51. *In:* Acta Oceanologica Sinica (ed.), *Proceedings of international symposium on "Sedimentation on the continental shelf, with special reference to the East China Sea"*, April 12-16, 1983, Hangzhou, China. China Ocean Press, Beijing, China.

Chough, S. K. 1981. Further evidence of fine-grained sediment dispersal in the southeastern Yellow Sea. pp. 173-193. *In:* Y. A. Park, J. Y. Chung, J. H. Shim, K. Kim and S. K. Chough (ed.), *A Basic Study on Oceanographic Environments off the West Coast of Korea—the Keum Estuary and the Adjacent Continental Shelf.* Res. Rep., Dept. of Oceanography, Seoul National University, Seoul, Korea.

Chung, J. Y. 1981. Estuarine dynamics of the Keum Estuary I. Flow characteristics and a circulation model. pp. 58-115. *In:* Y. A. Park, J. Y. Chung, J. H. Shim, K. Kim and S. K. Chough (eds.), *A Basic Study of Oceanographic Environments off the West Coast of Korea—the Keum Estuary and the Adjacent Continental Shelf.* Res. Rep., Dept. of Oceanography, Seoul National University, Seoul, Korea.

Korea Industrial Site Development Corporation. 1971. *Report on Investigation of the Han River Basin, Appendix I.* Ministry of Construction, Seoul, Korea. 584 pp. (In Korean).

Korea Industrial Site Development Corporation. 1979. *Report on Investigation of the Keum River Basin.* Ministry of Construction, Seoul, Korea. 530 pp. (In Korean).

Korea Ocean Research and Development Institute. 1978. *Survey of the Water Quality, Flow and Pollutant Sources of the Han and the Nagdong Rivers.* Seoul, Korea. (In Korean).

Lee, C. B. 1984. Sedimentation and geo-chemical characteristics of suspended sediments in the Keum Estuary. Res. Rep., Dept. of Oceanography, Seoul National University. Seoul, Korea. 52 pp. (In Korean).

Lee, S. W. 1972a. Characteristics of tidal currents near Asan Bay and Port Incheon. pp. 7-15. *In:* Tech. Rep., Korea Hydrographic Office, Seoul, Korea. (In Korean).

Lee, S. W. 1972b. Tide, tidal current and tidal prism at Incheon. *J. Oceanol. Soc. Korea* 7:86-97. (In Korean).

Lee, S. W. and M. J. Park. 1982. Sediment transport near Buan. Unpubl. Rep. Korea Ocean Sci. Eng. Corp., Seoul, Korea. 19 pp.

Milliman, J. D. and R. H. Meade. 1983. World-wide delivery of river sediment to the oceans. *J. Geol.* 91:1-21.

Milliman, J. D., Zuo-Sheng Yang and Robert H. Meade. 1983. Flux of Suspended Sediment in the Changjiang Estuary. pp. 382-399. *In:* Acta Oceanologica Sinica (ed.), *Proceedings of international symposium on "Sedimentation on the continental shelf with special reference to the East China Sea",* April 12-16, 1983. China Ocean Press, Beijing, China.

Ministry of Construction. 1974. *Report on Investigation of Korean Rivers.* Seoul, Korea. 1900 pp. (In Korean).

National Institute for Environmental Research. 1982. *Report on Basic Investigation on Major Korean Rivers* (2nd year report). Seoul, Korea. 1061 pp. (In Korean).

Oh, J. K. 1982. Investigation of depositional sedimentary environment around Gyeong-gi Bay by the survey of primary sedimentary structures and textures. Res. Rep., Dept. of Oceanography, Inha Univ., Incheon, Korea (In Korean). 34 pp.

Oh, J. K. 1984. Estuarine sedimentological study in a mouth of the Han River estuary: Yumha channel, west coast of Korea. American Geophysical Union Fall meeting Abstracts, San Francisco, Calif.

Pang, Jia-zhen and Shu-heny Si. 1979. The Estuary Changes of the Huang He River (Yellow River). I. Changes in Modern Time. *Oceanologia et Limnologia Sinica.* 10(2):136-141 (in Chinese).

Pang, Jia-zhen and Shu-heny Si. 1980. The Estuary Changes of the Huanghe River (Yellow River). II. Hydrographical character and the Region of sediment silting. *Oceanologia et Limnologia Sinica* 11(4):295-305 (in Chinese).

Park, Y. A. 1981. Marine geological study in the Keum Estuary and the adjacent continental shelf-clay minerals of the bottom sediments and suspended material in surface and near-bottom waters. pp. 3-57. *In:* Y. A. Park, J. Y. Chung, J. H. Shim, K. Kim and S. K. Chough (eds.), *A Basic Study on Oceanographic Environments off the West Coast of Korea—the Keum Estuary and the Adjacent Continental Shelf.* Res. Rep., Dept. of Oceanography, Seoul National University, Seoul, Korea.

Pritchard, D. W. 1967. What is an estuary: Physical viewpoint. pp. 3-5. *In:* G. H. Lauff (ed.), *Estuaries.* Amer. Asso. Adv. Sci. Pub. Spec. Paper No. 83, Washington, DC.

Schubel, J. R. and H. H. Carter. 1984. The estuary as a filter for fine-grained sediment. pp. 81-105. *In:* V. S. Kennedy (ed.), *The Estuary as a Filter.* Academic Press, New York.

Shen, Huan-ting and Ding-an Pan. 1979. The characteristics of tidal current and its effect on the channel changes of the Yangtze Estuary. *Journal of East China Normal University (Natural Science)* 2:131-144 (in Chinese).

Shen, Huan-ting, Hui-fang Zhu and Zhi-Chang Mao. 1982. Circulation of the Changjiang Estuary and its effects on the transport of suspended sediment. pp. 667-691. *In:* V. S. Kennedy (ed.), *Estuarine Comparisons.* Academic Press, New York.

Shen, Huan-ting, Jiu-fa Li, Hui-fang Zhu, Ming-bas Han, and Fu-gen Zhou. 1983. Transport of the Suspended Sediments in the Changjiang Estuary. pp. 359-381. *In:* Acta Oceanologica Sinica (ed.), *Proceedings of international symposium on "Sedimentation on the continental shelf with special reference to the East China Sea"*, April 12-16, 1983, China Ocean Press, Beijing, China.

Wells, J. T., Y. A. Park and J. H. Choi. 1984. Storm induced fine sediment transport, west coast of Korea. Marine Geology and Physical Processes of the Yellow Sea. pp. 309-313. *In:* Y. A. Park, O. H. Pilkey and S. W. Kim (eds.), *Proceedings of Korea-U.S. Seminar and Workshop.* Korea Institute of Energy and Resources, Seoul, Korea. 328 pp.

COMPARATIVE OCEANOGRAPHY OF COASTAL LAGOONS

Björn Kjerfve
Belle W. Baruch Institute
for Marine Biology and Coastal Research
Department of Geology and Marine Science Program
University of South Carolina
Columbia, South Carolina

Abstract: Coastal lagoons are shallow water bodies, separated from the adjacent ocean by a barrier. Some lagoons have only a narrow entrance channel, which at times may be completely closed off. These are the *choked* lagoons, which are characterizied by dynamic wind forcing, highly variable circulation response, and lack of significant tides. Whereas these lagoons may be fresh water systems in areas of high rainfall and runoff, they sometimes become salt flats in arid regions where they can be closed off from the sea for prolonged time periods. On the other end of the spectrum are the *leaky* lagoons. Such lagoons are either connected by several entrance channels to the adjacent ocean or are separated from it by an incomplete barrier, consisting of a series of sand islands or coral reefs. Leaky lagoons are usually strongly affected by tidal action, have oceanic salinity characteristics, exhibit persistent tidal circulation patterns, and are impacted by wind forcing. *Restricted* lagoons are intermediary to the choked and leaky extremes. Oceanographic features in ten lagoon systems are compared and classified.

Introduction

Coastal lagoons are among the most common coastal environments, occupy 13 percent of the world's coastline (Barnes 1980), and have a number of common features summarized by Phleger (1969, 1981). Yet very little systematic, scientific work has been carried out in coastal lagoons in comparison to estuaries. One reason for this might be that coastal lagoons are shallow, less suitable as harbors, and thus often without major population centers. Estuaries, on the other hand, usually have deeper channels, make better harbors, encourage population growth, and are located on historical river-transportation routes. Another probable reason for the lack of scientific studies of lagoons is that the majority of coastal lagoons are located either in lesser developed countries or in sparsely populated areas of developed countries.

Coastal lagoons are found on all continents from the tropics to polar regions but are less common on emergent high-latitude coasts (Nichols and Allen 1981). They are particularly prominent in the low latitudinal zone (Davies 1980). They occupy shallow coastal depressions and are separated from the ocean by a barrier. Like estuaries, they are ephemeral coastal features of recent origin. They were formed during the eustatic rise of sea level between the time of the Wisconsin glaciation 18,000 years ago and the present, and stand the risk of being completely infilled by sediments or closed off from the sea by littoral drift (Lankford 1976).

Most previous system-level studies of coastal lagoons have focused on biological/ecological characteristics (e.g. Day *et al.* 1973; Lara-Lara *et al.* 1980; Nixon and Less 1981; Millan-Nuñez *et al.* 1982; Yáñez-Arancibia and Day 1982; Day and Yáñez-Arancibia 1982; Day *et al.* 1982; and Farfan and Alvarez-Borrego 1983). Sedimentological regimes in coastal lagoons and processes of lagoon formation have been analyzed by Orme (1975), Lankford (1976) and in an excellent summary paper by Nichols and Allen (1981). The cumulative knowledge of the functioning of these systems has led to management recommendations and some analyses and syntheses (e.g. Lasserre 1979; Lee and Olsen 1985).

Our ability to predict future changes in coastal lagoons depends on an integrated understanding of hydrological and physical-dynamic lagoon processes. A few individual lagoons have been investigated from a hydrographic-hydrodynamic point of view: e.g. Wadden Sea, Denmark, Germany and The Netherlands (Postma and Dijkema 1982; Zimmerman 1976); Laguna de Terminos, Mexico (Graham *et al.* 1981); Indian River, U.S.A. (Smith and Kierspe 1981); Laguna Caimanero-Huizache, Mexico (Moore and Slinn 1984); Mississippi Sound, U.S.A. (Eleuterius 1976; Kjerfve 1983); Lake Pontchartrain, U.S.A. (Sakou 1983; Swenson 1981; Chuang and Swenson 1981; Swenson and Chuang 1983; Sikora and Kjerfve 1985); Lagoa dos Patos, Brazil (Herz 1977); Ninigret Pond, U.S.A. (Isaji and Spaulding 1985); and the lagoons of Texas (Collier and Hedgepeth 1950; Copeland *et al.* 1968; Smith 1977). In other instances, the physical processes in lagoons have been the focus of studies, e.g. wind-induced effects (Beer and Black 1979; Noye and Walsh 1976); and tidal mixing, dispersion, and flushing (Zimmerman 1978, 1981; Dronkers and Zimmerman 1982). Few previous studies have focused on common physical and oceanographic features of coastal lagoons.

The purpose of this paper is to describe and analyze relevant geographical, hydrological, and oceanographic characteristics of coastal lagoons in an attempt to formulate some generalizations. My main hypothesis is that physical lagoon characteristics and variabilities depend primarily on the nature of the channel(s) connecting the lagoon to the adjacent coastal ocean. The channel types vary along a continuum resulting in lagoons which can be classified as choked, restricted, and leaky. At a later time, it may be desirable to parameterize lagoon processes and present the results in a more quantitative fashion than is done here (cf. O'Brien 1969).

Characteristics of Ten Lagoons

Initially, I will describe salient physical characteristics of 10 selected lagoon systems (Table 1).

Laguna Joyuda (Fig. 1) on the west coast of Puerto Rico, is an example of a choked system (Levine 1981). The lagoon is microsized, measures only 1.4 km^2, and has a maximum depth of 4 m. The 500-m entrance channel

Table 1. Comparative data on ten coastal lagoons.

Lagoon	Latitude	Entrance Type	Ocean Entrances	Surface Area (km^2)	Mean Depth (m)	Ocean Tide at Entrance (m)	Predominant Salinity Characteristics	R Annual Rainfall (m)	E Annual Evaporation (m)	Q Mean River ($m^3 s^{-1}$)	Offshore Wave Energy
Laguna Joyuda (Puerto Rico)	N18°	Choked	1	1	1	0.3 - Diurnal	Estuarine	1.73	R ~ E	0	Low
Coorong (Australia)	S36°	Choked	1	260	2	0.5 - Mixed	Hypersaline	0.30	E > R	0	High
Lake St. Lucia (South Africa)	S28°	Choked	1	312	1	1.5 - Semidiurnal	Estuarine	1.00	1.32	20	High
Gippsland Lakes (Australia)	S38°	Choked	1	340	4	1.0 - Semidiurnal	Estuarine/ Brackish	0.67	1.00	100	High
Lake Songkla/ Thale Luang (Thailand)	S7°	Choked	1	1,040	2	0.5 - Semidiurnal	Estuarine/ Brackish	2.16	1.84	160	Medium
Lagoa dos Patos (Brazil)	S31°	Choked	1	10,360	5	0.2 - Diurnal	Fresh/ Brackish	1.25	R > E	~4,000	High
Lake Pontchartrain (USA)	N30°	Restricted	3	1,630	4	0.5 - Diurnal	Brackish	1.52	R > E	188	Low
Laguna de Términos (Mexico)	N19°	Restricted	2^+	2,500	4	0.5 - Mixed	Estuarine	1.70	R > E	~600	Medium
Mississippi Sound (USA)	N30°	Leaky	Multiple	2,130	3	0.5 - Diurnal	Oceanic/ Estuarine	1.49	R > E	1,400	Low
Belize Lagoon/ Chetumal Bay (Belize/Mexico)	N17°	Leaky	Multiple	12,700	15	0.2 - Mixed	Oceanic	5.0-0.5	1.80	~400	Medium

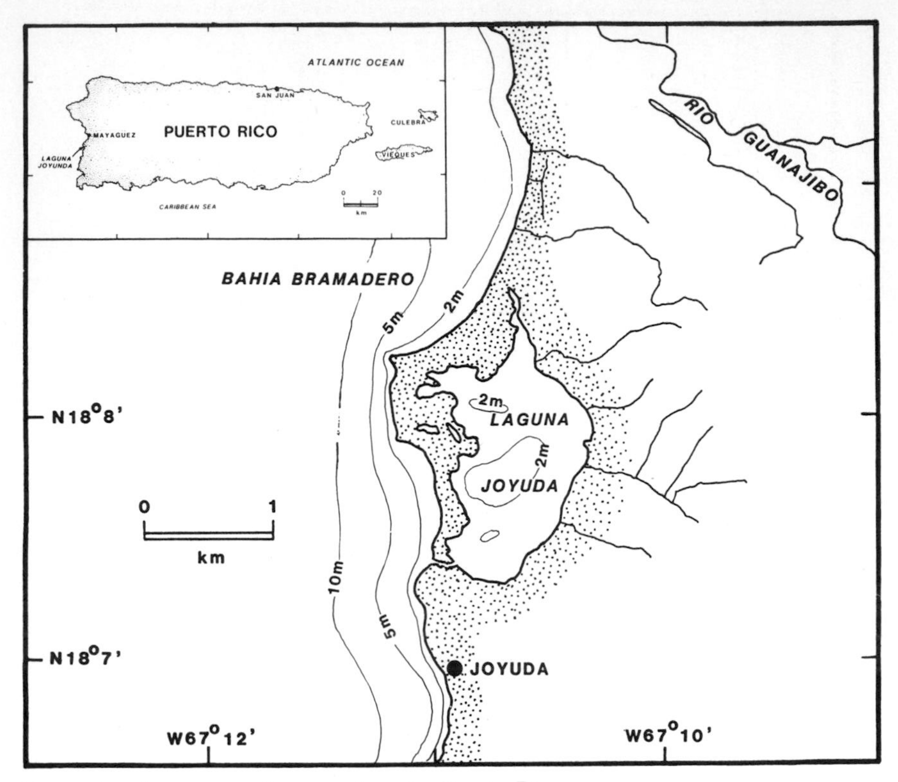

Figure 1. Area map of Laguna Joyuda, Puerto Rico.

winds through dense mangrove vegetation (*Rhizophora mangle*). The channel is typically only 0.4 m in depth and 6 m in width, and is occasionally blocked off by sediments accumulating on the ocean side. The diurnal microtide in the lagoon measures 0.1 m or 30 percent of the tide in the adjacent Mona Passage. Fresh water to the lagoon is derived from ground water and diffuse surface runoff from rainfall over the 6-km² drainage basin. The salinity of the lagoon is largely uniform, varies slowly, and ranges from 8 to 44‰ (Laurence Tilly, pers. comm.).

The Coorong (Fig. 2) is a narrow, shore-parallel, choked lagoon system in South Australia, separated from the Southern Ocean by a dune field 50-m high. The average surface area is 260 km². The maximum depth averages 1.8 m in the summer but regularly increases by 1.2 m in the winter (Noye 1973). The lagoon is located in an arid region where evapotranspiration exceeds runoff and rainfall. Accordingly, the lagoon is hypersaline. The northern basin connects to the Southern Ocean via a lake system at the mouth of the Murray River. This is also the main source of occasional fresh water flow into the Coorong, causing salinity variations in the northern basin. The southern basin is entirely hypersaline with an average salinity of 90‰ (Noye 1973). In

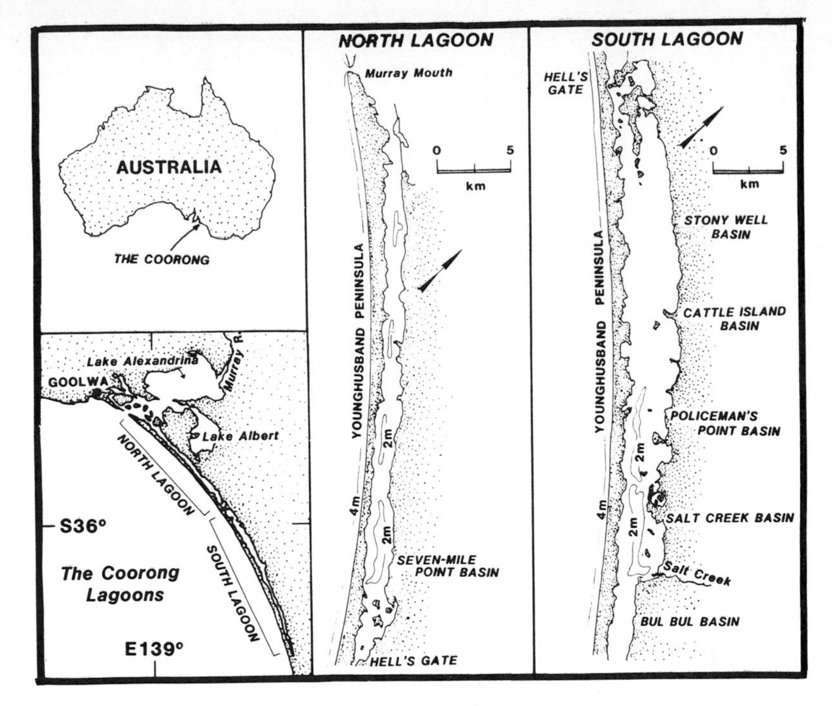

Figure 2. Area map of Coorong, South Australia.

the extreme south, lagoon salinities commonly exceed 200‰ in summer, and little plant and fish life can be sustained (Noye 1973). No water exchange occurs between the northern and southern basins in the summer. The coastal ocean experiences a 0.5-m mixed tide and an extreme wave climate with 4-m (15-s period) waves. These effects do not propagate into the Coorong. On the other hand, the wind stress parallel to the lagoon forces a 0.2-m longitudinal oscillation of water level in each of the two basins with a period of approximately 24 hours (Noye and Walsh 1976).

Lake St. Lucia (Fig. 3), South Africa, is another choked lagoon in a semiarid region, separated from the Indian Ocean by massive coastal barriers 180 m in height. The water-covered area is 312 km^2 but increases to 417 km^2 during the rainy summer period, when the mean water level rises one m (Orme 1975). The average water depth is less than two m, and a long (20 km) entrance channel (the Narrows) connects the lagoon to the high energy wave conditions in the coastal ocean. As a result the strong littoral drift can close off the entrance. The semidiurnal ocean tide measures 1.5 m but is completely filtered out north of the Narrows (Orme 1975). The salinity regime of the lagoon is highly variable on subtidal time scales and also changes seasonally. As an example, the northern end of the lagoon experiences 15‰ salinity during normal rainy seasons but salinities are as high as 50‰ during periods of severe drought (Orme 1975). Sedimentation in Lake St. Lucia is very high and has

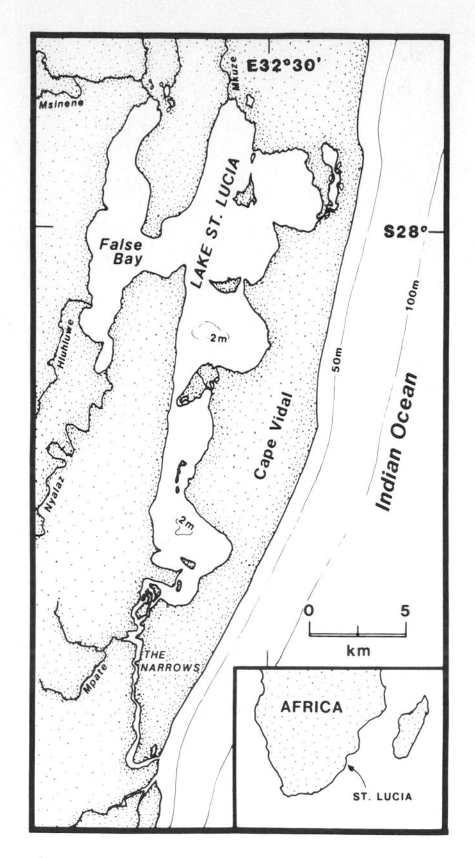

Figure 3. Area map of Lake St. Lucia, South Africa.

been accelerated by human activities. Over the past 5,000 years, the lagoon has experienced a 66 percent reduction in area and a 95 percent reduction in water volume as a result of sediment infilling (Orme 1975).

The *Gippsland Lakes* (Fig. 4) in southeastern Australia are a series of choked lagoons: Lake King (92 km^2), Lake Victoria (110 km^2), Lake Wellington (138 km^2), and a number of smaller lakes. Five rivers feed the lagoons, draining the 10,000 km^2 of catchment basin that consists mostly of farmlands (Bird 1978). The three connected lagoons are rather deep with average/maximum depths of 3/6 m for L. King, 4/10 m for L. Victoria, and 3/4 m for L. Wellington. The high coastal wave power formerly resulted in intermittant closure of the entrance to the lakes. However, the present entrance is an artificially cut and maintained navigation channel that remains open at all times. The semidiurnal ocean tide has a range of 1.0 m but is diminished to less than 0.1 m within 10 km of entering the lagoon channel. The combined fresh water inflow averages 100 m^3 s^{-1} with average September high flows of 180 m^3 s^{-1}. During times of flood discharge, the water level in the Gippsland Lakes may

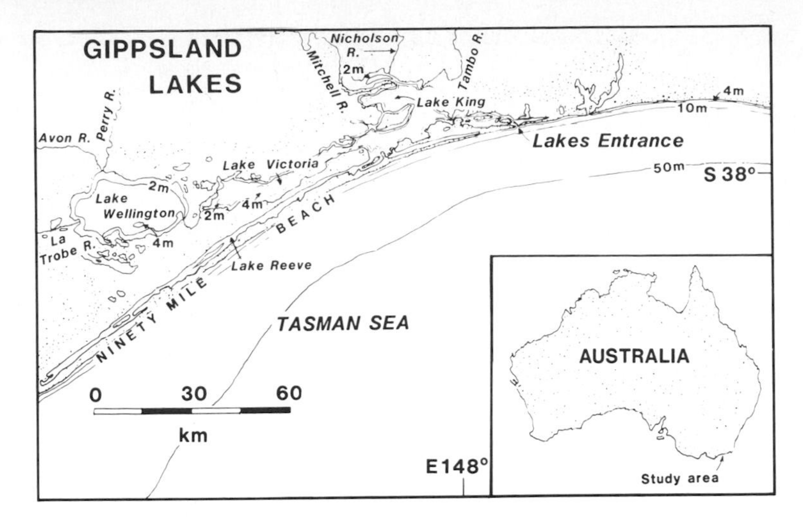

Figure 4. Area map of the Gippsland Lakes lagoon system in Victoria, Australia.

rapidly increase by two m and flood adjacent lowlands (Bird 1978). The salinities vary seasonally with spring lows in November and fall highs in May. The typical spring/fall salinities are 20/30‰ for L. King, 5/20‰ for L. Victoria, and 0/7‰ for L. Wellington (Bird 1978). Predominant and resultant westerly winds blow along the length of the lagoon system, causing choppy, 1-m waves on the lakes (Bird 1978).

Lake Songkla/Thale Luang/Thale Noi (Fig. 5) is a 1,040-km² system of choked coastal lagoons in southern Thailand. The mean depth of the lagoon system is 1-2 m, while the channels connecting Lake Songkla to the coastal ocean and to Thale Luang are 5-8 m deep. The surrounding land is low-lying, dominated by extensive rice fields and mangrove vegetation. The 0.5-m semidiurnal tide in the lower Gulf of Thailand is filtered by the entrance channel and measures less than 0.2 m in Lake Songkla. Fresh water input measures 160 $m^3 s^{-1}$ on the average and is derived from rain falling over the 70,000 km² drainage basin. The rainy season is August-December during the SW monsoon and the dry season is January-July during the dry NE monsoon. Lake Songkla, Thale Luang, and Thale Noi have average salinities of 19, 14, and 0.3‰, respectively (Asian Institute of Technology 1979; Sojisuporn 1984). The upper part of the system is largely covered by water hyacinths, *Eichornia spp*, which impairs the circulation and water exchange.

Lagoa dos Patos (Fig. 6) in southern Brazil is the world's largest choked lagoon. It measures 10,360 km², has a 140,000-km² drainage basin, and connects in its southern extreme to the South Atlantic via a 20-km entrance channel 0.5 to 3 km in width. Porto Alegre, Brazil's fifth largest city and a major port, is situated on the shores of Rio Guiaba in the innermost region of Lagoa dos Patos, and can be reached by seagoing ships navigated 320 m along the

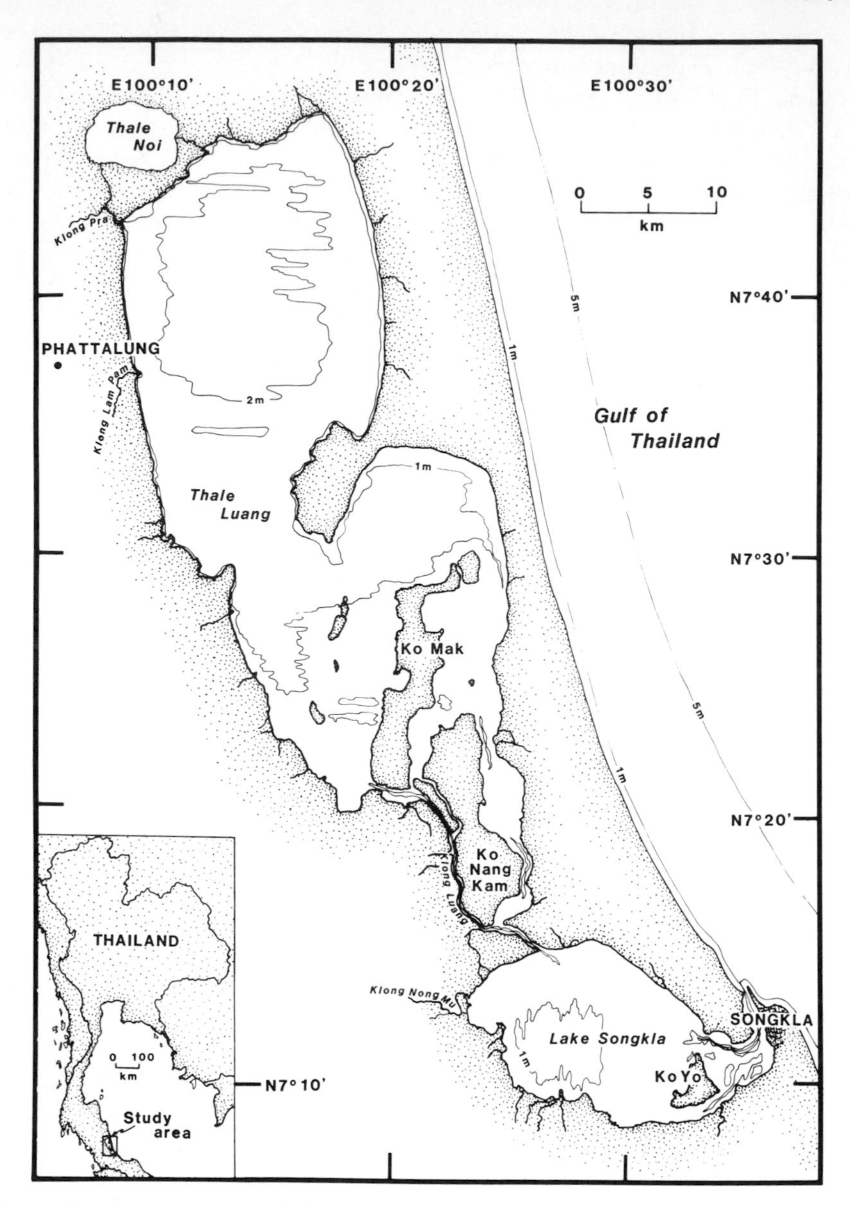

Figure 5. Area map of the Lake Songkla lagoon system in southern Thailand.

length of the 5-m (depth) lagoon. Lagoa dos Patos is usually a fresh water system with an estuarine region limited to the entrance channel and extreme southern lagoon. With winds from the north, the entire system is fresh, but with prolonged winds from the south, brackish water (1-5‰) may extend inland to the mouth of Rio Guiaba (Delaney 1963). Rio Guiaba supplies 86 percent of the average total fresh water input of approximately 4,000 $m^3 s^{-1}$. The peak winter fresh water input exceeds 1,500 $m^3 s^{-1}$ annually with extremes exceeding 25,000 $m^3 s^{-1}$ (Herz 1977). The low diurnal tide range in the adjacent South Atlantic vanishes within the confines of the entrance channel. Currents and water level variations are dominated by wind forcing, which causes a highly variable internal circulation (Bonilha 1974). Winds are predominantly from the northeast during all seasons. However, occasional southerly winds are common in June, associated with winter-time frontal passages (Herz 1977). Because of the large area of the lagoon and the low elevation of the surrounding dune-oriented topography, the wind is the major factor controlling circulation and dispersion in Lagoa dos Patos.

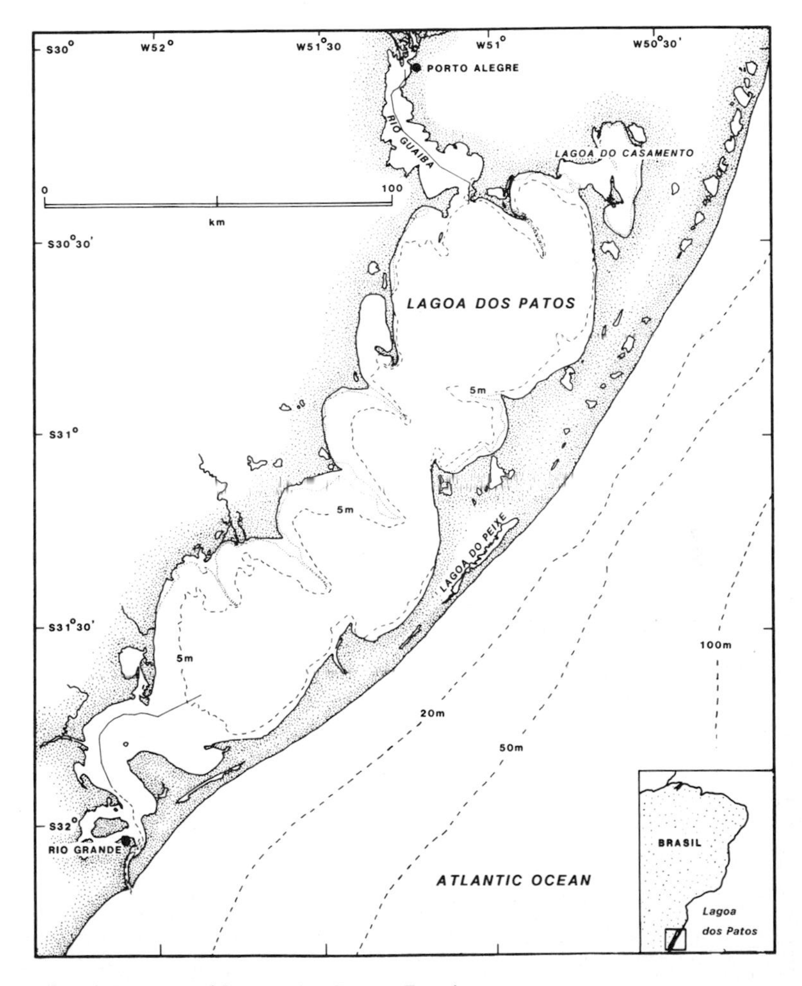

Figure 6. Area map of Lagoa dos Patos, Brazil.

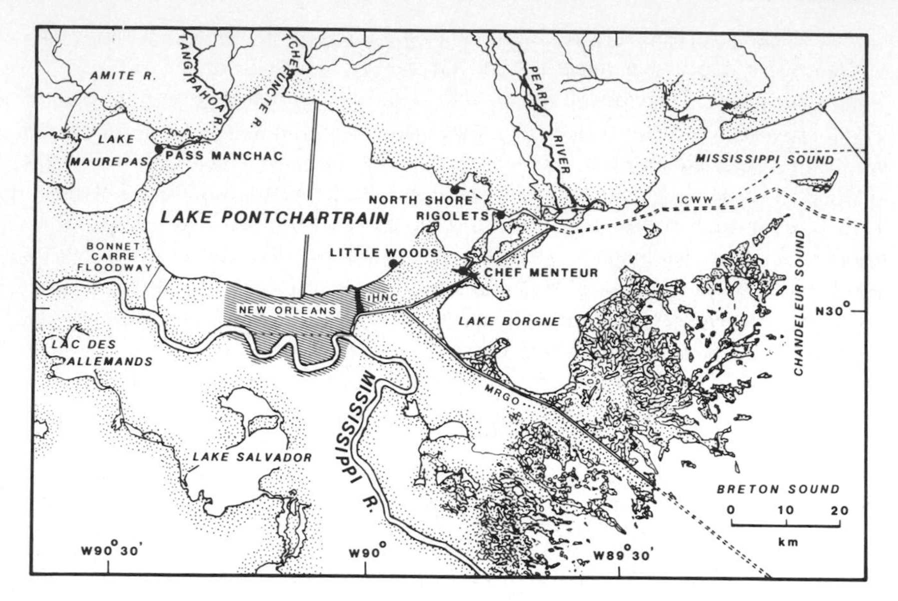

Figure 7. Area map of Lake Pontchartrain, Louisiana, U.S.A.

Lake Pontchartrain (Fig. 7) is a 1,630-km² saucepan-shaped, restricted lagoon in Louisiana with a depth of 3.7 m. It was formed as a result of two Mississippi River delta-building periods during the last 5,000 years. The lagoon connects to the Gulf of Mexico via three routes: 1) Mississippi River Gulf Outlet (MRGO), a 120-km navigation channel dredged to 11-m depth and completed in 1963; 2) the Rigolets; and 3) Chef Menteur Pass. The 0.5-m diurnal tide range is reduced to 0.1 m within the lagoon. Tidal currents are pronounced at all three entrances. Subtidal water-level changes and currents account for approximately 50 percent of the flow variance (Swenson and Chuang 1983). The tidal prism measures 1.56×10^8 m³ (Swenson and Chuang 1983), and flushing is achieved in 20–105 days (Swenson 1980). Substantial subtidal currents and water-level changes result from wind forcing and far-field Gulf effects (Chuang and Swenson 1981). On the average, Lake Pontchartrain receives 188 m³ s^{-1} of fresh water from numerous streams plus a fraction of the 305 m³ s^{-1} Pearl River discharge, which debouches into Lake Borgne near the entrance to the Rigolets. Lake Pontchartrain is a well-mixed brackish lagoon with a mean salinity of 5.4‰ near the Rigolets and 1.2‰ in the channel between Lakes Pontchartrain and Maurepas but with annual salinity variations of 8‰ (Sikora and Kjerfve 1985). As a result of the construction of MRGO, the mean salinity increased by 1 to 2‰ (Sikora and Kjerfve 1985). To protect New Orleans from Mississippi River floods, the Bonnet Carré Floodway was completed in 1931. It has been operated seven times (from 13 to 75 days each time) with discharges varying from 1,950 to 6,343 m³ s^{-1} (Sikora and Kjerfve 1985). Lake Pontchartrain became a fresh water system within two days of the opening of the Bonnet

Carré release structure and returned to average conditions within two months of the end of floodwater release (Sikora and Kjerfve 1985).

Laguna de Términos (Fig. 8) with a 2,500-km² surface area, including fluvial lagoon systems and marshes, is the largest lagoon in Mexico. It is an example of a restricted system, connecting to the Gulf of Mexico via two main inlets, El Carmen and Puerto Real, and several smaller inlets. The lagoon is 4 m deep and surrounded by low-lying mangrove swamps. The tide is mixed with a 0.3-0.7 m range and corresponding oscillatory tidal currents throughout the lagoon. The Palizada, a distributary of the Usumacinta river system, discharges an average 500 $m^3 s^{-1}$ of freshwater into Laguna de Terminos. During September floods, the Palizada discharge may exceed 3,000 $m^3 s^{-1}$ and is then responsible for the input of vast quantities of terrigenous silts and clays to the southwestern end of the lagoon. Two smaller rivers, Chumpan and Candelaria, add an average 100 $m^3 s^{-1}$ to the fresh water input. The typical salinity is 27‰, but it varies seasonally both with distance from the Palizada and with depth (Botello 1978). At the Gulf entrances, the salinity is typically 32-35‰ (Botello 1978; Yáñez-Arancibia et al. 1983) except during floods when salinity in El Carmen Inlet may be depressed to 20‰ or less. Most of the rains occur from June to October. During the dry season (February to June), trade winds from the southeast and east predominate and cause net water movement entering through Puerto Real and exiting through El Carmen inlet. This is the common circulation pattern which exists into the fall. During "norte"

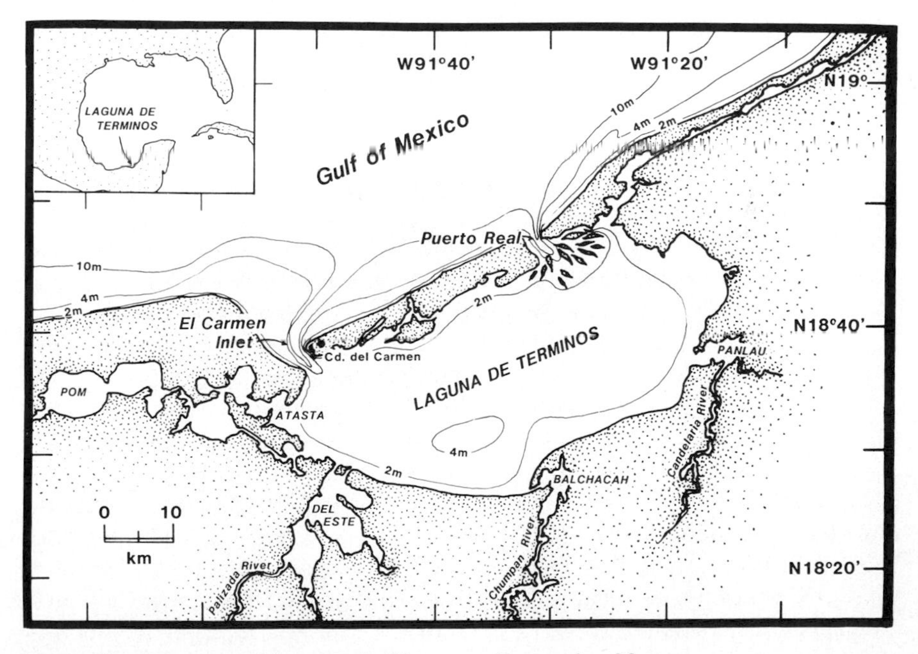

Figure 8. Area map of Laguna de Términos, Campeche, Mexico.

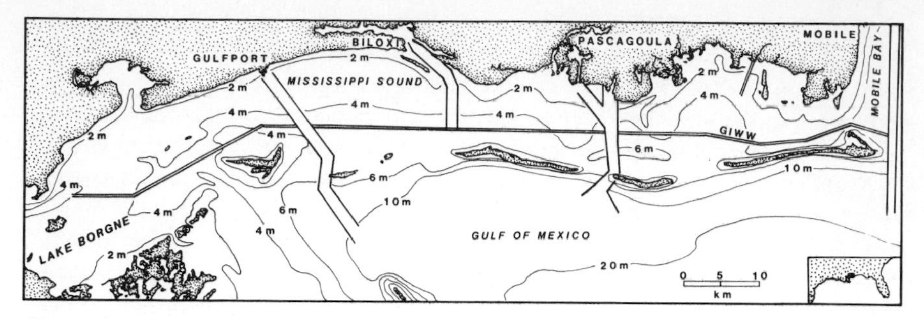

Figure 9. *Area map of Mississippi Sound, along the shores of Alabama and Mississippi, U.S.A.*

wind events from October to February, however, the circulation pattern may reverse directions.

Mississippi Sound (Fig. 9) measures 2,130 km^2 and is a leaky lagoon with a 3-m mean low-water depth. It extends 130 km along the Alabama and Mississippi coastlines and is separated from the Gulf of Mexico by a series of low, sandy barrier islands. It connects to Lake Borgne in the west and Mobile Bay in the east. Half a dozen or more wide tidal passes readily permit water exchange between the lagoon and the Gulf. Because of the leaky nature of the barrier, waves from the Gulf propagate unimpeded into Mississippi Sound. During passages of hurricanes, storm waves often strike the low-lying, densely-populated inner shoreline. The tide is diurnal and has a range of 0.5 m. Although tidal currents account for more than 50 percent of the flow variance, the lagoon responds rapidly to wind forcing (Kjerfve 1983). This is evidenced by sub-tidal sea level variations up to one m and persistent net currents in the tidal passes. Circulation is usually developed in the passes where currents may reach 1.0 m s^{-1}. The circulation within the lagoon is weak and variable, and the system is coherent and vertically well-mixed (Kjerfve 1983). Two major rivers, Pascagoula and Pearl, discharge into the lagoon an average 417 and 362 m^3 s^{-1}, respectively. During peak floods, each river may discharge more than 3,000 m^3 s^{-1}. In addition, a fraction of the discharges from the Mobile and Mississippi River systems enter Mississippi Sound. As a result, the salinity regime is variable and characterized by multiple, sharp fronts. Most commonly, the salinity varies from 20 to 35‰ (Kjerfve 1983; Eleuterius 1976).

The *Belize Lagoon/Chetumal Bay* (Fig. 10) system is a 12,700 km^2 leaky lagoon along the east coast of the Yucatan peninsula. The region is a shallow, reef-rimmed carbonate platform (James and Ginsburg 1979). Numerous tidal passes intersect sections of the shore-parallel barrier reef, the most extensive coral reef in the Americas, and connect the lagoon to the adjacent Caribbean Sea. Sandy cays and mangrove-covered islands are scattered throughout the lagoon, especially in the middle section. This tropical lagoon is effectively divided at the Belize River delta into two separate lagoons. The northern system is uniformly 4 m in depth, largely land-bound, and surrounded by mangroves. The southern system varies in depth from an average 10 m over most of the area to 65 m in the south. The main connection to the Caribbean Sea is an open entrance 25 km wide from which a trough 35 m deep stretches 45 km northward into the lagoon. Sediment profiles across the lagoon indicate a gradient from fine, organic-rich, terrigenous silts and clays at the coast to calcareous silts and clays adjacent to the barrier reef (Miller and Macintyre 1977). The annual rainfall measures only 0.5 m in the semi-arid north to five m annually in the humid tropical southern extreme. Salinity exceeds 30‰ over most of the lagoon except for a band of low-salinity coastal water along the mainland coast in the south, where high-discharge rivers enter the lagoon (Purdy 1974). Belize Lagoon, in particular, is influenced by northeasterly trade winds, experiences occasional hurricanes (Kjerfve and Dinnel 1983), and is susceptible to high, choppy waves. The tide is mixed diurnal with a 0.2-m range and progresses the length of the lagoon system from south to north in 2.1 hours (Kjerfve 1981).

Discussion

Coastal lagoons can conveniently be classified into choked, restricted, and leaky systems. This appears to be a useful division, as each type exhibits a number of functional similarities. The nature of the channel(s) connecting the lagoon to the adjacent ocean controls, more than any other parameter, how the system functions. Choked lagoons are characterized by a single entrance channel and a small ratio of entrance channel cross-sectional area to surface area of the lagoon. They are dominated by the hydrologic/riverine cycles; have long residence times; are wind forced; and experience limited short-term marine variability. They are most common along coastlines with medium to high wave energy. Leaky lagoons, on the other hand, are characterized by multiple entrance channels and a relatively large ratio of entrance-channel cross-sectional area to surface area of the lagoon. They are dominated by marine influence, near-oceanic salinities, strong tidal variability, and occasional significant wave energy. Restricted lagoons represent the middle of the spectrum of lagoons between the choked and leaky extremes.

Choked lagoons are most common on coasts with high wave energy and low tidal range. Strong coastal wave action in combination with available sediment produces littoral drift, which builds the barrier that separates lagoon and ocean, e.g. Lagoa dos Patos. If the littoral drift is substantial, the lagoon may temporarily become closed off from the ocean, e.g. Gippsland Lakes and Lake St. Lucia. On the west coast of Mexico, the closing and opening of lagoon entrance channels is a seasonal cycle (Moore and Slinn 1984). These choked lagoons are closed off from the sea during the dry season. As they are filled by runoff during the rainy season, the channel reopens.

In arid climates, e.g. South and Western Australia, closed-off lagoons may turn into salt flats for periods up to ten years (Patrick Hesp, pers. comm.). Occasional heavy rains and flood runoff can refill the lagoon basin, percolate through the barricaded entrance, and eventually weaken and break through

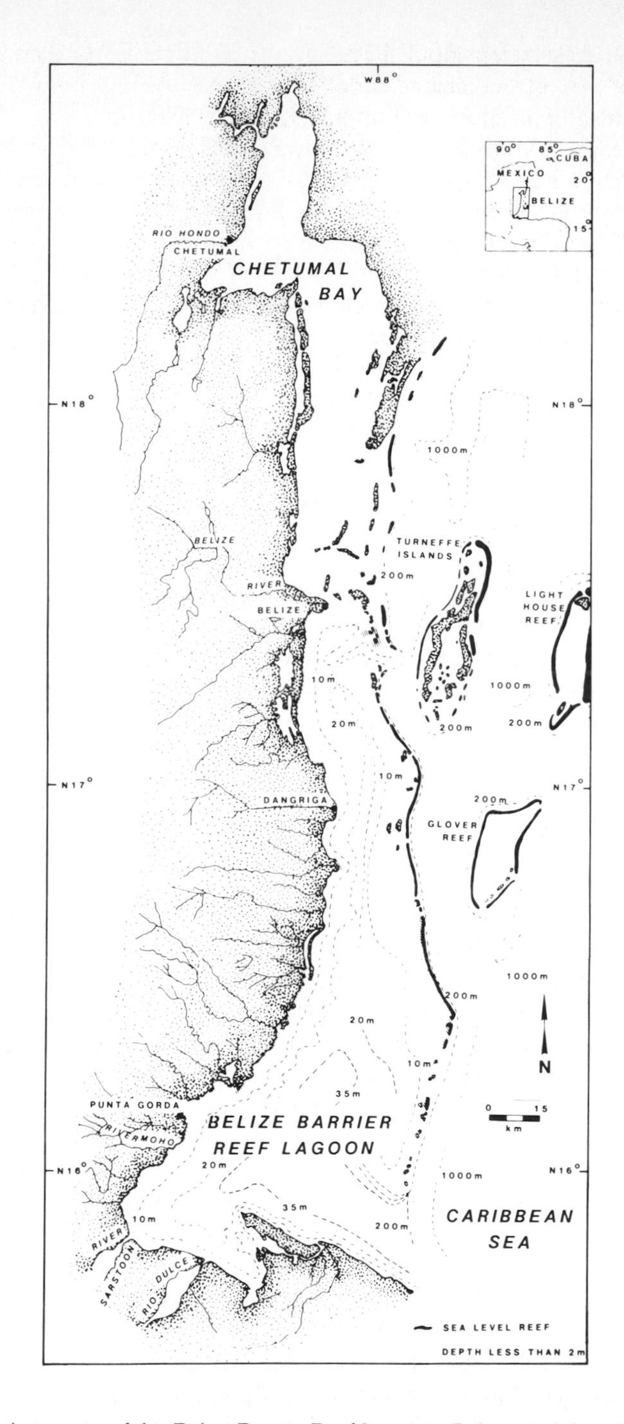

Figure 10. Area map of the Belize Barrier Reef Lagoon, Belize, and the adjacent Chetumal Bay, Mexico, along the Caribbean shores of the Yucatan peninsula.

the barrier. As fresh water input dwindles, salt water will intrude and evaporation will greatly exceed combined runoff and rainfall. As the channel is blocked off by littoral drift, the lagoon will again turn into a salt flat.

Tidal influence in choked lagoons is usually confined to the entrance channel. As a consequence, variability of current and water level is largely related to wind-forcing over a spectrum of frequencies from minutes to weeks. Seiching and set-up/set-down cycles are particularly intense in response to frontal passages (Copeland *et al.* 1968; Kjerfve 1983; Schwing *et al.* 1983). In general, systematic wind-driven circulation patterns are poorly developed and highly variable. Many choked lagoons experience seasonal water level changes with a range exceeding one m in response to onset of the rainy season, e.g. the Coorong and Lake St. Lucia. Similarly, the salinity distribution lacks tidal variability but responds rather to changes in fresh water input on scales from days to months, e.g. Laguna Joyuda and the Lake Songkla system. The climate and hydrological cycles control the magnitude and frequency of the low-frequency salinity variability. A system can change seasonally over the entire range from fresh water lagoon to hypersaline lagoon, to salt flat in extreme cases. Choked coastal lagoons are typically not susceptible to oceanic far-field forcing (Wang and Elliott 1978). Flushing times are usually on the order of months.

Restricted lagoons are usually located on low/medium wave energy coasts with a low tidal range. They are usually connected to the ocean by two or more channels which typically remain open at all times. Tidal water level and current variability are readily transmitted into the lagoon without excessive filtering, e.g. Laguna de Terminos. As a result, restricted lagoons exhibit well-defined tidal circulation, which is modified by wind forcing and fresh water runoff. Low frequency sea level changes in restricted lagoons are less a function of runoff than far-field oceanic forcing, transmitted into the lagoon, e.g. Lake Pontchartrain (Chuang and Swenson 1981). Restricted lagoons are less likely to undergo dramatic fluctuations in salinity as compared to choked lagoons. For the most part, restricted lagoons have rather homogeneous salinity, e.g. Laguna de Terminos (Botello 1978), ranging anywhere from 1-35‰, depending on the fresh water input. They are usually well mixed vertically. Fresh to brackish water is found near the river mouths. During flood discharge the entire restricted lagoon may turn fresh or brackish. Because of the small tidal range and the occurrence of near uniform salinity over large areas, semidiurnal/diurnal tidal salinity variations are minimal except in the entrance channels and tidal passes, e.g. Lake Pontchartrain (Sikora and Kjerfve 1985).

Leaky lagoons are connected to the ocean by wide tidal passes (rather than entrance channels) that transmit oceanic effects into the lagoon with a minimum of resistance. The separating barrier can be either a series of sand islands e.g. Mississippi Sound, or sections of a barrier reef. A reef can be either of coral origin as in the case of the Belize Lagoon/Chetumal Bay system and the Great Barrier Reef in Australia or of sandstone origin as in the case of the reefs along the coasts of northeastern Brazil (Mabesoone 1964) or western Australia (Allison and Grassia 1979). Leaky lagoons are located along coasts with variable tidal and wave characteristics. However, tidal currents must be sufficiently energetic to keep wave-generated littoral drift from closing off the passes. In areas of high tidal range, leaky lagoons exist despite high wave energy conditions, e.g. Great Barrier Reef and Wadden Sea. On coasts with low tidal range, leaky lagoons may still exist as long as wave energy conditions are also low, e.g. Mississippi Sound.

Leaky lagoons are usually characterized by near-oceanic salinities. However, in regions of a lagoon where river runoff is significant in comparison to local tidal exchange, estuarine salinities may persist. The occurrence of a complex pattern of sharp salinity and turbidity fronts is then common, e.g. in Mississippi Sound (Kjerfve 1983) and Belize Lagoon/Chetumal Bay. Tidal circulation patterns in leaky lagoons are usually well-defined but are sometimes modified dramatically by wind forcing. Frontal passages may cause lagoon-wide water level seiching and high wave energy conditions. Leaky lagoons are readily flushed. Most large leaky lagoons are oceanic, rather than estuarine in regard to salinity.

Final Comments

I propose that coastal lagoons (rather than bar-built estuaries) be considered as one of the three major classes of estuaries in addition to drowned river valleys and fjords. This requires an extension of the standard, narrow definition of estuaries (Cameron and Pritchard 1963) to include systems with salinity ranging from fresh to hypersaline. An estuary is then a semi-enclosed body of water, at least intermittently in open connection to the coastal ocean (c.f. Kjerfve, In press).

Coastal lagoons can conveniently by subdivided into choked, restricted, and leaky systems. The type of channel connecting the lagoon to the ocean defines the salient oceanographic characteristics of the system. Dynamic processes due to tides, wind, and density currents are often of equal magnitude, making it difficult to separate these effects.

Future scientific investigations of lagoon processes could benefit from techniques employed in the study of drowned river valley estuaries. Coastal lagoons lend themselves particularly well to numerical circulation/dispersion modeling. Because of weak vertical stratification, two-dimensional tidal models usually work well. In addition to conventional study approaches with moored instruments and hydrographic survey vessels, remote sensing techniques (c.f. Girloff-Emden 1976) are likely to yield useful results in coastal lagoons because of the large surface areas, lack of significant vertical stratification, and sharp density and turbidity fronts. For example, the satellite-sensed turbidity distribution in a coastal lagoon can provide a good synoptic clue to the circulation (c.f. Herz 1977).

Acknowledgments

This synthesis was made possible as a result of NSF grants INT 79-11180 and INT-8215374, NASA grant NAS5-28741, contract DACW01-82-Q-0022 with Army Corps of Engineers, Mobile

District; and other grants, allowing me an opportunity to visit all described lagoons with exception of Lake St. Lucia.

References Cited

Allison, H. and A. Grassia. 1979. Sporadic sea-level oscillations along the Western Australian coastline. *Aust. J. Mar. Freshwater Res.* 30:723-730.

Asian Institute of Technology. 1979. Effect of channel dredging on discharge pattern and salinity distribution of Songkla Lagoon. Songkla Port Project. Report 104. Vol. II. Bangkok.

Barnes, R.S.K. 1980. *Coastal Lagoons.* Cambridge University Press. Cambridge. 106 pp.

Beer, T. and R. Black. 1979. Water exchange in Peel Inlet, Western Australia. *Aust. J. Freshwater Res.* 30:135-141.

Bird, E.C.F. 1978. The Geomorphology of the Gippsland Lakes Regions. *Publ. 186.* Ministry for Conservation, Victoria, Australia. 158 pp.

Bonilha, N.F. 1974. Circulacão em lagos-um modelo matematico simplificado aplicado au Rio Guiaba e Lagoa dos Patos. M.S. thesis. Department of Civil Engineering. Universidade Federal do Rio Grande do Sul. Porto Alegre, R.S., Brazil. 74 pp.

Botello, A.V. 1978. Variación de los parámetros hidrológicos en las épocas de sequía y lluvias (Mayo y Noviembre de 1974) en la Laguna de Términos, Campeche, México. *An. Inst. Cienc. del. Mar.* y Limnol. *Univ. Nal. Autón. México* 5:159-178.

Cameron, W.M. and D.W. Pritchard. 1963. Estuaries. pp. 306-324. *In:* M.N. Hill (ed.), *The Sea,* Vol. 2. John Wiley & Sons, New York.

Chuang, W.-S. and E.M. Swenson. 1981. Subtidal water level variations in Lake Pontchartrain, Louisiana. *J. Geophys. Res.* 86:4198-4206.

Collier, A. and J.W. Hedgepeth. 1950. An introduction to the hydrography of tidal waters of Texas. *Publ. Inst. Mar. Sci. Univ. Texas* 1:125-194.

Copeland, B.J., J.H. Thompson, Jr. and W.B. Ogletree. 1968. The effects of wind on water levels in the Texas Laguna Madre. *Texas J. Sci.* 20:196-199.

Davies, J.L. 1980. *Geographical Variation in Coastal Development.* Second Edition. Langman, London. 212 pp.

Day, J.W., Jr., W.G. Smith, P.R. Wagner, and W.C. Stowe. 1973. Community structure and carbon budget of a salt marsh and shallow bay estuarine system in Louisiana. *Pub. LSU-SG-72-04.* Center for Wetland Resources, Louisiana State University. Baton Rouge. 80 pp.

Day, J.W., Jr. and A. Yáñez-Arancibia. 1982. Coastal lagoons and estuaries: ecosystem approach. *Ciéncia Interamericana* 22:11-26.

Day, J.W., R.H. Day, M.T. Barreiro, F. Ley-Lou, and C.J. Madden. 1982. Primary production in the Laguna de Terminos in the Southern Gulf of Mexico. *Oceanol. Acta* SP:269-276.

Delaney, P.J.V. 1963. Quaternary geologic history of the coastal plain of Rio Grande do Sul, Brazil. *Tech. Rep. 7,* Coastal Studies Institute, Louisiana State University. Baton Rouge. 63 pp.

Dronkers, J. and J.T.F. Zimmerman. 1982. Some principles of mixing in tidal lagoons. *Oceanol. Acta* SP:107-117.

Eleuterius, C.K. 1976. Mississippi Sound salinity distribution and indicated flow patterns. *MASP-76-023.* Mississippi-Alabama Sea Grant Consortium. Ocean Springs, MS. 128 pp.

Farfan, B.C. and S. Alvares-Borrego. 1983. Variability and fluxes of nitrogen and organic carbon at a mouth of a coastal lagoon. *Estuar. Coastal Shelf Sci.* 17:599-612.

Gierloff-Emden, H.G. 1976. Manual of interpretation of orbital remote sensing satellite photography and imagery for coastal and offshore environmental features (including lagoons, estuaries, and bays). Munchener Geographische Abhandlungen. Institute für Geographie der Universität München. Band 20. IOC-UNESCO contract SC/RP 600/341.

Graham, D.S., J.P. Daniels, J.M. Hill, and J.W. Day, Jr. 1981. A preliminary model of the circulation of Laguna de Terminos, Campeche, Mexico. *An. Inst. Cienc. del Mar. y Limnol. Univ. Nal. Autón. México* 8:51-62.

Herz, R. 1977. Circulacão das aguas de superficie da Lagoa dos Patos. Ph.D. dissertation. Departmento de Geografia. Universidade de São Paulo, Brazil. 312 pp.

Isaji, T. and M.L. Spaulding. 1985. Tidal exchange between a coastal lagoon and offshore waters. *Estuaries* 8:203-216.

James, N.P. and R.N. Ginsburg. 1979. The seaward margin of Belize barrier and atoll reefs. *Spec. Pub. 3.* Int. Assoc. Sedimentologists. Oxford. 191 pp.

Kjerfve, B. 1981. Tides of the Caribbean Sea. *J. Geophys. Res.* 86: 4243-4247.

Kjerfve, B. 1983. Analysis and synthesis of oceanographic conditions in Mississippi Sound. April-October 1980. *Final Rep.* U.S. Army Engineer District, Mobile (Alabama). DACW01-82-Q-0022. 438 pp.

Kjerfve, B. and S.P. Dinnel. 1983. Hindcast hurricane characteristics on the Belize barrier reef. *Coral Reefs* 1:203-207.

Kjerfve, B. In press. Estuarine characteristics and physical process. *In:* J.W. Day, Jr. *et al.* (eds.), *Estuarine Ecology.* John Wiley & Sons, New York.

Lankford, R.R. 1976. Coastal lagoons of Mexico: their origin and classification. pp. 182-215. *In:* M. Wiley (ed.), *Estuarine Processes.* Vol. II. Academic Press. New York.

Lara-Lara, J.R., S. Alvarez-Borrego, and L.F. Small. 1980. Variability and tidal exchange of ecological properties in a coastal lagoon. *Estuar. Coastal Shelf. Sci.* 11:613-637.

Lasserre, P. 1979. Coastal lagoons: Sanctuary ecosystems, cradles of culture, targets for economic growth. *Nature and Resources* 15(4):1-21.

Lee, V. and S. Olsen. 1985. Eutrophication and management initiatives for the control of nutrient inputs to Rhode Island coastal lagoons. *Estuaries* 8:191-202.

Levine, E.A. 1981. Nutrient cycling by the red mangrove, *Rhizophora mangle,* L., in Joyuda Lagoon on the west coast of Puerto Rico. *CEER-M-128.* Center for Energy and Environment Research, University of Puerto Rico, Mayaguez. 103 pp.

Mabesoone, J.A. 1964. Origin and age of the sandstone reefs of Pernambuco (Northeastern Brazil). *J. Sed. Pet.* 34:715-726.

Millan-Nuñez, R., S. Alvarez-Borrego, and D.M. Nelson. 1982. The effects of physical phenomena on the distribution of nutrients and phytoplankton productivity in a coastal lagoon. *Estuar. Coastal Shelf Sci.* 15:317-335.

Miller, J.A. and I.G. Macintyre. 1977. Field guidebook to the reefs of Belize. Atlantic Reef Committee. University of Miami. 36 pp.

Moore, N.H. and D.J. Slinn. 1984. The physical hydrology of a lagoon system on the Pacific coast of Mexico. *Estuar. Coastal Shelf Sci.* 19:413-426.

Nichols, M. and G. Allen. 1981. Sedimentary processes in coastal lagoons. pp. 27-80. *In: Coastal Lagoon Research, Present and Future.* UNESCO Technical Papers in Marine Science 33. Paris.

Nixon, S.W. and V. Lee. 1981. The flux of carbon, nitrogen and phosphorus between coastal lagoons and offshore waters. pp. 325-248. *In: Coastal Lagoon Research, Present and Future.* UNESCO Technical Papers in Marine Science 33. Paris.

Noye, B.J. 1973. *The Coorong—Past Present and Future.* Department of Adult Education. The University of Adelaide, S.A., Australia. 47 pp.

Noye, B.J. and P.J. Walsh. 1976. Wind-induced water level oscillations in shallow lagoons. *Aust. J. Mar. Freshwater Res.* 27:417-430.

O'Brien, M.P. 1969. Dynamics of tidal inlets. pp. 397-406. *In:* Ayala-Castañares and Phleger (eds.), *Coastal Lagoons, a Symposium.* Univ. Nal. Autón. México Press. Mexico, D.F.

Orme, A.R. 1975. Ecological stress in a subtropical coastal lagoon: Lake St. Lucia, Zululand. pp. 9-22. *In:* H.J. Walker (ed.), *Geoscience and Man: Coastal Resources.* Vol. XII. Louisiana State University Press. Baton Rouge.

Phleger, F.B. 1969. Some general features of coastal lagoons. pp. 5-25. *In:* A. Ayala Castañares and F.B. Phleger (eds.), *Coastal Lagoons a Symposium.* Univ. Nal. Autón. México Press. Mexico, D.F.

Phleger, F.B. 1981. A review of some general features of coastal lagoons. pp. 7-14. *In: Coastal Lagoon Research, Present and Future.* UNESCO Technical Papers in Marine Science 33. Paris.

Postma, H. and K.S. Dijkema. 1982. Hydrography of the Wadden Sea: Movements and Properties of Water and Particulate Matter. Rep. 2. Final report on "Hydrography" of the Wadden Sea Working Group. Stichting Veth tot Steun aan Waddenonderzoek. Leiden. 75 pp.

Purdy, E.G. 1974. Karst-determined facies patterns in British Honduras: holocene carbonate sedimentation model. *Am. Assoc. Petr. Geol. Bull.* 58:825-255.

Sakou, T. 1963. The Salinity Regime and Exchange Characteristic of a Shallow Coastal Bay System. *Tech. Rep. 63-21T.* Department of Oceanography. Texas A & M Univ. College Station. 155 pp.

Schwing, F.B., B. Kjerfve, and H.E. Sneed. 1983. Sea level oscillations in a salt marsh lagoon system, North Inlet, South Carolina. *An. Inst. Cienc. del Mar y Limnol. Univ. Nal. Autón. México* 10:231-236.

Sikora, W.B. and B. Kjerfve. 1985. Factors influencing the salinity regime of Lake Pontchartrain, Louisiana, a shallow coastal lagoon: analysis of a long-term data set. *Estuaries* 8:170-180.

Smith, N.P. 1977. Meteorological and tidal exchanges between Corpus Christi Bay, Texas, and the northwestern Gulf of Mexico. *Estuar. Coastal Shelf Sci.* 13:159-167.

Smith, N.P. and G.H. Kierspe. 1981. Local energy exchanges in a shallow coastal lagoon: winter conditions. *Estuar. Coastal Shelf Sci.* 13:159-167.

Sojisuporn, P. 1984. Estuarine oceanography of Songkla Lagoon. Unpublished manuscript. University of South Carolina. Columbia. 16 pp.

Swenson, E.M. 1980. General hydrography of the tidal passes of Lake Pontchartrain, Louisiana. pp 157-215. *In:* J.H. Stone (eds.), Environmental Analysis of Lake Pontchartrain, Louisiana, Its Surrounding Wetlands, and Selected Land Uses. Vol. 1. Louisiana State University, Baton Rouge, L.A. Prepared for U.S. Army Engineer District, New Orleans. Contract No. DACW29-77-C-0253. 591 pp.

Swenson, E.M. and W.-S. Chuang. 1983. Tidal and subtidal water volume exchange in an estuarine system. *Estuar. Coastal Shelf Sci.* 16:229-240.

Wang, D.P. and A.J. Elliott 1978. Non-tidal variability in the Chesapeake Bay and Potomac River: evidence for non-local forcing. *J. Phys. Oceanogr.* 8:225-232.

Yáñez-Arancibia, A. and J.W. Day, Jr. 1982. Ecological characterization of Términos Lagoon, a tropical lagoon-estuarine system in the southern Gulf of Mexico. *Oceanol. Acta* SP:431-440.

Yáñez-Arancibia, A., A.L. Lara-Dominquez, P. Chavance, and D. Flores Hernandez. 1983. Environmental behavior of Términos Lagoon ecological system, Campeche, Mexico. *An. Inst. Cienc. del Mar y Limnol. Univ. Nal. Autón. México* 10:137-176.

Zimmerman, J.T.F. 1976. Mixing and flushing of tidal embayments in the western Dutch Wadden Sea. Part I: Distribution of salinity and calculation of mixing time scales. *Netherlands J. Sea Res.* 10:149-191.

Zimmerman, J.T.F. 1978. Dispersion by tide-induced residual current vortices. pp. 207-216. *In:* J.C.J. Nihoul (ed.), *Hydrodynamics of Estuaries and Fjords.* Elsevier Amsterdam.

Zimmerman, J.T.F. 1981. The flushing of well-mixed tidal lagoons and its seasonal fluctuation. pp. 15-26. *In: Coastal Lagoon Research, Present and Future.* UNESCO Technical Papers in Marine Science 33. Paris.

RELATIONSHIPS AMONG PHYSICAL CHARACTERISTICS, VEGETATION DISTRIBUTION AND FISHERIES YIELD IN GULF OF MEXICO ESTUARIES

Linda A. Deegan, John W. Day, Jr., James G. Gosselink

Coastal Ecology Institute
Center for Wetland Resources
Louisiana State University
Baton Rouge, Louisiana

and

Alejandro Yáñez-Arancibia, G. Soberón Chávez and P. Sánchez-Gil

Universidad Nacional Autónoma de México
Instituo de Ciencias del Mar y Limnología
México

Abstract: The relationship of physical factors to vegetation distribution and fishery harvest was analyzed using data from the Gulf of Mexico. Data from 64 estuaries were used to investigate the relationships with vegetation. Fishery harvest in the southern Gulf of Mexico was analyzed using harvest statistics, estuarine area, and river discharge by state. Results show that the fishery harvest and area of an estuary are strongly related to freshwater input and physiography. Intertidal area is correlated to coastal land slope, length of coastline occupied by the estuary, and inshore open water area. The area of emergent vegetation is related to intertidal area and rainfall, but not to riverflow, because there are large areas of emergent vegetation with low riverflow (e.g., south Florida). Salt flat area is related to intertidal area and rainfall deficit. Fishery harvest per unit open water area in the southern Gulf is highly correlated to river discharge ($r = 0.98$).

Introduction

During the past several years comparative estuarine studies have been undertaken to identify characteristics and relationships that are common among divergent estuarine systems. Of particular interest are those factors controlling productivity (Turner 1977; Cross and Williams 1981; Nixon 1980, 1981, 1982; Rayburn 1981; Armstrong 1982; Boynton *et al.* 1982; Welsh *et al.* 1982). None of these studies have addressed the interaction of physical (climate and river flow), geomorphic and biological factors in determining the gross structure and productivity of estuarine systems.

For a number of reasons the Gulf of Mexico is an excellent region for comparisons among estuaries: (1) there are a large number of estuaries; (2) the climate ranges from tropical to temperate and from humid to arid; (3) the area encompasses a wide range of riverine influence, from systems with almost no riverine input to the Mississippi River; and (4) the size of estuarine areas (in terms of both open water and intertidal area) varies from very small to the largest in North America. Several studies have summarized selected aspects of the natural features (e.g., geology) of the Gulf of Mexico; however, a comprehensive description of the natural resources of the Gulf and their relationship to climatic and geomorphic factors has not been attempted previously even though a large quantity of pertinent information is available. The purpose of this paper is two-fold: (1) to describe in a general way the variability in climate, river discharge, geology, estuarine dimensions and vegetation and fisheries yield around the coast of the Gulf of Mexico; and (2) to analyze relationships among these variables. This paper combines original observation with a review of the literature on the dimensions, vegetation, geology, river discharge, tide range, rainfall, evapotranspiration, climate, slope of the land, area of adjacent continental shelf, and fisheries yield in the southern Gulf of Mexico. It is a summary of two more extensive reports which are available from the authors (Deegan *et al.* 1984; Yáñez-Arancibia *et al.* in press).

Methods

Information on climate, geology, and areal extent of vegetation was compiled for sixty four major estuarine areas in the Gulf of Mexico region (Fig. 1, Table 1).

For each estuary we determined: the width of the estuary (km) along the coast, including open water and wetland areas; areas of emergent vegetation (marsh or mangrove) and submergent vegetation (seagrass); total saltflat, open water and intertidal areas; and river discharge as the sum of the average annual discharge ($m^3\ s^{-1}$) of all the gauged streams and rivers discharging into the estuary. These were obtained from the literature for Florida (McNulty *et al.* 1972; Florida Department of Environmental Regulation 1978), Mississippi (Eleuterus 1972; Christmas 1973), Alabama (Swingle 1971; O'Neil and Mette 1982), Louisiana (Perret *et al.* 1971; Chabreck 1972), and Texas (Fisher *et al.* 1972; Diener 1975; Shew *et al.* 1981, Brown *et al.* 1977, 1980; McGowen *et al.* 1976a, b). Equivalent data for Mexico were planimetered from Direccion General de Geografica del Territorio Nacional topographic maps, scale 1:250,000. Submergent vegetation area could not be determined from the maps for Mexico. Information on submerged vegetation in Mexico was available only for Laguna de Términos (Day *et al.* 1982; Yáñez-Arancibia and Day 1982).

Average climatic water balance information (precipitation, potential and actual evapotranspiration, average daily water surplus, average daily water deficit) was from Thornthwaite Associates (1964a, b), and the average freeze-free period was from U.S. Department of Commerce (1968) and Garcia (1973). Tide range was obtained from standard tide tables (U.S. Department of Commerce 1982).

Slope of the land was estimated as the average of 5 measurements of the distance from the inland edge of the open water body to the 50-m upland contour on United States Geological Survey and Direccion General de Geografica del Territorio Nacional topographic maps, scale 1:250,000. Continental shelf

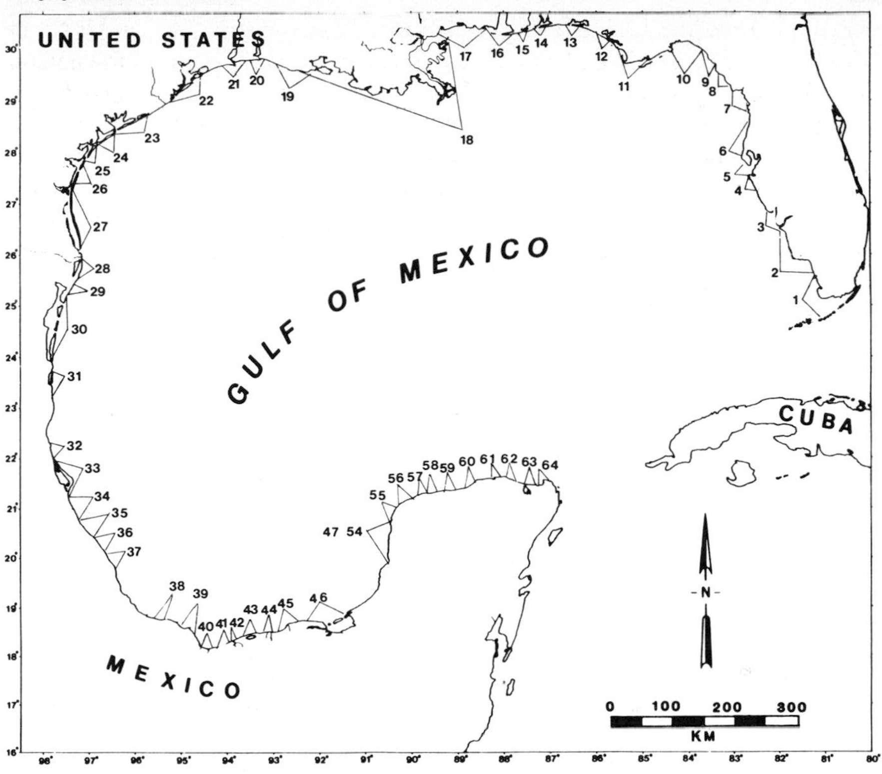

Figure 1. Map of the Gulf of Mexico estuaries included in this study. Numbers on map refer to estuaries listed in Table 1.

width was measured as the distance to the 10- and 100-fathom depth contours on a line perpendicular from the outer edge of the estuary on a bathymetric map of the Gulf of Mexico (U.S. Department of Commerce 1979).

Simple and stepwise multiple linear regression analyses (Draper and Smith 1981; SAS Institute Inc. 1985) were used to determine relationships among biotic and physical variables. For stepwise multivariate models the change in F-test value, used to determine predictive variable entry into the model, was set at 0.15. The Cp statistic was used in selection of the final regression model (Daniel and Wood 1980). Pearson product-moment correlations were used to test for interdependence of the predictive variables. Predictive variables were considered independent if the correlation value was below 0.6 (Draper and Smith 1981). Significance level of 95% ($p<0.05$) was used for all analyses.

To determine if relationships existed between fishery yield and the parameters described above, we conducted regression analyses on fish catch, river discharge, estuarine area, and several other environmental parameters for several Mexican coastal areas (Yáñez-Arancibia *et al.* in press). Data on riverflow were obtained directly from the Secretaria de Agricultura y Recursos

Hidraulicos, Direccion General de Estudios, Subdireccion de Hidrologia, Mexico, D.F. Fish catch data were from Secretaria de Pesca (1978-1982). Data for both riverflow and fish catch are for 1978-1982. Fish catch, river flow, and estuary area were all grouped by state.

Variability of Gulf of Mexico Estuaries

The estuarine systems bordering the Gulf of Mexico exhibit a broad range of characteristics (Table 1). Regions with little relief and high freshwater input are characterized by large areas of inshore open water and intertidal vegetation. Data from this report and from the literature also suggest that these areas support high fisheries yield.

Geomorphology

There is no completely satisfactory way to describe the geomorphology of coastal systems, primarily because many estuaries are formed by combinations of tectonic, coastal processes and riverine deposition. The basic foundation of the Gulf of Mexico was formed over millions of years by crustal plate movements and volcanic eruptions. During the last 5000-7000 years, however, sea level change has been very small and the tectonic shoreline has been modified by riverine sediment deposition and marine reworking and erosion of coastal sediments. These three factors, superimposed and operating on different time scales, have formed the present intertidal and inshore open water areas bordering the Gulf of Mexico.

In a general sense, the relief around the Gulf of Mexico can be inferred from Fig. 2a, which shows the distance from the inland edge of the estuary to the 50-m upland contour. As the distance to the upland contour increases, the slope of the land decreases. Two major areas of broad flat coastal landscapes are the calcareous lacustrine plateau in southern Florida and the Mississippi River Deltaic Plain. Other flat areas are associated with the deltaic plains of the Apalachicola River in Florida and the Grijalva River, Tabasco-Campeche Mexico. The middle coast of Florida has (for the Gulf of Mexico) a steep, rugged topography similar to the coast along the Yucatan peninsula in Mexico. These areas are both drowned calcareous shelf regions. The slopes of the Alabama, Mississippi and Texas coasts have resulted from marine reworking of sediments and are intermediate between the flatness of the deltaic plains, and the ruggedness of the calcareous plateau.

Climate And Riverflow

The water budget around the Gulf is controlled primarily by rainfall. Average annual rainfall varies from a high of 2800 mm in southern Mexico near the Grijalva River to a low of 466 mm in the Yucatan peninsula and averages 1334 mm gulfwide (Fig. 2b). Potential evapotranspiration, which integrates the effects of temperature and solar radiation, increases with decreasing latitude. Evapotranspiration is not as variable as rainfall, averaging 1300 mm and ranging from a low of 1025 mm in Louisiana to a high of 1586 mm in the Yucatan

Table 1. Dimensions, climatic, geologic, and River discharge characteristics of Gulf of Mexico estuaries. Dashes indicate no data.

Map	Estuary Name	Intertidal Area: Submerged vegetated area (ha)	Intertidal Area: Emergent vegetated area (ha)	Intertidal Area: Saltflat area (ha)	Open water area (ha)	Mean annual river discharge (CMS)	Width along coast (Km)	Climatic water Budget[1] (mm) PE	P	AE	D	S	Distance to contour (Km) 50 m upland	10 fathom	100 fathom	Mean tide range (m)	Freeze free period (days)	Mean depth (m)
1	Florida Bay	105104	86473	0	245518	283.1	46.0	1295	1309	1257	38	52	600.00	86.4	259.2	0.84	365	1.3
2	Tenthousand Islands	1955	72095	0	42000	9.5	117.0	1253	1368	1234	19	134	450.00	54.0	270.0	1.17	365	1.4
3	Charlotte Harbor	21558	26181	0	112463	86.0	43.0	1253	1368	1234	19	134	150.00	43.2	226.8	0.55	365	2.3
4	Sarasota Bay	3080	1647	0	14067	2.3	14.0	1127	1359	1112	15	247	67.50	32.4	205.2	0.67	330	1.7
5	Tampa Bay	11985	8517	0	123855	43.8	49.5	1186	1278	1171	15	107	30.00	43.2	205.2	0.77	330	3.3
6	St. Joseph Sound	30569	21573	0	49146	33.4	43.0	1186	1278	1171	15	107	37.50	64.8	216.0	1.00	300	1.2
7	Waccassa Bay	36857	36310	0	56981	20.6	19.8	1078	1455	1076	2	379	78.75	75.6	237.6	1.00	270	1.6
8	Suwanne Sound	3277	13114	0	16084	311.2	14.0	1078	1455	1076	2	379	112.50	64.8	226.8	1.00	270	1.6
9	Deadman Bay	3971	29337	0	17182	9.5	16.5	1078	1455	1076	2	379	75.00	64.8	237.6	1.00	270	1.2
10	Apalachee Bay	12806	28337	0	28440	86.2	33.2	1025	1465	1022	3	433	82.50	64.8	237.6	1.00	270	1.3
11	Apalachicola Bay	3796	8623	0	68814	763.6	23.3	1080	1432	1071	9	361	142.50	43.2	172.8	0.72	300	2.9
12	St. Andrew Bay	5245	21017	0	47651	15.1	39.7	1070	1487	1062	8	425	37.50	10.8	97.2	0.47	270	3.6
13	Choctawahatches	1251	1140	0	34937	204.6	50.0	1065	1519	1057	6	462	11.25	21.6	97.2	0.18	270	4.1
14	Pensacola Bay	3202	4216	0	61300	268.0	49.3	1063	1519	1057	6	462	7.50	21.6	64.8	0.44	270	5.9
15	Perdido Bay	0	433	0	6989	26.5	13.0	1063	1519	1057	6	462	22.50	21.6	86.4	0.15	300	2.6
16	Mobile Bay	2024	8693	0	115255	1664.0	28.0	1040	1614	1039	1	575	52.50	32.4	108.0	0.36	270	2.5
17	Mississippi Sound	12000	27087	0	175821	715.0	78.0	1079	1578	1077	2	471	52.50	43.2	151.2	0.52	300	3.0
18	Deltaic Plain	100	771193	0	15[illegible]5814	22897.7	201.0	1098	1533	1095	2	437	425.00	43.2	108.0	0.36	300	2.0
19	Mermentau	0	182571	0	55072	169.0	39.7	1051	1415	1034	23	381	300.00	54.0	194.4	0.73	300	1.5
20	Calcasieu	0	102073	0	53815	157.8	52.9	1073	1448	1054	19	394	217.50	64.8	216.0	0.61	300	1.5

Table 1. (Continued)

Map	Estuary Name	Intertidal Area: Submerged vegetated area (ha)	Intertidal Area: Emergent vegetated area (ha)	Intertidal Area: Saltflat area (ha)	Open water area (ha)	Total river discharge (CMS)	Length along coast (Km)	Climatic water Budget[1] (mm): PE	P	AE	D	S	Distance to contour (Km): 50 m upland	10 fathom	100 fathom	Tide range (m)	Freeze free period (days)	Depth (m)
21	Sabine	0	17199	0	22605	474.0	26.4	1086	1339	1061	25	278	210.00	64.8	216.0	0.67	300	1.4
22	Galveston	7327	93684	179	143210	73.1	39.8	1125	1137	1040	85	97	75.00	43.2	172.8	0.30	300	2.3
23	Matagorda	2850	48582	4532	118057	68.8	12.9	1173	888	888	285	0	75.00	21.6	108.0	0.18	300	1.1
24	San Antonio Bay	6619	10121	5723	56161	60.8	59.6	1173	888	888	285	0	52.50	21.6	108.0	0.18	300	1.1
25	Copana-Aransas Bay	8552	18218	—	46279	5.7	36.3	1173	888	888	285	0	45.00	32.4	108.0	0.42	300	1.1
26	Corpus Cristi Bay	5161	18218	—	44451	24.5	36.6	1118	677	677	441	0	71.25	21.6	108.0	0.39	300	1.2
27	Laguna Madre	77327	101214	66400	150060	1.3	109.3	1243	686	686	557	0	105.00	10.8	86.4	0.42	330	1.0
27	Las Animas	—	0	—	1760	0.0	17.0	1243	686	686	557	0	112.50	10.8	86.4	0.42	330	—
29	Laguna et Barrill	—	2724	—	5587	0.0	28.0	1243	686	686	557	0	35.00	10.8	86.4	0.42	330	—
30	Lower Laguna Madre	—	8461	—	200978	25.0	206.0	1283	761	761	522	0	25.00	10.8	75.6	0.42	330	—
31	L. San Andres	—	3804	—	9654	—	47.0	1320	1168	1168	152	0	2.50	10.8	43.2	0.42	330	—
32	L. el Chairel	—	0	—	14026	488.0	28.5	1320	1168	1168	152	0	2.50	10.8	43.2	0.42	365	—
33	Laguna Tamiahua	—	908	—	63430	—	60.0	1327	1239	1209	118	31	2.50	10.8	32.4	0.42	365	—
34	L. Tuxapan	—	2393	0	5852	613.0	7.5	1334	1311	1249	85	62	11.25	10.8	43.2	0.52	365	—
35	Barra de Cazones	—	0	0	1000	45.0	2.5	1358	1489	1230	178	258	10.00	10.8	43.2	0.52	365	—
36	Barra de Tecoluta	—	851	—	1000	17.0	2.5	1358	1489	1230	178	258	6.25	10.8	43.2	0.52	365	—
37	Barra de Nautla	—	6530	—	1000	—	2.5	1358	1489	1230	178	258	6.25	10.8	43.2	0.52	365	—
38	L. Alvardo	—	56550	0	1428	—	53.0	1392	2322	1269	123	1053	65.00	10.8	32.4	0.39	365	—
39	L. Sontecompan	—	0	0	454	—	5.0	1338	2110	1172	166	938	2.50	10.8	32.4	0.39	365	—
40	L. Ostion	—	0	0	795	—	5.0	1401	2879	1356	45	1523	17.50	5.4	43.2	0.39	365	—
41	Coatzacoalcos	—	7552	0	2895	431.0	15.0	1401	2879	1356	45	1523	27.50	5.4	54.0	0.46	365	—
42	Chicozapore	—	5338	0	1079	—	7.5	1401	2879	1356	45	1523	22.50	5.4	43.2	0.46	365	—
43	L. del Carmen	—	23508	0	12323	—	70.0	1401	2879	1356	45	1523	37.50	10.8	43.2	0.50	365	—
44	L. Mecoaca	—	23511	0	5963	—	35.0	1401	2879	1356	45	1523	31.25	21.6	64.8	0.50	365	—

Table 1. (Continued)

Map	Estuary Name	Intertidal Area: Submerged vegetated area (ha)	Intertidal Area: Emergent vegetated area (ha)	Intertidal Area: Saltflat area (ha)	Intertidal Area: Open water area (ha)	Total river discharge (CMS)	Length along coast (Km)	Climatic water Budget[1] (mm) PE	P	AE	D	S	Distance to contour (Km) 50 m upland	10 fathom	100 fathom	Tide range (m)	Freeze free period (days)	Depth (m)
45	Grijalva		26634	0	10224	1900.0	10.0	1401	2879	1356	45	1523	70.00	21.6	64.8	0.50	365	—
46	Laguna terminos	10000	130000	0	160000	200.0	108.0	1586	1738	1471	115	267	52.50	43.2	129.6	0.50	365	3.5
47	Punta Sanita	—	119	1073	0	0.0	20.2	1476	1019	1019	457	0	12.50	64.8	171.8	0.50	365	—
48	La Ensenada	—	119	1073	0	0.0	12.2	1476	1019	1019	457	0	12.50	64.8	172.8	0.50	365	—
49	MOA	—	255	2300	0	0.0	20.2	1476	1019	1019	457	0	15.00	64.8	172.8	0.50	365	—
50	Santa Juana	—	102	920	0	0.0	12.2	1476	1019	1019	457	0	18.75	64.8	172.8	0.50	365	—
51	Huaymil 1	—	62	562	0	0.0	6.2	1476	1019	1019	457	0	40.00	54.0	205.2	0.50	365	—
52	Isla Piedra	—	107	971	0	0.0	7.5	1476	1019	1019	457	0	42.50	54.0	194.4	0.50	365	—
53	Huaymil 2	—	272	2453	396	0.0	15.0	1476	1019	1019	457	0	28.00	54.0	205.2	0.50	365	—
54	El Nemate	—	141	1277	0	0.0	6.2	1476	1019	1019	457	0	33.75	43.2	205.2	0.50	365	—
55	Estero Yaltun	—	346	3118	1135	0.0	18.7	1476	1019	1019	457	0	33.75	32.4	216.0	0.50	365	—
56	Estero Celestun	—	550	4957	0	0.0	32.5	1444	446	446	978	0	35.00	43.2	205.2	0.50	365	—
57	Parque Celestun	—	3379	30411	0	0.0	35.0	1444	466	466	978	0	35.00	43.2	205.2	0.55	365	—
58	Estero Yukalpeten	—	2460	22148	2089	0.0	37.5	1444	466	466	978	0	36.25	32.4	183.6	0.55	365	—
59	L. Rosada	—	2420	21781	1901	0.0	21.2	1444	466	466	978	0	36.25	32.4	194.4	0.55	365	—
60	San Crisanto	—	1033	9311	0	0	21.2	1444	466	466	978	0	36.25	32.4	194.4	0.55	365	—
61	Bocas Dzilan	—	19291	192	2755	0	31.2	1502	898	898	604	0	42.50	21.6	194.4	0.55	365	—
62	Punta Nichili	—	2066	206	6660	0	30.0	1502	898	898	604	0	50.00	32.4	226.8	0.55	365	—
63	Estero Rio Lagarto	—	35062	3895	26066	0	52.5	1502	898	898	604	0	50.00	32.4	237.6	0.55	365	—
64	Laguna Conil	—	45740	27304	108278	0	45.0	1502	898	898	604	0	50.00	43.2	237.6	0.55	365	—

[1]PE = Potential Evapotranspiration, P = Precipitation, AE = Actual Evapotranspiration, D = Average Daily Water Deficit, S = Average Daily Water Surplus.

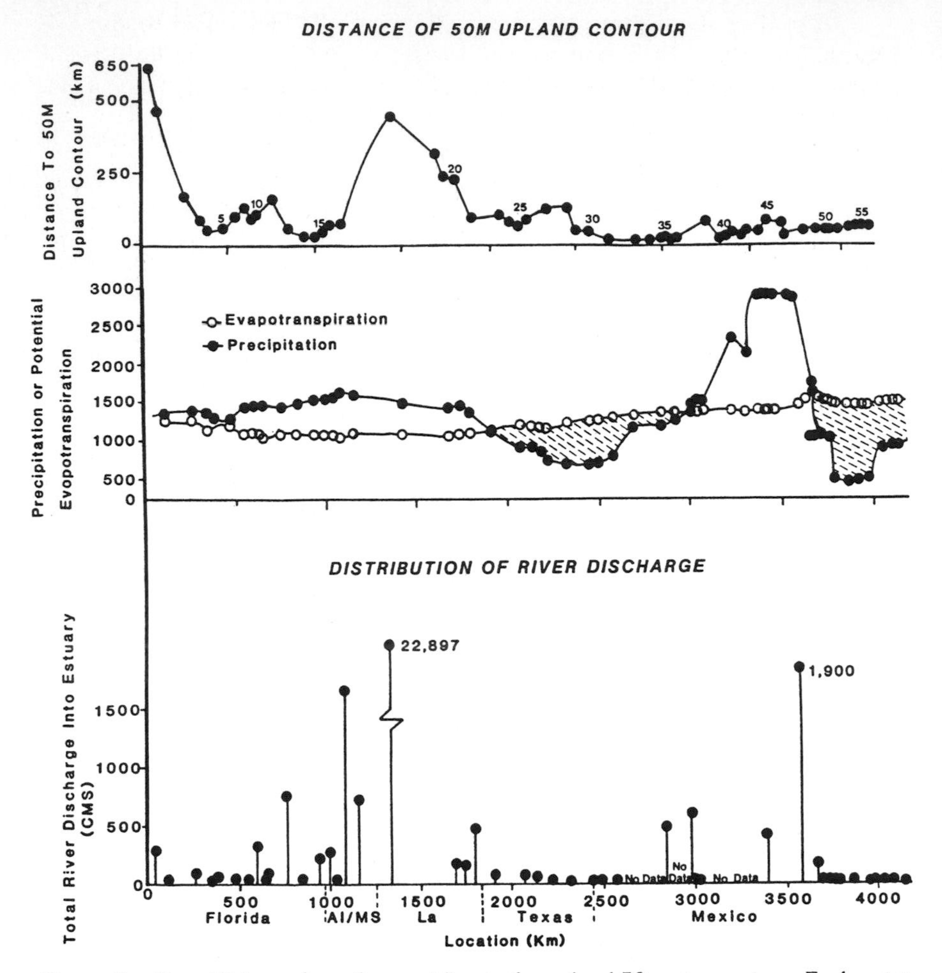

Figure 2. Top: *Distance from the coast line to the upland 50-meter contour. Each point corresponds to an estuary defined in Table 1. Numbers above the points identify each fifth estuary.* Middle: *Climatic water budget. Shading represents areas where annual potential evapotranspiration exceeds rainfall.* Bottom: *Variation of River Discharge. River discharge is the sum of the annual average discharge (cubic meters per second) of all gauged rivers and streams flowing into an estuary. The horizontal axis is the distance (km) from the southern tip of Florida to the eastern tip of the Yucatan (Fig. 1).*

peninsula. Yearly rainfall exceeds potential evapotranspiration, giving a water surplus, from Florida to northern Texas and along the south central Mexican coast. Arid areas, where annual potential evapotranspiration exceeds rainfall (net annual water deficit), extend along the south Texas and north Mexican coast and in the Yucatan peninsula in southern Mexico.

Riverflow also varies greatly over the Gulf of Mexico and tends to be highest in areas with strong rainfall surpluses (Fig. 2c). The largest river discharge is from the Mississippi River system. The combined discharge of the Mississippi and Atchafalaya Rivers (22,897 m^3 s^{-1}) is an order-of-magnitude larger than the discharge of the next largest river, the Rio Grijalva (1,900 m^3 s^{-1}). Areas with high rainfall surpluses have more rivers and streams and higher discharges than areas with low rainfall or water deficits, as a result of the generally restricted regional nature of the watersheds of these rivers (Moody 1967). Only the Mississippi River, which drains two-thirds of the United States, has a discharge pattern different from the regional rainfall pattern.

Areal Distribution of Vegetation

The shoreline and estuaries of the Gulf are characterized by tidal marshes, woody swamp forests, submerged grassbeds and tidal flats. Coastal marshes, mangroves and tideflats form the interface between aquatic and terrestrial habitats, while seagrass beds occupy the transition zone between the intertidal zone and unvegetated bottoms. These habitats may occupy narrow bands or vast expanses and may consist of sharply delineated zones of different species, monotonous stands of a single species, or mixed plant communities.

Relationships Among Physical and Biotic Factors

Reaches of the coast with high open water area generally also have large areas of emergent and/or submergent vegetation (Fig. 3a, b). The vegetated areas are characterized by low relief and high freshwater input. Saltflats occupy most of the intertidal area in arid regions (Fig. 3c). In the following discussion, we consider ways that the climate and geology of the region may interact to influence the areal distribution of vegetation types.

The amount of intertidal area is related to the physiography of the area. Stepwise linear regression showed that intertidal area was related directly to the 50-m upland contour, the length along the coast, and the area of open water (Fig. 4). Differences in intertidal area were not significantly related to tide range, probably because tide range shows little variation around the Gulf, and accounts for only a small portion of the water level variation in the Gulf (Marmer 1954; Baumann 1980).

Areal development of different vegetation types within the available intertidal area is related to the water budget of the estuary. There are two principal sources of freshwater to most estuaries, rivers and local rainfall. It is important to distinguish between these sources of freshwater because their effects are different. Rainfall is spread relatively evenly over an estuary, while river discharge is more nearly a point source of freshwater that often does not affect all areas of the estuary equally. The percent of total freshwater input into estuaries from rivers and streams averages 42.1% over the Gulf, and ranges from zero to 99%. Obviously estuaries with large rivers, such as Apalachicola Bay or the Mississippi River deltaic plain, receive most of their freshwater from rivers,

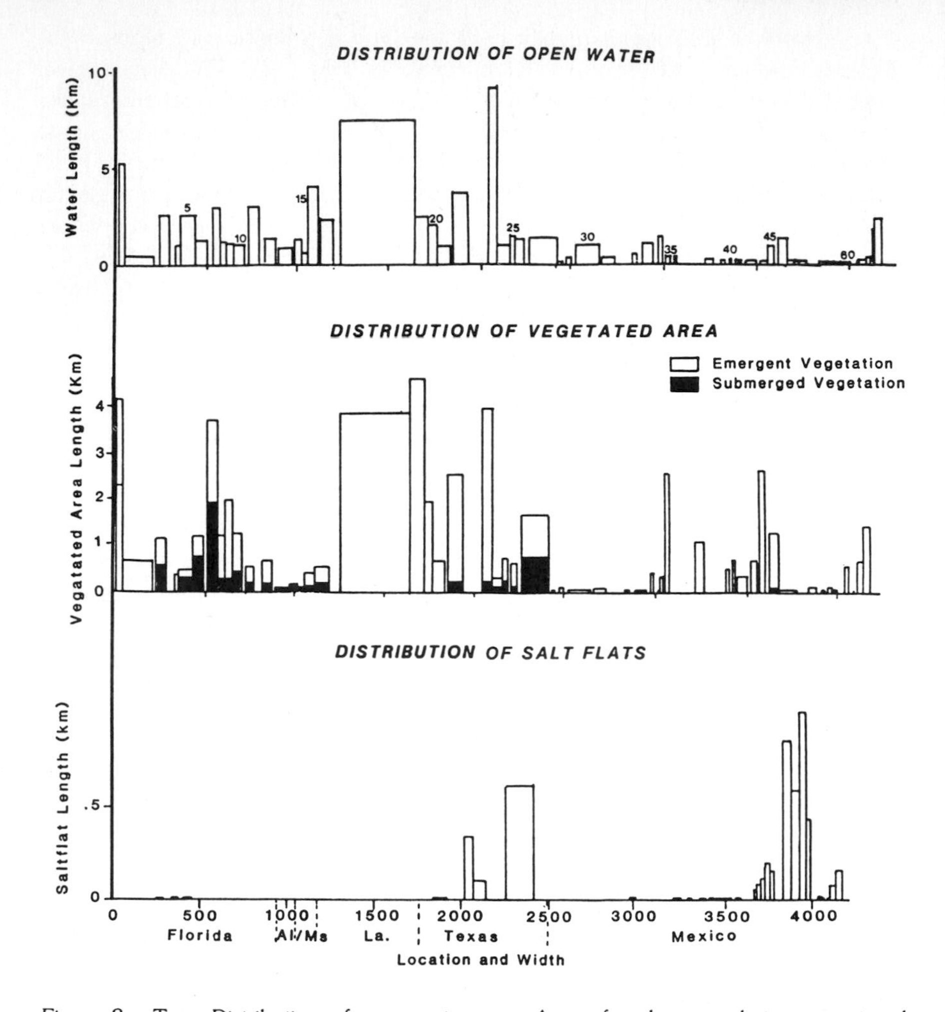

Figure 3. Top: *Distribution of open water area. Area of each rectangle is proportional to the area of the open water in each estuary. Width of each rectangle is the measured length along the coast. Height of the bar was obtained by dividing the measured area by the measured width of the system along the coast.* Middle: *Distribution of emergent and submergent vegetation area. Area of the rectangle is proportional to the area of the vegetation in the system. Width of the rectangle is the measured width along the coast. Height of the bar was obtained by dividing the measured area by the measured width of the system along the coast.* Bottom: *Distribution of saltflats. Area of each rectangle is proportional to the area of saltfalt within each estuary. Width of each rectangle is the measured length along the coast. Height of the bar was obtained by dividing the measured area by the measured length of the system along the coast. The horizontal axis is the distance (km) from the southern tip of Florida to the eastern tip of the Yucatan (Fig. 1).*

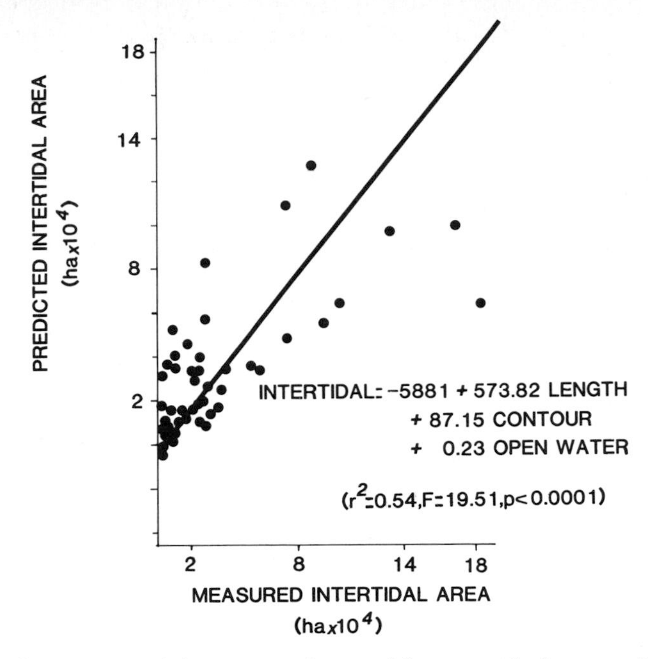

Figure 4. Comparison of the measured intertidal areas with those predicted by the model shown. The line represents a 1:1 correspondence, and the points represent individual estuaries in Table 1, excluding the Mississippi River Deltaic Plain. Length = length along the coast (km), contour = distance (km) to the 50-m upland contour, open water = area of open water (ha).

whereas estuaries in arid areas (where river discharge is usually very low) receive most of their freshwater from rainfall.

The area of emergent vegetation (either marsh or mangrove) is directly dependent on intertidal area and rainfall (Fig. 5). Areas with little rainfall develop fewer hectares of marsh or mangrove than those with high rainfall. In areas with little rain the marsh species tend to be small salt- and temperature-tolerant plants such as *Salicornia*. Other wetland plants are usually stunted and have limited distribution. The importance of rainfall is probably related to the influence of rain on interstitial water salinities and the osmotic water balance of the vegetation (Mendelssohn *et al.* 1982).

In the stepwise linear regression, the area of emergent vegetation was not significantly related to river discharge. However, all areas with high river discharge also have high rainfall, thus complicating the statistical relationship. While estuaries with large rivers (e.g., the Mississippi River Deltaic Plain, the Rio Grijalva, Mobile Bay) seem to have large emergent areas, there are also large areas of estuarine vegetation in some areas without large rivers (e.g., Florida Bay). Groundwater and overland sheet flow are important sources of freshwater to coastal systems in southern Florida and influence the distribution of emergent

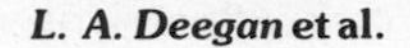

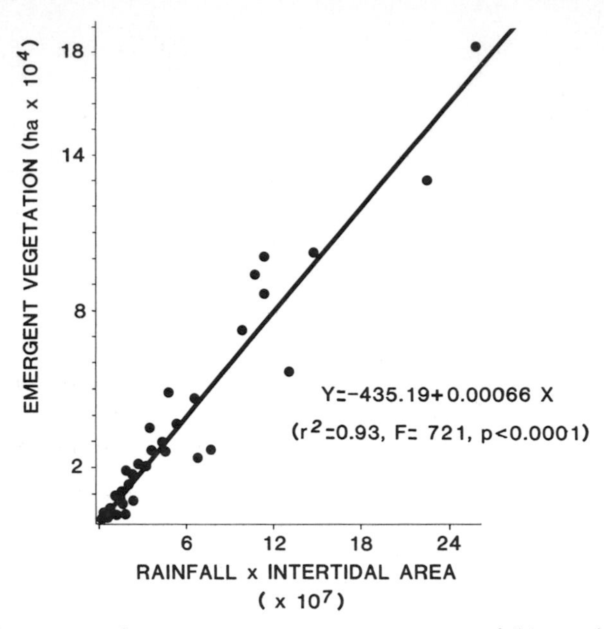

Figure 5. *Regression of emergent vegetation area on a rainfall/intertidal area index. The Mississippi River Deltaic Plain was excluded from the regression. Rainfall = precipitation in the estuary.*

vegetation (McNulty *et al.* 1972). Groundwater was not included in our analyses because the available regional data were insufficient. A long-term role of rivers in building emergent vegetation area is expressed by changing the land slope through sediment deposition over periods of hundreds to thousands of years. This effect would not necessarily be related, however, to the present annual river discharge.

The seasonal balance of rain and potential evapotranspiration is also of great importance in predicting intertidal vegetation. Saltflat vegetation can occur in estuarine areas with net annual water surpluses if there is a deficit at some time in the year; saltflats develop when the average daily water deficit is greater than zero and increases in extent as average daily deficit increases (Fig. 6).

Fisheries Yield

A number of investigations, of the Gulf of Mexico and elsewhere, suggest relationships between fisheries yield and such factors as freshwater input and estuarine area. For example, Turner (1977) reported that worldwide penaeid shrimp production was related to the area of intertidal vegetation worldwide. Nixon (1980) found that, except for Chesapeake Bay, commercial fishery landings per unit area along the Gulf and Atlantic coasts increased as the relative area of marsh increased. Sutcliffe (1972) reported that catch of several species in two eastern Canadian bays was related to river discharge. Similar findings have been

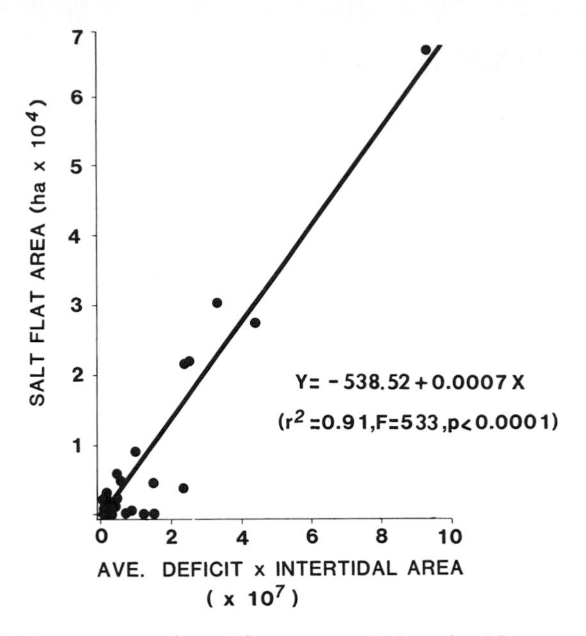

Figure 6. Multiple regression for saltflat area excluding the Mississippi River deltaic plain. For each estuary, the predicted saltflat area is plotted against the measured saltflat area. Average deficit = average daily water deficit (D).

reported for the Gulf of Mexico. Average annual river discharge and average annual fishery harvest were related in Texas estuaries, and overall fishery harvest was highest in wet years (Chapman 1966). Moore *et al.* (1970) concluded that abundance of these species was related to estuarine area and river discharge. Shrimp catch in the Northern Gulf of Mexico was related to the area of intertidal vegetation, and oyster landings in Mobile Bay were related to riverflow index (Turner 1979).

These studies suggested that such relationships might exist for the southern Gulf of Mexico. The fishery analysis showed that fish capture per unit area was positively related to river discharge (Fig. 7). This suggests that the factors which are related to fish catch in other regions are also important in the southern Gulf of Mexico. The high correlation ($p < 0.001$) partially reflects the fact that most fish captured are less than one year old. Approximately 60% of the total catch is made up of the following fisheries groups: oysters 24.8%, shrimp 11.1%, mojarras 7.5%, sierras 4.4%, snook 3.8%, crabs 2.8%, mullet 2.1%, and sharks 1.9% (Secretaria de Pesca, 1978-82). Of these species, only sharks and (in some instances) snook are taken when they are older than one year. It is interesting to note that the correlation shown in Fig. 7 is with open water area, rather than with area of emergent vegetation as has been demonstrated for a number of other areas. This could be due to a number of factors.

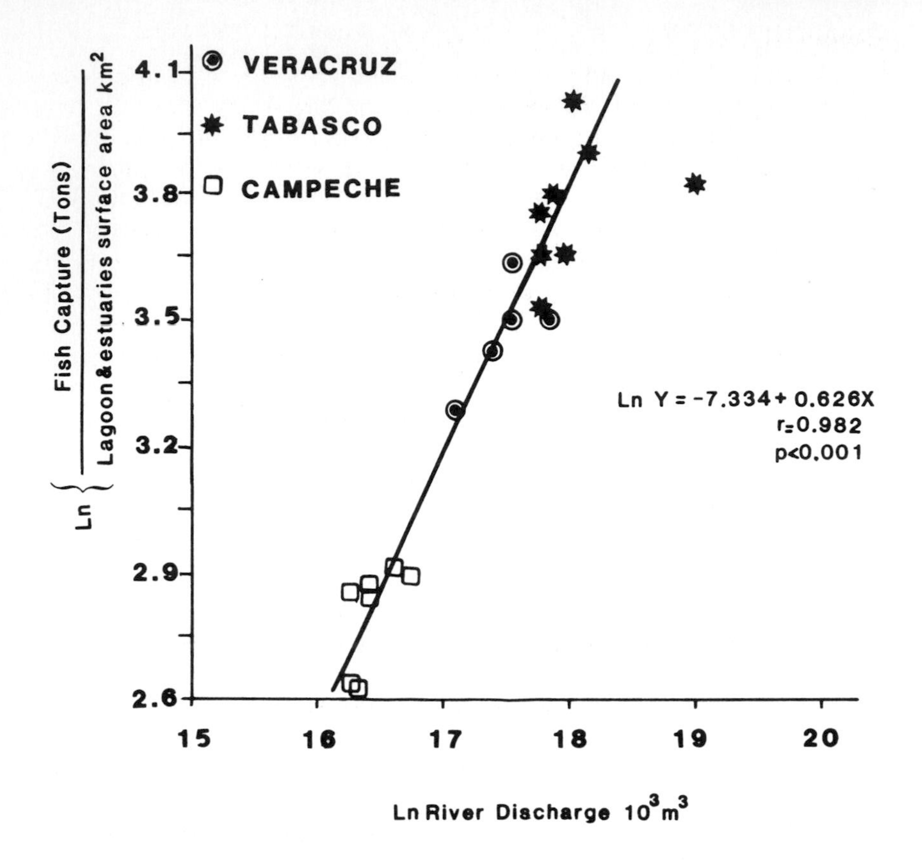

Figure 7. Linear logarithmic regression between fishery harvest per unit open water and average river discharge. The data are from the states of Veracruz, Tabasco and Campeche.

Perhaps the broad open shallow water systems which characterize many of these systems are important as habitat for fishery species, as Nixon's (1980) work suggested for Chesapeake Bay. Open water includes grassbeds, which are important nursery areas for many species. There is also a strong correlation between open water area and emergent vegetation area which could obscure the importance of emergent vegetation areas. Field studies by Yáñez-Arancibia and his coworkers showed that larval and juvenile species are also abundant in river mouths, protected shores, and mangrove channels (Amezcua Linares and Yáñez-Arancibia 1980; Aguirre León *et al.* 1982; Bravo-Nuñez and Yáñez-Arancibia 1979; Díaz-Ruiz *et al.* 1982; Lara-Dominguez *et al.* 1981; Sanchez-Gil *et al.* 1981; Vargas *et al.* 1981; Yáñez-Arancibia *et al.* 1982, 1983). While the mechanisms remain uncertain, there seems little doubt that both estuarine area and freshwater input are related to fishery harvest.

Synthesis

We have so far considered the relationship among physical and biological variables as simple linear functions. Interaction among different parameters may be obscured when they are considered one at a time. Estuarine events are cast within a series of time frames or time hierarchies from tectonic and sedimentary geological events that occur over thousands to millions of years, through ecological scales of tens to hundreds of years, to seasonal variability. The physical driving forces in estuaries act at all these levels, and the resulting picture can be confused. The intertidal and open water areas are determined by the geologic template of the estuary. The type and extent of intertidal vegetation are determined by the intertidal-area and by climate. The fishery type and yield are determined in turn by area of open water and vegetation. These general relationships are modified by the interaction of fluvial and oceanic processes on each time scale. For example, the worldwide extent of the intertidal area is determined in general by the oceanic tidal range and the sediment volume deposited by rivers. Variation in tidal range is not significant in the Gulf of Mexico, but otherwise the largest intertidal areas are found in active river deltas. The southwest coast of Florida is an exception, with broad expanses of intertidal vegetation and little influence from fluvial processes. This is true because the long-term geologic processes in this region were marine in origin, producing a broad, low platform subsequently modified by groundwater flow.

The type and extent of vegetation in an intertidal zone are determined by oceanic forces and the freshwater balance. Changes in the balance of these two forces lead to vegetation changes on an ecological time scale of tens to hundreds of years. An example is the shift of salinity zones and increase in freshwater vegetation area at the mouth of the Atchafalaya River because of increasing freshwater input. Finally, as we have shown, the volume of fisheries varies with the extent of open water area.

Acknowledgments

This project was funded by UNESCO Man and The Biosphere grant number 18-81-39 Part IIIBa, and by the National Oceanographic and Atmospheric Administration, Office of Marine Pollution Assessment, through a memorandum of understanding with the Louisiana Universities Marine Consortium. We acknowledge the help of Diane Baker, Isis Miler, Fransico Flores Verdugo, Suzanne Hautot, Dr. L. Rouse, and Chris Neill. Many of the ideas in this paper stem from conversations with our colleagues in the Coastal Ecology Institute and The Institute of Marine Sciences and Limnology. This is publication number LSU-CEI-85-15 of the Coastal Ecology Institute and number 425 from The Instituto de Ciencias del Mar y Limnologia.

References Cited

Aguirre Leon, A., A. Yáñez-Aranicibia and F. Amezcua Linares. 1982. Taxonomia, diversidad, distribucion y abundancia de las mojarras, de la Laguna de Terminos, Campeche (Pisces: Gerreidae). *An. Inst. Cienc. del Mar y Limnol., Univ. Nal. Auton. Mexico.* 9(10):213-250.

Amezcua Linares, F. and A. Yáñez-Arancibia. 1980. Ecologia de los sistemas fluvio-lagunares asociado a la Laguna de Terminos. El habitat y estructura de las comunidades de peces. *An. Centro Cienc. de Mar y Limnol., Univ Nal. Auton Mexico,* 7(1):69-118.

Armstrong, N. E. 1982. Response of Texas estuaries to freshwater inflows. pp. 103-120. *In:* V. S. Kennedy (ed.), *Estuarine Comparisons.* Academic Press, New York, 709 pp.

Bauman, R. 1980. Mechanisms of maintaining marsh elevation in a subsiding environment. M. S. Thesis. Louisiana State University, Baton Rouge. 91 pp.

Boynton, W. R., W. M. Kemp and C. W. Keefe. 1982. A comparative analysis of nutrients and other factors influencing estuarine phytoplankton production. pp. 69-90. *In:* V. S. Kennedy (ed.), *Estuarine Comparisons.* Academic Press. New York. 709 pp.

Bravo-Nuñez, E. and A. Yáñez-Arancibia, 1979. Ecologia de la boca de Puerto Real, Laguna de Terminos. Descripción del area y análisis estructural de las comunidades de peces. *An. Centro Cienc. del Mar y Limnol., Univ. Nal. Auton. Mexico.* 6(1):125-182.

Brown, L. F., Jr., J. H. McGowen, T. J. Evans, C. G. Groat and W. L. Fisher. 1977. *Environmental Geologic atlas of the Texas coastal zone—Kingsville area.* Bureau of Economic Geology. Univ. of Texas, Austin. 131 pp.

Brown, L. F., Jr., J. L. Brewton, T. J. Evans, J. H. McGowen, W. A. White, C. G. Groat and W. L. Fisher. 1980. *Environmental Geologic atlas of the Texas coastal zone—Brownsville-Harlingen area.* Bureau of Economic Geology. Univ. of Texas, Austin. 140 pp.

Chabreck, R. H. 1972. *Vegetation, water and soil characteristics of the Louisiana Coastal Region.* Bull. No. 664. Louisiana State University Agricultural Experiment station. Baton Rouge. 72 pp.

Chapman, C. R. 1966. The Texas basins project. *In:* R. F. Smith, A. Swartz, and W. Massmann (eds.), *Symposium on Estuarine Fisheries.* Amer. Fisheries Soc. Spec. Publ. 3 (suppl.), 95(4):83-92.

Christmas, J. Y., (ed.). 1973. *Cooperative Gulf of Mexico estuarine inventory and study, Mississippi.* Gulf Coast Research Laboratory. Ocean Springs, Mississippi. 434 pp.

Cross, R. D. and D. Williams (eds.). 1981. *Proceedings of the National Symposium of Freshwater Inflow to Estuaries.* Vol 1 and Vol 2. FWS/OBS-81/04. U.S. Fish and Wildlife Service. U.S. Dept. of Interior. Washington, D.C. 525 pp and 528 pp.

Daniel, C. and Wood, F. S. 1980. *Fitting Equations to Data.* John Wiley and Sons. New York. 549 pp.

Day, J., R. Day, M. Barreiro, F. Ley-Lou and C. Madden. 1982. Primary production in the Laguna de Terminos, a tropical estuary in the southern Gulf of Mexico. *Oceanologica Acta.* Special Volume: 269-276.

Deegan, L., J. W. Day, Jr. and A. Yáñez-Arancibia. 1984. Relationships of vegetation to climate, river discharge, and geomorphology in Gulf of Mexico estuaries. Final Report to Man and Biosphere Program. Center for Wetland Resources, Louisiana State University, Baton Rouge. 58 p.

Diaz-Ruiz S., A. Yáñez-Arancibia, and F. Amezcua Linares. 1982. Taxonomia, diversidad, distribucion y abundancia de los Pomadasidos de la Laguna de Términos, Sur del Golfo de Mexico (Pisces: Pomadasyidae). *An. Inst. Cienc. del Mar y Limnol., Univ. Nal. Autón. México.* 9(1):251-278.

Diener, R. A. 1975 *Cooperative Gulf of Mexico estuarine Inventory and Study—Texas: Area description.* National Oceanic and Atmospheric Administration, National Marine Fisheries Service Circ. 393. Galveston, Texas. 129 p.

Draper, N. R. and H. Smith. 1981. *Applied Regression Analysis.* John Wiley and Sons, New York. 699 pp.

Eleuterus, L. N. 1972. The Marshes of Mississippi. *Castanea* 37:153-168.

Fisher, W. L., J. H. McGowen, L. F. Brown, Jr. and C. G. Grout. 1972. *Environmental geologic atlas of the Texas coastal zone—Galveston-Houston area.* Bureau of Economic Geology. The University of Texas, Austin. 91 pp.

Florida Department of Environmental Regulation. 1978. *Statistical inventory of key biophysical elements in Florida's coastal zone.* Div. of Environmental Programs, Bureau of Coastal Zone Planning. Tallahassee, Florida. 115 pp.

Garcia, E. 1973. *Modificaciones al sistema de clasificacion climatica de Koppen*. Universidad Nacional Autónoma de México, Instituto de Geografica. Mexico City, Mexico. 246 pp.

Lara-Dominguez A. L., A. Yanez-Arancibia and F. Amezeua Linares. 1981. Biologia y ecologia del bagre *Arius melanopus* Gunther, en la Laguna de Terminos, Sur del Golfo de Mexico. *An. Inst. Cienc. del Mar y Limnol., Univ. Nal. Autón. México.* 8(1):267-304.

Marmer, H. A. 1954. Tides and sea level in the Gulf of Mexico. Pp. 101-118. *In:* P. S. Galtsoff (ed.), *Gulf of Mexico, Its Origin, Waters and Marine Life.* Fishery Bull. 89. U.S. Fish and Wildlife Service, Washington, D.C. 604 pp.

McGowen, J. H., L. F. Brown, Jr., T. J. Evans, W. L. Fisher, and C. G. Groat. 1976a. *Environmental geologic atlas of the Texas coastal zone—Bay City-Freeport area.* Bureau of Economic Geology. The University of Texas, Austin. 98 pp.

McGowen, J. H., C. V. Proctor, Jr., L. F. Brown, Jr., T. J. Evans, W. L. Fisher, and C. G. Groat. 1976b. *Environmental Geologic atlas of the Texas coastal zone-Port Lavaca area. Bureau of Economic Geology., Univ. of Texas, Austin. 107 pp.*

McNulty, J. K., W. N. Lindall, Jr., and J. E. Sykes (eds.). 1972. Cooperative Gulf of Mexico estuarine inventory and study, Florida: Phase I. Area Description. Circular 368. National Marine Fisheries Service, St. Petersburg, Florida. 126 pp.

Mendelssohn, I. A., K. L. McKee and M. T. Postke. 1982. Sublethal stresses controlling *Spartina alterniflora* productivity, pp. 223-242. *In:* B. Gopal, R. E. Turner, R. G. Wetzel, and D. F. Whigham (eds.), *Wetlands Ecology and Management.* National Institute of Ecology and International Scientific Publications. Lucknow Publishing, India.

Moody, C. L. 1967. Gulf of Mexico Distributional Province. *Am. Assoc. Petroleum Geologists Bull.* 51(2):179-199.

Moore, D., H. Brusher, and L. Trent. 1970. Relative abundance, seasonal distribution, and species composition of demersal fishes off Louisiana and Texas, 1962-1964. *Contr. Mar. Sci.* 15:45-70.

Nixon, S. 1980. Between coastal marshes and coastal waters—A review of twenty years of speculation and research on the role of salt marshes in estuarine productivity and water chemistry. pp. 437-525. *In:* P. Hamilton and K. MacDonald (eds.), *Estuarine and Wetland Processes.* Plenum Publishing Corp., New York.

Nixon, S. W. 1981. Freshwater inputs and estuarine productivity. pp. 31-57. *In:* R. Cross, and D. Williams (eds.), *Proceedings of the national symposium on freshwater inflow to estuaries.* Vol. I. FWS/OBS-81/04. U.S. Fish and Wildlife Service, U.S. Dept. of Interior, Washington, D.C.

Nixon, S. 1982. Nutrient dynamics, primary production and fisheries yields of lagoons. *Oceanologica Acta.* Special Volume: 357-371.

O'Neil, R. E. and M. F. Mettee (eds.). 1982. *Alabama Coastal Region Ecological Characterization. Vol. 2. A synthesis of environmental data.* FWS/OBS-82/42. Fish and Wildlife Service, U.S. Dept. of Interior, Washington, D.C. 346 pp.

Perret, W. S., B. B. Barrett, W. R. Lataple, J. F. Pollard, W. R. Mock, G. B. Adkins, W. J. Guidry, and C. J. White. 1971. *Cooperative Gulf of Mexico Estuarine Inventory and Study, Louisiana. Phase I. Area Description.* La. Wildlife and Fish. Comm. New Orleans. 175 pp.

Rayburn, R. 1981. Texas shrimp fisheries and freshwater inflow. pp. 431-438. *In:* R. D. Cross, and D. L. Williams (eds.), *Proceedings of the National Symposium on Freshwater Inflow to Estuaries. Vol. I.* U.S. Fish and Wildlife Service, U.S. Dept. of the Interior. Publ. No. FWS/OBS-81/04.

Sánchez-Gil P., A. Yáñez-Arancibia, and F. Amezcua Linares. 1981. Diversidad, distribucion y abundancia de las especies y poblaciones de peces demersales de la Sonda de Campeche (Verano, 1978). *An. Inst. Cienc. del Mar y Limnol., Univ. Nal. Autón. México.* 8(10):209-240.

SAS Institute Inc. 1985. *SAS Users Guide: Statistics.* Version 5 Edition. SAS Inst. Inc., Cary, North Carolina. 956 pp.

Secretaria de Pesca. 1978-82. Anuarios Estadisticos de Pesca. Direccion General de Informacion y Estadistica. Mexico. D.F. Multiple pagination.

Shew D. M., R. H. Baumann, T. H. Fitts, L. S. Dunn. 1981. *Texas Barrier Islands Region Ecological Characterization: Environmental synthesis papers.* Biological Services Program. U.S. Fish and Wildlife Services. FWS/OBS-81/32. 413 pp.

Sutcliffe, W. H. 1972. Some relations of land drainage, nutrients, particulate material, and fish catch in two eastern Canadian bays. *J. Fish. Res. Bd. Canada.* 29:357-412.

Swingle, H. 1971. *Biology of Alabama estuarine areas cooperative Gulf of Mexico estuarine inventory.* Alabama Marine Resource Bulletin. No. 5. Alabama Marine Resources Laboratory. Dauphin Island, Alabama. 123 pp.

Thornthwaite Associates. 1964a. Average climatic water balance data of the continents. Part VII: The United States. *Publications in Climatol.* 17(3):419-615.

Thornthwaite Associates. 1964b. Average climatic water balance data of the continents. Part VI: North America. *Publications in Climatol.* 17(3):372-403.

Turner, R. E. 1977. Intertidal vegetation and commercial yields of Penaeid shrimp. *Trans. Amer. Fish. Society* 106(5):411-416.

Turner, R. E. 1979. Louisiana's coastal fisheries and changing environmental conditions. pp. 368-370. *In:* J. Day, Jr. *et al.* (eds.), *Proc. Third Coastal Marsh and Estuary Management Symposium.* Louisiana State University Division of Continuing Education, Baton Rouge. 511 pp.

United States Dept. of Commerce. 1968. *Climatic Atlas of the United States.* U.S. Dept. of Commerce. Environmental Science Services Administration. Environmental Data Service. Washington, D.C. 80 pp.

United States Dept. of Commerce. 1979. Bathymetric map of the Gulf of Mexico, Scale 2,160,000. U.S. Dept. of Commerce, National Oceanic and Atmospheric Administration. Map Number 411.

United States Dept. of Commerce. 1982. *Tide Tables 1983. East coast of North and South America, including the coast of Greenland.* U.S. Dept. of Commerce, NOAA, National Ocean Survey. 285 pp.

Vargas Maldonado, I., A. Yáñez-Arancibia and F. Amezcua Linares, 1981. Ecologia y estructura de las comunidades de Peces en areas de *Rhizophora mangle y Thalassia testudinum* de la Isla del Carmen, Laguna de Términos, sur del Golfo de Mexico. *An. Inst. Cienc. del Mar y Limnol., Univ. Nal. Autón. México.* 8(1):241-266.

Welsh, B. L., R. B. Whitlatch and W. F. Bohlen. 1982. Relationship between physical characteristics and organic carbon sources as a basis for comparing estuaries in southern New England. pp. 53-67. *In:* V. S. Kennedy (ed.), *Estuarine Comparisons.* Academic Press. New York, 709 pp.

Yáñez-Arancibia, A. and J. W. Day, Jr. 1982. Ecological characterization of Terminos Lagoon, tropical lagoon-estuarine system in the southern Gulf of Mexico. *Oceanologica Acta.* Special Volume: 431-440.

Yáñez-Arancibia, A., A. L. Lara-Domínguez, P. Chavance and D. Flores Hernandez, 1983. Environmental behavior of Términos Lagoon ecological system, Campeche, Mexico. *An. Inst. Cienc. de Mar y Limnol., Univ. Nal. Autón. México.* 10(1):137-176.

Yáñez-Arancibia, A., A. Lara-Domínguez, P. Sánchez-Gil, I. Vargas Maldonado, P. Chavance, F. Amezcua Linares, A. Aguirre Leon and S. Diaz Ruiz, 1982. Ecosystem dynamics and nichthemeral and seasonal programming of fish community structure in a tropical estuarine inlet, Mexico. *Oceanologica Acta.* Special Volume: 417-430.

Yáñez-Arancibia, A., G. Soberón Chavez and P. Sánchez-Gil. In press. Ecology of control mechanisms of fish production in the coastal zone. *In:* A. Yáñez-Arancibia (ed.), *Fish Community Ecology in Estuaries and Coastal Lagoons: Towards an Ecosystem Integration.* Editorial Universitaria. Universidad Nacional Autónoma de México, Mexico, D.F.

TEMPORAL VARIABILITY IN ESTUARIES

Charles B. Officer, Convenor

TEMPORAL VARIABILITY OF ELEVATIONS, CURRENTS AND SALINITY IN A WELL-MIXED ESTUARY

R.J. Uncles, M.B. Jordan and A.H. Taylor
Natural Environment Research Council
Institute for Marine Environmental Research
Prospect Place, The Hoe, Plymouth, United Kingdom

Abstract: Data are presented for currents, salinity, surface elevations and meteorological variables over a three month period in the Severn Estuary, U.K. Currents and salinity were measured 5 m above the sea bed at a site 23 m deep. Low pass (residual) currents, salinity and surface elevations showed a pronounced temporal variability. Using maximum-entropy spectral analysis, this variability was interpreted in terms of the major driving mechanisms of atmospheric pressure, wind-stress, freshwater runoff and tidal range. Residual surface elevations were highly coherent with tidal range, showing a strong spring-neap variability indicative of non-linear tidal forcing. The elevations also showed strong coherence with both atmospheric pressure (inverse barometer effect) and wind-stress; winds from the south or west led to an increase in residual water levels in the estuary. Salinity exhibited small amplitude, short-period fluctuations which appeared to result from fluctuations in the residual current; these were superimposed on much larger, long-period (greater than one month) variations which were caused mainly by freshwater inputs and wind-stress. Residual currents were mainly topographically induced, and showed a strong spring-neap variability. Generally, maximum wind-induced currents were directed across the estuary, and were associated with oppositely directed winds. Depth-averaged, two-dimensional hydrodynamical models were used to evaluate wind and tide-induced residual elevations and currents in the estuary. Computed elevations compared well with observations. A depth-averaged model was used in conjunction with a model of the vertical structure of residual currents in order to interpret observed features of the current variability.

Introduction

Residual (tidally averaged) motion in macrotidal, well-mixed estuaries occurs on many time-scales, and can be generated by several mechanisms. The non-linearity of the equations describing tidal flows in such estuaries leads to residual flows and residual surface elevations. These exhibit their major temporal variability with the spring-neap cycle (Uncles and Jordan 1980). Density-driven residual currents and elevations are also influenced by the spring-neap cycle, because of variations in tidal viscosity and mixing. However, these currents and elevations respond primarily to variations in the salinity field, which in turn depend on temporal variations in the freshwater discharges and the flushing time of the estuary. Variations in residual flows and elevations also result from atmospheric pressure gradients and wind-stress, which vary on a wide range of time-scales.

The objectives of this work were to investigate the temporal variability of residual surface elevations, currents and salinity in the Severn Estuary (Fig. 1).

The investigation was a natural extension of earlier work in this macrotidal, vertically well-mixed estuary, in which observations were restricted to a few tidal cycles, and steady conditions were assumed (Uncles and Jordan 1979, 1980; Uncles 1982a,b, 1983). The methods employed were similar to those used in studies of wind-driven currents in mesotidal and microtidal systems (Elliott 1978, 1979a,b, 1981; Schwing *et al.* 1983; Wang 1979).

Observations and Data

Velocity and salinity were recorded at 10-min intervals over the period 18 April to 19 July 1978 at the site shown in Fig. 1. The combined current and salinity meters were moored 5 m above the seabed, and the mean depth of water was 23 m.

Half-hourly values of tidal height were observed at Avonmouth (Fig. 1). Hourly values of atmospheric pressure, wind speed and direction were observed at Cardiff (Fig. 1), situated roughly midway between the current-meter site and Avonmouth. Daily-averaged values of the freshwater inputs to the estuary were available from Water Authority records.

Daily-averaged values of low-pass (residual) velocity, salinity and surface elevation were extracted from the data and interpreted in terms of the daily averaged tidal range at Avonmouth, freshwater run-off, wind-stress and atmospheric pressure.

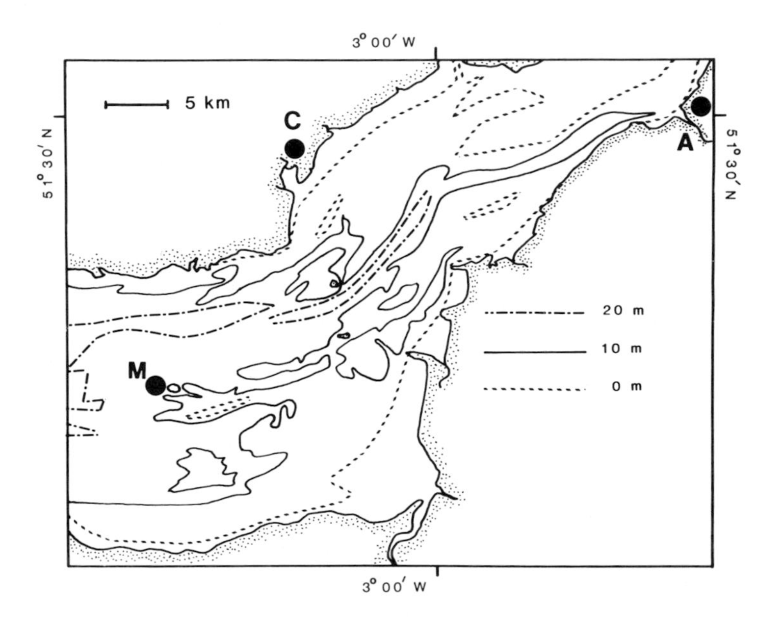

Figure 1. Severn Estuary, showing the locations of Avonmouth (A), Cardiff (C) and the meter site (M). See Randerson (this volume) Fig. 1 for a larger perspective of the Severn Estuary. Contoured depths are relative to a datum positioned at the depth corresponding to the lowest astronomical tide.

The daily-averaged tidal range, residual surface elevation (mean tide level) velocity and salinity are shown in Fig. 2 as functions of time in days. The tidal range over this period varied between the extremes of about 13.5 m at peak springs to 6.5 m at smallest neaps (Fig. 2a). The northerly component of residual velocity, v_N, was typically 0.23 m s^{-1} (Fig. 2c), and showed a strong spring-neap variability (through its correlation with tidal range). Spring tide residual currents were much higher than those during neap tides. The east component of residual current, v_E, was typically 0.08 m s^{-1} and showed little spring-neap variability (Fig. 2c).

The residual surface elevation varied between about 6.6 m and 7.2 m (Fig. 2b). Spring-neap variations were dominant, with spring tide elevations exceeding those at neap tides. The major variability in salinity occurred with time-scales much longer than the spring-neap cycle (Fig. 2d). The inverse relationship between salinity, S, and freshwater run-off into the estuary, Q, is evident from the time-series for run-off given in Fig. 2f. Time-series data for meteorological variables are given in Figs. 2e,g,h. We define the quadratic wind to be $W|W|$, where W is the wind-speed vector. East and north components of quadratic wind are given by $W_E|W|$ and $W_N|W|$, respectively (Figs. 2g,h). There were no strong winds over the period. The mean and standard deviation (S.D.) of the east and north components of the quadratic wind over the period were 7 ± 25 and -3 ± 9 $m^2 s^{-2}$, respectively. The mean and S.D. of the atmospheric pressure was 1015 ± 8 mb. Although not obvious from Figs. 2b and 2e, a strong negative correlation existed between atmospheric pressure and surface elevation because of the inverse barometer effect (Proudman 1953).

Time-Series Analysis

Maximum-entropy spectral analysis (Taylor 1983; Lacoss 1971; Ulrych and Bishop 1975) was used to derive coherence spectra between various pairs of variables. In addition to supplying a measure of the coherence between variables at a particular frequency, the technique also yielded the phase relationship between the variables and their linear regression equation at that frequency. Maximum entropy spectral analysis achieves a high frequency resolution because it involves minimal assumptions about how the time-series extends outside the sampled interval.

Using the regression relationships between surface elevation and tidal range, and v_N and tidal range, the strong spring-neap signal can be removed from these variables. Also, the strong inverse barometer effect can be removed from the surface elevation. The results of doing this are shown in Fig. 3. Fluctuations in surface elevation (Fig. 3b) clearly correlated with the north and east components of quadratic wind (Figs. 2g,h). Winds to the north and the east produced an increase in surface elevation (a set-up) at Avonmouth (see Fig. 1 for location). Fluctuations in v_N were negatively correlated with the north component of quadratic wind (Figs. 3a, 2h), such that a wind to the north generated a south-flowing current.

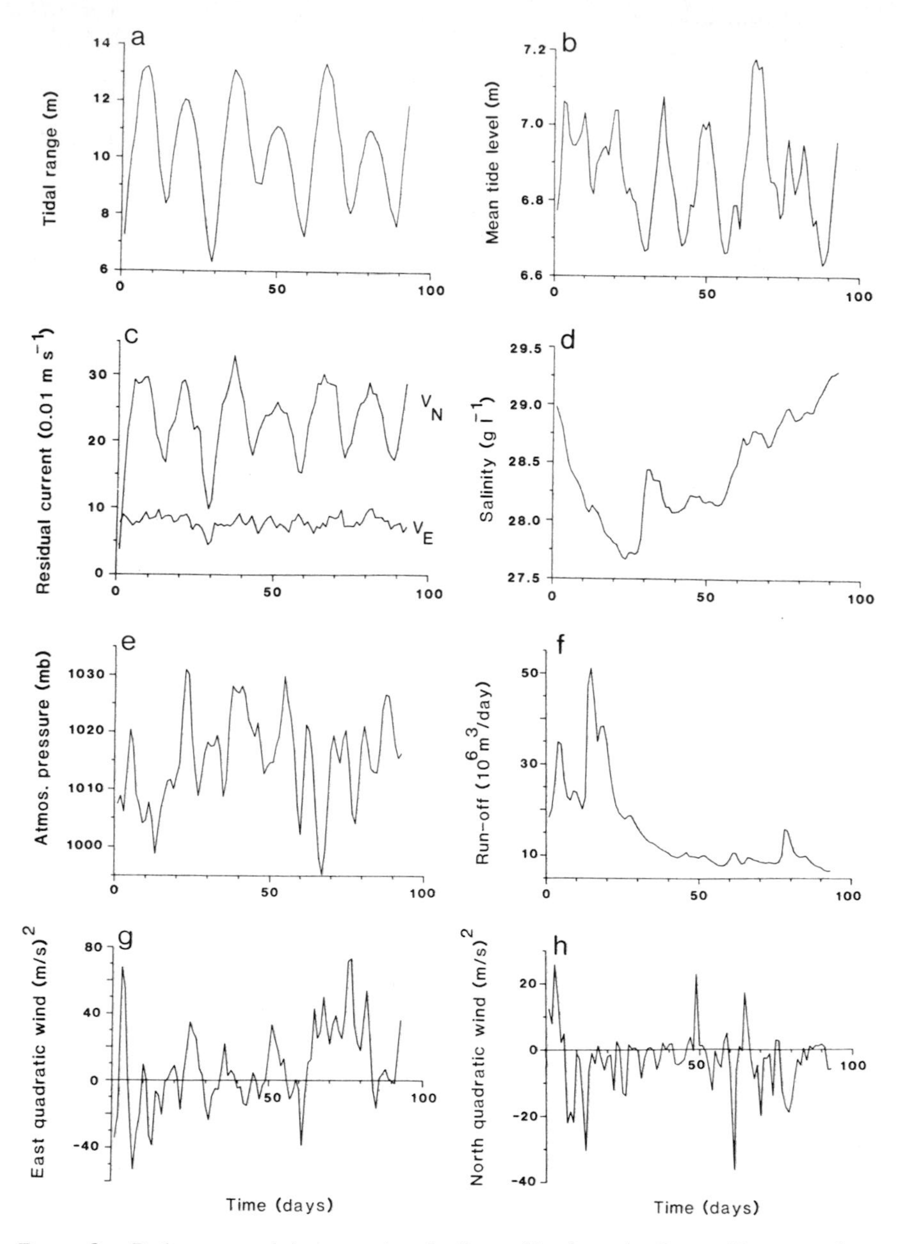

Figure 2. *Daily-averaged, low pass (residual) variables from the Severn Estuary as functions of time in days. (a) Tidal range, (b) surface elevation (mean tide level), (c) components of residual velocity, (d) salinity, (e) atmospheric pressure, (f) freshwater run-off, (g) east and (h) north components of quadratic wind.*

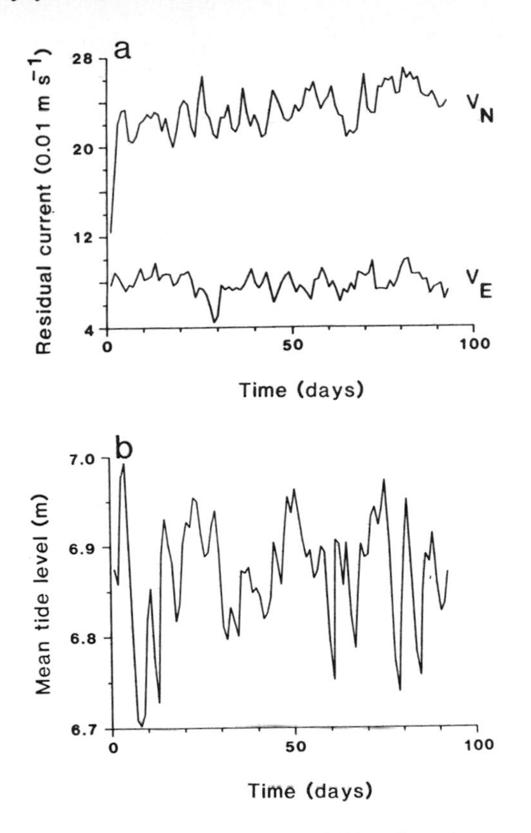

Figure 3. *Components of (a) residual current and (b) surface elevation (mean tide level), after removal of tidally-induced fluctuations, as functions of time in days. The inverse barometer effect has also been removed from the elevations.*

Surface Elevation

The dependence of surface elevation, ζ, on tidal range, R, is a very distinctive feature of the data (Figs. 2a,b). Coherence spectra between ζ and R show bands of coherence over the periods 25-29 days (representing forcing at the M_2-N_2 interaction period of 27.6 days) and 13-15 days (representing forcing at the spring-neap, M_2-S_2 interaction period of 14.8 days). M_2, S_2 and N_2 (in descending magnitude) are the three principal semi-diurnal tidal constituents in the Severn Estuary. The regression relationship between ζ and R at the spring-neap periodicity was:

$$\zeta = 0.06 \text{ R} \quad (1)$$

Surface elevation and tidal range were in phase. The mean and S.D. of R were 10.3 ± 1.8 m; this led to a S.D. in ζ due to tidal forcing of approximately 0.1 m. Therefore, according to eqn. 1 the residual surface elevation at mean spring tides (R = 12.3 m) is approximately 0.35 m higher than at mean neap tides (R = 6.5 m).

Another distinctive feature of the data at low frequencies was the dependence between surface elevation, ζ, and atmospheric pressure, P. The regression relationship was:

$$\zeta = -0.01 \, P \tag{2}$$

where the units are meters for ζ and millibars for P. Equation 2 is the inverse barometer effect. The S.D. in P was 8 mb, which led to a S.D. in ζ due to the inverse barometer effect of about 0.1 m.

The strong dependence of surface elevation on tidal range and atmospheric pressure tended to mask effects due to other mechanisms. Because of this, eqns. 1 and 2 were used to filter these effects from the surface elevation data. The resulting coherence spectra (coherence squared against $\log_{10}$ frequency, where frequency is in cycles per day) are shown in Fig. 4. The 95% confidence limits for the coherence functions are drawn as horizontal dashed lines. Relationships between pairs of variables are considered to be significant only if the coherence function exceeds the 95% confidence limit.

The dependence of filtered surface elevation on atmospheric pressure and spring-neap tidal range is reduced to an insignificant level. The elevation was not significantly coherent with freshwater run-off (Fig. 4), i.e., the direct effect of run-off on water levels was small. However, the surface elevations were significantly affected by salinity fluctuations, which were themselves affected by run-off, especially at low frequencies (periods longer than about 30 d). The regression relationship between surface elevation ζ and salinity S at low frequencies was:

$$\zeta = 0.04 \, S \tag{3}$$

where units are m and g l^{-1}. The S.D. of the fluctuations in elevation due to those in salinity were roughly 0.02 m. At low frequencies, S lagged ζ by roughly one third of the cycle period. This dependence of elevation on salinity was a manifestation of the set-up which was associated with axial density gradients along the estuary.

The surface elevation was highly coherent with north and east components of the wind-stress (proportional to the quadratic wind) over several frequency bands (Fig. 4). Maximum elevations corresponded with maximum north or east components of the quadratic wind; ignoring any coherence between these components, the averaged regression relationships were:

$$\zeta = 3.6 \times 10^{-3} \, W_E |W| \tag{4}$$

$$\zeta = 6.6 \times 10^{-3} \, W_N |W| \tag{5}$$

where units are m and $m^2 \, s^{-2}$. This led to a S.D. in the elevations due to fluctuations in winds of roughly 0.1 m.

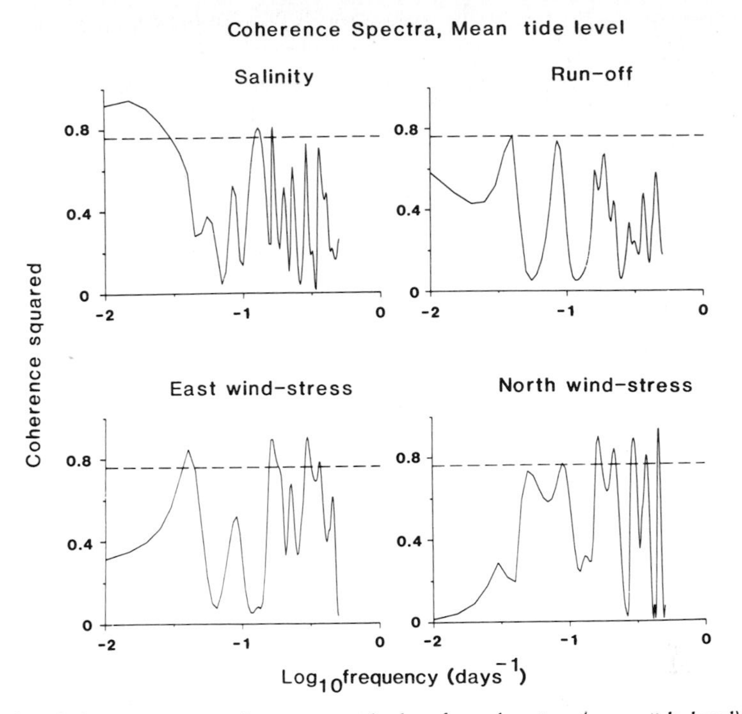

Figure 4. Coherence spectra between residual surface elevation (mean tide level) in the Severn Estuary and salinity, run-off and wind-stress components. The inverse barometer effect and tidally-induced fluctuations have been removed from the surface elevation.

Residual Currents

The east component of residual current, v_E, showed no significant coherence with atmospheric pressure, wind-stress or tidal range. While the speed of the residual current vector at 5 m above the bed increased with tidal range, the direction also changed, becoming more northerly with increasing tidal range. This behavior reduced the spring-neap variability in v_E to an insignificant level. The insignificance of wind-stress to v_E is difficult to understand. It is possibly a steering effect of local topography. East-west gradients in sea bed are very steep (Fig. 1), which might deflect the near-bed, wind-driven currents northward or southward.

High frequency fluctuations in v_E (S.D. about 0.01 m s^{-1}) had the appearance of random noise (Fig. 2c). Small, low frequency (period longer than 30 d) fluctuations were coherent with run-off and local salinity, indicating the presence of a density-driven component of current.

The dependence of the north component of residual current, v_N, on tidal range, R, is a very distinctive feature of the data (Figs. 2a,c). The regression relationship between v_N and R at the spring-neap periodicity was:

$$v_N = 0.029\,R \tag{6}$$

where units are m s^{-1} and m. The mean and S.D. of R were 10.3 ± 1.8 m; this led to a S.D. in v_N due to tidal forcing of approximately 0.05 m s^{-1}.

The strong dependence of v_N on tidal range tended to obscure other processes. Because of this, eqn. 6 was used to filter tidal effects from v_N (Fig. 3a). This reduced the S.D. to 0.02 m s^{-1}, and the dependence of filtered current on tidal range was insignificant. There was no coherence with atmospheric pressure. The resulting coherence spectra are shown in Fig. 5. The peaks of coherence with run-off probably reflect a dependence on wind-stress, rather than direct dependence on run-off itself (coherence existed between run-off and wind-stress). Similarly, the coherence with salinity at 6 d (Fig. 5) is probably a reflection of coherence with wind-stress, rather than a direct dependence on salinity. However, the coherence between current and salinity at the longest periods (roughly 100 d) appears to have been due to a density current component of v_N. The northerly component of this density current increased over the period due to the adjustment of the salinity field (see Fig. 3a, which shows the upward trend in v_N). This, in turn, was due to the increasing east-heading winds and decreasing run-off over the period (Figs. 2f,g), which re-oriented the isohalines in the area from NE-SW to N-S.

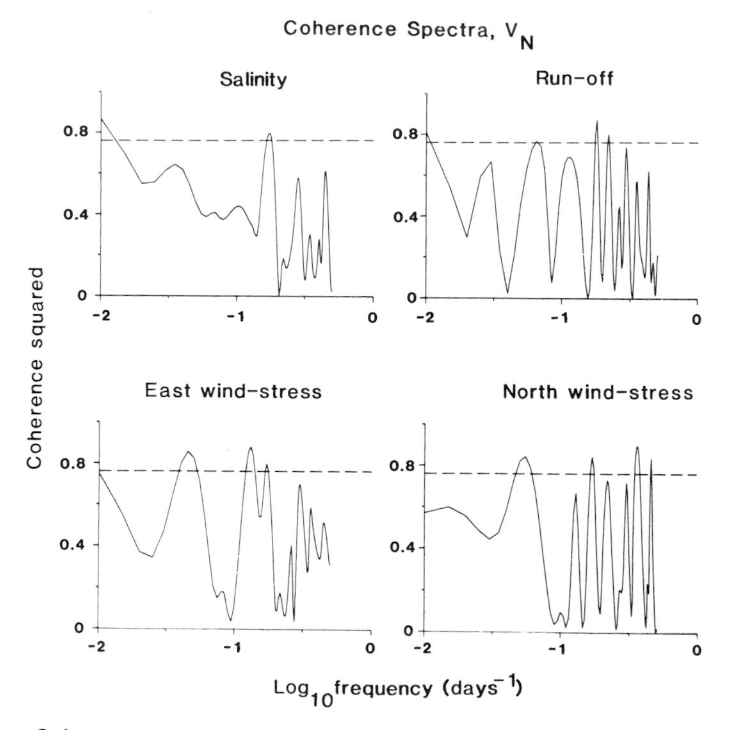

Figure 5. Coherence spectra between north component of residual current v_N in the Severn Estuary and salinity, run-off and wind-stress components. Tidally-induced fluctuations have been removed from the residual current.

Peaks of strong coherence existed between the north component of filtered residual current and wind-stress at certain frequencies (Fig. 5). For each of these peaks the quadratic wind components were evaluated at the instant when the filtered residual current reached its maximum speed (the semi-major axis of the filtered residual current ellipse). The maximum wind-driven current at 5 m above the seabed was directed across the estuary, and was associated with oppositely directed winds. The relationship between current speed and quadratic wind speed, averaged over the significant peaks, was:

$$v = -(2 \times 10^{-3} \pm 50\%) \times W|W| \tag{7}$$

The S.D. in the across-estuary (north component) quadratic wind was 9 $m^2\ s^{-2}$, so that this led to a S.D. in the residual current due to wind fluctuations of 0.02 $m\ s^{-1}$.

It is not known why the east component of residual current did not respond significantly to wind-stress, or why the east component of wind stress had little effect on the wind-driven current. It may be that the steep slope of the sea bed in the east-west direction steered the currents responding to east-west winds along the depth contours, and that these currents were too small to be separated from noise in our data.

Salinity

The salinity showed significant coherence with tidal range, especially at the spring-neap and monthly periodicities (Fig. 6). However, this merely reflected coherence with the north component of unfiltered residual current, v_N, which itself had strong tidal dependence.

The significant coherence with unfiltered v_N extended over the periodicity band 10-30 d (Fig. 6). Within this band, the salinity led the north component of current by approximately 90° of phase. Therefore, the increasing v_N between neap and spring tides advected lower salinity water from the south of the estuary through the site. The reverse occurred on the decreasing northward-flowing current between springs and neaps. In the long term, the transport of salt by the average residual current over the spring-neap period must have been approximately balanced by mixing of salt. Our data exhibited salinity perturbations from this balance due to tidally-induced current fluctuations. The regression relationship between salinity, S, and v_N was:

$$S = 2\ v_N \tag{8}$$

where units are $g\ l^{-1}$ and $m\ s^{-1}$. The S.D. of 0.05 $m\ s^{-1}$ in v_N led to a S.D. of 0.1 $g\ l^{-1}$ in S, which is the scale of the short-period fluctuations shown in Fig. 2d.

At long periods (40-100 d) the salinity was strongly coherent with freshwater run-off. This minimum time-scale of 40 d was a consequence of the long flushing time of the estuary (flushing time being a measure of the time taken for freshwater inputs to pass through and be removed from the estuary). The

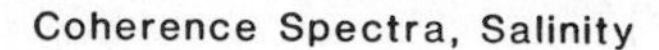

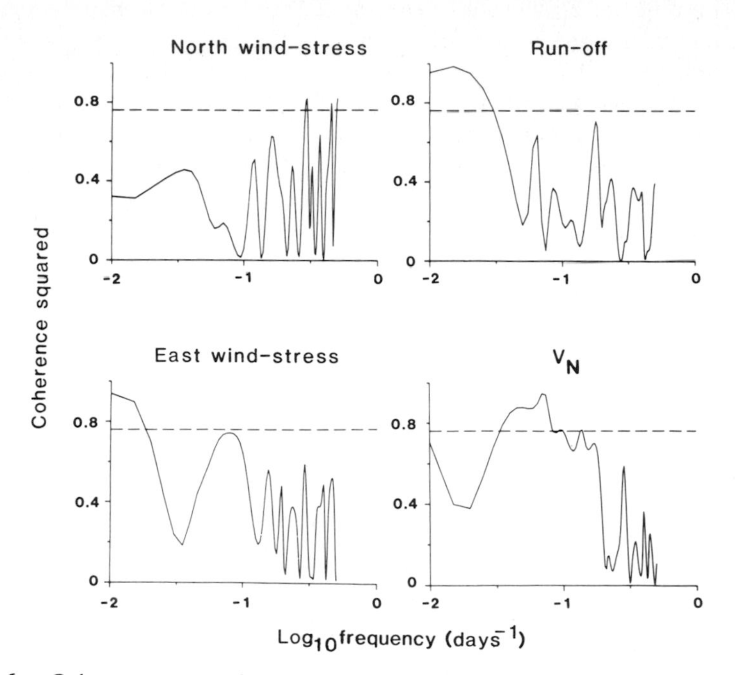

Figure 6. *Coherence spectra between salinity in the Severn Estuary and run-off, north component of residual current and wind-stress components.*

regression relationship between amplitudes of the salinity, S, and run-off fluctuations, Q, were:

$$S = 0.04\ Q \tag{9}$$

where units are g l^{-1} and 10^6 m^3 d^{-1}. The S.D. of Q was 10×10^6 m^3 d^{-1}, which led to a S.D. in salinity due to run-off fluctuations of 0.4 g l^{-1} (see Fig. 2d).

The salinity was also strongly coherent with the east component of wind stress at long periodicities (longer than 70 d). Maximum salinity occurred about 1 week after the maximum east component of wind-stress. Therefore, the observed long-term fluctuations in salinity at the site were the result of fluctuations in freshwater inputs and the east component of wind-stress. Currents associated with these long-term fluctuations in wind-stress would have been small. This is confirmed by the lack of significant coherence between measured residual currents and wind-stress (Fig. 5) for long periodicities.

Theoretical Analysis

We will consider models of surface elevations and currents in order to provide theoretical insight into the various mechanisms which have been identified through time-series analysis of the data.

Surface Elevations

The observed relationship between residual surface elevation ζ at Avonmouth and tidal range R is given by eqn. 1. Residual surface elevations were higher at spring tides than at neap tides. Figure 7 shows the modeled range of residual surface elevations over a spring-neap cycle. Data were computed using a depth-averaged, numerical hydrodynamical model (Uncles 1982a). Residual elevations were modeled for mean spring and neap tidal conditions. Residual surface elevations were then subtracted to give the contours drawn in Fig. 7. The range of elevations at the seaward boundary is roughly 0.05 m; it increases to approximately 0.35 m near Avonmouth. This value is in agreement with observations (eqn. 1).

The large increase in ζ with distance up-estuary is a non-linear effect. It is generated mainly by the Stokes drift which is associated with the large-amplitude tides, but is affected to some extent by the presence of overtides, such as M_4 (Uncles and Jordan 1980). A theoretical analysis of this phenomenon has been given by LeBlond (1979) for the St. Lawrence Estuary. The axial surface slope along the estuary shown in Fig. 7 drives a down-estuary directed residual current.

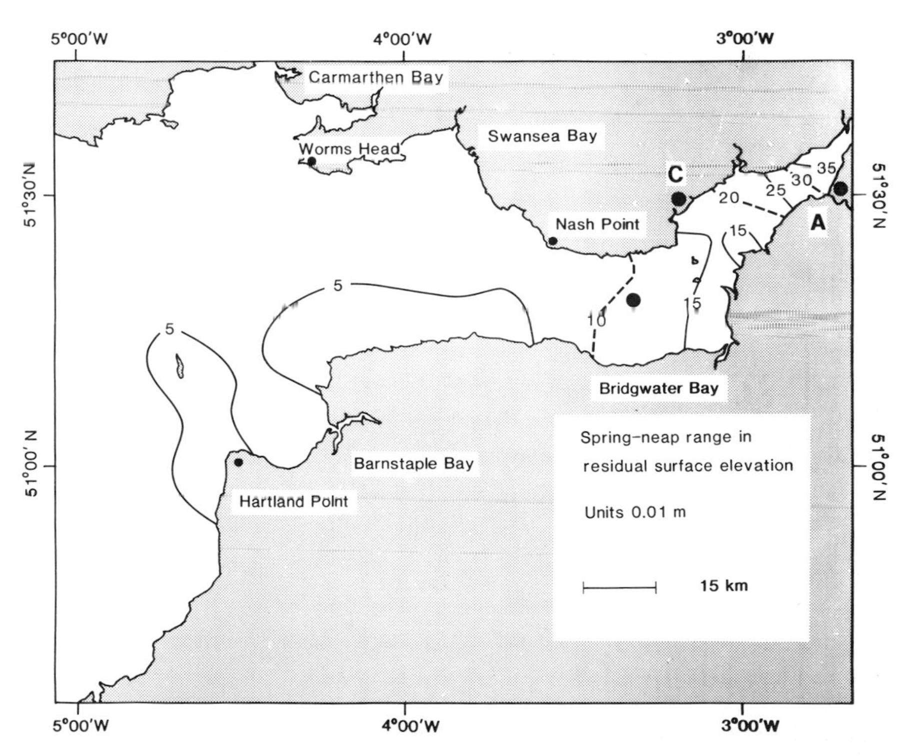

Figure 7. Range (units 0.01 m) in residual surface elevation between mean spring and mean neap tides. Avonmouth (A), Cardiff (C), current meter site (●). Data according to a numerical, hydrodynamical model.

This current has a typical, spatially averaged value of about 0.03 m s^{-1} in the Severn (Uncles and Jordan 1980). In the absence of freshwater inputs and meteorological perturbations it balances the Stokes drift that is directed up-estuary.

The observed residual surface elevation depended upon salinity, S, at low frequencies (eqn. 3). Amplitudes of these elevation fluctuations were 0.02 m. A rough estimate of the set-up in elevation at Avonmouth due to axial salinity gradients can be derived in the following way. If ΔS is the difference in salinity between Avonmouth and the ocean (where S = 35 g l^{-1}), $\Delta\zeta$ is the set-up due to ΔS, and h is the depth (spatially-averaged between the mouth of the Bristol Channel and Avonmouth to give h ~ 30 m), then Uncles (1982 a) has shown that an approximate balance exists between surface slopes and axial density gradients:

$$\Delta\zeta \sim -\frac{1}{2}\,ha\,\Delta S \qquad (10)$$

where a = 7.8×10^{-4}. Observations of salinity throughout the estuary during July 1978 gave $\Delta S = -10$ g l^{-1}, whereas those during April gave $\Delta S = -15$ g l^{-1}. The observed range in ΔS of 5 g l^{-1} between April and July gives a range in $\Delta\zeta$ of 0.06 m, from eqn. 10. Therefore the amplitude of surface elevation variations due to salinity fluctuations is estimated to be 0.03 m. This is comparable with the observed value of 0.02 m (eqn. 3).

Equations 4 and 5 relate the observed surface elevations at Avonmouth to the east and north component of quadratic wind, respectively. In reality, the set-up will depend on a weighted summation of these components:

$$\zeta = \gamma\zeta^{(1)} + \delta\zeta^{(2)} \qquad (11)$$

Where $\zeta^{(1)}$ and $\zeta^{(2)}$ are the surface elevations at Avonmouth due to the east and north components of wind-stress in isolation, $\tau^{(1)}$ and $\tau^{(2)}$. The total wind-stress is τ:

$$\tau = \gamma\, e\, \tau^{(1)} + \delta\, n\, \tau^{(2)} \qquad (12)$$

in which e and n are unit vectors pointing east and north, respectively, and τ is related to the quadratic wind by:

$$\tau = \varrho_a C_D W|W| \qquad (13)$$

The drag coefficient has been estimated as $C_D \approx 1.3 \times 10^{-3}$ (Phillips 1966), $C_D \approx 2.5 \times 10^{-3}$ (Proudman 1953) and $C_D \approx 4.5 \times 10^{-3}$ (Bowden 1956); ϱ_a is the density of air, and winds are measured at a height of 10 m.

Figures 8a,b show the modeled surface elevations due to an east and north component of wind-stress, respectively, equal to 0.078 N m^{-2}. The computed elevations have had the inverse barometer effect removed. The seaward boundary conditions for residual elevations were derived from data given for the

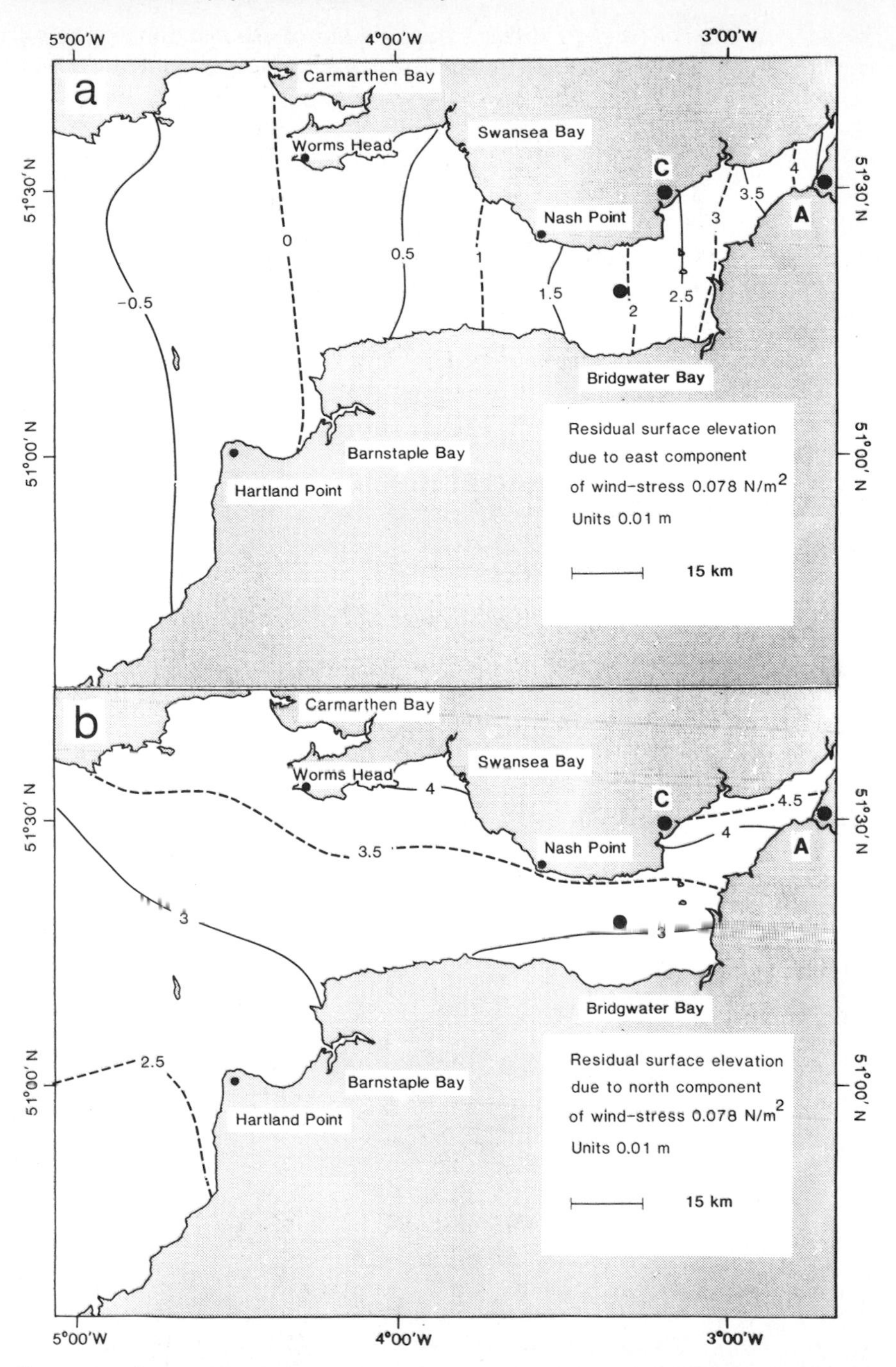

Figure 8. *Surface elevation (units 0.01 m) due to east and north components of wind-stress (i.e. westerly and southerly winds) equal to 0.078 N m^{-2}. (a) East component. (b) North component. Data according to a numerical, hydrodynamical model.*

shelf by Pingree and Griffiths (1980). The elevations at Avonmouth are, coincidently, 0.045 m for both wind-stresses. Therefore, from eqns. 11 and 12:

$$\tau_E = \tau^{(1)} = \tau_N = \tau^{(2)} = 0.078 \text{ N m}^{-2}$$

and $\zeta^{(1)} = \zeta^{(2)} = 0.045$ m

The set-up for any other wind-stress, τ, can be derived from eqn. 11 using values of γ and δ given by eqn. 12:

$$\gamma = \tau_E/\tau^{(1)} \text{ and } \delta = \tau_N/\tau^{(2)}.$$

Using these modeled data it is possible to estimate C_D from the observations. Coherence spectra between ζ and the east component of quadratic wind give a regression relation at each significant peak of coherence (Fig. 4). The corresponding regression relation between east and north components of quadratic wind can be derived from their coherence spectra in order to evaluate δ in eqn. 12. This procedure gives a mean value for C_D of 4×10^{-3}. The same procedure applied to the coherence spectra for ζ and the north component of quadratic wind gives a mean value for C_D of 2×10^{-3}. A mean value from all these observations is $C_D = 3 \times 10^{-3} \pm 50\%$. The error encompasses all derived values.

Tidally-induced Residual Currents

The regression relationship between north component of residual current, v_N, and tidal range, R, at the spring-neap periodicity is given by eqn. 6. This dependence on tidal range is the strongest feature of the residual current data, and implies that these currents are primarily tidally-induced. Their generation has been investigated using a depth-averaged, numerical hydrodynamical model of the Severn (Uncles 1982b).

Computed, depth-averaged, tidally-induced residual currents are given in Fig. 9 for the case of average (M_2) tidal flows. These are generated mainly by the interaction of tidal flow with local sea bed and coastline topography. A secondary down-estuary component of order 0.03 m s^{-1} also exists as a compensation current for the Stokes drift. At the current meter site the modeled, depth-averaged residual current is directed predominantly to the north, in agreement with our observations at 5 m above the seabed. However, the modeled east and north components of residual current are -0.005 and 0.015 m s^{-1}, respectively; these components are an order of magnitude smaller than the observed mean values of 0.08 and 0.23 m s^{-1} (Fig. 3a).

The disagreement between modeled and observed currents is possibly a consequence of inadequate spatial resolution in the model (1 km), and the neglect of variations through the water column. To investigate vertical structure we consider the following steady-state, linearized equation for the residual current (Uncles 1982a):

$$\partial_t v = 0 = -f x v - g\nabla\zeta - gha\nabla S(1-\eta) + C.F. + \alpha h^{-1}\partial_\eta^2 v \quad (14)$$

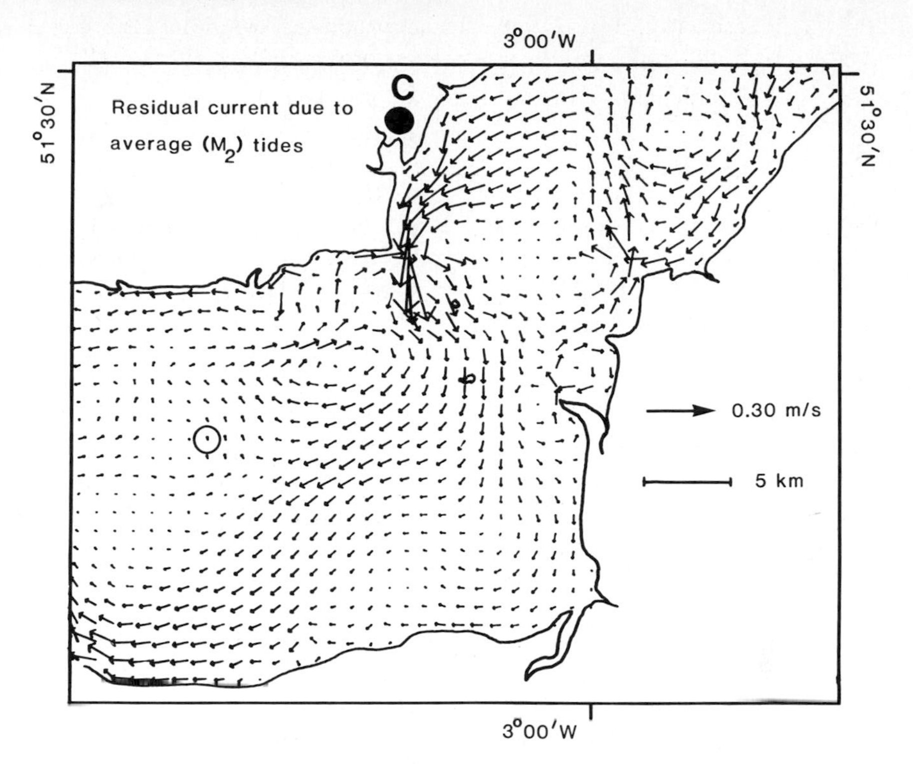

Figure 9. Topographically-induced, depth-averaged residual circulation in the Severn at mean tides. Data according to a numerical, hydrodynamical model.

with

$$\partial_\eta v = \tau/\varrho\alpha \text{ at } \eta = 1 \text{ (sea surface)} \tag{15}$$

and

$$\partial_\eta v = kv/\alpha \text{ at } \eta = 0 \text{ (sea bed)} \tag{16}$$

in which η is non-dimensional depth, k is a seabed friction coefficient, α the uniform vertical eddy viscosity divided by depth, f the Coriolis parameter, C.F. is centrifugal force, and g the acceleration due to gravity. Other symbols have been defined. For average (M_2) tides, k and α have been taken as 2×10^{-3} and 1×10^{-3} m s^{-1}, in accordance with previous modeling work (Heaps 1972, p. 423). The surface slope in eqn. 14 is assumed to take into account all of the forcing mechanisms except those due to density gradients and centrifugal forces. It is evaluated by solving eqn. 14 analytically, averaging the derived currents over the depth, and substituting for these the depth-averaged residual currents from the hydrodynamical numerical model (Fig. 9).

To investigate the tidally-induced current we omit wind-stress and density gradients ($\tau = 0$, $\nabla S = 0$) and model the centrifugal force (C.F.) as being perpendicular to the tidal currents, and of magnitude:

$$\text{C.F.} = (1-\eta)U^2/r \tag{17}$$

where U is the tidal current speed and r the radius of curvature. U^2/r is taken to be constant throughout the water column. Therefore, the centrifugal force is modeled as being zero for surface currents, and a maximum for near-bed currents. This simulates topographic steering of the near-bed tidal currents.

A typical value of U for average tides, U_o, is 1.4 m s^{-1}. The solution of eqn. 14 for average tides and for r = 5 km gives modeled currents which are comparable with observed values. The modeled residual current at 5 m above the bed is directed predominantly to the north. At the surface the modeled current is in the opposite direction, although we have no data to support this. Data can also be modeled for spring and neap tides by increasing or decreasing U in eqn. 17, and by increasing or decreasing k and α (both of which depend approximately linearly on tidal current speed) by U/U_o. Modeled values of the north component of residual current, v_N, vary with tidal range, R, in a manner which is comparable with observed data when r = 5 km (eqn. 6). Moreover, the east component of modeled residual current shows almost no spring-neap variability. This is because the computed near-bed currents tend to align closer to the direction of centrifugal forcing (N-S) at spring tides, owing to the greater influence of friction compared with Coriolis force. Therefore, although the speed of the computed current vector at 5 m above the bed increases with increasing tidal range, it also rotates towards the north. The overall effect is that v_E changes little during the simulated spring-neap cycle. The similarity between modeled and observed data implies that there is substantial physical validity in the way we have modeled centrifugal forces (eqn. 17).

Density-induced Residual Currents

Equation 14 has been solved solely for density currents by omitting centrifugal force and wind-stress (τ = C.F. = 0). Depth-averaged density-driven currents in the area of the current meter site are very small (Uncles 1982a), and have been put equal to zero in the solution of eqn. 14. Density gradients (proportional to salinity gradient vectors) and density currents were computed for three sets of salinity fields corresponding to observations during April, May and July 1978. The computed density currents at the surface and at 5 m above the sea bed, together with the associated horizontal salinity gradients, are shown in Fig. 10.

The computed southward component of density-driven current at 5 m above the sea bed decreased by roughly 0.02 m s^{-1} over the period. This accounts for much of the observed increase in the north component of residual current over the period (Fig. 3a). The salinity gradient vector decreased over the period (Fig. 10) as a consequence of the general decrease in freshwater run-off

(Fig. 2f). Rotation of the salinity gradient over the period was partly effected by the general increase in the east component of quadratic wind (Fig. 2g).

Wind-induced Residual Currents

Equation 7 gives the relationship between the speed of the wind-driven component of residual current at 5 m above the sea bed and the associated quadratic wind. Generally, these currents are directed across the estuary (the north-south axis) with the wind opposed to the current.

Depth-averaged, wind-driven currents have been computed using the numerical, hydrodynamical model. Figure 11a shows the wind-driven currents resulting from an east component of wind-stress (westerly wind) equal to 0.078 N m^{-2}. Currents are directed into the estuary in shallow areas, where the effects of wind forcing are largest, and directed out of the estuary in the deeper areas. Maximum currents for this wind-stress amount to 0.02 m s^{-1}. Depth averaged, wind-driven currents in the vicinity of the current meter site are 10^{-3} m s^{-1}, which are negligible.

Figure 11b shows the wind-driven current pattern for a north component of wind-stress (southerly wind) equal to 0.078 N m^{-2}. This pattern is very similar to that for the east component of wind-stress. Current speeds are similar, and depth-averaged, wind-driven currents in the vicinity of the current meter site are again 10^{-3} m s^{-1}.

When the vertical distribution of wind-driven current is computed from eqn. 14 by omitting density gradients and centrifugal force (∇S = C.F. = 0), the modeled current at 5 m above the sea bed is opposed to the surface wind-stress, as

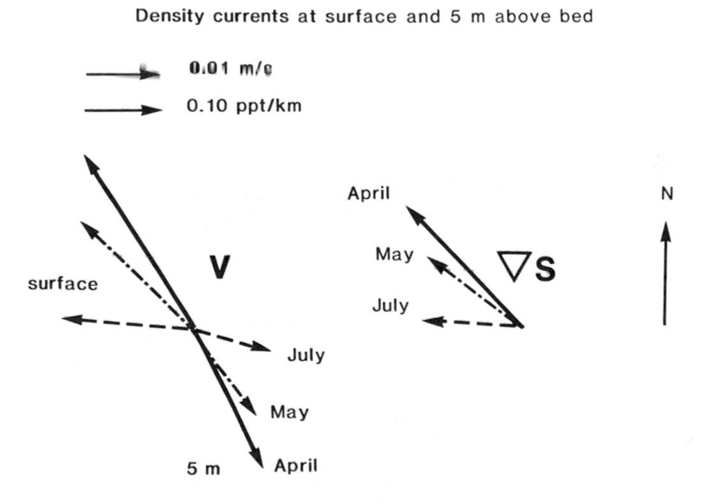

Figure 10. Currents, v, generated by density gradients at surface and 5 m above the sea bed in the Severn Estuary. Horizontal salinity gradient vectors ∇S are also shown. Data according to a linear model of the depth structure (eqn. 14) for April (———), May (-•-•-) and July (---) 1978.

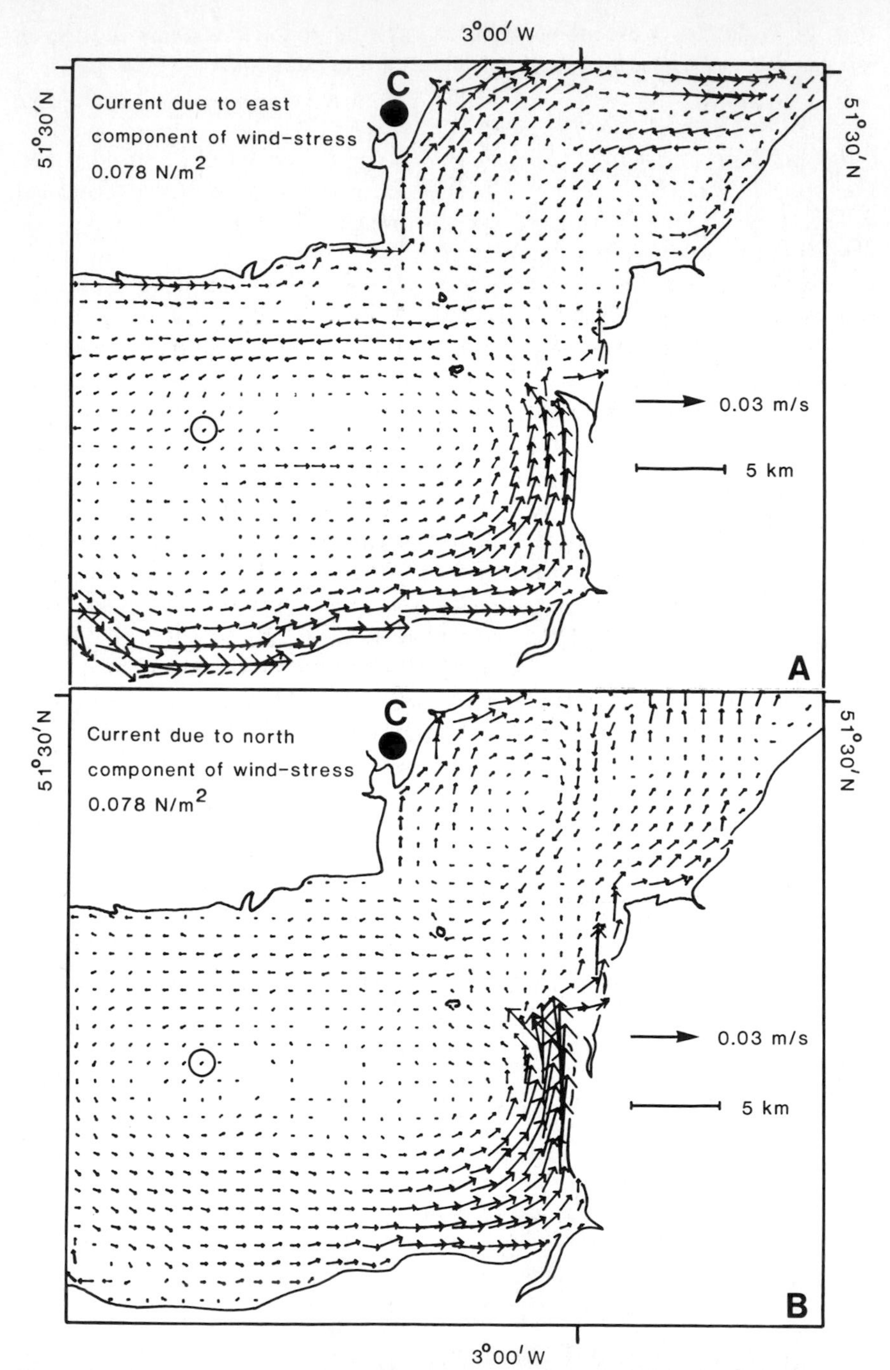

Figure 11. *Wind-driven, depth-averaged circulation due to east (A) and north (B) components of wind-stress (i.e. westerly and southerly winds) equal to 0.078 N m^{-2}. Data according to a numerical, hydrodynamical model.*

was observed. However, the modeled current (approximately 0.01 m s^{-1} for a wind-stress of 0.078 N m^{-2}) is only about a third of what one would expect from eqn. 7. Moreover, this simple model does not explain why the east component of residual flow near the sea bed was effectively blocked by the steep east-west gradient in topography (Fig. 1).

Modeling Conclusions

Observed temporal variability in residual surface elevations resulted from the inverse barometer effect, non-linear tidal forcing, wind-stress and, to a much smaller extent, salinity (density) forcing. A two-dimensional, depth-averaged hydrodynamical model of the Bristol Channel and Severn Estuary could accurately simulate the non-linear tidal forcing. The model also simulated fluctuations in surface elevation due to wind-stress with reasonable accuracy, provided a factor of 3×10^{-3} was assumed in the quadratic drag law for wind-stress. The dependence of surface elevation on salinity could be explained in terms of a simple balance between axial surface-slope forcing along the estuary, and axial salinity (density) gradient forcing.

A higher resolution model of the Severn demonstrated the existence of strong, tidally-induced residual currents in the estuary. However, while the modeled, depth-averaged currents at the mooring site were in the same general direction as observed currents at 5 m above the sea bed, they were much slower. A linear model of the vertical structure of residual flow (which was driven by centrifugal forcing due to topographic steering of the near-bed tidal currents) compared well with observations when a 5-km radius of curvature was assumed for the near-bed currents. The linear model was also used to compute the vertical structure of density-driven currents. The observed increase in the north component of residual current at 5 m above the sea bed could be attributed to a density current response to the changing salinity field.

The depth-averaged, hydrodynamical model of the Severn was used to evaluate wind-driven currents in the estuary. These followed the expected pattern, with currents in the general direction of wind forcing in shallow water, and return flows in the deeper water. Modeled currents were very small at the current meter site. The linear model of vertical current structure showed that the wind-driven current at 5 m above the sea bed was oppositely directed to the surface wind-stress, as was observed. This model could not explain why the wind-driven current responded only to cross-estuary winds (parallel to the local seabed depth contours); a three dimensional model would be necessary to investigate this feature of the flow.

Acknowledgments

We are grateful to J.A. Stephens and T.Y. Woodrow for assistance with data analysis. This work forms part of the Physical Processes programme of the Institute for Marine Environmental Research, a component body of the Natural Environment Research Council (NERC). It was partly supported by the Department of the Environment on Contract No. DGR 480/48.

References Cited

Bowden, K.F. 1956. The flow of water through the Straits of Dover related to wind and differences in sea level. *Phil. Trans. R. Soc. London, ser. A.* 248:517-551.

Elliott, A.J. 1978. Observations of the meteorologically induced cirulation in the Potomac Estuary. *Estuar. Coastal Mar. Sci.* 6:285-299.

Elliott, A.J. 1979a. Low frequency sea level and current fluctuations along the coast of North-West Italy. *J. Geophys. Res.* 84:3752-3760.

Elliott, A.J. 1979b. The effect of low frequency winds on sea level and currents in the Gulf of Genova. *Oceanol. Acta.* 2:429-433.

Elliott, A.J. 1981. Low frequency current variability off the West coast of Italy. *Oceanol. Acta.* 4:47-55.

Heaps, N.S. 1972. Estimation of density currents in the Liverpool Bay area of the Irish Sea. *Geophys. J.R. Astron. Soc.* 30:415-432.

Lacoss, R.T. 1971. Data adaptive spectral analysis methods. *Geophys.* 36:661-675.

LeBlond, P.H. 1979. Forced fortnightly tides in shallow rivers. *Atmosphere-Ocean.* 17:253-264.

Phillips, O.M. 1966. *The dynamics of the upper ocean.* Cambridge University Press, London. 261 pp.

Pingree, R.D. and D.K. Griffiths. 1980. Currents driven by a steady uniform wind stress on the shelf seas around the British Isles. *Oceanol. Acta.* 3:227-236.

Proudman, J. 1953. *Dynamical Oceanography.* Methuen, London. 409 pp.

Schwing, F.B., B. Kjerfve, and J.E. Sneed. 1983. Nearshore coastal currents on the South Carolina Continental Shelf. *J. Geophys. Res.* 88 (C8):4719-4729.

Taylor, A.H. 1983. Fluctuations in the surface temperature and surface salinity of the North-East Atlantic at frequencies of one cycle per year and below. *J. Climatol.* 3:253-269.

Ulrych, T.J. and T.N. Bishop. 1975. Maximum entropy spectral analysis and autoregressive decomposition. *Rev. Geophys. Space Phys.* 13:183-200.

Uncles, R.J. 1982a. Computed and observed residual currents in the Bristol Channel. *Oceanol. Acta.* 5:11-20.

Uncles, R.J. 1982b. Residual currents in the Severn Estuary and their effect on dispersion. *Oceanol. Acta.* 5:403-410.

Uncles, R.J. 1983. Modelling tidal stress, circulation, and mixing in the Bristol Channel as a prerequisite for ecosystem studies. *Can. J. Fish. Aquat. Sci.* 40(S1):8-19.

Uncles, R.J. and M.B. Jordan. 1979. Residual fluxes of water and salt at two stations in the Severn Estuary. *Estuar. Coastal Mar. Sci.* 9:287-302.

Uncles, R.J. and M.B. Jordan. 1980. A one-dimensional representation of residual currents in the Severn Estuary and Associated Observations. *Estuar. Coastal Mar. Sci.* 10:39-60.

Wang, D.-P. 1979. Wind-driven circulation in the Chesapeake Bay, Winter 1975. *J. Phys. Oceanogr.* 9:564-572.

INTERANNUAL VARIABILITY IN BIOGEOCHEMISTRY OF PARTIALLY MIXED ESTUARIES: DISSOLVED SILICATE CYCLES IN NORTHERN SAN FRANCISCO BAY

David H. Peterson

U.S. Geological Survey
Menlo Park, California

Daniel R. Cayan

Scripps Institution of Oceanography
La Jolla, California

and

John F. Festa

National Oceanic and Atmospheric Administration
Miami, Florida

Abstract: Much of the interannual variability in partially mixed estuaries in dissolved inorganic nutrient and dissolved oxygen patterns results from an enhancement or reduction of their annual cycle (generally via climatic forcing). In northern San Francisco Bay estuary the annual cycle of dissolved silicate supply peaks in spring and the effect of phytoplankton removal peaks in fall. Because riverine silicate sources are enhanced in wet years and reduced in dry years, the annual silicate cycle is modified accordingly. Effects of phytoplankton removal are reduced and delayed in wet years and enhanced and advanced (seen earlier) in dry years. Similar reasoning can apply to interpreting and understanding other mechanisms and rates.

Introduction

A knowledge of interannual variability in riverine and estuarine plant nutrient chemistry provides insight into understanding how estuaries operate and helps to define the role of nutrients in estuaries (c.f., Neilson and Cronin 1981; Boynton *et al.* 1982; Nixon and Pilson 1983). This paper considers interannual variability of dissolved silicate concentrations and salinity distributions in northern San Francisco Bay from 1960 to 1980. For a variety of reasons, unfortunately, there are very few observations in the open literature on the nature of interannual variability in estuarine chemistry. We therefore start with some perspective on this topic in the context of distinguishing between human and natural effects in riverine and estuarine chemical records.

To understand the biogeochemistry of estuaries it is necessary to study their chemical variability on all time scales. Long-term (e.g., decadal) trends in riverine and estuarine chemistry are often attributable to human causes, and in

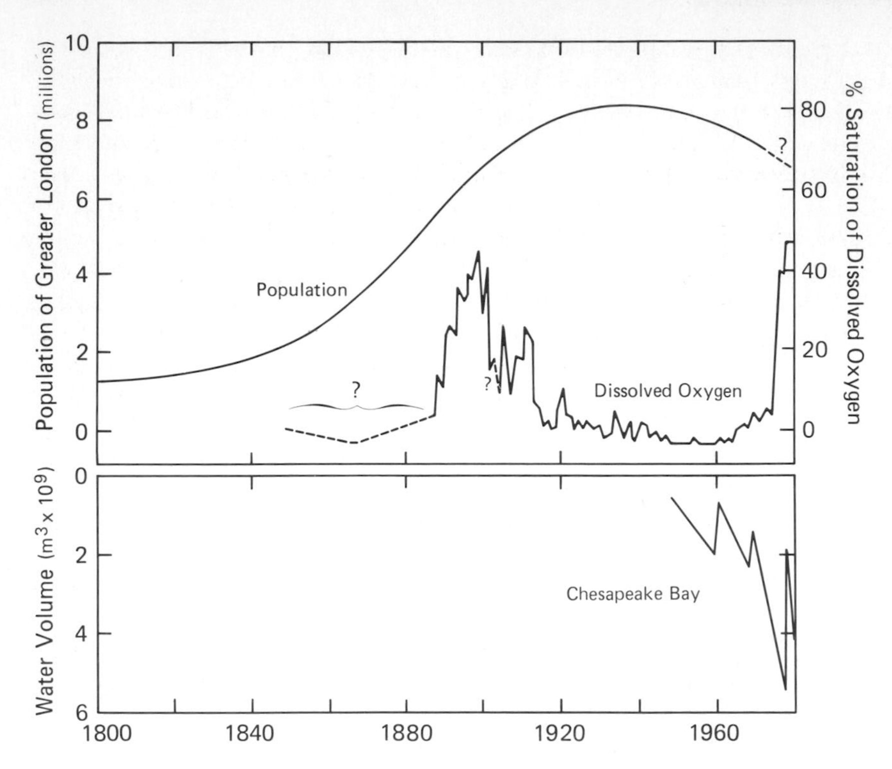

Figure 1. *Examples of long-term trends in riverine and estuarine biogeochemistry.* Top: *History of Thames estuary dissolved oxygen from circa 1850 and the population of greater London. Adapted from Andrews (1984).* Bottom: *Record of the volume of water in Chesapeake Bay with dissolved oxygen levels less than 45 μg-at l^{-1} during the summer from circa 1950. Adapted from Officer* et al. *(1984).*

this regard Europe has a longer history of urban and industrial activity than North America. Understandably then, estuarine research in Europe often is concerned with the longer term changes in chemistry due to human activities. A classic example is the history of the chronic hypoxia in the Thames near London (Fig. 1, Andrews 1984). Related problems include eutrophication and input of toxic substances such as heavy metals (Förstner 1983).The Rhine River, for example, transports more dissolved inorganic nitrogen than the Amazon River (Edmond *et al.* 1981; Wollast 1983; Zobrist and Stumm 1982). These and other long-term changes are probably typical of many European rivers. Superimposed on these lower frequency changes are the higher frequency (e.g., interannual) variations. Riverine and estuarine pollution effects can differ in wet years and dry years. For example, hypoxia increases in the Thames in years of low river flow (low dilution), and was of particular concern during the severe drought of 1975 to 1976 (Davies 1978).

Studies of riverine and estuarine chemistry along the eastern seaboard of the United States also reveal long-term trends caused by human activities (Meade 1982). One example, just now emerging above background interannual variability (Fig. 1), is the increasing frequency and extent of anoxia in the Chesapeake Bay (Officer *et al.* 1984). The effect of climatic forcing on anoxia development is important on an interannual and longer time scale in Chesapeake Bay (Seliger *et al.* 1985) as are prominent hurricane episodes (Chesapeake Bay Consortium 1976). In regions where several causes are important, distinguishing human from natural causes is not easy. A now classic example of such difficulties is the interpretation of the 1976 anoxia development in New York Bight, which was subsequently diagnosed as the result mainly of an anomalous climatic regime during the winter and spring of 1976 (Mooers 1978; Falkowski *et al.* 1980).

The western United States has an even shorter history of anthropogenic alterations, and a clear picture of long-term effects on estuarine and riverine chemistry along the west coast has not yet emerged. Major exceptions are those rivers, such as the San Joaquin in California, which have been strongly influenced by agricultural activity and others, such as the Colorado, which have been reduced in volume by water consumption and modified by impoundments (Bredehoeft 1984). The west coast temporal variability in riverine and estuarine chemistry is probably still associated more with natural causes such as climatic forcing than with human factors.

To illustrate this suggested progressive difference from Europe to the eastern seaboard to the west coast of the United States we examine the differences in source and distribution of two dissolved nutrients, silicate and nitrate. The principal source of dissolved silicate in rivers (Wollast and MacKenzie 1983) and estuaries (DeMaster 1981) is weathering of rocks and soils. In contrast to this natural process for silicate, dissolved inorganic nitrate concentrations are considerably increased by human activity (Wollast, 1983). Thus, for many rivers and estuaries, a very simplified measure of the effects of human activity is the ratio of (largely natural) silicate concentration to (human-enhanced) nitrate concentration. As a point of reference, the global mean value of riverine silicate concentrations is 200 μg-at l^{-1} (DeMaster 1981) and an estimate of a pristine river nitrate concentration is 5 μg-at l^{-1} (Wollast 1983). Thus, a global mean "pristine ratio" of dissolved silicate to nitrate is 40:1 (neglecting the possible small contribution from nitrite). In the winter of 1974-1975, this ratio was well above one (Fig. 2) for eleven rivers and estuaries of western U.S. (Alsea, Coos Bay, Coquille, Duwamish, Grays Harbor, Nehelem, San Francisco Bay, Siuslaw, Willapa Bay, and Yaquina). This ratio is clearly higher than what is found for some of the major rivers of eastern U.S. including the Delaware (Sharp *et al.* 1982; Nixon and Pilson 1983), Susquehanna (McCarthy *et al.* 1977; D'Elia *et al.* 1983; Lang 1982) and Potomac (Lang 1982; Smith and Herndon 1980), where the ratio of silicate to nitrate is often close to 1. In some of the major European rivers and estuaries this ratio can be consistently less than 1 (e.g., Helder *et al.* 1983; van Bennekom *et al.* 1985; Wollast 1983; Wollast and MacKenzie 1983; Zobrist

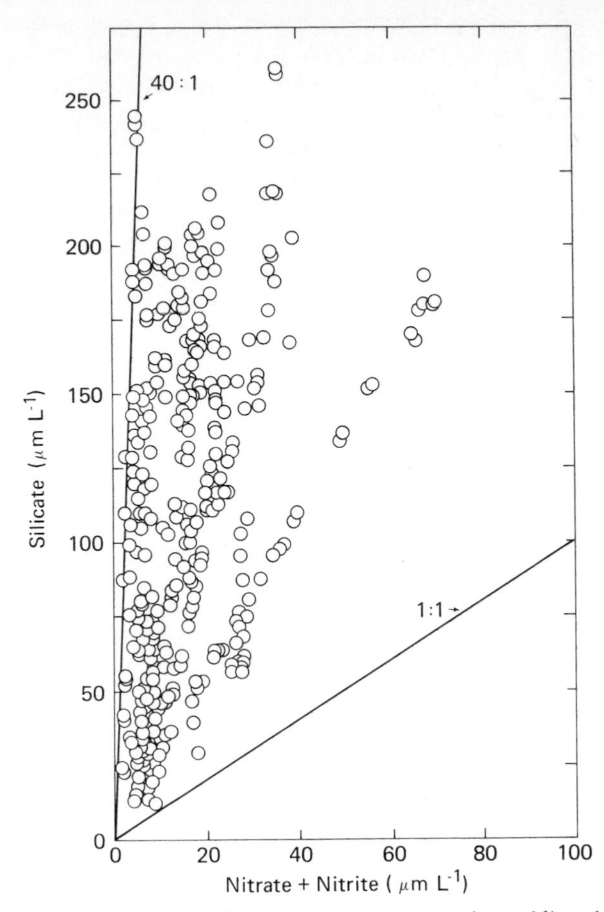

Figure 2. *Silicate versus nitrate plus nitrite concentrations (μg-at l^{-1}) in 11 estuaries (see text) from western United States. The solid line corresponds to a ratio of one. Unpublished data from U.S. Geological Survey for Mar.-Apr. 1975.*

and Stumm 1982). In polluted rivers dissolved inorganic nitrogen may be largely in the form of ammonium.

In addition to the history of human activity, the temporal chemical records of estuaries depend on their assimilative capacity. Our ability to detect the human signal in the chemical record depends, in part, on the dilution rate of the system. Large systems with short water residence times have a higher assimilative capacity than small systems with long water residence times (Officer and Ryther 1977). Another consideration might be the location where waste enters the estuary. Presumably when waste releases are located near the interior of an estuary (e.g., from London to the Thames; from Harrisburg to the Susquehanna/Chesapeake; Wilmington to the Delaware; and from Washington, D.C. to the Potomac) the chemical signal is stronger than for similar-sized releases located near the mouth (e.g., from San Francisco to San Francisco Bay).

Based on the history of human activities and the location of major waste sources, therefore, the human factor will be less obvious in the long-term chemical trends in northern San Francisco Bay estuary than in European and eastern seaboard estuaries. From this broad perspective we now focus on the interannual variability in chemistry of northern San Francisco Bay as illustrated by the nonconservative dissolved silicate-salinity distributions.

Methods

The data base for this paper covers the period 1960 through 1980 and represents four independent studies (Peterson *et al.* 1985). The nearly two decades of observations (no data are available for 1965-1967) are from an urbanized estuary (Conomos 1979; Cloern and Nichols 1975) and represent a variety of sampling and analytical methods (Smith *et al.* 1985).

Silicate concentrations were usually measured on a near-monthly or bimonthly frequency. Spatial coverage was limited to the main channel of northern San Francisco Bay and does not include waters overlying the shoals (for our purposes this limitation is not important, Peterson *et al.* 1985). Despite the gaps in the record, however, there are sufficient data in most instances to characterize silicate variability in relation to the interannual variability in river flow. The general extremes in silicate distributions in northern San Francisco Bay area are represented by nonlinear distributions in late summers of dry years and near-linear distributions in early summers of wet years (Fig. 3).

Assumptions and difficulties in characterizing conservative/nonconservative behavior in estuaries have been discussed elswhere (Officer 1979; Kaul and Froelich 1984). Our approach was kept very simple, the silicate defect (a deficiency as opposed to excess) was defined as the maximum concentration difference between an assumed linear (conservative) mixing concentration and the observed concentration. For example, such an estimate is 170 μg-at l^{-1} (the defect at a salinity of about 10‰) in the lower panel of Fig. 3. Interannual (nonseasonal) variability in river flow, total dissolved solids concentrations, riverine silicate concentrations and silicate defect were estimated by subtracting the mean value (the annual cycle) from the observed value (Chelton 1982). This difference was plotted in units of standard deviation from the mean.

The Annual and Seasonal Cycle

Annual cycles of river flow, total dissolved solids, riverine silicate concentration and estuarine silicate defect are illustrated in Fig. 4. To characterize more closely the annual variation of river flow during the times of the chemical oceanographic observations, we utilized the mean river flow values only from the months for which there were corresponding estuarine dissolved silicate observations.

River flow reflects the annual pattern of precipitation. In general, river flow peaks by early spring (March) and drops precipitously to a minimum in mid- to late-summer (July and August). Riverine silicate concentrations are slightly higher in winter than summer, probably reflecting effects of biogenic removal

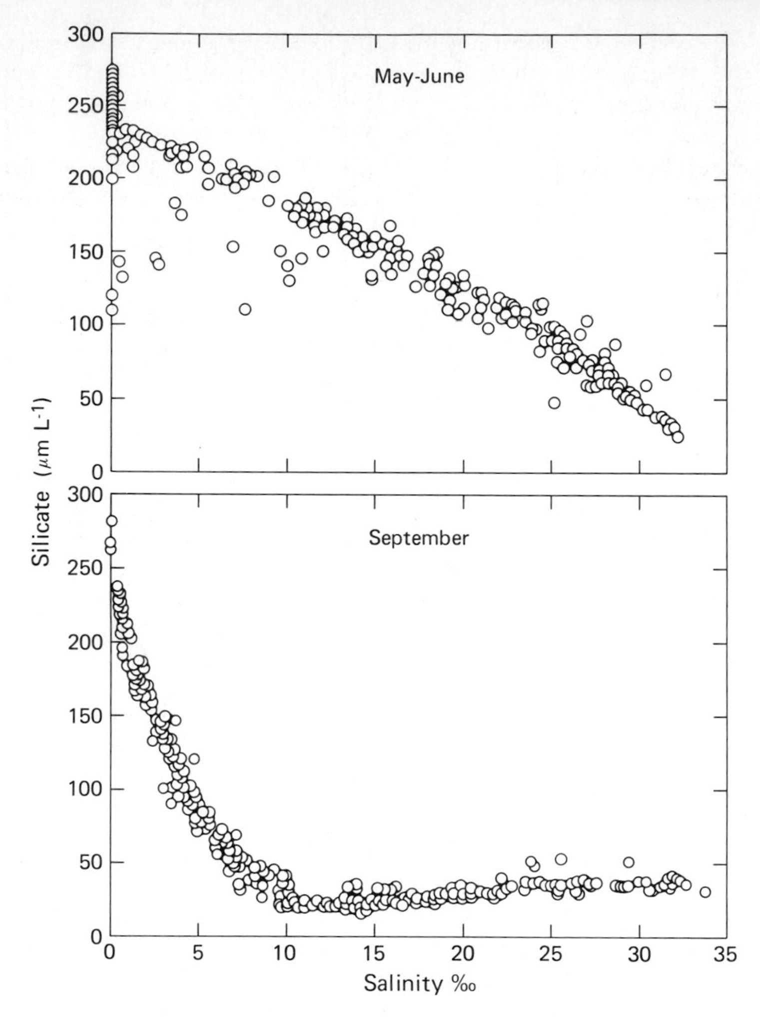

Figure 3. *Seasonal relationships between dissolved silicate (μg-at l^{-1}) and salinity (‰) in northern San Francisco Bay estuary. Upper panel: May-June composite of wet summers: 1963, 1969, 1974, 1975, 1978. Lower panel: September composite of dry summers: 1961, 1964, 1968, 1972, 1979.*

during the low summer flow as well as other processes (Fig. 4). The concentration of total dissolved solids (TDS) near the confluence of the Sacramento and San Joaquin Rivers depends in winter on the background TDS concentration in the river and in summer on the salinity which results from mixing river and sea water. Relations between silicate-salinity are conservative or near-conservative in the winter (November-March) and nonconservative during summer (May-September). The nonconservative behavior of silicate in September apparently reflects removal of the nutrient by diatoms (Peterson *et al.* 1978).

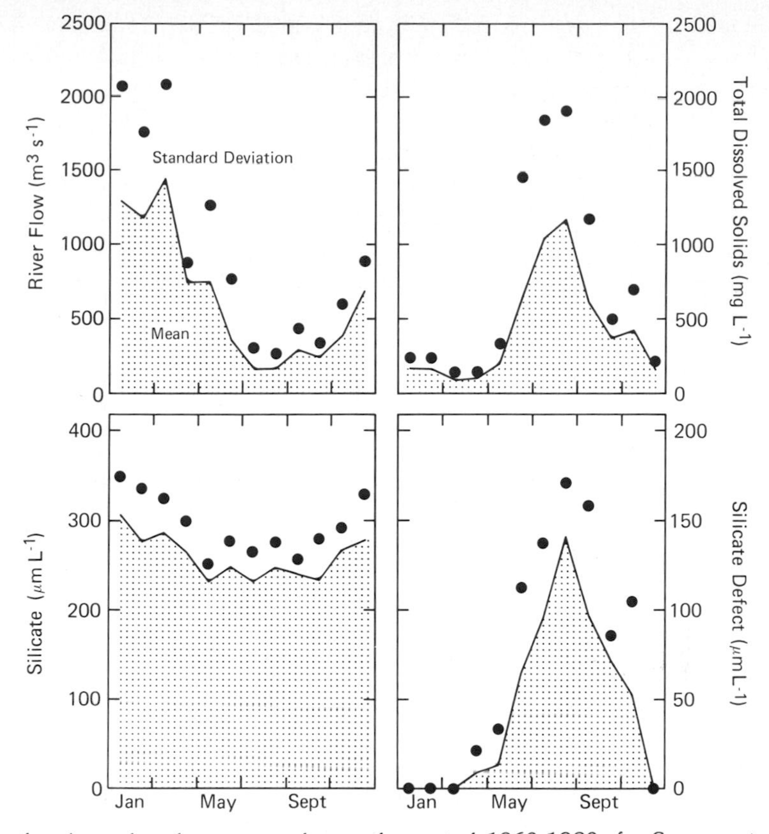

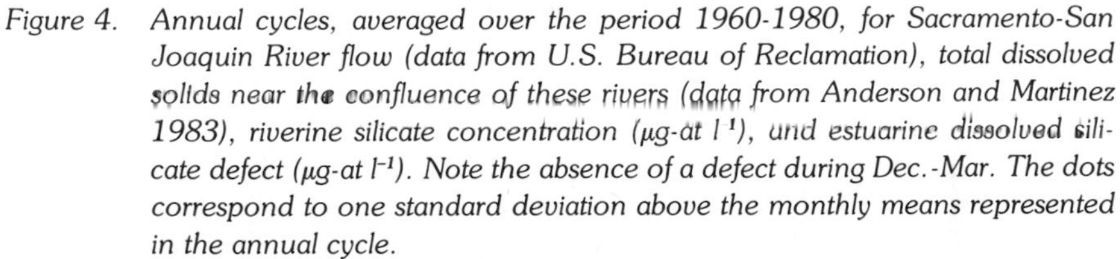

Figure 4. *Annual cycles, averaged over the period 1960-1980, for Sacramento-San Joaquin River flow (data from U.S. Bureau of Reclamation), total dissolved solids near the confluence of these rivers (data from Anderson and Martinez 1983), riverine silicate concentration (μg-at l^{-1}), and estuarine dissolved silicate defect (μg-at l^{-1}). Note the absence of a defect during Dec.-Mar. The dots correspond to one standard deviation above the monthly means represented in the annual cycle.*

Because the seasonal variation in phytoplankton production is small (Cole and Cloern 1984), we have assumed that the seasonal variation in rate of biogenic removal is also small. The seasonal variation in river flow, however, is large. River flow, then, is probably the major factor controlling the annual variation in dissolved silicate-salinity patterns (conservative versus nonconservative). Thus, the annual (deterministic) cycle of the dissolved silica defect will be enhanced in dry years during low river flow and reduced in wet years.

Some Causes and Characteristics of Interannual Variability

Interannual Variability in the Climate Regime

As discussed above, seasonal variations in silicate/salinity distributional patterns appear to be linked to seasonal patterns in precipitation (Fig. 5) and runoff. Thus, our interest is directed immediately to the major causes of variability between winter and spring precipitation specifically with regard to the region of the Sacramento-San Joaquin River basin.

Two dominant features of the Pacific atmospheric circulation are key factors in the winter and spring climate regimes for this area. The subtropical high pressure pattern to the south is relatively weak in winter and stronger in spring, while the Gulf of Alaska-Aleutian low pressure pattern to the north is relatively strong in winter and weaker in spring (Fig. 6). The deepening of the low and contraction of the high in winter allow north Pacific storms to invade the west coast, resulting in the rainy season. The year-to-year variations in strength and position of these features is a key to the associated variation in rainfall over California. The precipitation dynamics associated with these major high and low pressure cells can often be inferred from interannual variability in mean sea level pressure patterns. Some patterns are more prevalent than others, and it is not difficult to diagnose many of the inferred wind patterns and resulting downstream precipitation and runoff patterns for western United States (Cayan and Roads 1984). For example, a sea level pressure pattern associated with a "dry" winter in central and northern California is often more characteristic of the spring-like pattern (Fig. 6 lower panel), whereas the sea level pressure pattern associated with a "wet" spring in California is often more characteristic of the winter-like pattern (Fig. 6 upper panel). Similarly, the pressure field is more "winter-like" to the north and "spring-like" to the south (storm tracks and precipitation are usually more prevalent to the north).

One of the strongest climate signals in an interannual time scale is the tropical El Nino-Southern Oscillation (ENSO) phenomenon (Namias and Cayan 1983). Even though the thermal histories in the tropical ocean frequently exhibit a remarkably uniform temporal pattern during these episodes (Fig 7), its

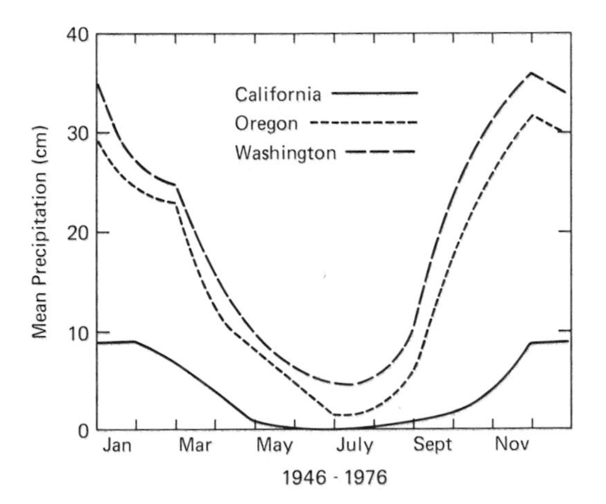

Figure 5. Annual cycle of precipitation based on 1946 to 1976 monthly means. Adapted from Cayan and Roads (1984).

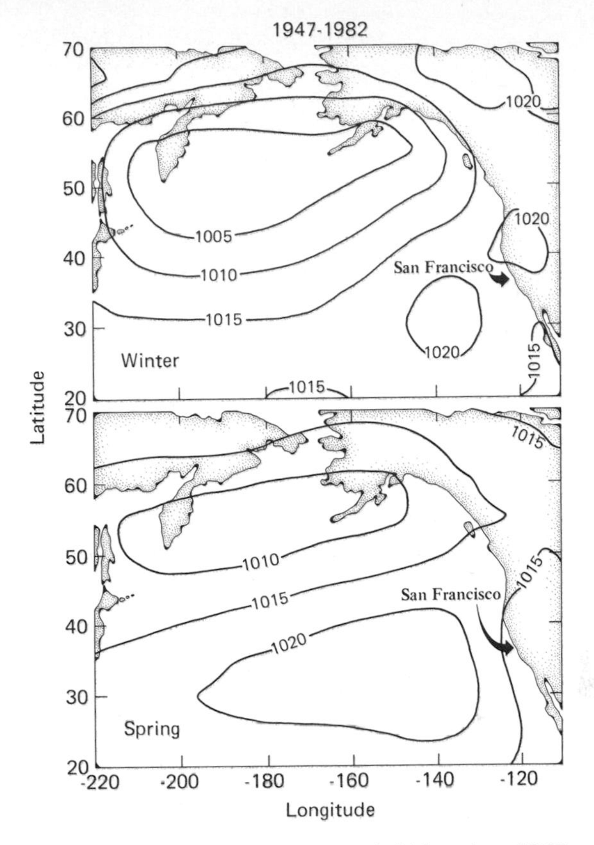

Figure 6. Climatological mean sea level pressure (mb) based on 1947 to 1982 monthly means for winter (Dec., Jan., Feb.) and spring (Mar., Apr., May). Adapted from Namias (1985) and Emery and Hamilton (1986).

extratropical atmospheric counterparts are somewhat surprisingly associated with some of California's wettest (1982-1983) and driest (1976-1977) winters (Namias and Cayan 1983). These extremes in response are partly related to the eastern (wet) or western (dry) location of the atmospheric low pressure anomaly (Namias 1985; Cayan and Douglas 1985; Fig. 8).

An Estuarine Response to Climate Variability

Interannual variability is described as the standard deviation in monthly value from its mean annual cycle (Fig. 9). In general when river flow is above its annual mean flow (positive anomaly), the total dissolved solids and silicate anomalies are below their annual values (negative anomalies); and when river flow is below its annual flow (negative anomaly), the anomaly patterns for total dissolved solids and silicate are positive. The late spring and early summer silicate anomaly of 1964 is a exception that may be at least partly explained by the

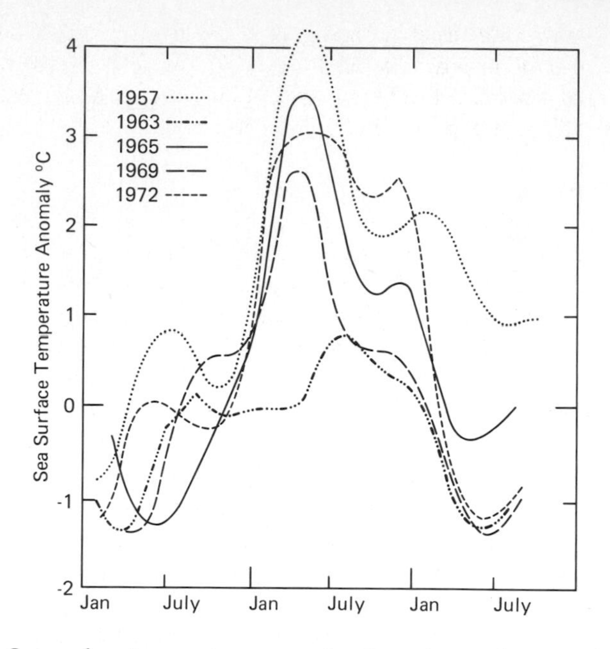

Figure 7. *Sea surface temperature anomalies from the regular annual cycle during El Nino-Southern Oscillation (ENSO) episodes. The 1976 episode, not shown, was similar to the 1972 event. Adapted from Philander (1983). The extra-tropical ENSO perturbation in the northern hemisphere is in winters following the years listed (e.g., an ENSO winter in California ws 1972-1973).*

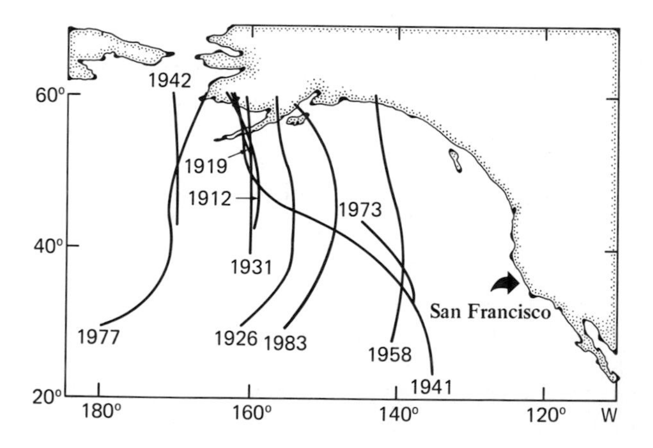

Figure 8. *Axes of the maximum anomaly of winter (Dec., Jan., Feb.) average sea level pressure during 10 strong El Nino episodes from 1899 to 1983. From Cayan and Douglas (1985).*

anomalously low riverine silicate concentration values during that period (Fig. 9). Also, data from the unprecedented 1976-1977 drought period were not included in this analysis because of the unusual changes in the entire chemical and biological system (Cloern *et al.* 1983; Peterson and Festa 1984; Nichols in press).

Implications

Observed annual and interannual silicate anomalies are similar to what has been previously derived from estuarine numerical (Peterson *et al.* 1978) and analytical (Rattray and Officer 1979) models and, therefore, are expected. This relatively simple behavior provides a framework for interpreting more complex patterns of chemical variability. In attempting to do this, however, two factors should be considered. First, there is apparently a balance between river flow (silicate supply) and the silicate defects (phytoplankton removal) during the summer and fall seasons in northern San Francisco Bay. This relation may not be so close in other estuarine systems. For example, in some estuaries at high latitudes in western North America the delayed seasonal peak in runoff (Fig. 10) probably dominates the seasonal dissolved silicate distribution. In this instance summer river flow is probably too strong to reflect interannual variability. Another extreme imbalance is found in Chesapeake Bay where the area of the river basin is only 15 times larger than the estuary, and the annual silicate cycle is dominated by exchange across the benthic surface of the estuary rather than by

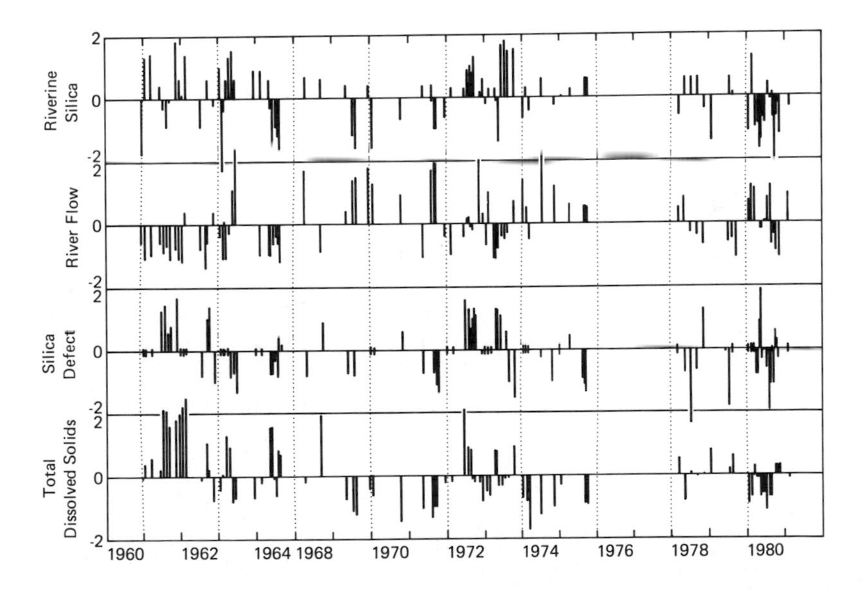

Figure 9. Nonseasonal anomalies of riverine silica, river flow, silica defect and total dissolved solids over the sampling period 1960-1980. Departures from the mean values shown in Fig. 4 are represented as standard deviations.

river flow (D'Elia *et al.* 1983). In this instance river flow is relatively weak except during winter and spring.

Insofar as the northern San Francisco Bay system is concerned, a long record (e.g., > 20 years) of estuarine behavior is needed to provide a statistical basis for characterizing interannual estuarine variability. Nevertheless, the less than 20-year record studied here in combination with proxy variables offers some insight into how the estuarine system operates on annual and interannual time scales and provides a basis for future research. One future task, for example, is to characterize estuarine behavior in more detail and in the context of climate variability and to test that understanding on historical scenarios (e.g., 1958 and 1941 via 1983, c.f. Schemel and Hager in press; and possibly, 1931 via 1977, see Fig. 8). Confidence developed from these exercises would enable us to anticipate how the system might respond with regard to future scenarios.

In closing, although we have restricted our analysis to dissolved silicate over this limited period, much work remains to be done with other records of associated parameters. How closely history repeats itself in years with similar proxy records is a very general, yet important question. Another example of a proxy record is the circa 1850 to present tidal record (sea surface elevation) near the mouth of San Francisco Bay (Fig. 11). Note, for instance, the strong 1940-1941 and 1957-1958 events, recorded (Chelton *et al.* 1982; Simpson

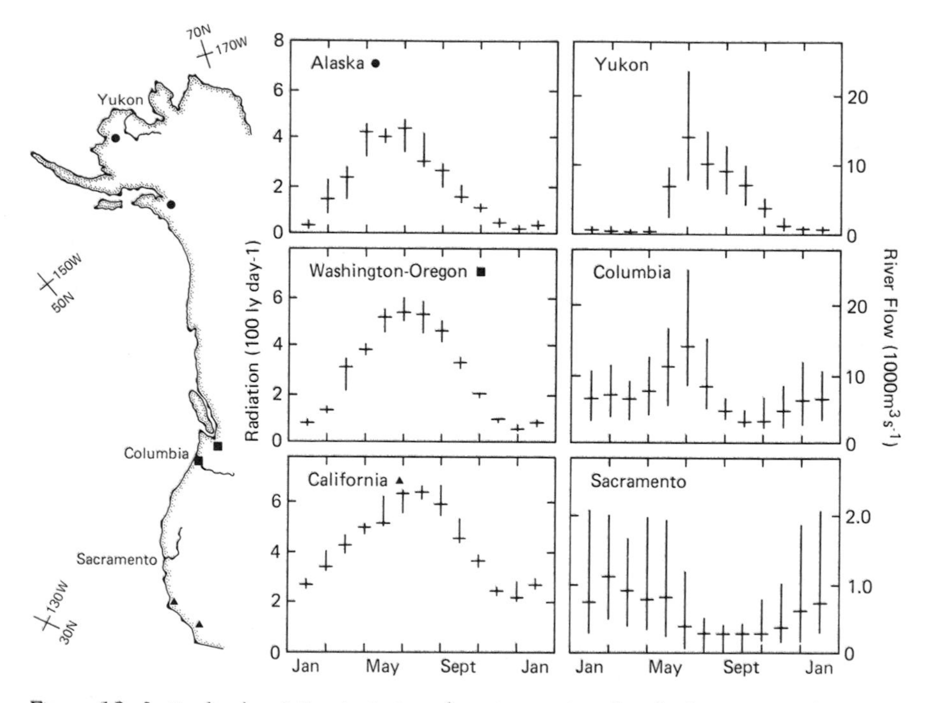

Figure 10. Latitudinal variation in timing of maximum river flow for large rivers of western North America, relative to solar radiation. Radiation data from National Climate Program; river flow from U.S. Geological Survey.

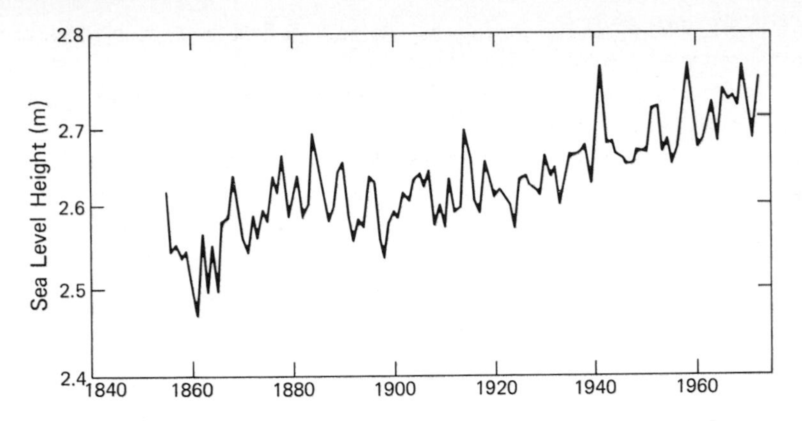

Figure 11. Yearly mean sea level, San Francisco. Data from National Oceanic and Atmospheric Administration, Rockville, Maryland.

1984) as increased sea surface elevation. The 1982-1983 episode (not shown) behaved similarly as oceanographic research suggests (e.g., Simpson 1984; Ebbesmeyer in press; Emergy and Hamilton 1985). One of several scientific and practical benefits of such analysis might be an historical understanding of erosion associated with the unfavorably high sea surface heights common with ENSO episodes (Cayan and Flick 1985). Inside San Francisco Bay the times when delta levees are breached and farm land is flooded apparently often correspond to major ENSO episodes (compare Fig. 11 herein with Fig. 6 in Josselyn and Atwater 1982). Such local inundations often cost millions of dollars to repair and the advantages in anticipating such events in upcoming ENSO winters are obvious. Of broader interest, the value of understanding and anticipating climate-related events in coastal and estuarine research and management has already been suggested (Seliger *et al.* 1985; Falowski *et al.* 1980; Tyler and Seliger 1978).

Acknowledgments

We thank C. Mooers, L. Schemel, D. Wolfe and two anonymous reviewers for their comments, Jean Delo Stevens for technical assistance, and the San Francisco Bay Estuarine study group for various observations.

References Cited

Anderson, G. M. and L. M. Martinez. 1983. Historical Sacramento-San Joaquin Delta salinity summary. *California Dept. of Water Resources,* Central District, Sacramento. 57 pp.

Andrews, M. J. 1984. Thames estuary: Pollution and recovery. pp. 195-227. *In:* P. J. Sheehan, D. R. Miller, G. C. Butler and Ph. Bourdeau (eds.), *Effects of Pollutants at the Ecosystem Level,* John Wiley and Sons, Chichester.

Boynton, W. R., W. M. Kemp and C. W. Keefe. 1982. A comparative analysis of nutrients and other factors influence estuarine phytoplankton production. pp. 69-90. *In:* V. S. Kennedy (ed.), *Estuarine Comparisons,* Academic Press, New York.

Bredehoeft, J. 1984. Physical limitations of water resources. pp. 17-44. *In:* E. A. Engelbert and A. F. Scheuring (eds.), *Water Scarcity Impacts on Western Agriculture,* University of California Press, Berkeley.

Cayan, D. R. and A. V. Douglas. 1985. Variations in U.S. west coast precipitation in response to major El Nino episodes. Unpublished manuscript.

Cayan, D. R. and R. E. Flick. 1985. Extreme sea levels in San Diego, CA winter 1982-1983. *Scripps Institution of Oceanography Reference No. 85-3.* 58 pp.

Cayan, D. R. and J. O. Roads. 1984. Local relationships between United States west coast precipitation and monthly mean circulation parameters. *Monthly Weather Review* 12:1276-1282.

Chelton, D. B. 1982. Statistical reliability and the seasonal cycle: Comments on "Bottom pressure measurements across the Antarctic Circumpolar Current and their relation to the wind." *Deep-Sea Res.* 29:1381-1388.

Chelton, D. B., P. A. Bernal and J. A. McGowan. 1982. Large-scale interannual physical and biological interaction in the California Current. *J. Mar. Res.* 40:1095-1125.

Chesapeake Research Consortium, Inc. 1976. *The Effects of Tropical Storm Agnes on the Chesapeake Bay Estuarine System.* CRC Publication No. 54. 639 pp.

Cloern, J. E. and F. H. Nichols (eds.) 1985. Temporal dynamics of an estuary: San Francisco Bay. *Hydrobiologia* 129:1-237.

Cloern, J. E., A. E. Alpine, B. E. Cole, R. L. Wong, J. F. Arthur and M. D. Ball. 1983. River discharge controls phytoplankton dynamics in northern San Francisco Bay estuary. *Estuar. Coastal Mar. Sci.* 16:415-429.

Cole, B. E. and J. E. Cloern. 1984. Significance of biomass and light availability to phytoplankton productivity in San Francisco Bay. *Mar. Ecol. Prog. Series* 17:15-25.

Conomos, T. J. (ed.) 1979. *San Francisco Bay: The Urbanized Estuary.* Pacific Division American Association for the Advancement of Science, San Francisco, California. 493 pp.

Davies, A. W. 1978. Pollution problems arising from the 1975-76 drought. *Proc. R. Soc. Lond. A.* 363:97-107.

D'Elia, C. F., D. M. Nelson and W. R. Boynton. 1983. Chesapeake Bay nutrient and plankton dynamics: III. The annual cycle of dissolved silicon. *Geochim. Cosmochim. Acta.* 47:1945-1955.

DeMaster, D. J. 1981. The supply and accumulation of silica in the marine environment. *Geochim. Cosmochim. Acta* 45:1715-1732.

Ebbesmeyer, C. C., C. A. Coomes, J M. Cox, G. A. Cannon, R. J. Stewart, J. R. Holbrook, D. E. Bretschneider and C. A. Barnes. In press. Current structure of Puget Sound 1972-1984: Long-term changes induced by a climatic shift. *J. Geophys. Res.*

Edmond, J. M., E. A. Boyle, B. Grant and R. F. Stallard. 1981. The chemical mass balance in the Amazon plume I: The nutrients. *Deep-Sea Res.* 18A:1339-1374.

Emery, W. J. and K. Hamilton. 1985. Atmospheric forcing of interannual variability in the northeast Pacific Ocean: Connections with El Nino. *J. Geophys. Res.* 90:857-868.

Falkowski, P. G., T. S. Hopkins and J. J. Walsh. 1980. An analysis of factors affecting oxygen depletion in the New York Bight. *J. Mar. Res.* 38:479-506.

Förstner, U. 1983. Assessment of metal pollution in rivers and estuaries. pp. 395-423. *In:* I. Thornton (ed.), *Applied Environmental Geochemistry,* Academic Press, London.

Helder, W., R. T. P. de Vries, and M. M. Rutgers van der Loeff. 1983. Behavior of nitrogen nutrients and silica in the Ems-Dollard Estuary. *Can. J. Fish Aquat. Sci.* 40 (Suppl. 1): 188-200.

Josselyn, M. N. and B. F. Atwater. 1982. Physical and biological constraints on man's use of the shore zone of the San Francisco Bay estuary, pp. 57-84. *In:* W. J. Kockelman, T. J. Conomos and A. E. Leviton (eds.), *San Francisco Bay Use and Protection,* Pacific Division American Association for the Advancement of Science, San Francisco, California.

Kaul, L. W. and P. N. Froelich, Jr. 1984. Modeling estuarine nutrient geochemistry in a simple system. *Geochimica Cosmochimica Acta* 48:1417-1433.

Lang, D. J. 1982. Water-quality of the three major tributaries to the Chesapeake Bay, the Susquehanna, Potomac and James Rivers, January 1979-April 1981. *U.S. Geol. Survey Water Resource Investigations* 82-32. Reston, Virginia 64 pp.

McCarthy, J. J., W. R. Taylor and J. L. Taft. 1977. Nitrogenous nutrition of the plankton in the Chesapeake Bay. 1. Nutrient availability and phytoplankton preferences. *Limnol. Oceanogr.* 22:996-1011.

Meade, R. H. 1982. Man's influence on the discharge of fresh water, dissolved material and sediment by rivers to the Atlantic coastal zone of the United States. pp. 13-17. *In: River Inputs to Ocean Systems,* UNESCO and UNEP, New York.

Mooers, C. N. K. 1978. Physical properties and their scales of possible significance to fisheries, especially off the N.E. United States. pp. 7-24. *In: Climate and Fisheries,* Center for Ocean Management Studies, University of Rhode Island, Kingston.

Namias, J. 1985. New evidence for relationships between north Pacific atmospheric circulation and El Nino. *Tropical Ocean/Atmosphere Newsletter* March: 2-3. NOAA Atlantic Oceanographic & Meteorological Laboratory. Miami, Florida.

Namias, J. 1975. Northern hemisphere seasonal sea level pressure and anomaly charts, 1947-1974. *California Cooperative Oceanic Fisheries Investigations, Atlas No. 22.* 243 pp.

Namias, J. and D. R. Cayan 1984. El Nino: The implications for forecasting. *Oceanus* 27:41-47.

Neilson, B. J. and L. E. Cronin. (eds.) 1981. *Estuaries and Nutrients.* Humana Press, Clifton, New Jersey. 643 pp.

Nichols, F. H. In press. Increased benthic grazing: An alternative explanation for low phytoplankton biomass in northern San Francisco Bay during the 1976-77 drought. *Estuar. Coastal Shelf Sci.*

Nixon, S. W. and M.E. Pilson 1983. Nitrogen in estuarine and coastal marine ecosystems. pp. 565-648. *In:* E. J. Carpenter and D. G. Capone (eds.), *Nitrogen in the Marine Environment.* Academic Press, New York.

Officer, C. B. 1979. Discussion of the behavior of nonconservative dissolved constituents in estuaries. *Estuar. Coastal Mar. Sci.* 9:91-94.

Officer, C. B. and J. H. Ryther. 1977. Secondary sewage treatment versus ocean outfalls: An assessment. *Science* 197:1056-1060.

Officer, C. B., R. B. Biggs, J. L. Taft, L. E. Cronin, M. A. Tyler and W. R. Boynton. 1984. Chesapeake Bay anoxia: Origin, development, and significance. *Science* 223:22-27.

Peterson, D. H., J. F. Festa and T. J. Conomos. 1978. Numerical simulation of dissolved silica in the San Francisco Bay. *Estuar. Coastal Mar. Sci.* 7:99-116.

Peterson, D. H. and J. F. Festa. 1984. Numerical simulation of phytoplankton productivity in partially mixed estuaries. *Estuar. Coastal Shelf Sci.* 19:563-589.

Peterson, D. H., R. E. Smith, S. W. Hager, D. D. Harmon, R. E. Herndon and L. E. Schemel. 1985. Interannual variability in dissolved inorganic nutrients in northern San Francisco Bay estuary. *Hydrobiologica* 129:37-58.

Philander, S. G. 1983. El Nino southern oscillation phenomena. *Nature* 302:295-301.

Rattray, M., Jr. and C. B. Officer. 1979. Distribution of a nonconservative constituent in an estuary with appilication to the numerical simulation of dissolved silica in the San Francisco Bay. *Estuar. Coastal Mar. Sci.* 8:489-494.

Schemel, L. E. and S. W Hager. In press. Chemical variability in the Sacramento River and in northern San Francisco Bay. *Estuaries.*

Seliger, H. H., J. A. Boggs and W. H. Biggley. 1985. Catastrophic anoxia in the Chesapeake Bay in 1984. *Science* 228:70-73.

Sharp, J. H., C. H. Culberson and T. M. Church. 1982. The chemistry of the Delaware estuary. General considerations. *Limnol. Oceanogr.* 27:1015-1028.

Simpson, J. J. 1984. A simple model of the 1982-83 California "El Nino". *Geophys. Res. Letters* 11:243-246.

Smith, R. E. and R. E. Herndon. 1980. Physical and chemical properties of Potomac River and environs, January 1978. *U.S. Geological Survey Open File Report* 80-742. Reston, Virginia.

Smith, R. E., D. H. Peterson, S. W. Hager, D. D. Harmon, L. E. Schemel and R. E. Herndon. 1985. Seasonal and interannual nutrient variability in biogeochemistry of northern San Francisco Bay estuary. pp. 137-159. *In:* A. C. Siglio and A. Hattori (eds.), *Marine and Estuarine Geochemistry*. Lewis Publishers, Inc., Chelsea, Michigan.

Tyler, M. A. and H. H. Seliger. 1978. Annual subsurface transport of a red tide dinoflagellate to its bloom area: Water circulation patterns and organism distributions in the Chesapeake Bay. *Limnol. Oceanogr.* 23:227-246.

van Bennekom, A. J., W. W. Gieskes and S. B. Tijssen. 1975. Eutrophication of Dutch coastal waters. *Proc. R. Soc. Lond. B.* 189:359-374.

Wollast, R. 1983. Interactions in estuaries and coastal waters. pp. 385-410. *In:* B. Bolin and R. B. Cook (eds.), *The Major Biochemical Cycles and Their Interactions. SCOPE 21*. John Wiley and Sons, Chichester.

Wollast, R. and F. T. MacKenzie. 1983. The global cycle of silica. pp .39-76. *In:* S. R. Aston (eds.), *Silicon Geochemistry and Biogeochemistry*. Academic Press, London.

Zobrist, J. and W. Stumm 1982. Chemical dynamics of the Rhine catchment area in Switzerland, extrapolation to the "pristine" Rhine River input to the ocean. pp. 52-63. *In: River Inputs to Ocean Systems*. UNESCO and U.N. Environment Programme, Unipub, New York.

CONVERGENT FRONTS IN THE CIRCULATION OF TIDAL ESTUARIES

J. H. Simpson and W. R. Turrell

Department of Physical Oceanography
Marine Science Laboratories
University College of North Wales
Bangor, Gwynedd, United Kingdom

Abstract: The interaction of tidal flow with freshwater buoyancy input produces a variety of frontal structures in estuaries. These fronts are important, not only as indicators of significant circulation mechanisms, but also as rather specialized features of the estuarine regime in which strong convergence and sinking lead to the concentration of buoyant material, surface-active components and swimming organisms.

In estuaries where tidal flow is limited, the freshwater outflow forms a buoyant layer that spreads on the surface as a density current with a plume front at the leading edge. Strong flood tide currents may force the plume front back into the estuary to produce a tidal intrusion front whose movement within the estuary may be explained in terms of density current dynamics. This type of front exhibits a characteristic V configuration with an isolated point convergence at the apex and an associated gyre system.

In still more vigorous tidal flows the estuary becomes well-mixed and the density current structure breaks down. In these circumstances the vertical and lateral shear in the flood current may interact with the longitudinal density gradient to produce a transverse circulation with an intense convergence along the axis of the tidal flow. A diagnostic model has been used to determine the form and amplitude of this circulation but still requires direct validation. An improved method for the quasi-synoptic measurement of the transverse velocity field is described. Initial results confirm the existence of full-depth transverse circulation cells in the flow.

Introduction

Regions of intensified horizontal gradients, termed fronts, occur widely in the estuarine environment. They are readily apparent even to the casual observer, because of the surface accumulation of buoyant material in the convergence zone invariably associated with a front. The resulting foam lines often stretch for many kilometers forming prominent surface features. In spite of these obvious manifestations, the mechanisms involved in estuarine frontogenesis have not yet been fully explored, nor has the relationship of frontal structure to larger scale circulation and mixing been elucidated.

Some progress has been made, however, in documenting particular classes of fronts in estuaries and it is increasingly recognized that fronts will need to be assimilated into our models of estuarine dynamics and mixing processes if the models are to correspond more closely to reality. Certainly the occurrence, extent and nature of fronts can give vital clues as to the form of organized circulations in estuaries.

Fronts in estuaries may also be of practical and ecological importance. The frontal convergence will act to concentrate surface materials providing a favorable feeding ground for particular organisms. At the same time they may serve to concentrate buoyant contaminants and act as indicators of incipient pollution (Klemas and Polis 1977).

In this contribution we present some recent studies of fronts in the strongly tidal estuaries of North Wales with particular emphasis on the axial convergence front which we argue is the result of an important secondary circulation characteristic of well-mixed estuaries. We consider the difficulties of observing such circulations and report on a new measurement technique and its application to the validation of models of transverse flow.

Frontogenesis by Buoyant Forces

The buoyancy forces associated with freshwater input are mainly responsible for driving the circulations that sustain estuarine fronts. The central role of the buoyancy forces is clearly apparent in the well-documented case of the plume front. The Connecticut River outflow, for example, flows into Long Island Sound as a thin brackish surface layer during the ebb phase of the tide (Garvine and Monk 1974). The subsequent spreading of the surface plume takes the form of a density current with the leading edge of brackish water propagating as a clearly defined front characterized by strong salinity gradients with obvious surface color changes and accumulations of surface material.

The front propagates into the ambient seawater at a velocity $U_g \sim \sqrt{g'D}$ where g' is the reduced gravity and D is the depth of the brackish layer (Fig. 1).

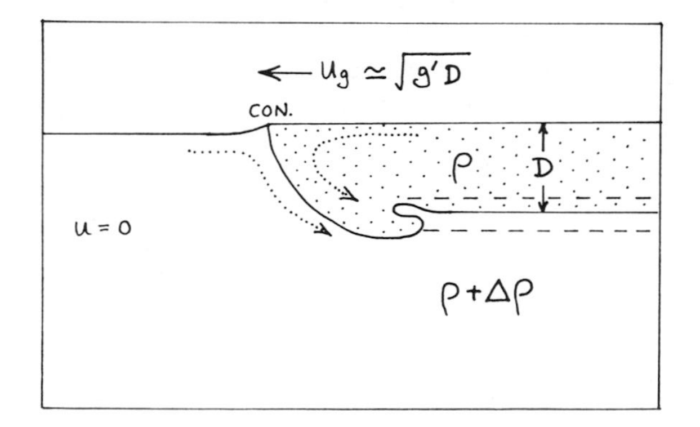

Figure 1. Schematic of a density current in which lighter water advances at speed U_g into denser stationary fluid. Flow within the surface layer towards the head of the density current is required to replace water entrained in mixing waters behind the head. The partially mixed waters produced are left behind the advancing flow in a layer indicated by the dashed line.

The spreading speed is somewhat less than that given by the hydraulic theory of density currents (Brooke-Benjamin 1968) because of the mixing processes that occur just behind the front boundary. The mixing in this region, which has been studied in laboratory experiments by Simpson and Britter (1979), produces mixed water of intermediate density behind the advancing front. To replace fluid lost from the surface layer, the brackish water overtakes the front at a speed of ~0.16 U_g so that the front exhibits a double-sided convergence with strong sinking motion at the boundary. Stigebrandt (1980) presents a simplified model of these processes.

Tidal movement of the ambient fluid may modify the intensity of the frontal boundary, producing sharp fronts when the spreading is opposed by tidal flow, and weak or negligible discontinuities when the tidal flow is in the same direction as the spreading (Garvine In Press). Where ambient motion is minimal and the discharge is effectively from a point source, a radial theory of spreading may be applied (Garvine 1984). This approach has been used by Garvine to explain the "multi-ring fronts" sometimes observed in plume discharge (McClimans 1978). A combination of rapid radial spreading together with the reflection of internal waves from the leading edge of the plume is necessary for the evolution of a ring structure.

The reflection of fronts from coastal boundaries has been studied by Thorpe *et al.* (1983) who observed the production of internal wave trains when a density front, propagating out of a Scottish sea loch, was "bounced off a promontary". About 3 to 7 waves (with wavelength ~60 m), generated in the reflection process, returned into the stratified water behind the advancing front.

Tidal intrusion fronts

Where tidal currents are sufficiently strong, the advancing plume front may be swept back into the estuary by the flood current, to form a tidal intrusion front (Simpson and Nunes 1981). The movement of such fronts can be understood to first order in terms of competition between the down-estuary spreading at U_g and the upstream tidal flow U_t (Fig. 2a). As the brackish layer is forced back into the estuary its thickness increases so that U_g becomes larger until an equilibrium is reached when $U_g \approx U_t$. In the Seiont estuary (N Wales), the intrusion front has been observed to take the form of a pronounced V with a strong point convergence at the apex of the V together with an associated gyre system (Fig. 2b). Time-lapse filming has revealed that the frontal boundary shows considerable variability but tends always to revert to the V configuration with one or the other of the gyres dominating. The mechanisms responsible for the V structure are not clear; there is some analogy apparent with the lobe and cleft structure observed in atmospheric density currents (Simpson 1972) but laboratory studies point to the importance of bottom topography (Nunes 1982).

At the time of maximum intrusion into the estuary, which occurs during the peak flood current, the density current flow is in equilibrium with the tidal inflow. In these circumstances the advance of the frontal interface at U_g is balanced by the local component of the tidal flow perpendicular to the front (Fig. 2b), i.e.

$$U_g = U_t(y) \cos \Theta$$

so that the angle Θ between the front and a line across the estuary is just

$$\Theta = \mathrm{Cos}^{-1} \frac{U_g}{U_t} \tag{1}$$

Equation 1 has two solutions for $-\pi/2 < \Theta < \pi/2$ corresponding to the two branches of the front. For a fixed U_g, Θ increases with U_t so that we would expect the maximum inclination in mid-channel where U_t is greatest—which is just what is observed.

Similar frontal features have been observed on flood tide in other estuaries (Tully and Dodimead 1957; Godfrey and Parslow 1976; Huzzey 1982; Nunes 1982), although in many estuaries in the U.K., vertical mixing by the tidal stream limits the development of the tidal intrusion front. This is certainly the case in the estuary of the Conway (Lailey 1980), where a marked boundary, often in the form of a V, forms during the early part of the flood flow where the incoming saline water plunges under a fresher surface layer. As the flood currents increase, however, the density current structure fails to adjust and there is a catastrophic breakdown of stratification, which advances rapidly up the estuary.

Such observations suggest that tidal intrusion fronts will be prominent only in estuaries which exhibit the right range of tidal inflow in relation to the freshwater discharge, i.e., large enough currents to drive the freshwater outflow back

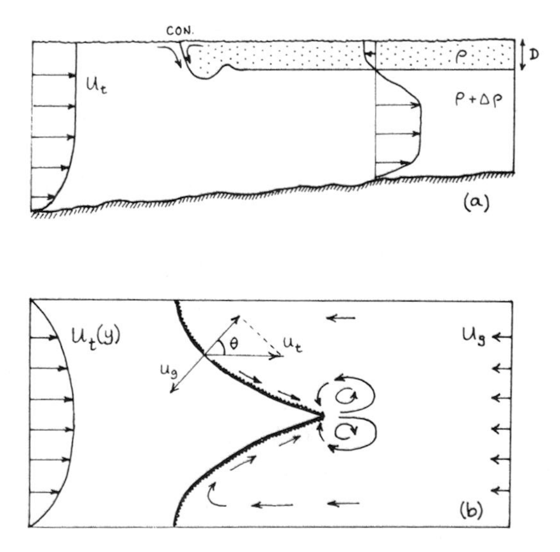

Figure 2. *Schematic of tidal intrusion fronts: (a) the competition of the gravity current with the tidal current into the estuary; (b) interaction of tidal intrusion front with the lateral shear to produce a V front with a single point convergence at the apex and attendant gyres.*

into the estuary ($U_t > U_g$) but not so large that the density structure is disrupted by vertical mixing. The latter requirement could be stated in terms of an estuarine Richardson number (Fischer 1972) but it appears that the geometrical configuration of the estuary may also be important. For example, the complete draining of the Seiont estuary on each tide (because of the large tidal range) may promote the formation of a strong tidal intrusion front by removing all the products of mixing on the previous tidal cycle.

Axial Convergence

In estuaries where strong tidal stirring prevails and vertical gradients are minimal, the buoyancy forces are involved in driving a transverse circulation that produces an extensive convergence line along the axis of the tidal flow. This type of front is observed in the Conway estuary (Fig. 3) during the second half of the flood after the collapse of the density current structure described above. The convergence appears 2-3 hours after low water and is clearly apparent because of the large accumulation of flotsam in the convergence. In summer large quantities of foam may be present, thought to be due to organic materials derived from phytoplankton blooms in the northern Irish Sea.

The convergence is evident on most flood tides—being disrupted only by severe windstress. It may extend (sometimes without interruptions) for as much as 10 km up the estuary, as we have been able to verify by means of aerial photography (Fig. 4a,b). The location of the front varies with the rate of freshwater discharge through the estuary and the range of the tide, and there are indications that the strongest convergences occur on spring tides. Time-lapse photography has demonstrated clearly that the convergence is present only on the flood; no persistent surface convergence is apparent on the ebb when the mid-channel flow appears to be divergent with surface material being transported to the banks of the estuary. Direct indications of sinking in mid-channel were reported by West and Cotton (1981) who found that rhodamine dye released at the surface in midchannel during the flood could not be detected upstream at the surface but was present in the near-bottom samples.

Salinity measurements show that the convergence occurs only in regions where there is a significant longitudinal salinity gradient. Upstream of the limit of salt penetration, for example, we see no indication of surface convergence. On the basis of the above circumstantial evidence, Nunes and Simpson (1985) proposed that the front is maintained by a transverse circulation which results from an interaction between the vertical and lateral shear in the flood current and the longitudinal density gradient. The essential mechanism (Fig. 5) involves the creation of a pattern of transverse density gradients by the tidal current, which is swiftest in the central part of the channel. This central region therefore tends to show a positive salinity anomaly relative to the rest of the section. The corresponding pressure gradients will then drive a transverse 2-cell circulation with convergence and sinking in mid-channel.

On the assumption that the cross-section of density *anomalies* is steady in time, a diagnostic model of the circulation may be developed. (Nunes and

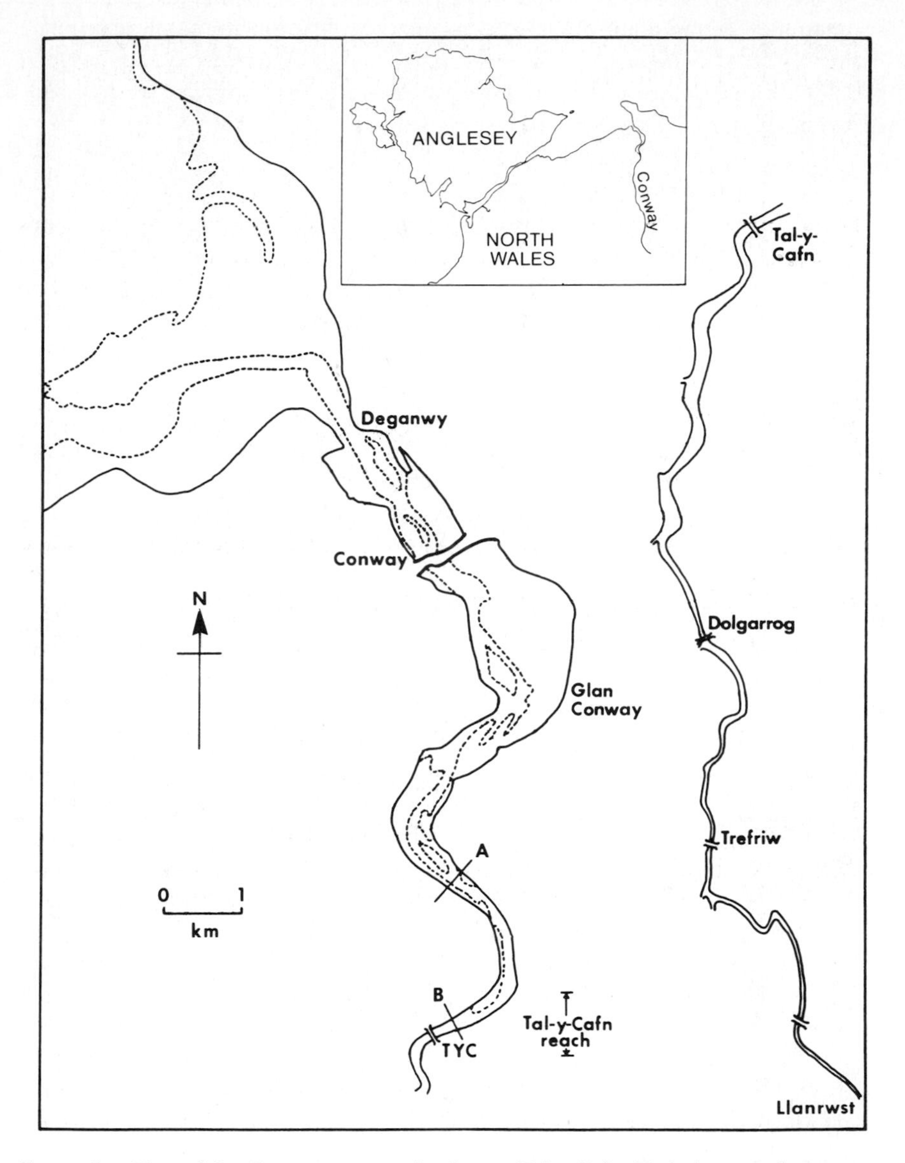

Figure 3. *Plan of the Conway estuary (broken at Tal-y-Cafn, T_{YC}) showing the sections (A + B) where the transverse flow has been determined. The axial convergence is commonly observed between Glan Conway and Dolgarrog.*

Figure 4. *The axial convergence in the Conway: (a) downstream view of foamline from Tal-y-cafn bridge; (b) aerial view of the estuary between Tal-y-cafn and Dolgarrog.*

Simpson 1985). The dynamic balance between the pressure gradients and frictional stresses may be written as:

$$\frac{\partial^2 v}{\partial z^2} = \frac{g}{\overline{\varrho} N_z} \left(\overline{\varrho}_s \frac{\partial \zeta}{\partial y} - \int_0^z \frac{\partial \varrho}{\partial y} dz \right) \tag{2}$$

where v = transverse velocity; g = acceleration due to gravity; $\overline{\varrho}_s$ = mean surface density; N_z = eddy viscosity (assumed constant); ζ = surface elevation; and y,z = transverse and vertical (positive upwards) coordinates.

Equation 2 may be integrated to determine the transverse velocity v(y, z), with the form of the result depending on the surface and bottom boundary conditions. Here we take the bottom stress as given by an approximation to the quadratic drag law:

$$\tau_b = k \overline{\varrho} U v \qquad \text{at } z = -H \tag{3}$$

where U is the longitudinal tidal current and k is the quadratic drag law constant.

For no surface stress $\left(\frac{\partial v}{\partial z} = 0 \text{ at } z = 0\right)$ the solution may then be written:

$$v(y, z) = \frac{g}{\overline{\varrho} N_z} \left\{ B \left(\frac{z^2}{2} - \frac{H^2}{2} - \frac{N_z H}{kU} \right) - I_3^z - I_3^H - \frac{N_z}{KU} I_2^H \right\} \tag{4}$$

where:

$$B = \frac{3}{H^3} \left(I_4^H + H I_3^H - \frac{N_z H}{kU} I_2^H \right) \left(1 + \frac{3N_z}{kHU} \right)^{-1}$$

and

$$I_n^z = \int_0^z \int_0^z \cdots \int_0^z \frac{\partial \varrho}{\partial y} dz_1 \ldots dz_n$$

and H = total depth

From the observed density distribution we use $\left(\frac{\partial \varrho}{\partial y}\right)$ to evaluate the integrals I_n^z and hence determine v (y, z). The vertical velocity field may then be obtained from continuity.

Two examples of the transverse circulation computed by this approach from observed density distributions in the Conway are shown in Fig. 6a,b. Horizontal differences of ~1 kg m^{-3} in density drive significant transverse flow that may take the form of two almost symmetrical cells separated by a pronounced double-sided surface convergence in mid-channel—with the strongest

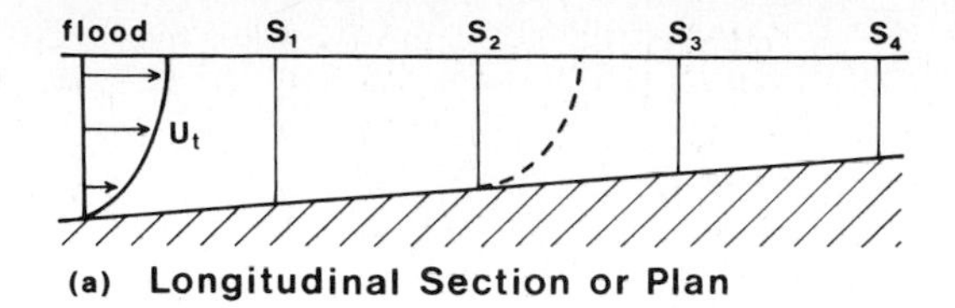

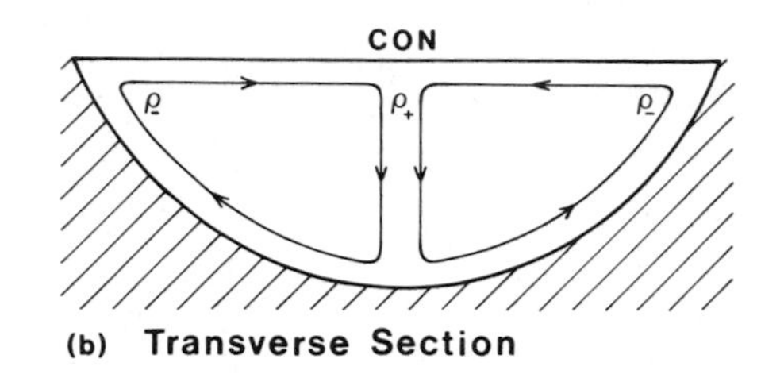

Figure 5. *Conceptual model of a transverse circulation in a well-mixed estuary: (a) Shear in the flood current distorts the isohalines (the isohaline S_2 is distorted as shown) to produce density gradients which drive a two-celled convergent lateral flow. This occurs both in the vertical and horizontal. (b) The secondary flow with associated density anomalies.*

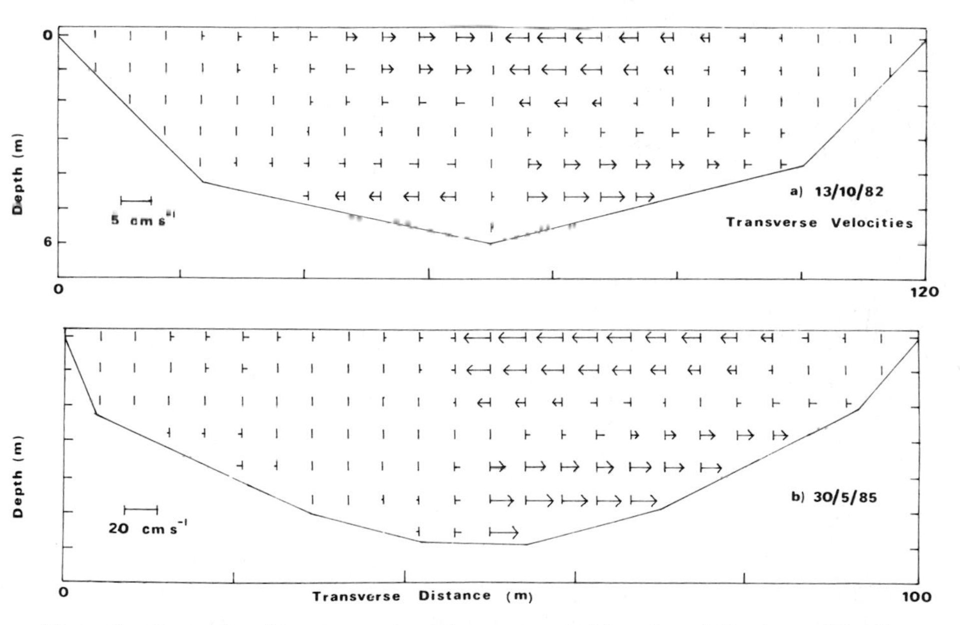

Figure 6. *Examples of transverse circulations computed from lateral density gradients by the diagnostic model for Section A of the Conway Estuary (Fig. 3). (a) Almost symmetrical density field producing two-cell circulation, October 13, 1982. (b) Asymmetrical density distribution producing a single cell with one-sided convergence, May 30, 1985.*

convergent surface flow occurring towards the center. A corresponding divergence occurs in the lower layer with negligible horizontal motion at mid-depth. In many cases the observed distribution of density is markedly asymmetric so that one cell predominates. Figure 6b illustrates an extreme example where the second cell is almost completely suppressed and the surface convergence will be one-sided.

Direct Observations of Transverse Motions

To test the above inferences from the diagnostic model, we clearly need precise data sets on the observed transverse circulations. Estimates of the surface currents may be obtained by tracking the motion of surface floats released on either side of the front. Measurements of this kind (Nunes 1982) confirm that there is convergent flow at the surface with speeds of ~ 10 cm s^{-1} increasing towards the front. Float measurements are not practical, however, for the subsurface circulation and it became clear from a number of trials with current meters that the full determination of the rather small transverse component of the flow represented a considerable observational challenge in the presence of large longitudinal currents.

Nunes (1982) made an initial attempt based on the use of high-resolution directional sensors attached to a mast which was fixed with stays in the estuary at low water. The results (Nunes and Simpson 1985) provided support for the notion of a full-depth two-cell circulation. However, because the data were confined to one fixed vertical profile per tide the method was rather inflexible and could not furnish fully convincing evidence of convergence and divergence in the flow. To provide for much more flexible, quasi-synoptic measurements of transverse flow, a new boat-mounted sensor system has recently been developed. It consists of four direction-sensing vanes attached to a tubing framework which is rigidly fixed to the hull (Fig. 7a). The vertical spars can be raised and lowered to deploy sensors at depths of 0.5–3.5 m while maintaining a precise relative directional reference. The direction of each vane is measured by a precision plastic-film potentiometer which is fed by a constant current source that generates a voltage output proportional to angular displacement. Mechanical coupling of the vane to the potentiometer is achieved via an oil-filled bearing which prevents loading of the potentiometer shaft by the drag forces on the vane.

The directional data together with the output of two speed sensors are logged on a microcomputer adapted to run from a 12V power supply. The system samples each directional output at 8Hz with a resolution of $<0.2°$. Mean values of the vane directions over a 10-s period are computed and displayed on a visual display unit in analog form to provide a convenient monitor while recording (Fig. 7b). Output from the current speed sensors are likewise averaged over 10 s to provide estimates of the current modulus U at two depths. Direction and speed data are then merged and logged with the time of recording on floppy disks.

Operation of the system requires careful attention to the bottom depth. The esuary section to be profiled is first surveyed by using the craft's

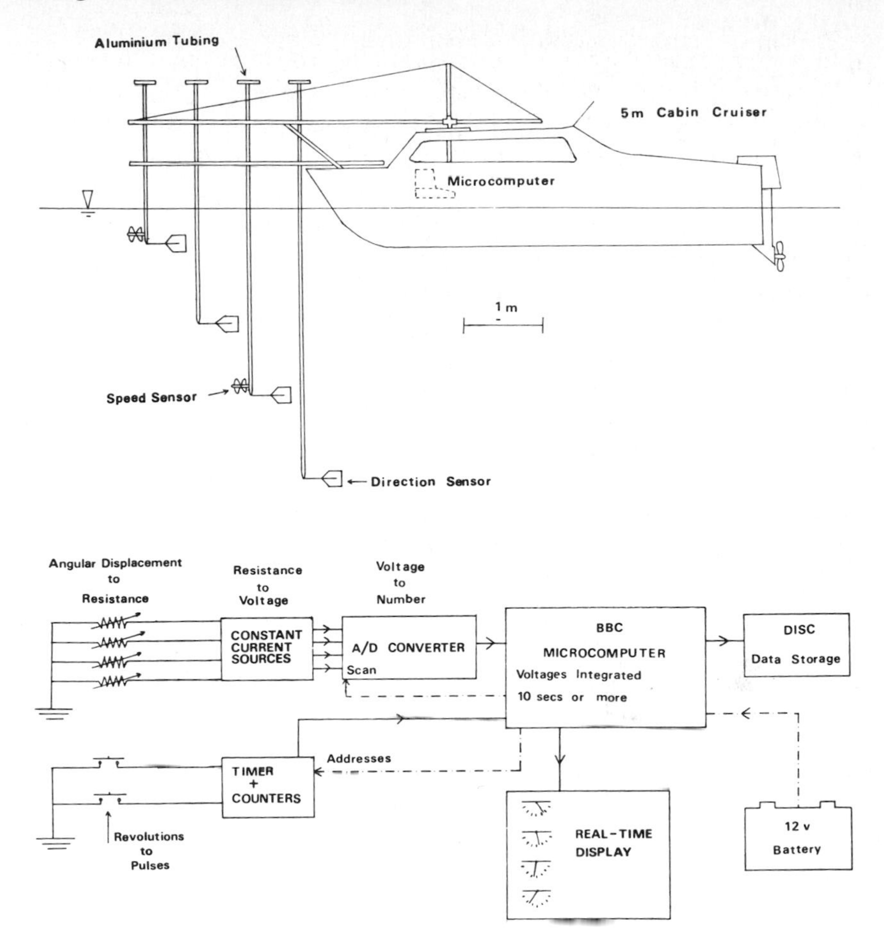

Figure 7. System for observing transverse flow components: (a) adapted cabin cruiser with sensor framework; (b) schematic of data acquisition system.

echo-sounder (resolution 0.1 m). The sensors are then lowered to the appropriate depths and the vessel is maneuvered onto the section which is slowly traversed using the engine and helm to balance the longitudinal tidal flow. Marker buoys deployed at known points across the estuary assist in fixing lateral position. Each directional profile takes only 10 s so that the 100-m section can be profiled at high resolution (~2 m) in 8 minutes.

In spite of the extensive structure immersed ahead of the bow, the boat responds well to the controls and no difficulty has been experienced in this dynamic mode of operation apart from occasional damage to the deepest vane system through impact with the bottom.

The directional information provided by the vanes is relative to the axis of the vessel, and must be corrected for lateral motion of the boat. This is done by subtracting the mean direction $\overline{\Theta}$ from the individual sensor readings Θ, so the transverse velocities are given by

$$v_i = U_i \sin(\Theta_i - \overline{\Theta}) \tag{5}$$

where U_i is the modulus of the current interpolated from the two speed sensors.

This procedure is equivalent to defining the transverse flow direction as that which is normal to the mean flow direction as measured by the four sensors. Under most conditions this will be a fair approximation to choosing the coordinate frame so that the net lateral transport is zero i.e.,

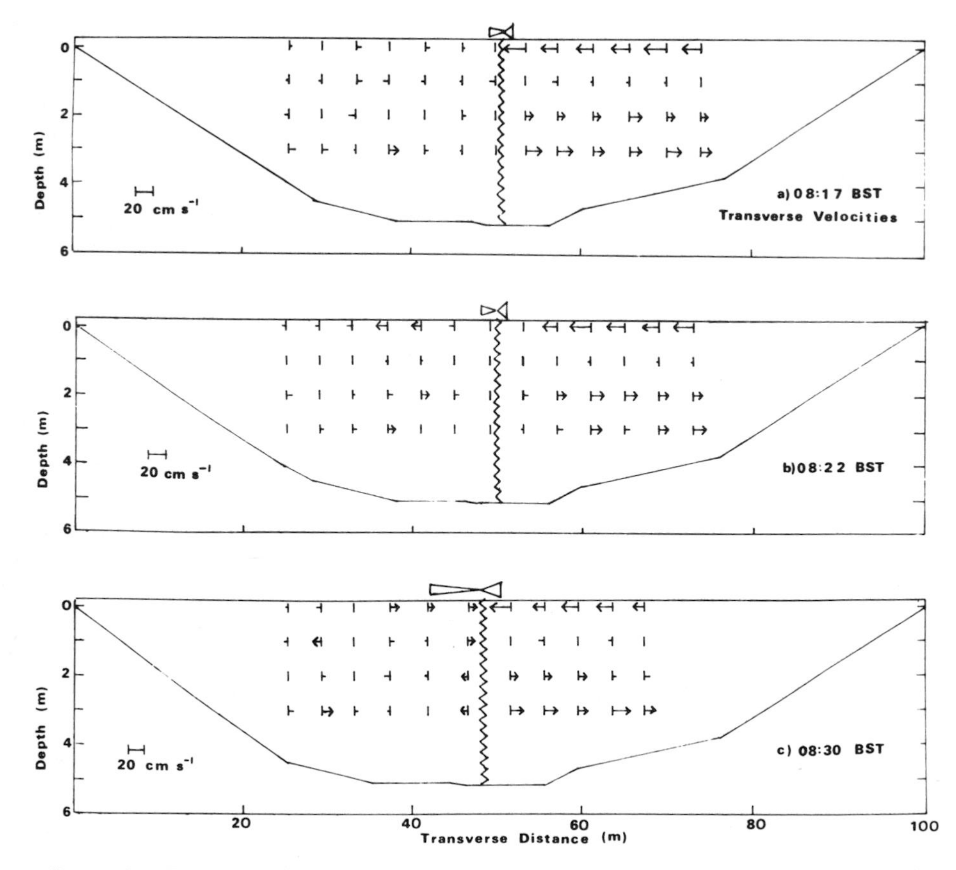

Figure 8. *Examples of observed transverse velocities on three successive traverses of section B (Fig. 3) on May 31, 1985 (Conway 0914 BST). The components are measured relative to the mean flow direction. Axial velocity U = 100.3 cm s⁻¹ at z = 0.5 m and U = 93.1 cm s⁻¹ at z = 3.5 m. The predominant cell on the eastern side of the estuary where convergent surface flow is 20 percent of the axial speed. The symbol marks the observed position of the foam line with the small black triangle indicating the side with visibly stronger convergence.*

$$\int_{-H}^{0} v dz = 0 \tag{6}$$

A sample of the transverse flow obtained in this way from an initial survey of the lateral convergence in the Conway estuary is shown in Fig. 8. The three panels are based on data at 2-m intervals from 3 successive traverses of the same section (B in Fig. 3). All show evidence of a strong circulation cell on the eastern side of the estuary extending over the full depth with surface speeds up to 20 cm s^{-1}. The cell extends exactly as far as the observed surface convergence, which in this case was mainly single-sided. Transverse flow on the western side of the channel was much weaker with rather variable velocities which showed significant surface flow towards the axis only in the 0830 transect.

Visual observation of the surface convergence supported the conclusion that, in this case, it was markedly asymmetric. Whereas the eastern side of the foam line was sharply defined, there was a more diffuse distribution of surface material on the western side.

Discussion

These first results from our system have demonstrated its capacity for measuring the transverse flow component over much of the estuarine cross-section. They have established positively the link between the axial convergence and the transverse circulation and confirmed that the cell structure extends over the full depth of water. Future observational studies will be designed to fully exploit the system by generating parallel sets of transverse flow and density data to test the diagnostic and other models of the circulation.

In a broad perspective of estuarine flow we should recognize both the tidal intrusion front and the axial convergence as products of the interaction between the tidal flow and the density gradients. While the associated circulations are kinematically quite different, a link between them may be identified. With increasing tidal flow, the V of the intrusion front becomes increasingly acute and the transverse component of the circulation increases. Further increase in the axial flow stretches the interface even more but at some point induces the breakdown of vertical structure with the abrupt transition to the axial convergence.

While these buoyancy-related mechanisms have been observed to predominate in a range of estuaries in the British Isles, they are not responsible for all convergent flow in estuaries. Flow separation, due to sharp features in the estuary shoreline or islands, for example, can also lead to the formation of strong foam lines. They occur at the boundary between the free stream and the almost stationary horizontal eddy which occupies the separated region. Such features are readily identified by their topographic association and the lack of systematic density contrast.

Acknowledgment

We are pleased to acknowledge the important contribution of Dave Boon and Alan Nield in the development of the shear sensor system. Jeremy Lowe and Cath Hudson helped with the observations. We are also grateful to John West and Reg Uncles for useful discussions on some of the ideas presented here.

References Cited

Brooke-Benjamin, T. 1968. Gravity currents and related phenomena. *Fluid Mech.* 32(2):209-248.

Fischer, H. B. 1972. Main transport mechanisms in partially stratified estuaries. *Fluid Mech.* 53:671-687.

Garvine, R. W. and J. D. Monk. 1974. Frontal structure of a river plume. *Geophys. Res.* 79(15):2251-2259.

Garvine, R. W. 1984. Radial spreading of buoyant, surface plumes in coastal waters. *Geophys. Res.* 89(C2):1989-1996.

Garvine, R. W. In Press. The role of brackish plumes in open shelf waters. *In:* S. Skreslet (ed.), Workshop on Influence of Freshwater Run-off on Coastal Ecosystems. May 21-25, 1985. NATO, Bodo, Norway.

Godfrey, J. S. and J. Parslow. 1976. Description and preliminary theory of circulation in Port Hacking estuary. C.S.I.R.O. Division of Fisheries and Oceanography Report No. 67, Sidney, Australia. 25 pp.

Huzzey, L. M. 1982. The dynamics of a bathymetrically arrested estuarine front. *Estuar. Coastal Shelf Sci.* 15:537-552.

Klemas, V. and D. F. Polis. 1977. Remote sensing of estuarine fronts and their effects on pollutants. *Photogrammetric Eng. Remote Sens.* 43(5):599-612.

Lailey, R. S. 1980. Tidal convergence in the Conway estuary. M.Sc. Thesis; UCNW, University of Wales, Bangor. 81 pp.

McClimans, T. A. 1978. Fronts in fjords. *Geophys. Astrophys. Fluid Dynamics* 11:23-34.

Nunes, R. A. 1982. Dynamics of small-scale fronts in estuaries, Ph.D. thesis, University of Wales, Bangor. 175 pp.

Nunes, R. A. and J. H. Simpson. 1985. Axial convergence in a well-mixed estuary. *Estuar. Coastal Shelf Sci.* 20(5):637-649.

Simpson, J. E. and R. E. Britter. 1979. The dynamics of the head of a gravity current advancing over a horizontal surface. *Fluid Mech.* 94(3):477-495.

Simpson, J. E. 1972. Effects of the lower boundary on the head of a gravity current. *Fluid Mech.* 53:759-768.

Simpson, J. H. and R. A. Nunes. 1981. The tidal intrusion front: an estuarine convergence zone. *Estuar. Coastal Shelf Sci.* 13:257-266.

Stigebrandt, A. 1980. A note on the dynamics of smallscale fronts. *Geophys. Astrophys. Fluid Dynamics* 16:225-238.

Thorpe, S. A., A. J. Hall and S. Hunt. 1983. Bouncing internal bores of Ardmucknish Bay, Scotland. *Nature, London.* 306 (5939):167-169.

Tully, J. P. and A. J. Dodimead. 1957. Properties of water in the Strait of Georgia, British Columbia and influencing factors. *J. Fish. Res. Bd. Canada.* 14:241-319.

West, J. R. and A. P. Cotton. 1981. The measurement of diffusion coefficients in the Conway estuary. *Estuar. 'Coastal Shelf Sci.* 12:232-336.

TIDAL RECTIFICATION IN THE PECONIC BAYS ESTUARY

Robert E. Wilson, Mario E. C. Vieira and J.R. Schubel

Marine Sciences Research Center
State University of New York
Stony Brook, New York

Abstract: Low-frequency currents from a total of fourteen moorings within the Peconic Bays Estuary were analyzed through frequency-domain empirical orthogonal functions with the objective of distinguishing spatially-coherent fluctuations at fortnightly periods associated with tidal rectification from synoptic-period fluctuations due to meteorological forcing. Results indicated that current fluctuations within the frequency band encompassing fortnightly periods were coherent only within a few distinct regions of the bay. These spatially coherent fluctuations were also highly coherent with demodulated sea level, apparently indicating tidal rectification in those areas. Current fluctuations within a frequency band centered on 0.275 cycles per day were spatially coherent over almost the entire bay and apparently forced by meteorologically-induced coastal sea level fluctuations.

Introduction

In shallow estuaries and bays with complicated coastline morphology, tidal rectification contributes to the production of residual flow. This nonlinear process is often associated with the advection of vorticity produced near the perimeter of the bay into the interior. (Kashiwai 1984a,b 1985; Oonishi 1977, 1978).

The detection of residual flow produced by tidal rectification within a given waterbody is not always straightforward. One method is to search for low-frequency current velocity fluctuations that are coherent with fluctuations in the amplitude of the tidal currents themselves. This study presents an analysis of current observations in the Peconic Bays Estuary, a shallow bay with strong tidal currents, with the objectives of:

1) distinguishing fortnightly fluctuations associated with tidal rectification from meteorologically-induced fluctuations; and
2) identifying those areas within the bay where appreciable tidally-rectified residual motion might be expected.

The Peconic Bays Estuary represents a system of very shallow (mean depth approximately 4.5 m) interconnected bays situated between the north and south forks of Eastern Long Island (Fig. 1). The total annual average fresh water inflow to the bays, including stream flow, groundwater seepage and precipitation, is extremely low ranging from approximately 3 to 5 $m^3 s^{-1}$. Tides within the bays are predominantly semidiurnal. The mean range at the mouth of the estuary is 0.78 m, and the tidal range increases within the estuary to 0.84 m at its western end (instead of suffering attenuation within the narrow channels

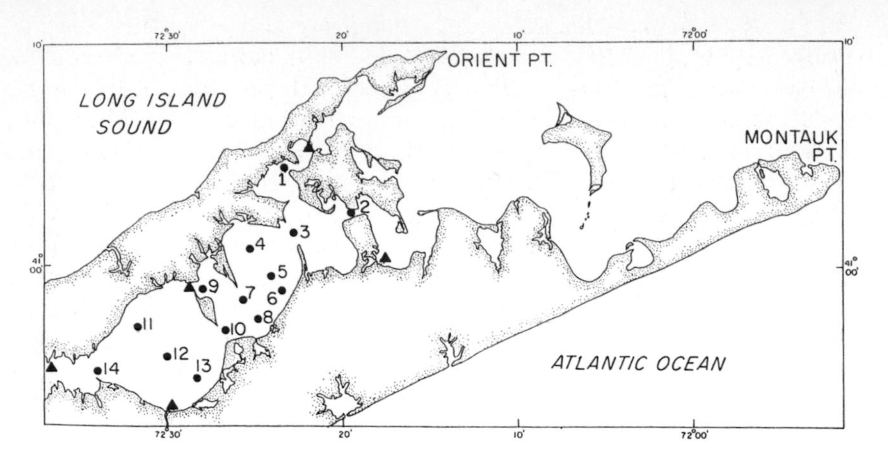

Figure 1. Peconic Bays Estuary showing locations of current moorings (circles) and sea level recorders (triangles).

and constructions). This effect could be due to Helmholtz resonance because the Helmholtz period for the bays is nearly equal to the semidiurnal period. Because of this effect semidiurnal tidal currents remain strong throughout much of the interior of the estuary. Tidal currents within the constrictions may exceed 1.2 m s^{-1} and some of the strongest currents encountered were those at mooring 9 (Fig. 1).

The data used for the present analyses were obtained in conjunction with a field survey of the Peconic Bays Estuary during 1984. The overall objective of the survey was to provide a data base for the evaluation of water movements and exchanges within the estuary. A more specific objective was to provide current velocity and sea level data for numerical model verification. The survey involved two one-month current meter deployments, one during March and another during July 1984, as well as the maintenance of sea level recorders around the perimeter of the bays for one year. Each current meter deployment involved a total of fourteen moorings with two or three instruments (ENDECO-174) per mooring. Positions of moorings and sea level recorders are shown in Fig. 1. Data from the July current meter deployment were unavailable for the analyses presented here.

Wilson and Vieira (1986) analyzed low frequency current fluctuations for the March 1984 deployment from moorings 1, 2 and 3 (Fig. 1) near the eastern end of the bay. They found that low-frequency current fluctuations at any given mooring did not vary significantly with depth. From the mean estimates they found no evidence for depth-dependent gravitational flow, although gravitational convection involving inflow on one side of the channel and outflow on the other might occur. At moorings 1 and 2, low frequency currents exhibited fortnightly fluctuations that were coherent with the fluctuations in the amplitude of the semidiurnal tide. At moorings 1, 2 and 3 currents exhibited synoptic period

(2-10 days) fluctuations that were coherent with fluctuations in coastal sea level. Wilson and Vieira (1986) also found evidence for fortnightly fluctuations in the low-pass-filtered sea level, which were coherent with fluctuations in the amplitude of the semidiurnal tide. These fluctuations represented a set up or set down of sea level in the estuary necessary to balance a residual bottom stress whose magnitude depended on the strength of the tidal stream. In this present study we extend the analyses of Wilson and Vieira (1986) to an analysis of the spatial variations of current response to forcing by coastal sea level fluctuations and by fortnightly fluctuations in tidal current amplitude throughout the entire bay. The latter response depends on the nonlinear interaction of the tidal stream with the basin topography and would be expected to vary in intensity over the basin depending on local bathymetry and coastline morphology.

Analysis

Our assessment of the spatial variations in the response of low-frequency currents to tidal and coastal sea-level forcing was based on Frequency-Domain Empirical Orthogonal Function (FEOF) analysis (Denbo and Allen 1984) applied to low-pass-filtered current velocity time series and to low-pass-filtered sea level and to demodulated (at the semidiurnal period) sea level. These latter series were chosen to be representative of coastal sea level and fortnightly tidal forcing functions. In order to keep the number of input series to a minimum and thereby simplify interpretation of the FEOF results, only the principal axis component of current velocity from the upper instrument at each mooring (Fig. 1) and the record from one sea level recorder (closest to mooring 1) were used.

Figures 2 and 3 illustrate examples of time series for low-pass-filtered, band-pass-filtered, and demodulated (at the semidiurnal period) principal-axis current for the upper instrument at moorings 1 and 3, respectively. Note that the principal axis direction is that determined for the low-pass-filtered current velocity. The low-pass-filtered series (Figs. 2a and 3a) exhibit mean velocities ranging from 10 to 20 cm s^{-1} and appreciable synoptic period fluctuations about the mean. The band-pass-filtered series (Figs. 2c and 3c) show clearly the strong fortnightly fluctuations in the amplitude of the semidiurnal currents. When the band-pass-filtered series are demodulated at the semidiurnal period, the time-varying amplitudes are obtained (Figs. 2b and 3b). Both band-pass-filtered series show that the filter used (Lanczos with half-power point at 34 hours) did not adequately filter out short period meteorological fluctuations near days 6 and 7. Figure 2c also shows evidence of distortion by overtides.

Comparison of the spectra for the low-pass-filtered series (Fig. 4) shows similarity at periods shorter than approximately 4 days, but significant differences at periods longer than 4 days. The low-pass-filtered current at mooring 1 (Fig. 2a) exhibits appreciable variance at periods longer than 4 days. Because the band-pass-filtered current (Fig. 2c) at this mooring showed distortion characteristic of overtides production, we suspect that the long-period variability in the low-pass-filtered current could be associated fortnightly fluctuations in the strength of a tidally rectified current. The FEOF analysis described below allows

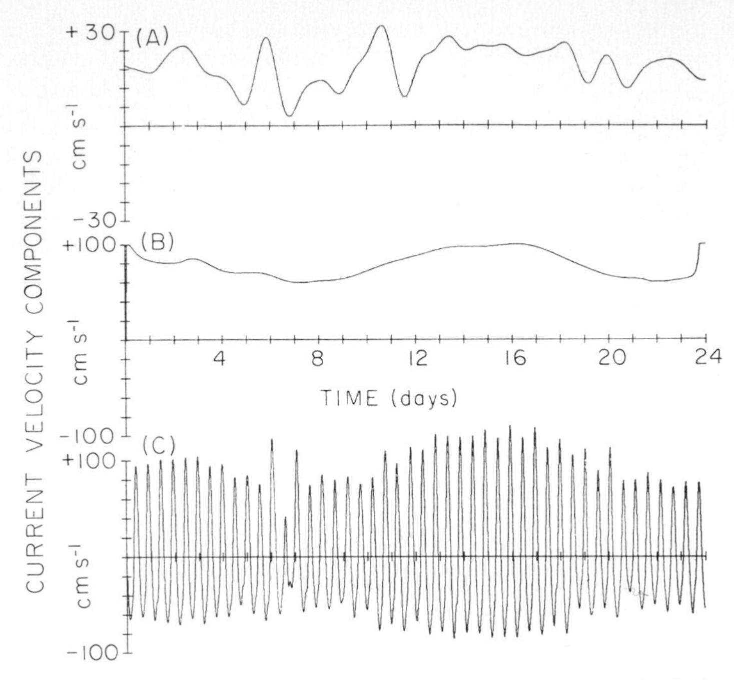

Figure 2. Times series for principal axis current at mooring 1 at 17-m depth showing the low-pass-filtered series (A), the amplitude of the band-pass-filtered series obtained from demodulation (B), and the band-pass-filtered series (C). The origin for the abscissa is 0000 hours on 3 March 1984.

us to describe frequency-dependent relationships between series as suggested by the spectra in Fig. 4.

Results of the FEOF analysis are presented (Table 1) for two frequency bands, one centered on 0.0 cycles per day (cpd) and the other centered on 0.275 cpd. On the basis of spectra in Fig. 4 and the spectrum for the time rate of change of coastal sea level (not shown), these bands were chosen to be most representative of relationships at fortnightly periods associated with tidal forcing, and relationships at synoptic periods associated with forcing by meteorologically-produced coastal sea level fluctuations. Note that our records were not long enough to resolve fortnightly fluctuations, and so this information is contained in the lowest frequency band. Table 1 presents the coherency-squared between each of the 14 input current series, the two inputs sea level series and the modal series for mode 1 and mode 2 in each of the two frequency bands. The reader is referred again to Denbo and Allen (1984) for the definition of the modal series for a given mode.

Discussion

The coherency-squared estimates between the input series and the series for the first mode (Table 1) are contoured in Figs. 5 and 6 for frequency bands centered on 0.0 and 0.275 cpd, respectively. Note that for these two bands the

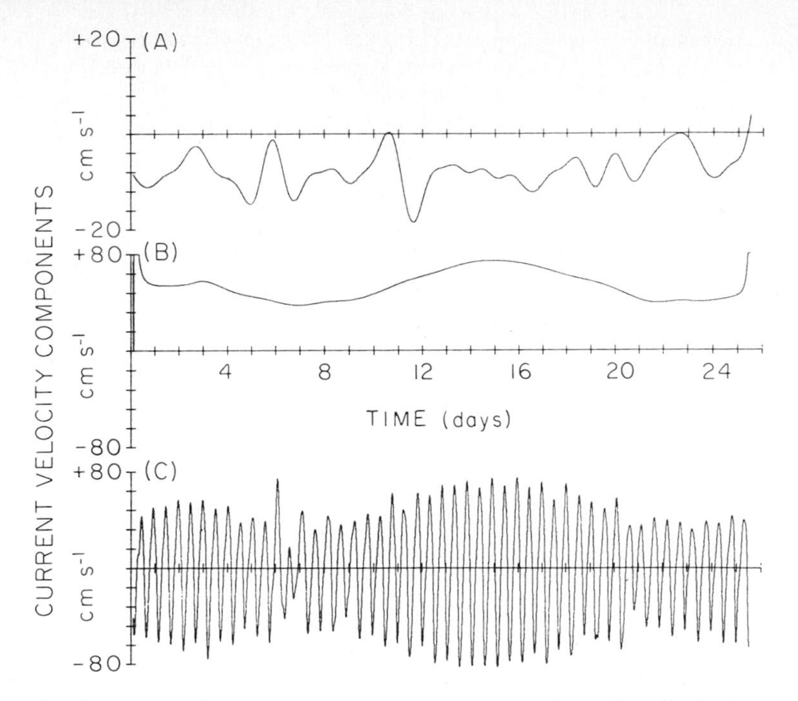

Figure 3. *Time series for principal axis current at mooring 3 at 11-m depth showing the low-pass-filtered series (A), the amplitude of the band-pass-filtered series obtained from demodulation (B), and the band-pass-filtered series (C). The origin for the abscissa is 0000 hours on 3 March 1984.*

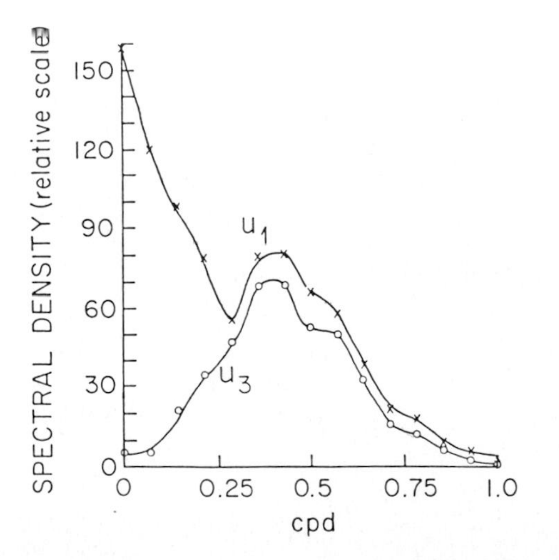

Figure 4. *Auto spectra for low-pass-filtered principal axis current at mooring 1 and mooring 3 (See Figs. 2 and 3). Scales are linear.*

Table 1. Results of Frequency-Domain Empirical Orthogonal Function (FEOF) analysis for two frequency bands (ω), showing percentage of total variance (in all series) accounted for by each mode, and the coherency-squared (γ^2) between each input series and the modal FEOF series. The first 14 series are currents at the moorings in Fig. 1 (in sequence); the last two series are demodulated sea level and low-pass-filtered sea level, respectively. Mode 1 coherency-squared estimates are contoured in Figs. 5 and 6.

	ω = 0.0 cpd		ω = 0.275 cpd	
Series	γ^2; Mode 1 51% of variance	γ^2; Mode 2 24% of variance	γ^2; Mode 1 64% of variance	γ^2; Mode 2 22% of variance
1	.84	.14	.92	.04
2	.40	.10	.63	.30
3	.28	.00	.95	.02
4	.11	.57	.53	.42
5	.44	.01	.67	.24
6	.54	.29	.74	.11
7	.78	.06	.59	.18
8	.80	.00	.79	.11
9	.41	.40	.89	.03
10	.53	.20	.77	.13
11	.76	.20	.84	.01
12	.00	.93	.05	.67
13	.42	.06	.37	.45
14	.79	.00	.52	.11
15	.81	.07	.05	.67
16	.16	.78	.95	.02

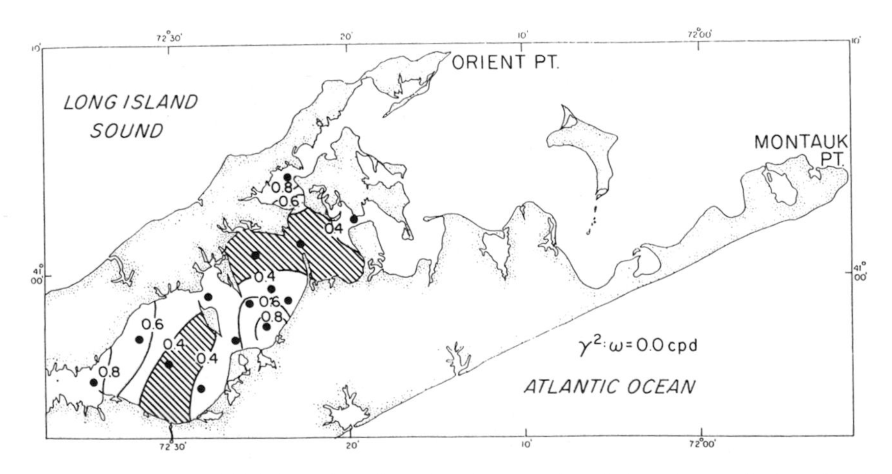

Figure 5. Results of FEOF analysis (for a band centered on 0.0 cycles per day) applied to low-pass-filtered currents at all moorings; contours are coherency-squared between each current series and the series for Mode 1 (see text). Shaded regions are below the significance level.

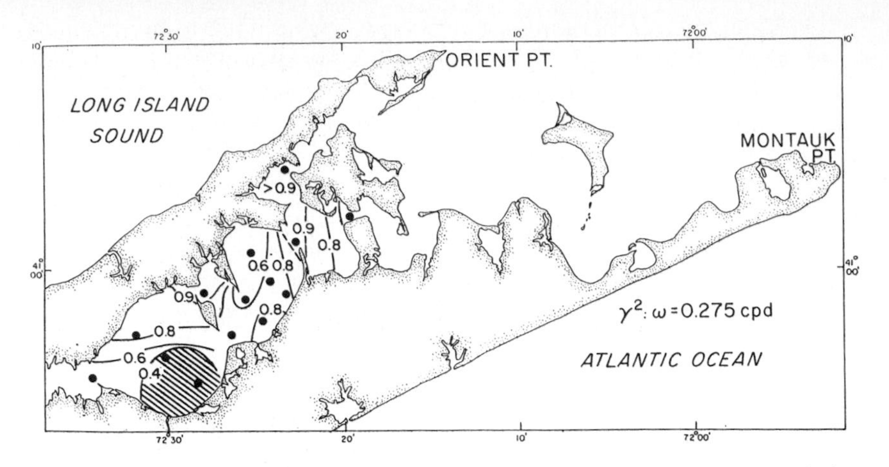

Figure 6. Results of FEOF analysis (for a band centered on 0.275 cycles per day) applied to low-pass-filtered currents at all moorings; contours are coherency-squared between each current series and the series for Mode 1 (see text). Shaded regions are below the significance level.

first mode accounts for 51% and 64%, respectively, of the total normalized variance in all 16 series. In general, current fluctuations are spatially coherent over much of the entire bay; they are coherent with the demodulated sea level in the 0.0 cpd band and with low-pass-filtered sea level in the 0.275 cpd band.

Spatial patterns of coherency-squared between each current series and the series for mode 1 in the 0.0 cpd band show that there are two regions of low coherency (below the significance level of approximately 0.4). Outside of these regions current fluctuations are spatially coherent although there may be phase shifts. The basic significance of these coherent fluctuations in this frequency band is that it implies that the residual flow pattern within these regions is due at least partially to tidal rectification. It is difficult to speculate why moorings 3, 4 and 12 would be little influenced by tidal rectification. Kashiwai (1984a,b, 1985) discusses the patterns of tidally rectified motion which might be expected for extremely simplified basin geometries. Mean flows recorded at moorings 4 and 12 were quite weak, but at mooring 3 mean flow was approximately 8 cm s^{-1} directed towards the east. This flow was nearly depth-independent and its origin is somewhat puzzling. One possible cause is gravitational convection associated with an inflow through the northern half of the section and an outflow to the south.

The mode 1 current fluctuations in the 0.275 cpd band were highly coherent over much of the entire bay with the exception of a region in the western bay (Fig. 6). This mode represents a simple emptying and filling of the bay in response to coastal sea level fluctuations which were presumably meteorologically induced. Current fluctuations within this band are approximately 90° out of phase with the sea level fluctuations and the spectrum for the time rate of change in sea level has a peak at precisely the same frequencies as the currents in Fig. 4.

In summary, results of the FEOF analysis applied to series for low-pass-filtered principal-axis current provide insight into the spatial patterns of tidally rectified residual motion within the bay through a description of the coherent fortnightly fluctuations. The FEOF analysis also effectively described the characteristics of the synoptic-period current fluctuations associated with coastal sea-level fluctuations in spite of the fact that a number of the current spectra were red and exhibited little or no spectral peak at 2-to-4-days period.

Additional insight into the patterns of tidally rectified residual motion could be obtained by applying FEOF analysis to the current vector time series and constructing the eigenvector components at each mooring. These patterns are likely to be quite complex. The results of Oonishi (1977, 1978), for example, lead us to expect strong vorticity generation near capes such as those near moorings 7, 9 and 10. This vorticity is then advected and diffused into the interior of the basin to produce a residual circulation consisting of counter-rotating vortices on either side of the cape. Oonishi (1977) has shown that the size of the residual vortex depends strongly on the ratio of tidal amplitude to basin depth, increasing as this ratio increases. In this basin with islands and multiple capes, numerical simulations could be used to obtain the detailed patterns of residual circulation.

One additional process of possible importance in this system is the interaction of the tidally-induced residual flow with the density-driven flows associated with freshwater input. In spite of the fact that current records did not show depth-dependent gravitational circulation, salinity surves do show variatons of several parts per thousand and a general increase in salinity seawards. Two-level tidal models which include freshwater forcing can begin to provide insight into the interaction of tidally rectified currents and density-driven currents in wide estuaries and bays.

Acknowledgments

This research was supported by New York State Sea Grant Institute under Grant #NA799AAD00053 to H. H. Carter and D. W. Pritchard who kindly made these data available to us. Contribution 506 of the Marine Sciences Research Center.

References Cited

Denbo, D. W. and J. S. Allen. 1984. Rotary Empirical Orthogonal Function Analysis of currents near the Oregon coast. *J. Phys. Oceanogr.* 14:35-46.

Kashiwai, M. 1984a. Tidal residual circulation produced by a tidal vortex, Part 1. *J. Oceanogr. Soc. Jap.* 40:279-294.

Kashiwai, M. 1984b. Tidal residual circulation produced by a tidal vortex, Part 2. *J. Oceanogr. Soc. Jap.* 40:437-444.

Kashiwai, M. 1985. Control of tidal residual circulation and tidal exchange in a channel basin system. J. Oceanogr. Soc. Jap. 41:1-10.

Oonishi, Y. 1977. A numerical study of the tidal residual flow. *J. Oceanogr. Soc. Jap.* 33:207-218.

Oonishi, Y. 1978. A numerical study on the tidal residual flow—vertical motion induced by tidal currents. *J. Oceanogr. Soc. Jap.* 34:140-159.

Wilson, R. E. and M. E. C. Vieira. 1986. Residual currents in the Peconic Bay Estuary. *In:* B. Neilson (ed.), *Circulation Patterns in Estuaries.* Humana Press Inc., Clifton, New Jersey. (In Press).

FLOW-INDUCED VARIATION IN TRANSPORT AND DEPOSITION PATHWAYS IN THE CHESAPEAKE BAY: THE EFFECT ON PHYTOPLANKTON DOMINANCE AND ANOXIA

Mary Altalo Tyler

College of Marine Studies
University of Delaware
Lewes, Delaware

Abstract: Comparison of data from the drought year 1981 and the moderately high flow year 1983 revealed the influence of streamflow on stratification, anoxia and phytoplankton dominance in the Chesapeake Bay. In 1981, the low flow resulted in a weakly stratified system with considerable mixing between surface and bottom waters. This produced: (1) increased dominance of the diatom *Rhizosolenia* baywide due to increased turbulent mixing; (2) abbreviation of the typical episodes of anoxia throughout the summer period; (3) up-estuary penetration of phytoplankton species normally confined to the southern bay (*Heterocapsa triquetra*) due to the abnormally high salinity in the northern bay; and (4) flow-mediated transfer of the dinoflagellate *Katodinium rotundatum* from its origin in the northern bay (Patapsco River) into the Delaware Bay through the Chesapeake and Delaware Canal. In 1983, the high flow intensified the stratification providing and effective barrier to vertical mixing in the system. This reduction in mixing led to (1) the early onset, and persistence of anoxia that originated just south of the bay bridge in mid-May and spread bay wide by mid-June; and (2) the early transport and deposition of the late winter *Rhizosolenia* bloom into benthic sediments in the northern and mid-bay due to the reduction in mixing in the system.

Introduction

The Chesapeake Bay is a partially-mixed coastal plain estuary, and the freshwater inflow provided by the Susquehanna River is the major driving force for the return flow of high salinity coastal waters along the bottom, thus establishing a two-layer system (Pritchard 1952). The degree of stratification varies seasonally with the steamflow cycle, and strongest density gradients are observed in the late spring. Concomitant with the development of a strong pycnocline is the establishment of both surface and bottom "corridors" that connect the upper and lower bay and through which planktonic organisms migrate or are transported during their seasonal cycle. Marine diatoms such as *Rhizosolenia fragilissima, Leptocylindrus danicus, Skeletonema costatum, Ditylum,* and *Coscinodiscus* which are resuspended in winter in the coastal region or from the mouth of the bay can move up the estuary in bottom waters in late spring (Tyler 1985; Tyler and Seliger 1986). Fresh water cyanobacteria *Oscillatoria* or nanoplankton such as *Cryptomonas* and *Chroomonas* may be inoculated into the estuary in the surface freshette. Bloomforming dinoflagellates in the Northern Bay may cycle annually throughout the estuary, alternating between surface and bottom corridors (Tyler and Seliger 1978, 1981). Other bloom species such as *Gyrodinium uncatenum, Gymodinium pseudopalustre* and *Heterocapsa triquetra* may be introduced into the corridors from the sediment on an annual basis when benthic cysts germinate into motile planktonic stages (Tyler *et al.* 1982; Tyler and Heinbokel 1985; Chin-Leo 1985). Transport of organisms up-estuary in bottom waters is usually curtailed by mid-May since anoxic bottom waters provide a physiological block to migration during the summer stability period (Officer *et al.* 1984; Seliger *et al.* 1985).

Because the continuity and duration of the transport "corridors" are dependent upon strong stratification, variability in yearly streamflow may produce changes in the transport pathway. In this paper we examine and document how the interannual variations in streamflow influence stratification in Chesapeake Bay and explain the impact of flow changes upon the formation of anoxia as well as on the transport and deposition of phytoplankton.

Methods

During the period from 1975 to 1984, more than 100 cruises were taken aboard the R/V Ridgely Warfield and R/V Cape Henlopen to establish the seasonal migration cycles of phytoplankton in the Chesapeake Bay. Here, we concentrate on two years (1981 and 1983) with extreme flow characteristics. Data from other years are also included to illustrate certain normal patterns of distribution. The study area encompassed the entire Chesapeake Bay and the major tributaries.

Vertical profiles of temperature, salinity and dissolved oxygen were taken with a Neil Brown CTD or Plessy Systems CTD. Profiles of *in vivo* fluorescence were taken with a Seamartec fluorometer mounted on the CTD. The dissolved oxygen probe values were calibrated with Winkler titrations performed for discrete depths at each station. Water samples for extracted chlorophyll *a* concentrations, species identification and enumeration, and dissolved nutrients were taken with a rosette sampler mounted on the CTD. Sediment samples were collected with a gravity corer and processed for benthic algal concentrations using methods outlined in Anderson *et al.* (1982).

Results

Cumulative mean streamflow for the entire Chesapeake Bay was 22,073 $m^3 s^{-1}$ for January through July 1983 and 13,646 $m^3 s^{-1}$ for the same time period in 1981 (U.S. Geological Survey, Towson, Maryland. Pers. Comm.) The most pronounced effect associated with this 40% reduction in flow was the decrease in stratification and the increased up-estuary penetration of high salinity waters. It was not uncommon to see bottom salinities at the Bay Bridge of 22‰ during January-February 1981. Watermen observed barnacles fouling their boats near the Susquehanna, attesting to the upbay invasion by salt water organisms. The difference in streamflow between 1981 and 1983 illuminated a number of regulatory mechanisms for the spatial and temporal variability of dinoflagellate and diatom blooms.

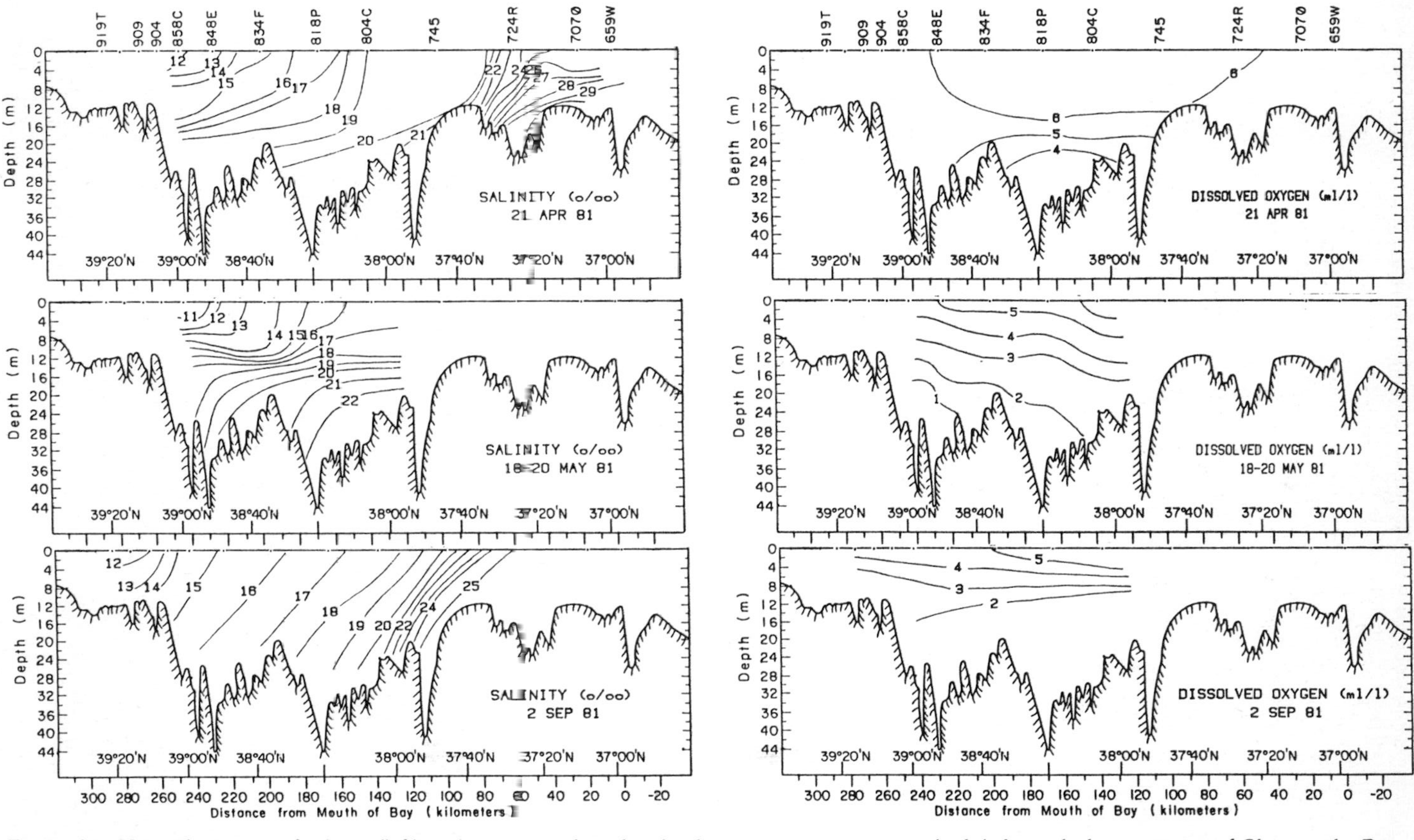

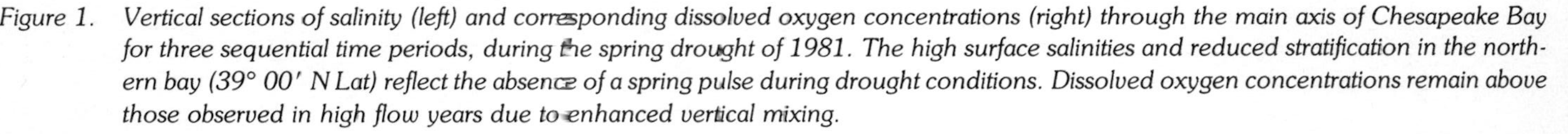
Figure 1. *Vertical sections of salinity (left) and corresponding dissolved oxygen concentrations (right) through the main axis of Chesapeake Bay for three sequential time periods, during the spring drought of 1981. The high surface salinities and reduced stratification in the northern bay (39° 00′ N Lat) reflect the absence of a spring pulse during drought conditions. Dissolved oxygen concentrations remain above those observed in high flow years due to enhanced vertical mixing.*

Drought and Decreased Stratification: Reduction of Anoxia and Prolongation of Diatom Blooms

A major manifestation of low runoff is the reduction in stratification throughout the estuary—particularly in the spring. Salinity profiles differed considerably in the drought year (1981) and high-flow year (1983). In the spring of 1981, a stream flow pulse was not evident (Fig. 1). Surface salinities in the Bay Bridge area (39°00′N) remained high (11‰) through early summer and stratification was reduced. By late August, the system was relatively homogeneous. In 1983, however, the spring streamflow was heavy and sustained (Fig. 2). By mid-April a 13‰ surface-to-bottom salinity difference existed in the northern bay.

The reduction in vertical exchange processes during 1983 was also reflected in the oxygen concentrations below the pycnocline. The development of anoxia can be related to the onset of stratification, which in turn is governed mainly by the spring flow "pulse" (Officer *et al.* 1984). One consequence of the layering is the curtailment of oxygen transfer vertically across the sharp halocline. However, if the water column does not stratify significantly, anoxia may not develop to the same degree. In 1981, oxygen concentrations decreased in bottom waters during the spring period (Fig. 1). However, the widespread anoxia typical of late May was not evident. We observed localized anoxia in July in the mid-bay region, which indicated that oxygen concentrations did go to zero, but for a short period of time. In 1983, oxygen concentrations decreased to 4 ml l^{-1} in the northern bay by mid-April (Fig. 2). As the flow peaked in early May, oxygen fell below 1 ml l^{-1}, and within a two-week period, anoxia had spread down-estuary. By 20 June the effect of stratification was reinforced by insolation, and anoxic waters had developed from the Bay Bridge south to the Potomac River with zero ml l^{-1} oxygen detected from the bottom up to the pycnocline at 9-m depth. An interesting hydrography-related phenomenon appears to have occurred from 38°48′ southward (Fig. 2, 20 June 1983). In this mid-bay location the dissolved oxygen minimum was found at mid-depth. Below about 16 m, the oxygen levels began to rise. This increase corresponded to salinities >18‰ which appeared to have been transported from the lower bay beneath the anoxia. If the streamflow is increased to an even greater extent, the severity of the anoxia (spatial as well as temporal extent) appears to increase as was reported for the anomolously high flow year of 1984 (Seliger *et al.* 1985).

The reduced stratification in 1981 had a significant impact upon the spatial as well as temporal distribution of the spring diatom bloom. Initially, the major inoculum for the spring diatom bloom is from the mouth of the bay (Tyler 1985). Seed populations are typically transported into the estuary in response to a wind event and begin to reproduce rapidly in the high-nutrient and high-light environment. As the spring streamflow peaks and stratification intensifies, the diatoms are carried upbay by the net non-tidal flow of bottom waters and are deposited in the deep trough along the axis of the bay. During drought years, however, the reduced stratification and increased up-estuary penetration of high salinity waters increase both the duration and areal extent of the diatoms.

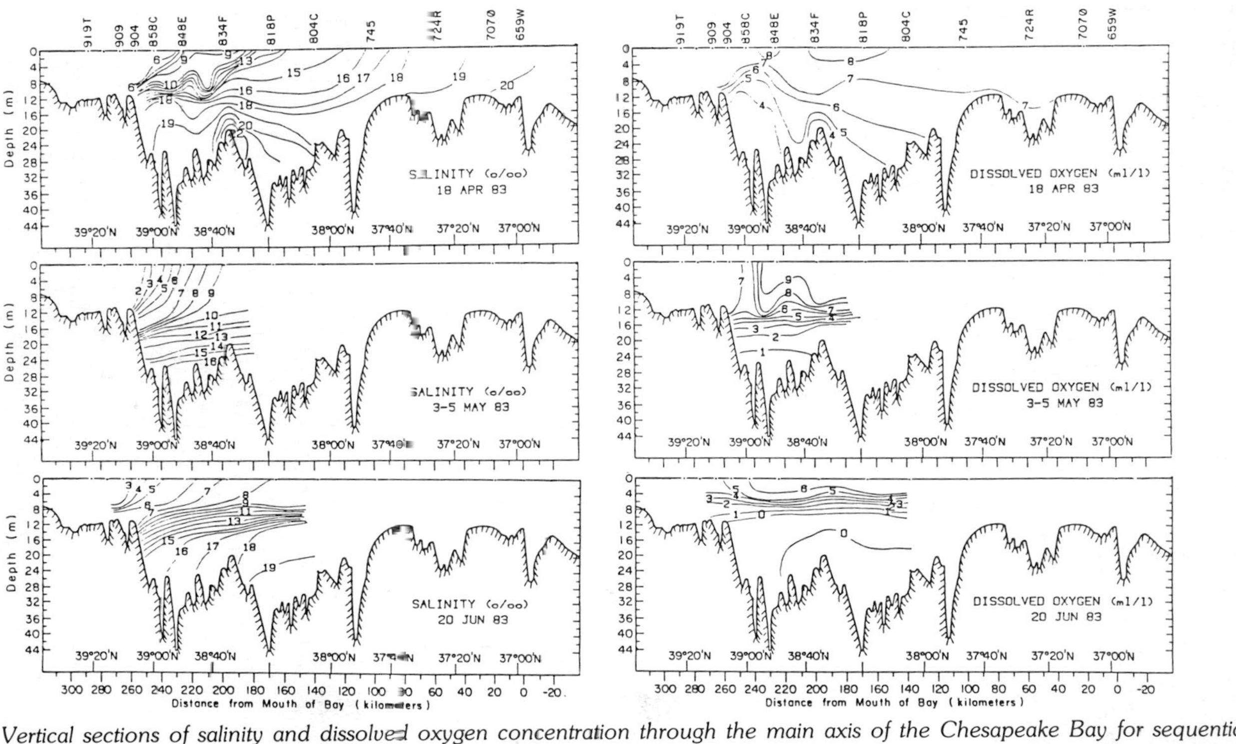

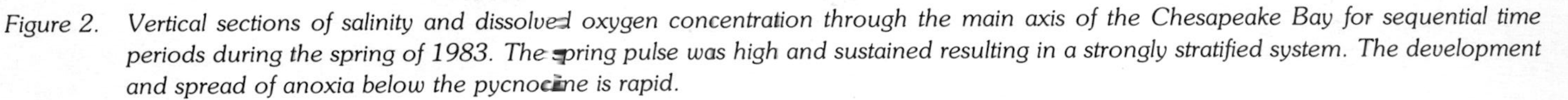

Figure 2. *Vertical sections of salinity and dissolved oxygen concentration through the main axis of the Chesapeake Bay for sequential time periods during the spring of 1983. The spring pulse was high and sustained resulting in a strongly stratified system. The development and spread of anoxia below the pycnocline is rapid.*

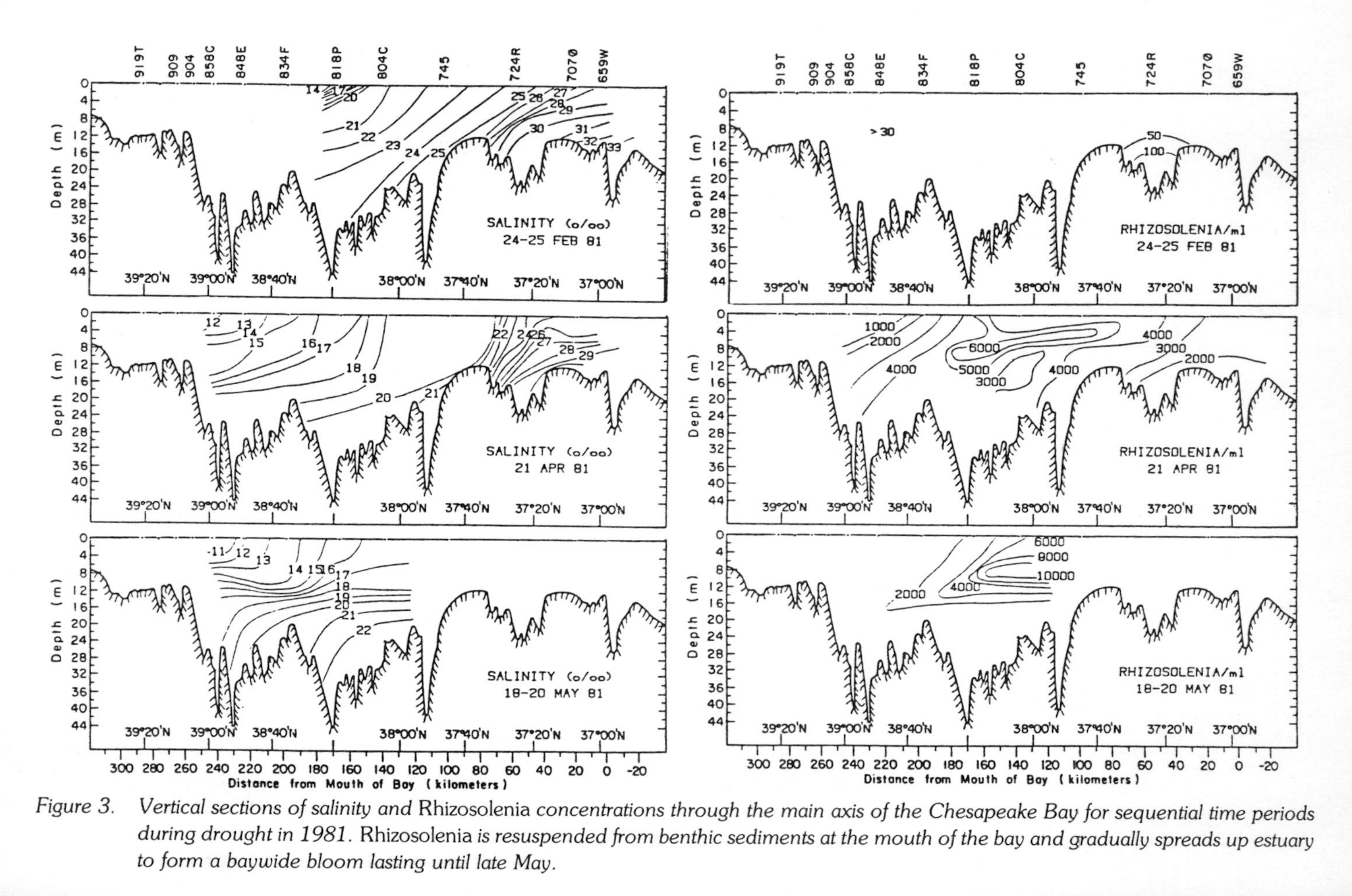

Figure 3. *Vertical sections of salinity and* Rhizosolenia *concentrations through the main axis of the Chesapeake Bay for sequential time periods during drought in 1981.* Rhizosolenia *is resuspended from benthic sediments at the mouth of the bay and gradually spreads up estuary to form a baywide bloom lasting until late May.*

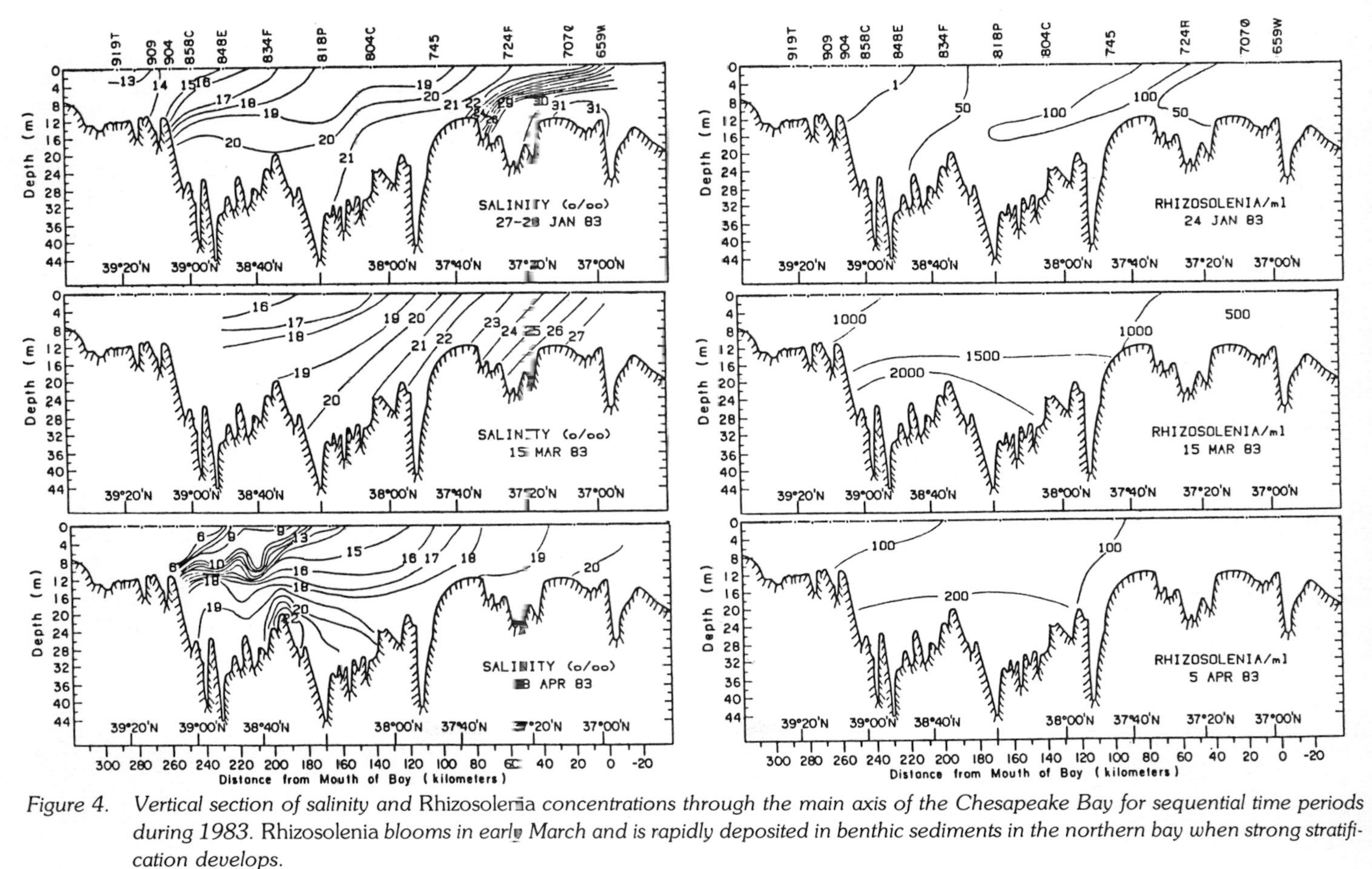

Figure 4. *Vertical section of salinity and* Rhizosolenia *concentrations through the main axis of the Chesapeake Bay for sequential time periods during 1983.* Rhizosolenia *blooms in early March and is rapidly deposited in benthic sediments in the northern bay when strong stratification develops.*

Figures 3 and 4 illustrate this change in distribution patterns of the diatom *Rhizosolenia fragilissima* due to the drought. In 1981, the bloom of *Rhizosolenia* originated in the southern bay during the late winter. The persistence of vertical advection until early summer coupled with a still operative net non-tidal flow of bottom waters up-estuary allowed the bloom to penetrate the upper Bay and persist through May. Although the mechanisms for bloom formation and deposition were similar in 1983, the bloom was short lived, terminating abruptly in March as the strong streamflow cut off the turbulent mixing of cells to the surface euphotic zone. The resultant light limitation lead to bloom dissipation and deposition into the sediments.

Drought and Home Range Extension

The dinoflagellate *Heterocapsa triquetra* also forms major winter/spring blooms in the Chesapeake Bay. Unlike *Rhizosolenia*, the cell is motile and does not rely upon vertical turbulence for suspension. Phytoplankton of this type appear to thrive under more stratified conditions. Life cycle stages of organism include a resting form, in addition to the motile cells that cause red water (Braarud and Pappas 1951; Campbell 1973; Chin-Leo 1985). Under adverse environmental conditions, the organism may shed its theca, settle into the sediment and remain there for an unknown period of time until some physiological or environmental factor stimulates the organism to regain its motility and commence the growth phase. Such a physiological mechanism allows the cell to become established in one locality for many years.

Prior to 1980, *Heterocapsa* was reported annually in the southern bay, particularly in the tributaries of the York and Rappahannock Rivers (Patten *et al.* 1963; Mackiernan 1968; Zubkoff and Warinner 1975). Sporadic reports of the cells are also recorded for the Patuxent River (Morse 1947; Marshall 1967; Mulford 1972). According to these observations, cell concentrations typically begin to increase in the late fall with red water conditions prevalent by late winter/early spring. Observations in the mid-to-late 70's indicate that the organisms are strongly affected by frontal circulation patterns that tend to accumulate the cells at a convergence and thus retain them within the rivers (Seliger *et al.* 1979). Figure 5 illustrates one such accumulation and possible recirculation within the York River in late spring. Cells in the bottom waters were observed to be shedding their thecas, an indication of the initiation of the encystment phase.

Beginning in 1981, a very different picture of the spatial extent of the organism began to appear. Red water of *Heterocapsa* appeared in the Potomac estuary as well as the upper bay region and we have observed them each year in the Potomac since that time. The change in areal distribution is depicted in Fig. 6, a plan view of the estuary outlining the location of blooms since 1978. *Heterocapsa* was an effective tracer for the invasion of the Southern Bay high-salinity water mass to the north, illustrating how a species can become introduced into a new location through a low streamflow event. To become a permanent resident, the organism must possess some defense against flushing. The

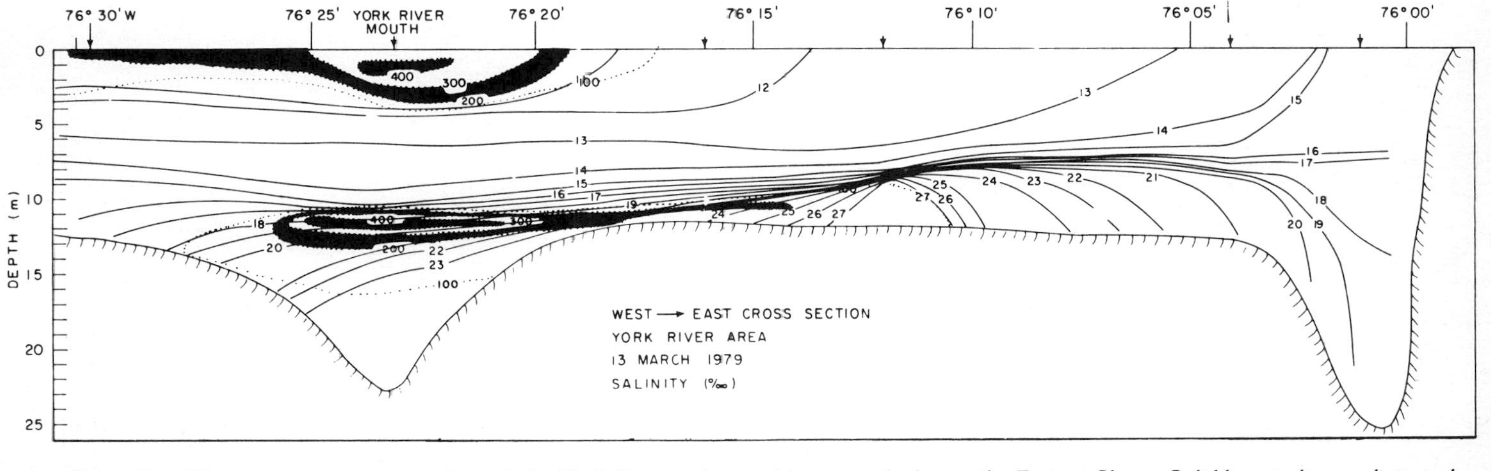

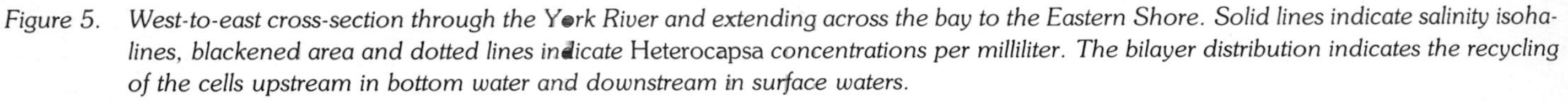

Figure 5. *West-to-east cross-section through the York River and extending across the bay to the Eastern Shore. Solid lines indicate salinity isohalines, blackened area and dotted lines indicate* Heterocapsa *concentrations per milliliter. The bilayer distribution indicates the recycling of the cells upstream in bottom water and downstream in surface waters.*

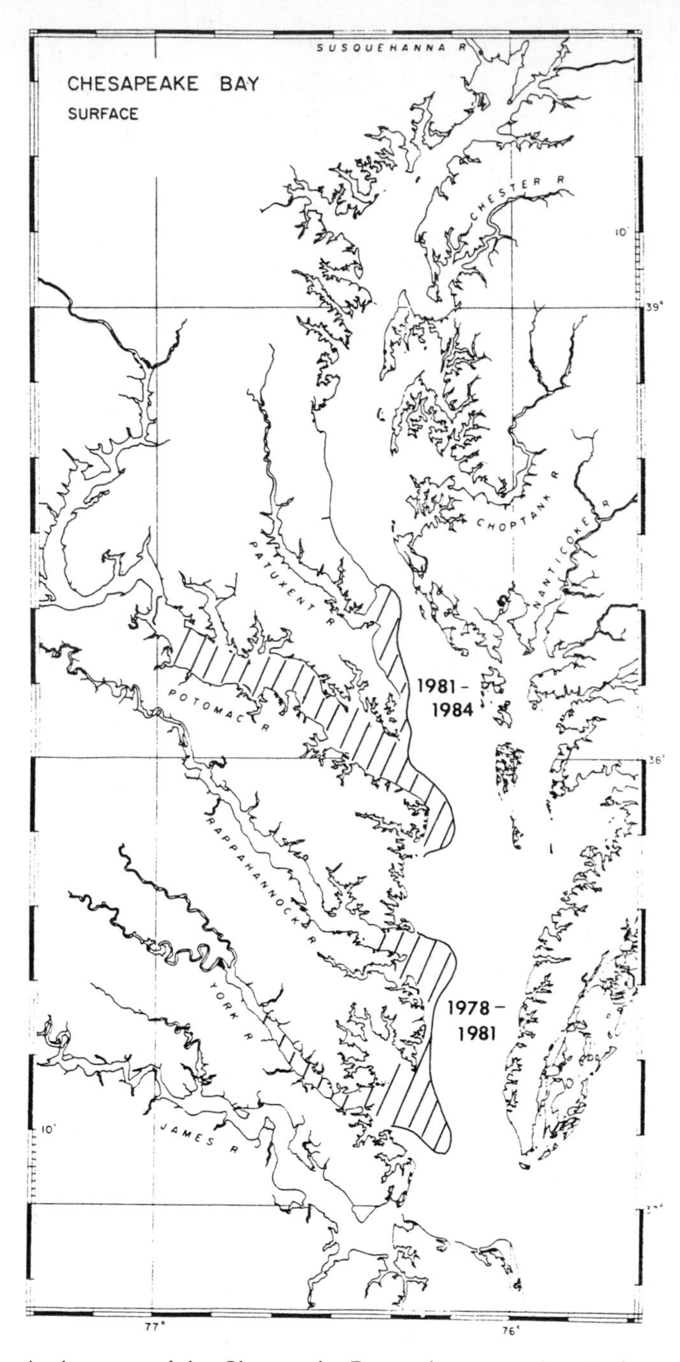

Figure 6. *A plan view of the Chesapeake Bay outlining approximate locations of observed* Heterocapsa *blooms since 1978. Note the up-estuary displacement of the blooms in 1981 due to drought conditions.*

formation of a resting stage of *Heterocapsa* in the sediments allows the cells to be retained and reappear in the same location periodically. A similar mechanism was used to explain the change in the bloom location of the summer species *Gyrodinium uncatenum* from 1979 to 1981 (Tyler *et al.* 1982). The same mechanism was proposed for the seeding and persistence of toxic red tides (Anderson and Wall 1978; Seliger *et al.* 1979).

Another instance of the displacement of an organism into a new location during a drought year was illustrated by the dinoflagellate *Katodinium rotundatum. Katodinium* annually forms blooms in the Chesapeake Bay—culminating in red water in the bay and tributaries in mid-autumn or early spring (Fig. 7; Seliger *et al.* 1975). *Katodinium* maintains high concentrations of biomass directly below the ice cover throughout the winter and is recycled annually in the bay. Its presence in the bay proper is due to a riverine introduction and it appears to form a cyst that allows it to oversummer in the sediments of the tributaries. Figure 8 represents the distribution of this organism along the length of the bay proper during the spring of 1982. The source of *Katodinium* to the Bay was the Patapsco River (39°10′) and the organism's inflow into the Bay at depth reflected the three-layer flow of that tributary (Stroup *et al.* 1961). In the region of the Potomac River just upstream of the area where the major pycnocline of the Bay breaks the surface, the population of *Katodinium* appeared to be recirculated into upstream-flowing bottom waters enroute to the northern bay. A bloom of this dinoflagellate was followed over a 3-month period in late 1980 during the early stages of the drought (Fig. 9). First signs of red water appeared in the Patapsco River and gradually spread to the northern Bay proper. These surface accumulations did not follow the usual route, however. In the fall of 1980, high surface patches entered the Bay directly at the surface (due to the absence of a freshwater layer from the Susquehanna) and moved in a northward (up-estuary) direction. Approximately three weeks after the red water was observed in the Chesapeake Bay, high concentrations of these organisms (2000 cells ml^{-1}) spread to the mouth of the Chesapeake and Delaware Canal. By the third week in November a plume of *Katodinium* (>100 cells ml^{-1}) was observed within the Delaware Bay. This was a localized distribution with concentrations rapidly dissipating as one moved away from the canal entrance. Thus, in this instance, the anomolous streamflow condition led to an unusual exchange of biota between estuaries.

Discussion

The data presented in this paper substantiate previous conclusions that major changes in flow volume have a significant impact on the degree of stratification. Although the increased mixing due to low runoff appears to reduce the degree and duration of anoxia, oxygen is depleted near the bottom even in drought years (Fig. 1). Thus the physiological processes responsible for anoxia are still operative. It is the seasonal increase in streamflow that puts the "lid" on the system, however, hampering the mixing of the low-oxygen bottom water

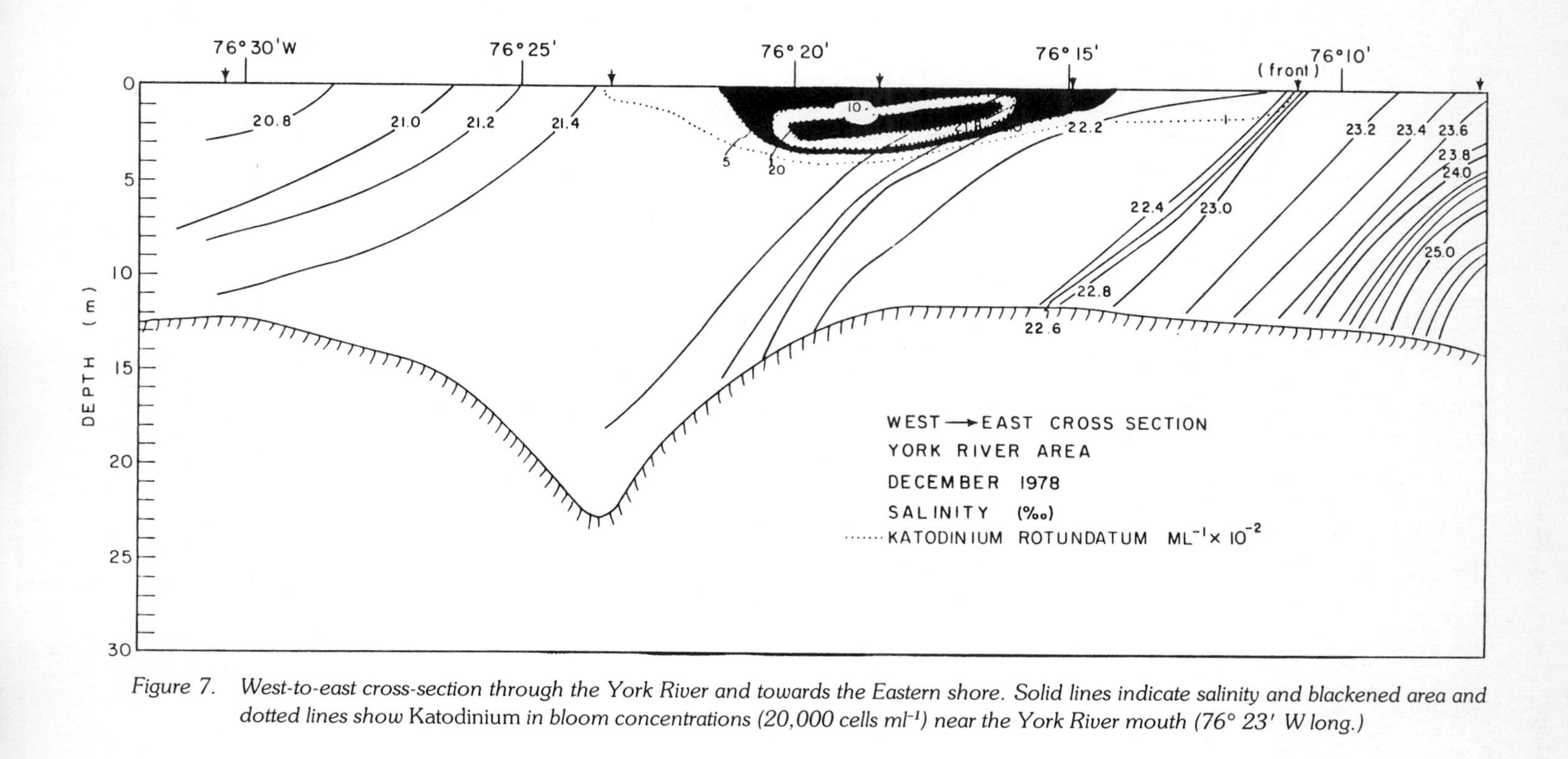

Figure 7. *West-to-east cross-section through the York River and towards the Eastern shore. Solid lines indicate salinity and blackened area and dotted lines show* Katodinium *in bloom concentrations (20,000 cells ml^{-1}) near the York River mouth (76° 23′ W long.)*

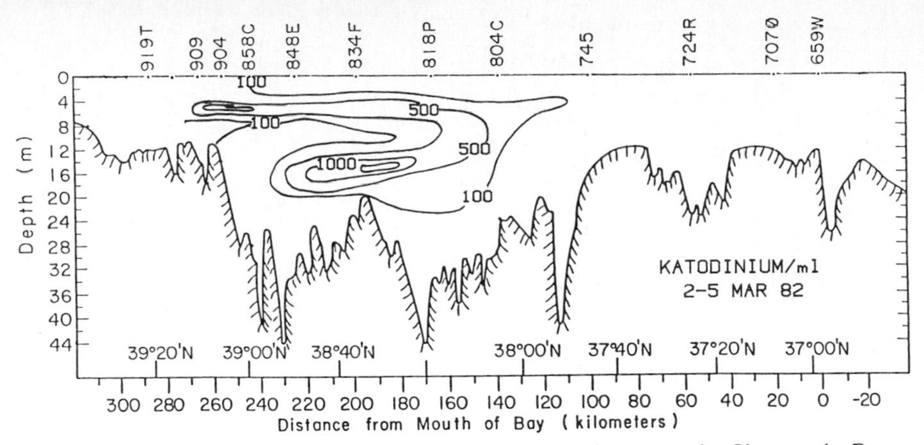

Figure 8. *Vertical section of salinity and* Katodinium *distributions in the Chesapeake Bay during March 1982 illustrating a recycling of the species in the region of the Potomac River.*

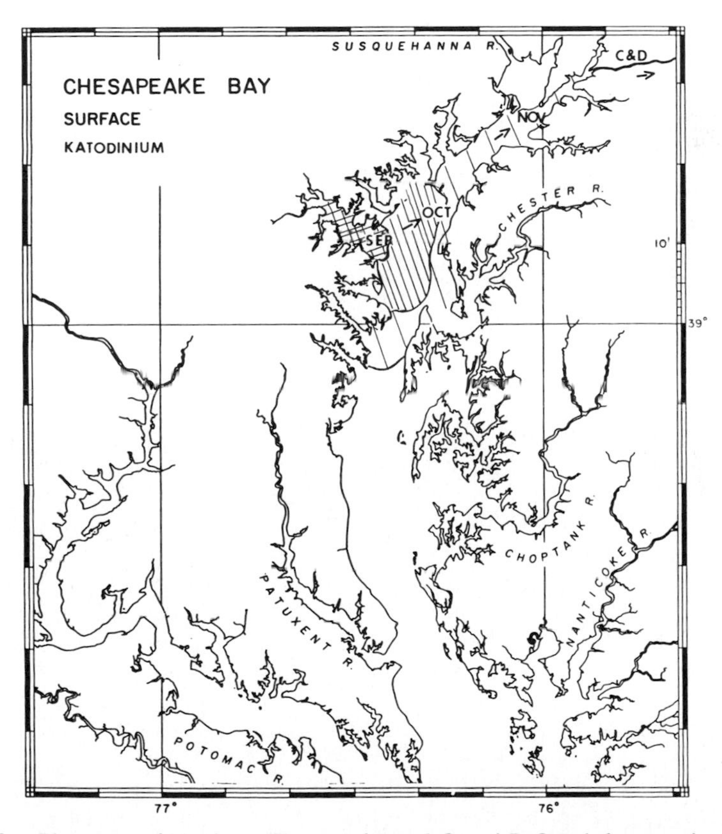

Figure 9. *Plan view of Northern Chesapeake and C and D Canal showing the up-bay movement of a surface bloom of* Katodinium *from the Patapsco River area in October to the C and D Canal and Delaware Bay in November 1980.*

and driving the system into the anoxic state. In a low flow year, this "lid" never materializes, and the low oxygen waters are mixed into surface layers.

In addition to capping the system, an increased flow tends to enhance the up-estuary transport of lower bay biomass. For many organisms, bottom water transport contributes to a yearly migration cycle. This may include higher trophic levels as well as the phytoplankton. The diatom component of the phytoplankton may die or sink out during transport, and much of that biomass is deposited in the northern bay, providing additional substrate for bacterial action and oxygen consumption. The drought conditions of 1981 were responsible for the introduction of phytoplankton species into new locations hundreds of kilometers from their origin. These processes of exchange within the bay and inoculation into adjacent estuaries are extremely important to the population dynamics of the species. If, in addition, the organism maintains its viability in the new location, it may become a permanent resident. Such an event may have a significant impact upon the new ecosystem—potentially devastating if the phytoplankton species happens to be toxic.

In conclusion, therefore, changes in streamflow and stratification patterns can potentially affect the biological as well as the physical characteristics of the system. Although this paper has focused on distributions of oxygen and certain phytoplankton species, streamflow variation will also affect nutrient and sediment distributions as well as higher levels of the food chain. Any study of planktonic systems in estuaries must consider interannual variations in streamflow to properly interpret observed organism distributions.

Acknowledgments

Support for this research was obtained from the National Science Foundation grants OCE 8310407, OCE 8109928 and OCE 8011039 to M. Tyler.

References Cited

Anderson, D. M. and D. Wall. 1978. Potential importance of benthic cysts of *Gonyaulax tamarensis* and *Gonyaulax excavata* in initiating toxic dinoflagellate blooms. *J. Phycol.* 14:224-234.

Anderson, D. M., D. G. Aubrey, M. A. Tyler and D. W. Coats. 1982. Vertical and horizontal distributions of dinoflagellate cysts in sediments. *Limnol. Oceanogr.* 27(4):757-765.

Braarud, T. and I. Pappas. 1951. Experimental studies on the dinoflagellate *Peridinium triquetrum* (EHRB) Lebour. *Norse Vidensk-Akad. Oslo Avh. I Nat-Natury. Kl.* No. 2:4-23.

Campbell, P. H. 1973. Studies on brackish water phytoplankton. Sea Grant Publications UNC-SG-73-07. School of Public Health, Univ. North Carolina. Chapel Hill. 407 pp.

Chin-Leo, G. 1985. Factors determining the distribution of the dinoflagellate *Heterocapsa triquetra* in the Potomac and Chesapeake Estuaries: 1983-1984. Masters Thesis. Univ. of Delaware, Lewes. 124 pp.

Mackiernan, G. B. 1968. Seasonal distribution of dinoflagellates in the lower York River, Va. Masters Thesis. College of William and Mary, Williamsburg, Virginia. 104 pp.

Marshall, H. G. 1967. Plankton in the James River Estuary. 1. Phytoplankton in Willoughby Bay and Hampton Roads. *Chesapeake Sci.* 8:90-101.

Morse, D. G. 1947. Some observation on seasonal variation in plankton population. Patuxent River, Maryland 1943-1945. Pub. No. 65. State of Maryland. Dept. Research and Education. Chesapeake Biological Laboratory. Solomons, Maryland. 31 pp.

Mulford, R. A. 1972. An annual plankton cycle on the Chesapeake Bay in the vicinity of Calvert Cliffs, Maryland, June 1969-May 1970. *Proc. Acad. Nat. Sci. Philadelphia.* 124(3):17-40.

Officer, C. B., R. B. Biggs, J. L. Taft, L. E. Cronin, M. A. Tyler and W. Boynton. 1984. Chesapeake Bay anoxia: Origin, development and significance. *Science* 223:22-27.

Patten, B. C., R. A. Mulford and J. E. Warinner, 1963. An annual phytoplankton cycle in the lower Chesapeake Bay. *Chesapeake Sci.* 4(1):1-20.

Pritchard, D. W. 1952. Salinity distribution and circulation in the Chesapeake Bay estuarine system. *J. Mar. Res.* 11:106-123.

Seliger, H. H., M. E. Loftus and D. V. Subba Rao. 1975. Dinoflagellate accumulations in Chesapeake Bay. pp. 181-206. *In:* V. R. LoCicero (ed.), *Proceedings of the First International Conference on Toxic Dinoflagellate Blooms.* Massachusetts Sci. Tech Foundation, Wakefield, Massachusetts.

Seliger, H. H., K. McKinley, and M. A. Tyler. 1979. Phytoplankton distributions and red tides resulting from frontal circulation patterns. pp. 239-248. *In:* H. H. Seliger and D. L. Taylor (eds.), *Toxic Dinoflagellate Blooms, Proc. 2nd Internatl. Conf.* Elsevier, New York.

Seliger, H. H., J. A. Boggs and W. H. Biggley. 1985. Catastrophic anoxia in the Chesapeake Bay in 1984. *Science* 228:70-73.

Stroup, E. D., D. W. Pritchard and J. H. Carpenter. 1961. Final Report, Baltimore Harbor Study. Chesapeake Bay Inst. Tech. Rep. 26 Johns Hopkins Univ. Baltimore, Maryland 79 pp.

Tyler, M. A. 1985. Fate of a shelf inoculum of phytoplankton into two east coast estuaries: productivity vs. anoxia. *EOS* 66:18.

Tyler, M. A. and J. F. Heinbokel. 1985. Cycles of red water of the dinoflagellate *Gymnodinium pseudopalustre* in the Chesapeake Bay: Effects of hydrography and predation. pp. 213-218. *In:* D. M. Anderson, A. W. White and D. G. Baden (eds.), *Toxic Dinoflagellates, Proc. 3rd Internatl. Conf.* Elsevier, New York.

Tyler, M. A. and H. H. Seliger. 1978. Annual subsurface transport of a red tide dinoflagellate to its bloom area: water circulation patterns and organism distribution in the Chesapeake Bay. *Limnol. Oceanogr.* 23(2):227-246.

Tyler, M. A. and H. H. Seliger. 1981. Selection for a red tide organism: physiological responses to the physical environment. *Limnol. Oceanogr.* 26(2):310-324.

Tyler, M. A. and H. H. Seliger. 1986. Significance of time scale variations on stratification-parameters and impact on biota of Chesapeake Bay. *In:* B. Neilson (ed.), *Estuarine Circulation (Proceedings of the Symposium of the College of William and Mary).* Humana Press, Clifton, New Jersey. In Press.

Tyler, M. A., D. W. Coats and D. M. Anderson. 1982. Encystment in a dynamic environment: deposition of dinoflagellate cysts by a frontal convergence. *Mar. Ecol. Prog. Ser.* 7:163-178.

Zubkoff, P. L. and J. E. Warinner. 1975. Synoptic sightings of red water of the lower Chesapeake Bay and its tributary rivers (May 1973-September 1974). pp. 105-119. *In:* V. R. LoCicero (ed.), *Proceedings of the First International Conference on Toxic Dinoflagellate Blooms.* Massachusetts Sci. and Technol. Foundation, Wakefield, Massachusetts.

EFFECT OF WATERSHED MODIFICATION ON A SMALL COASTAL PLAIN ESTUARY

J. F. Ustach, W. W. Kirby-Smith and R. T. Barber

Duke University Marine Laboratory
Beaufort, North Carolina

Abstract: A four-year study of physical, chemical, and biological properties of South River, a lateral estuary of the Neuse River estuary, N.C., characterized the mean conditions and the degree of variability during conversion of South River's watershed from swamp/forest to farm. Salinity, turbidity PO_4, NO_3, and chlorophyll-*a* concentrations were determined at 10 stations in the South River during (1) clearing and draining of the land and (2) farm operation; the results from three stations are discussed in this paper. There was little change in the variability of these parameters (as measured by coefficients of variation) over the two periods, but there were significant increases in mean turbidity, NO_3 and PO_4 attributed to farm runoff. There was no significant change in the mean salinity or chlorophyll concentration.

Introduction

The southeastern United States contains more than 55% of the wetlands in the U.S. (Geraghty *et al.* 1973). These wetlands, while not virgin, had been used mostly for timber production and wildlife, uses which have little impact on surrounding waters. Draining and clearing of these southeastern wetlands for agricultural production began on a large scale during the early 1970's. In order to use these wetlands, which have a relatively rich soil but high water table, the water balance must be altered with large-scale drainage projects. Such projects have been common in the southeastern coastal plain since the 19th century (Miles 1910; Lefler and Newsome 1954); but many of these early projects failed due to pump failure or low commodity prices (Carter 1975).

By 1972 approximately 30% of the coastal plain in North Carolina had been drained, and an additional 40% was believed to be suitable for agriculture if the surface water table could be lowered (Doucetic and Phillips 1978). Between 1973 and 1975 about 500,000 acres of North Carolina's coastal plain wetlands were drained and cleared (Carter 1975). These scrub bogs and pocosins are part of the Palustrine System in wetlands classification (Cowardin *et al.* 1979) or Type 8 scrub/scrub bog (Shaw and Fredine 1956). The Palustrine System includes all non-tidal wetlands dominated by trees, shrubs, persistent emergent vegetation, emergent mosses or lichens (Cowardin *et al.* 1979). These wetlands are not protected by federal or state law and drainage of such lands for agriculture has been actively assisted and promoted by the federal and state governments.

In January 1974, approximately 45,000 acres of low-lying land in Carteret County, eastern N.C., were purchased for agricultural development by Open Grounds Farm, Inc. (Fig. 1). Approximately half this acreage is in the drainage basin of the South River, a coastal-plain estuary with little fresh water runoff, and the other half is divided among numerous other small estuarine basins. The maximum elevation of the land is approximately four meters. The overall impact of this agricultural development on water quality of fresh water entering the estuary and on the receiving estuarine waters was the subject of a

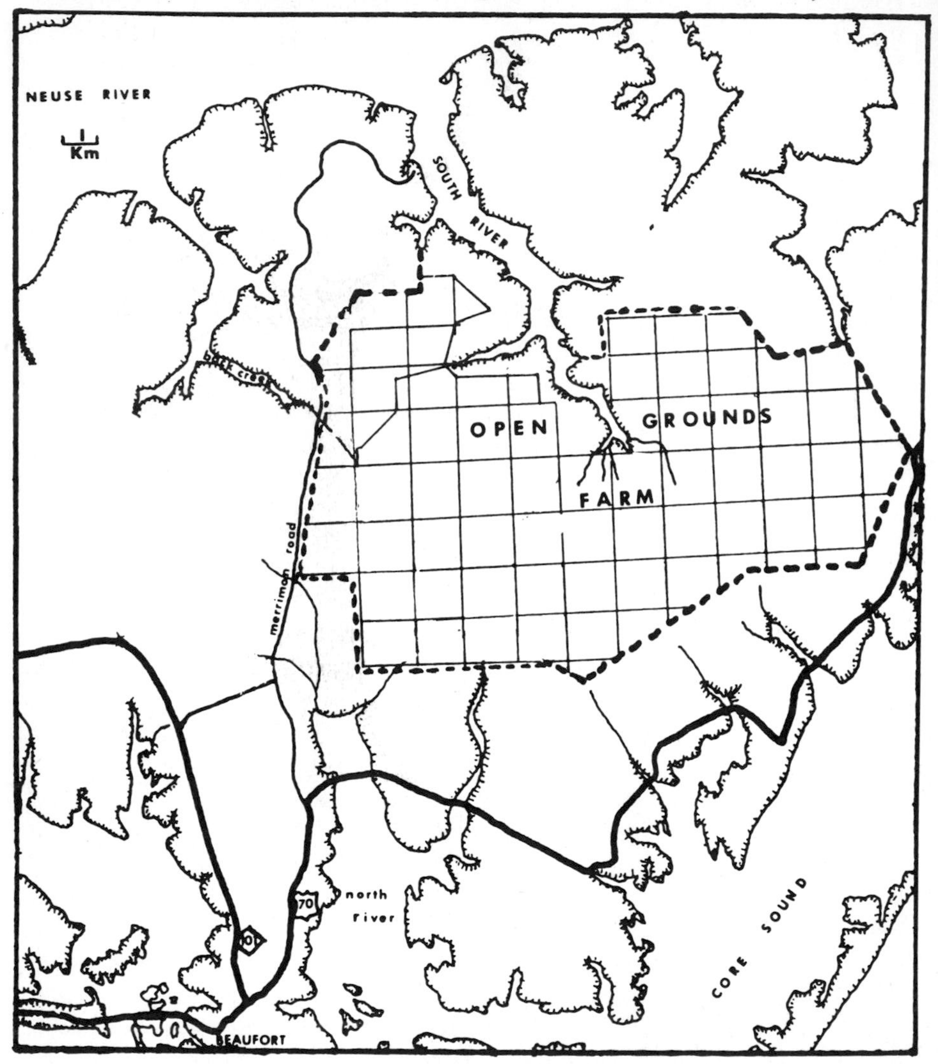

Figure 1. Map of Open Grounds Farm, Inc. in Carteret County, North Carolina. The cross hatch grid represents the location of major drainage ditches (1 mile square grid). The dashed line outlines the farm. The Neuse River is connected to the northeast with Pamlico Sound behind Cape Hatteras.

4-year study (Kirby-Smith and Barber 1979). In this paper we present and discuss the impacts of this changing land use on the variability of water quality parameters in the headwaters of the South River estuary.

Research Area

This study was conducted in the South River (Fig. 2) which is a coastal plain estuary connecting with the Neuse River estuary which in turn is part of the Pamlico Sound estuarine complex. The South River is a shallow, lateral embayment approximately 13-km long with an average depth of 2 m. The mouth of the estuary has a shallow sand bar sill, approximately 1-m deep, separating it from the Neuse River. Inside the sill the depth of the river is 3-4 m, slowly becoming shallower towards the headwaters of the tributary creeks where, even when the channel is only 5-m wide, the depth is still greater than 1 m. The South River has a very smooth, flat bottom with sediments composed of sand and highly organic silts (Berryhill *et al.* 1971). The slight lunar tide oscillations are masked by wind-driven tides that originate in Pamlico Sound and determine the water level in connecting areas (Pietrafesa 1985); so tidal height in the South River is largely a function of wind speed and direction. The upper half of the estuary has a large border of irregularly flooded marsh (*Juncus roemerianus*) between the open water and land and is typical of estuarine systems in the southeastern United States which receive fresh water from pine forest and coastal plain swamps. Coastal plain estuaries, such as the South River, that do not receive fresh water from large rivers make up a large percentage of the total estuarine area of the southeastern United States.

Prior to 1974 the drainage basin of the South River was covered by a mixture of pine swamp/forest, pocosin bog, and open grassland. A few old ditches remained from earlier attempts to drain part of the land and an abandoned dirt road crossed the land from east to west. The surface drainage entering the headwaters of the tributaries of the South River consisted of natural, "blackwater" swamp drainage (see Ash *et al.* 1983).

During 1974-1978 most of the South River watershed was converted into productive farmland. Development of the land followed a general pattern in which major drainage ditches were constructed through a swamp/forest on a one-mile square grid (Fig. 1). Smaller field ditches were then constructed at approximately one-eighth mile intervals, running north/south or east/west between the major ditches. Finally the land was cleared, shaped for drainage, heavily limed (to increase soil pH), and planted in pasture. The main ditches empty into the surrounding estuarine headwaters at several points through flood-gated culverts and short canals. Most, if not all, of these outfalls were constructed at the location of previously existing ditches which were cleaned out and enlarged. By the end of this study (November 1978), development of the western half of the farm had been completed while the process of ditching and clearing continued in the extreme eastern part. Planting of corn and other row crops began in 1976 in the southwest section of the farm, outside the South River watershed. In 1977 and 1978 extensive row cropping with concomitant fertilization was done throughout the western and central portions of the farm, including much of the South River watershed. Open Grounds Farm, Inc. is now raising cattle and growing corn, soybeans and other row crops. It has been and continues to be a very well managed operation. Modern farming practices include carefully controlled applications of fertilizers, herbicides, and pesticides,

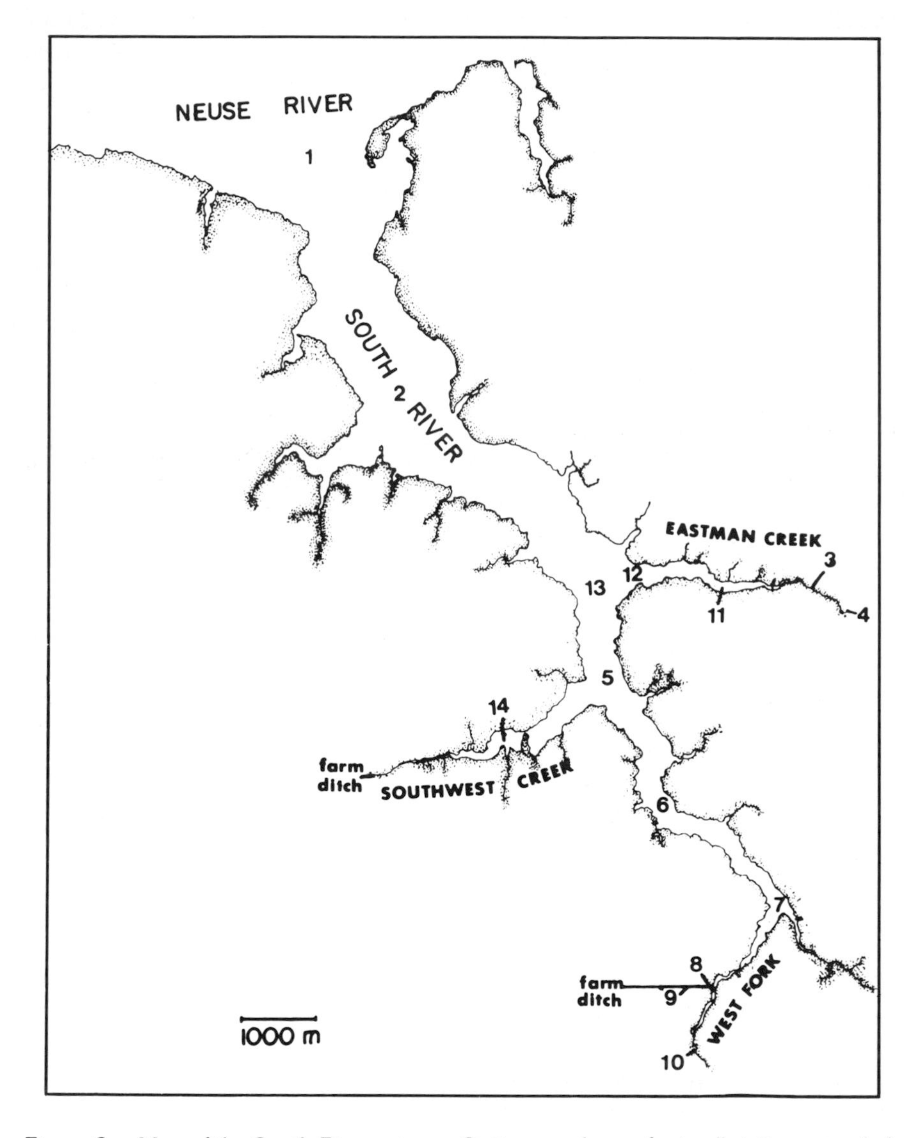

Figure 2. Map of the South River estuary. Station numbers refer to all stations sampled at some time during the project (Kirby-Smith and Barber 1979). The three stations discussed in this paper are: #3, in Eastman Creek; #7, at the head of the estuary; and #1, in the Neuse River.

maintenance of green strips along ditches and other soil conservation practices, and controlled release of water from ditches through floodgates.

The circulation of water in the South River is a classical estuarine circulation with transport of Neuse River water into the system along the bottom. These waters (averaging about 12‰ salinity) circulate to the headwaters of all the tributaries of the South River, mixing continuously with the surface waters. The water quality of the South River system as a whole depends in large part upon the water quality of the Neuse estuary which has a long history of receiving human impacts (Hobbie and Smith 1975). Even though the surrounding watershed and the quality of the surface waters of its tributaries were basically unaltered prior to 1974, the South River as a whole should not be considered a pristine estuary because of the Neuse River influence.

Methods

From March 1975 through September 1976 (Period 1) and again from June 1977 through October 1978 (Period 2), water samples were collected approximately twice each month from up to 14 stations located in the South River (Fig. 2). The first period coincided with clearing and pasture development while the second period was one of intensive cropping of corn and soybeans.

Surface water samples from stations on the South River were collected with a plastic bucket from a small boat. Part of the sample was placed in a one-gallon polyethylene carboy, kept in the shade at ambient temperature, and returned to the laboratory (generally within 4 h) where subsamples were analyzed for turbidity and chlorophyll-*a*. Chlorophyll-*a* was determined by fluorescence emission of an acetone extract in a Turner Model 111 fluorometer (Strickland and Parsons 1968), and turbidity was measured on a Hach Turbidimeter. Other aliquots were drawn into 250-ml polyethylene bottles, kept on ice in the dark, filtered immediately upon return to the laboratory, and stored on ice (24-48 h) until spectrophotometric analyses were completed for phosphate, nitrite, nitrate, and silicate (Strickland and Parsons 1968). During 1977-1978 these iced samples were also used (within 4 h) for chlorophyll-*a* and turbidity determinations, and the use of the one-gallon polyethylene carboys was discontinued.

Salinity was determined in the field with a YSI or Beckman portable salinometer. Water for nutrient analyses was filtered through glass fiber filters on the same day as collected to remove particulate matter. Limits of detection for nutrients were different from those published (Strickland and Parsons 1968) because we used 1-cm cells. Our limits of detection (μg-at l^{-1}) were calculated to be: phosphate, 0.20 and nitrate, 0.25.

Temperature, nitrite, silicate, ammonia, particulate organic carbon and nitrogen, seston, and pH were determined on both surface and bottom water samples at the stations shown in Fig. 2. (Kirby-Smith and Barber 1979). For the sake of brevity, we have not included all those data in this paper. Their inclusion, however, would not have altered our results or conclusions. Discussion in this paper is limited to five water quality parameters (salinity, turbidity, phosphate, nitrate, and chlorophyll-*a*) at Stations 1, 3, and 7 (Fig. 2). Means, standard deviations, and coefficients of variation (V) were calculated on data sets from Periods 1 and 2 for each of the five parameters. Means were compared by t-tests (Sokal and Rohlf 1981); V's were compared by a method similar to a t-test, after appropriate transformations, when necessary (Sokal and Braumann 1980).

Station 1 was located near marker 1 in the Neuse River (Fig. 2) and was about 3-m deep. The Neuse River is the main source of saline water for the South River and provides the downstream boundary conditions. Stations 3 and 7 (Fig. 2) were selected for comparison because both were located upstream in the headwaters of the estuary.

Station 7, 1.9-m deep, was located at the joining of two tributaries of the South River (West Fork and East Fork), both of which received farm drainage through cleared out ditches. During Period 1, most runoff from the farm came from a ditch entering West Fork at Station 8 (Fig. 2); additional drainage entered into East Fork during Period 2.

Station 3, 1.5-m deep, was located in the headwaters of a tributary of South River, Eastman Creek, which had no farm ditches along its entire length (Fig. 2). It represented the most "pristine" station in the sense of receiving no direct drainage via ditches. The watershed of Station 3 was not cleared and drained during Period 1 (1975-1976); by the end of Period 2 (November 1978) fields had been constructed in the watershed but no ditches entered the creek, and the area was separated from fields to the south and east by a forest/marsh border 100 m or greater and by a dike around the edge of the fields.

Analysis and Results

Estuaries are usually highly variable compared with other marine ecosystems (Bowden 1967). The basic question addressed here is how this characteristic variability is changed when the wetland watershed of a small estuary is cleared and drained, and agricultural runoff contributes to the nutrient flux. The pattern of sampling in the South River enabled us to determine how watershed modification changed the mean values and degree of variability for a series of properties that are related to different aspects of farm development.

The first line of analysis was a spatial comparison between the upstream station (Station 7) that receives drainage from the developed area and the downstream boundary conditions that prevailed in the Neuse River (Station 1). If "local" land clearing or fertilizer use modified the mean condition or degree of variability, that change should be most evident in this Station 7-versus-Station 1 comparison, but any difference could also result from a "natural" upstream-versus-downstream gradient in the estuary. The second line of analysis was based on the different drainage conditions at the two upstream sites and provided a control on the upstream gradient problem. Water from the cleared area was conveyed directly to the branch of South River where Station 7 is located while no farm drainage flowed directly in the branch where Station 3 is located. Changes resulting from land clearing or farm operations should therefore be more pronounced at Station 7.

A third line of analysis involved comparison of Period 1 (clearing and drainage) with Period 2 (intensive farming). This comparison is particularly focused on resolving estuarine transitions that involved nutrient changes. During Period 1 there was minimal use of fertilizers on the South River watershed because intensive farming had not yet started. If runoff of nitrate and phosphate from row crops is affecting nutrients or phytoplankton, there should be a clear difference between Period 1 and Period 2. If, on the other hand, land clearing *per se* is an important source of nutrients then we can expect to see a difference in Period 1 between Station 7 and Station 1. This pattern of one temporal and two spatial comparisons was used to examine the changes in each water quality property.

Turbidity

Turbidity did not change significantly between Period 1 and Period 2 at any stations in either surface (Table 1) or bottom (Table 2) water samples. Turbidity was clearly elevated at Station 7, however, relative to other stations. The comparison of the mean concentration upstream (Station 7) versus downstream (Station 1) in Table 3 shows a significant difference in turbidity in both Period 1 and 2 in the surface and a significant difference in the bottom in Period 1. This result was expected; clearing the thick pocosin forest ground cover would promote sediment runoff from the farm operation and cause a 3- to 4-fold increase in the quantity of sediment that was in suspension in the estuary closest to the ditches that drain the developed land.

Table 3 shows that Station 7 also had significantly higher levels of turbidity than Station 3. This result was also expected since at or near Station 3 the estuary receives essentially no drainage from disturbed land. Station 3 was also somewhat less turbid than Station 1 in the Neuse River (Tables 1, 2). During Period 1 the mean surface salinity at Station 3 was somewhat lower than at Station 1 (9.3‰ vs. 11.9‰) indicating that there is freshwater entering the estuary from the land around Station 3, but this freshwater must have had a relatively low suspended load. This decreased suspended load upstream probably reflects the sediment-sink character of the headwaters of small coastal plain estuaries. In the unmodified South River estuary greatest deposition probably occurred in the less energetic headwaters where wind mixing and tidal mixing are reduced.

Interestingly the degree of variability of turbidity was not significantly different between periods (Tables 1, 2) or stations (Table 4), despite the 3- to 4-fold difference in turbidity between Station 7 and Station 1. This result indicates that while land disturbance increased the mean turbidity it did not significantly alter the processes that introduce variability around that mean. This was an unexpected result. One interpretation that is consistent with this result is that the inherent variability of the weather (e.g., wind and rain) regulates the variability of turbidity while land use affects the magnitude of the flux of sediment to the receiving waters.

Salinity

At the start of this project we hypothesized that one effect of land clearing and drainage would be to increase the variability of salinity. Our conceptual

Table 1. Mean values ($\overline{X}$), coefficients of variation (V), and number of observations (n) for water quality parameters in surface water samples from Stations 1, 3, and 7 during land clearing (Period 1) and farming (Period 2); t_x and t_v are results of t-tests comparing means and coefficients of variation, respectively. * = test significant at $p < 0.05$; ** = test significant at $p < 0.01$.

	Turbidity (JTU)					Salinity ‰					Phosphate (μg-at l^{-1})					Nitrate (μg-at l^{-1})					chl *a* (μg l^{-1})				
	$\overline{X}$	t_x	V	t_v	n	$\overline{X}$	t_x	V	t_v	n	$\overline{X}$	t_x	V	t_v	n	$\overline{X}$	t_x	V	t_v	n	$\overline{X}$	t_x	V	t_v	n
Station 1																									
Period 1	3.6		60.7		35	11.9		29.5		37	0.68		92.0		37	0.80		210.5		36	12.8		177.3		37
		1.15		0.23			0.67		1.76			2.75*		0.35			2.26*		0.74			1.87		1.94	
Period 2	4.5		64.4		18	11.0		45.3		17	1.54		81.4		18	3.66		138.9		17	22.1		63.4		18
Station 3																									
Period 1	2.4		68.6		35	9.3		43.1		36	0.63		90.1		37	0.39		73.8		36	11.1		192.0		37
		0.48		0.38			0.65		0.89			2.19*		0.37			2.80*		1.33			1.18		1.66	
Period 2	2.6		62.0		24	10.1		52.9		26	1.04		80.9		27	1.16		121.0		27	16.1		77.8		27
Station 7																									
Period 1	11.8		85.7		35	8.5		40.3		36	0.69		112.0		37	0.91		106.0		36	17.4		204.0		37
		0.89		1.16			0.32		2.06*			3.03**		0.18			3.63**		0.58			0.18		1.36	
Period 2	16.1		140.0		26	8.1		71.7		27	2.32		119.0		28	9.44		131.0		28	18.6		99.1		28

Table 2. Mean values ($\overline{X}$), coefficients of variation (V), and number of observations (n) for water quality parameters in bottom water samples from Stations 1, 3, and 7 during land clearing (Period 1) and farming (Period 2); t_x and t_v are results of t-tests comparing means and coefficients of variation, respectively. * = test significant at $p < 0.05$; ** = test significant at $p < 0.01$.

	Turbidity (JTU)					Salinity ‰					Phosphate (μg-at l^{-1})					Nitrate (μg-at l^{-1})					chl *a* (μg l^{-1})				
	$\overline{X}$	t_x	V	t_v	n	$\overline{X}$	t_x	V	t_v	n	$\overline{X}$	t_x	V	t_v	n	$\overline{X}$	t_x	V	t_v	n	$\overline{X}$	t_x	V	t_v	n
Station 1																									
Period 1	4.5		70.8		35	12.5		28.0		37	0.59		87.2		36	0.61		253.4		36	10.5		144.4		37
		0.83		1.0			0.36		1.63			3.07**		0.06			2.15*		0.85			1.71		2.05*	
Period 2	5.7		99.8		18	12.0		43.9		17	1.68		84.8		17	2.61		145.2		18	16.3		59.8		18
Station 3																									
Period 1	3.3		60.2		35	10.7		27.4		36	0.79		103.8		37	0.38		87.9		36	7.6		101.6		37
		2.10*		0.37			0.40		2.49*			1.83		0.09			2.48*		1.25			3.15**		1.71	
Period 2	4.6		54.9		24	11.2		51.6		26	1.32		100.8		27	1.59		158.7		27	15.2		60.5		25
Station 7																									
Period 1	7.3		57.8		35	10.9		27.1		36	0.67		107.3		37	0.74		111.8		37	13.9		167.7		37
		1.63		1.78			0.31		2.83*			3.18**		0.06			3.17**		0.72			0.40		1.35	
Period 2	14.9		154.4		25	10.5		58.8		27	2.24		113.4		28	7.54		150.5		28	15.8		93.2		28

Table 3. Statistical comparison (t-test) of station differences in mean water quality parameters (Table 2). * = means are significantly different at $P < 0.05$; ** = means are significantly different at $P < 0.01$; *** = means are significantly different at $P < 0.001$.

	Station 1 vs. Station 7		Station 3 vs. Station 7	
	Period 1	Period 2	Period 1	Period 2
Turbidity				
Surface	4.83**	2.53*	11.14***	2.99**
Bottom	3.19**	2.99**	5.14***	2.22*
Salinity				
Surface	4.22**	1.74	0.90	1.27
Bottom	2.00*	0.87	0.30	0.40
Phosphate				
Surface	0.06	1.30	0.38	2.34*
Bottom	0.55	0.95	0.67	1.69
Nitrate				
Surface	0.34	2.18*	3.13**	3.51**
Bottom	0.44	1.95	2.14*	2.71**
Chlorophyll *a*				
Surface	0.37	0.74	0.93	0.57
Bottom	0.75	0.13	1.56	0.17

hydrological model was that an intact wetland acts as a sponge that can absorb rain and release it slowly to the estuary, whereas cleared land would release fresh water in pulses. Observations available from other studies indicate that there is a slight increase in the annual quantity of runoff from cleared coastal plain land (Daniel 1981; Konyha *et al.* 1985), but the largest effect is in the shortened temporal cross-section for runoff from cleared and drained land. The expected change in variability of runoff was not observed consistently, however, in the South River salinity observations. There was a significant increase in variability between Period 1 and Period 2 in surface (Table 1) and bottom (Table 2) salinity at Station 7. This result suggests that the operating farm released freshwater runoff in a more episodic manner, as hypothesized. There was no significant difference in the coefficients of variation, however, between Station 7 and other stations during either period (Table 4). This result indicates that clearing of the watershed did not increase the degree of variation of freshwater runoff.

Mean surface and bottom salinities were significantly lower at Station 7 than at Station 1 during Period 1 (Table 3). This result is probably not related to any man-made watershed change since small lateral estuaries typically show an upstream decrease in salinity. During Period 2, however, the salinity difference between Stations 1 and 7 was not significant (Table 3). One possible explanation

Table 4. Statistical comparisons (modified t-test) of coefficients of variation (V's in Table 1) for water quality parameters between stations. None of the comparisons are significantly different at $p < 0.05$.

	Station 1 vs. Station 7		Station 3 vs. Station 7	
	Period 1	Period 2	Period 1	Period 2
Turbidity				
Surface	1.31	1.59	0.85	1.75
Bottom	0.85	0.88	0.18	1.96
Salinity				
Surface	1.60	1.52	0.33	1.11
Bottom	0.19	1.03	0.06	0.51
Phosphate				
Surface	0.67	1.04	0.73	1.06
Bottom	0.71	0.75	0.11	0.32
Nitrate				
Surface	1.53	0.12	1.23	0.19
Bottom	1.23	0.06	0.79	0.12
Chlorophyll *a*				
Surface	0.29	1.28	0.12	0.75
Bottom	0.36	1.32	1.32	1.37

for this change is that during Period 1 (before farming) the watershed around Station 7 retained runoff more effectively than during Period 2. The slower, more continuous inflow of freshwater at Station 7 would result in a continuously lower salinity which was resolved statistically at our sampling frequency. In Period 2, farming may have caused shorter and faster pulses of freshwater runoff, more likely to be missed at our sampling frequency. Station 7 did not show a significant difference in mean salinity relative to Station 3 in either surface or bottom waters. This surprising result indicates that the direct drainage to Station 7 had no detectable local effect in terms of the mean annual salinity.

Nutrients

With regard to nutrients we hypothesized that fertilizer use during Period 2 would alter the nutrient concentration in the receiving waters of the estuary. Nitrate and phosphate increased significantly in surface waters in Period 2 at all three stations (Table 1); in bottom waters there was a significant increase at Station 7 and Station 1 (Table 2). The effect of direct runoff from the farm can be seen by the enhanced nitrate concentrations at Station 7, which were about 3 times the mean concentrations at Station 1. In sum, the mean nitrate and phosphate concentrations showed the same patterns during the two periods: they were low and relatively constant throughout the estuary during Period 1 and increased significantly during Period 2 with the largest increase at Station 7, nearest the drainage outfall and least at Station 3, the station least affected by farm runoff.

The spatial comparison of mean nutrient concentrations (Table 3) further reinforces the interpretation that direct farm runoff was responsible for increasing the mean concentration of surface and bottom nutrients. Additional information is provided, however, by the spatial comparison. In the temporal comparison (Tables 1, 2), significant increases were observed in both nitrate and phosphate in Period 2. Comparison of Station 7 with the other stations (Table 3) shows that the increase is considerably stronger and more significant for nitrate than for phosphate. These spatial differences were most pronounced between Stations 3 and 7 (Table 3). This result may reflect the naturally enhanced phosphate concentrations that characterize shallow coastal plain estuaries in the southeastern U.S. (Pomeroy *et al.* 1967). An increase in runoff phosphate is less likely to result in a significant increase because the mean concentration without runoff is maintained at relatively high stable concentrations due to rapid recycling from sediments (Pomeroy *et al.* 1967).

Despite the highly significant increases in mean concentrations from Period 1 to Period 2 and between stations, there was no significant change in the coefficients of variation either in time at each station (Tables 1, 2) or in space between stations (Table 4). This result is parallel to that observed in turbidity and emphasizes that while agricultural runoff increased the mean nutrient loading it did not significantly alter the processes that introduce variability.

Chlorophyll

The mean chlorophyll concentration increased somewhat at all stations in surface and bottom samples between Periods 1 and 2, but these increases were not statistically significant except at Station 3 in the bottom waters (Tables 1, 2). In view of the significant and highly significant increases in nitrate and phosphate it is surprising that there were not larger increases in chlorophyll. No significant differences between stations were seen in chlorophyll-*a* (Table 3). As with both turbidity and nutrients there was no significant change in the coefficient of variation either between periods (Tables 1, 2) or between stations (Table 4).

Discussion

This analysis of a small coastal plain estuary provided two clear insights into the consequences of watershed development. The first insight deals with the inherent variability of the estuary and the processes responsible for maintaining that variability. Variability of the measured parameters, as indicated by the coefficients of variation, was high; and except for salinity, modification of the watershed did not significantly alter that variability. The second insight was that watershed development and intensive farming did increase the mean turbidity and nutrient levels, but had no significant effect on the biomass of phytoplankton.

Returning to the first point, this analysis determined that variability of all five water quality parameters was high in the South River even during Period 1

(Table 1). Coefficients of variation were very similar at all three stations for five surface water quality measurements, ranging from about 30 to 210 during Period 1 and from 45 to 140 during Period 2 (Table 1). Modification of the watershed therefore did not substantially change the variability of most water quality parameters in South River. The one exception to this observation is salinity at Stations 3 and 7 where there was a significant increase in variability between Periods 1 and 2. This increase in the variability of salinity is interpreted to have resulted from the faster runoff of freshwater from the cleared and drained land. We caution, however, that the increased variability in Period 2 also could have been related to increased variability in the pattern of rainfall (Kirby-Smith and Barber 1979). Konyha *et al.* (1985) found that the average annual freshwater outflow rates for agricultural land use was about 10% higher than average annual rates for areas with native vegetation in coastal North Carolina. They also found, however, that between-year variations in outflow for both land uses are much larger than the differences caused by land use. The increased coefficients of variation for salinity at Stations 3 and 7 may not be due entirely to farming, therefore, but could also reflect larger variability in runoff in Period 2. The coefficient of variation for salinity increased at all stations and both depths in the South River during Period 2 (Tables 1, 2), the only instance in which all stations showed in increase (Kirby-Smith and Barber 1979).

The absence of a change in the coefficient of variation for all the properties except salinity (turbidity, phosphate, nitrate, chlorophyll) was an unexpected result, especially in view of the significant and highly significant changes in mean concentrations of some of these properties. We interpret the absence of effect to indicate that watershed development did not significantly alter the manner in which the weather cycle introduces variability into estuarine water quality properties.

Clearing of the watershed and intensive farming did alter the mean concentrations of turbidity and nutrients in the estuary, but did not alter the mean salinity or chlorophyll. Rainfall during the two periods was essentially identical: 108 inches in Period 1 and 111 inches in Period 2 (Kirby-Smith and Barber, 1979) so the changes in turbidity and nutrients can be related unambiguously to the watershed modifications. The significant increases in turbidity occurred in the portion of the estuary that directly received drainage (Table 2). Station 8, located in the mouth of the major farm drainage ditch less than 2 km upstream from Station 7 (Fig. 2), had mean turbidity values about 50% higher than Station 7 (Kirby-Smith and Barber 1979). At Station 5 (Fig. 2), the readings were similar to those obtained at Stations 1 and 2. Thus, the high turbidity due to direct runoff from the farm decreases rapidly down-river and the greatest effects are seen at the headwaters which are important nursery areas for juvenile fish (Miller 1985). Increases in sediment load in these regions can affect larvae and juvenile fishes both directly via damage to gill epithelium and indirectly via burial and alteration of benthic food organisms (Sherk 1971).

Mean phosphate and nitrate concentrations, both surface and bottom, increased significantly throughout the South River during Period 2 when fertilizers were in use on the farm. During Period 1, mean nitrate and phosphate concentrations were low and similar from Stations 1-7. Mean phosphate concentrations ranged from 0.5-0.7 μg-at l^{-1} and mean nitrate concentrations were from 0.3-0.9 μg-at l^{-1} (Kirby-Smith and Barber 1979). The greatest increases occurred at Station 7 where phosphate increased by a factor of 3 and nitrate increased by an order of magnitude (Table 1). That these large increases were due to farm runoff is obvious since the concentrations at Station 8, the ditch outfall, were even higher than those at Station 7 and the concentrations at Station 7 were always greater than at any downstream station (Kirby-Smith and Barber 1979). The decrease in concentrations downstream is probably due to a combination of algal uptake, dilution (Kunishi and Glotfelty 1985), and absorption by suspended sediments and their subsequent deposition (Pritchard and Schubel 1981).

Algal uptake of nutrients may not be the most important cause of dissolved nutrient decreases in the South River. There were no significant increases in algal biomass, as measured by chlorophyll *a*, accompanying the significant increases in mean nutrient concentrations (Table 1). The waters of South River, like those from other coastal estuaries which drain pocosins, are highly colored with dissolved organic materials (Ash *et al.* 1983). The secchi depth readings of Stations 1-7 average less than 1 m for both periods (Kirby-Smith and Barber 1979). Light limitation may have kept the phytoplankton from increasing significantly in the upstream portion of the estuary.

Grazing is another factor which may have kept the phytoplankton biomass from increasing significantly between Periods 1 and 2. It has been suggested that benthic filter feeders have the potential to control algal biomass in shallow waters (Cloern 1982; Officer *et al.* 1982). There are populations of *Crassostrea virginica* and *Rangia cuneata* in the South River. Our observations, however, indicate that these molluscs are confined to a narrow band of sandy sediments along the east and west shores of the estuary and that their populations are not extensive. Consequently we feel that it is doubtful that benthic grazers could regulate the phytoplankton biomass in the South River. So if grazing is a factor in controlling algal biomass in the South River, it most likely is due to zooplankton.

The most striking effect of the farm operation is the increase in mean concentrations of nutrients, yet those increases did not result in a commensurate increase in chlorophyll *a* concentration and there was no change in species composition of the phytoplankton community between the two periods (LaPennas 1982). The other striking result was that there was no change in the degree of variability among the water quality parameters except for salinity. Our sampling frequency may not have been adequate to detect the effects of short-term, pulsed events such as rain storms. Also coastal plain estuaries are highly variable (Gilliam *et al.* 1985) and the natural sources of variability may have absorbed and masked any added variability due to farming. The primary nursey function of estuaries can obviously be sustained over a broad spectrum of variability. It remains unclear, however how that nursery function is affected by either the amplitudes or rates of variation.

Acknowledgments

We thank L.A. Barling, W.L. Bretz, L.M. Garrigan and P.J. Whaling for field and laboratory assistance. D.A. Wolfe and other, anonymous reviewers provided criticisms and recommendations which improved the paper. Primary financial support was provided by the Water Resources Research Institute of the University of North Carolina. Additional support came from the N.C. Sea Grant Program, the Department of Natural Resources and Community Development of the State of N.C. and Duke University Marine Laboratory.

References Cited

Ash, A. N., C. B. McDonald, E. S. Kane and C. A. Pories. 1983. Natural and modified pocosins: literature synthesis and management options. FWS/OBS-83/04. U.S. Fish and Wildlife Service, Div. of Biol. Sci., Washington, D.C. 156 pp.

Berryhill, H. L., Jr., V. E. Swanson and A. H. Love. 1972. Organic and trace element content of holocene sediments in two estuarine bays, Pamlico Sound area, North Carolina. U.S. Geol. Surv. Prof. Paper No. 1314-E. Washington, D.C. 32 pp.

Bowden, K. F. 1967. Circulation and diffusion. pp. 15-36. *In:* G. H. Lauff (ed.), *Estuaries.* AAAS Press, Washington, D.C. 757 pp.

Carter, L. J. 1975. Agriculture: a new frontier in coastal North Carolina. *Science* 189:271-275.

Cloern, J. E. 1982. Does the benthos control phytoplankton biomass in south San Francisco Bay? *Mar. Ecol. Prog. Ser.* 9:191-202.

Cowardin, L. M., V. Carter, F. C. Golet and E. T. LaRoe. 1979. Classification of wetlands and deepwater habitats of the United States. FWS/OBS-79/31. U.S. Fish and Wildlife Service, Div. of Biol. Sci., Washington, D.C. 103 pp.

Daniel, C. C. 1981. Hydrology, geology, and soils of pocosins: a comparison of natural and altered systems. pp. 69-108. *In:* C. J. Richardson (ed.), *Pocosin Wetlands: An Integrated Analysis of Coastal Plain Freshwater Bogs in North Carolina.* Hutchinson Press Publ., Stroudsburg, Pennsylvania.

Doucetic, W. H., Jr. and J. A. Phillips. 1978. Overview: Agriculture and forest land drainage in North Carolina's coastal zone. Center for Rural Resource Development Rep. No. 8. North Carolina State University, Raleigh. 54 pp.

Geraghty, J. J., D. W. Miller, F. Van Der Leeden and F.L. Troise. 1973. *Water Atlas of the United States.* Water Information Center, Inc., Water Research Building, Manhasset Isle, Port Washington, New York. 219 pp.

Gilliam, W., J. Miller, L. Pietrafesa and W. Skaggs. 1985. Water management and estuarine nurseries. Sea Grant Working Paper UNC-SG-WP-85-2, North Carolina State University, Raleigh. 84 pp.

Hobbie, J. E. and N. W. Smith. 1975. Nutrients in the Neuse River estuary. Sea Grant Publication UNC-SG-75-21. North Carolina State University, Raleigh. 183 pp.

Kirby-Smith, W. W. and R. T. Barber. 1979. The water quality ramifications in estuaries of converting forest to intensive agriculture. Pub. UNC-WRRI-79-148. Water Resources Research Institute, North Carolina State University, Raleigh. 70 pp.

Konyha, K., R. W. Skaggs and J. W. Gilliam. 1985. Agricultural runoff and water quality in coastal areas, pp. 1-20. *In:* W. Gilliam *et al.* (eds.), *Water Management and Estuarine Nurseries.* Sea Grant Working Paper UNC-SG-WP-85-2, North Carolina State University, Raleigh.

Kunishi, H. M. and D. E. Glotfelty. 1985. Sediment, season, and salinity effects on phosphorus concentrations in an estuary. *J. Environ. Qual.* 14:292-296.

LaPennas, P. P. 1982. Phytoplankton community structure in a variable estuary. Ph.D. Thesis. Duke University, Durham, North Carolina. 174 pp.

Lefler, H. T. and A. R. Newsome. 1954. *North Carolina, the History of a Southern State.* The University of North Carolina Press, Chapel Hill 676 pp.

Miles, M. 1910. *Land Draining.* Organe Judd Co., New York and Chicago. 199 pp.

Miller, J. M. 1985. Effects of freshwater discharges into primary nursery areas for juvenile fish and shellfish: criteria for their protection. pp. 62-84. *In:* W. Gilliam *et al.* (eds.), *Water Management and Estuarine Nurseries.* Sea Grant Working Paper UNC-SG-WP-85-2, North Carolina State University, Raleigh.

Officer, C. B., T. J. Smayda and R. Mann. 1982. Benthic filter feeding: a natural eutrophication control. *Mar. Ecol. Prog. Ser.* 9:203-210.

Pietrafesa, L. J. 1985. Response of Rose Bay to freshwater inputs. pp. 21-61. *In:* W. Gilliam *et al.* (eds.), *Water Management and Estuarine Nurseries.* Sea Grant Working Paper UNC-SG-WP-85-2, North Carolina State University, Raleigh.

Pomeroy, L. R., L. R. Shenton, R. D. H. Jones and R. J. Reimold. 1972. Nutrient flux in estuaries. *Limnol. Oceanogr. Special Symposia* 1:274-281.

Pritchard, D. W. and J. R. Schubel. 1981. Physical and geological processes controlling nutrient levels in estuaries. pp. 25-34. *In:* B. J. Nielsen and L. E. Cronin (eds.), *Estuaries and Nutrients.* Humana Press, Clifton, New Jersey. 643 pp.

Shaw, S. P. and C. G. Fredine. 1956. Wetlands of the United States. Fish and Wildlife Circular 39. U.S. Dept. of Interior, Washington, D.C. 67 pp.

Sherk, J. A., Jr. 1971. Effects of suspended and deposited sediments on estuarine organisms. Contribution No. 443, Chesapeake Biol. Lab, Solomons, Maryland. 73 pp.

Sokal, R. R. and C. A. Braumann. 1980. Significance tests for coefficients of variation and variability profiles. *Systematic Zool.* 29:50-66.

Sokal, R. R. and F. J. Rohlf. 1981. *Biometry.* W. H. Freeman and Co., New York. 859 pp.

Strickland, J. D. H. and T. R. Parsons. 1968. A practical handbook of sea water analysis. *Fish. Res. Bd. Canada Bull.* 167. 311 pp.

DIURNAL VARIATION OF SURFACE PHYTOPLANKTON IN THE PATUXENT RIVER

Jerome Williams and John Foerster
Oceanography Department
U.S. Naval Academy
Annapolis, Maryland
and

Fred Skove
Computer Science Department
U.S. Naval Academy
Annapolis, Maryland

Abstract: Data were taken at two-hour intervals over a 24-hour period at a station near the mouth of the Patuxent River, a sub-estuary of the Chesapeake Bay. This station was occupied on July 13-14, 1983 and the sequence was repeated on June 25-26, 1984 with similar results. Measurements of surface beam transmittance indicated a diurnal variation with a minimum occuring at about 1800 and a maximum after midnight. The late afternoon transmittance was about half that during nighttime. Optical measurements were correlated with concentration of suspended particles, and changes in the size distribution of the suspended material were associated with changes in transparency. Through the use of optical measurements, particle counts, chlorophyll *a* measurements, and phytoplankton analyses, we determined that phytoplankton populations are probably changing by at least an order of magnitide over each diurnal cycle. Examination of tidal movement and the entire water column led us to conclude that these daily variations were due to grazing, and were not associated with horizontal advection or sinking. A mathematical model using a bimodal grazing pattern coupled with an exponential solar-controlled growth rate fitted the data reasonably well.

Introduction

The diurnal variations in feeding habits of zooplankton have been documented both in laboratory studies (Duval and Geen 1976) and in the field (Chisholm *et al.* 1975; Haney and Hall 1975; Mackas and Bohrer 1976; Dagg and Wyman 1983; Simard *et al.* 1985), but very little work has been done to relate these feeding habits to diel variations in phytoplankton populations. In this study we not only document the recurring daily variations in surface phytoplankton, but also develop a simplistic model to explain these variations.

The Patuxent River (Fig. 1a) has had many studies over the past twenty-five years. Of importance to this study were the works of Heinle (1966, 1974) and Herman *et al.* (1968) on the zooplankton composition and grazing. In addition, works by Stross and Stottlemeyer (1965) and Flemer and Olmon (1971) on phytoplankton and primary production were useful. However, much of the work concentrated on areas of the river upstream or downstream of our sampling area.

As part of a continuing effort to monitor and understand changes in the optical characteristics of Chesapeake Bay and its tributaries, a 24-hour station was occupied on July 13-14, 1983 near the mouth of the Patuxent River. Concentrations of suspended particles and chlorophyll *a* were measured. A marked decrease in light attenuation was noted starting in the late afternoon and lasting until the early morning hours. These data suggested the hypothesis that the increase in water clarity was directly related to a decrease in the surface phytoplankton population produced by the grazing of either zooplankton or fish. To substantiate this hypothesis the station was again occupied on June 25-26 1984, and phytoplankton counts were also made.

Methods

Station P-1 (Figure 1A) near the mouth of the Patuxent River (lat. 38° 20′ 08:: N, long. 76° 29′ 05:: W) was occupied for 24-hour periods on 13-14 July, 1983 and 25-26 June, 1984. Measurements were made every two hours at depth intervals of one meter for all parameters except chlorophyll *a* and particle size distribution (surface and bottom) and phytoplankton counts and identification (surface only). Phytoplankton counts and identification were performed only during the 1984 period.

The primary optical measurement made was that of beam transmittance, using a 25-cm pathlength beam transmissometer as described by Jerlov (1976). The 25-cm transmittance was converted to the beam attenuation coefficient (α) by:

$$\alpha = 4 \ln(100/\%T) \text{ m}^{-1}$$

where:
%T is the percent transmittance for a 25 cm path (Williams 1970).

A Coulter counter with a 100 micrometer (μm) aperture was utilized to determine the size distribution of the suspended particulates. Particles were counted within each of 16 channels. Those particles having diameters between about 1.8 μm and 4.2 μm arbitrarily were termed to be suspended sediments. The larger particles ranging in diameter from about 14 to 36 μm were termed phytoplankton. Particle size distributions with a large number of smaller particles and very few particles in the larger size ranges, were common in Chesapeake Bay. However, if these data are converted to cross-sectional area per unit volume of sample in each size channel, larger particles are shown to be present in significant numbers. Even though the number of larger particles is much less than the smaller, their individual cross-sectional areas are so much greater that they are important when optical measurements are made.

Larger particles have an effect on the total cross-sectional (scattering) area. These larger particles are more in evidence at the surface than near the bottom. This substantiates the assumption that the larger particles were primarily phytoplankton, since one would expect them to be in the photic zone of the water column.

Salinity and temperature were measured *in situ* using a CSTD (Inter-Ocean, Inc.). Chlorophyll *a* was measured by the fluorometric technique (Yentsch and Menzel 1963; Strickland and Parsons 1972). Samples were collected from the desired depths and determinations were made within ten minutes of collection. Phytoplankton were collected on a 47-mm diameter membrane filter of 0.45 μm porosity, preserved with 10% formalin solution, and enumerated according to McNabb (1960). All times are reported as Eastern Standard Time (EST).

Results

The beam attenuation coefficient, alpha (Fig. 1A), the algal cross-sectional scattering area (Fig. 1B) and the chlorophyll *a* concentration (Fig. 1C) all declined precipitously in the late afternoon in both years. The algal scattering area plotted (Fig. 1B) is the total particulate cross-sectional area of the particles in channels 11-15. The decline occurred at about the same time of day for both years and for all three measures.

In 1984 plankton samples were collected and phytoplankton counts were made coincidentally with other measurements. Three major species were found, representing two diatoms and one dinoflagellate. The diatoms in greatest abundance were *Cyclotella kuetzingiana* (about 10-20 μm in diameter) and *Melosira varians* (ranging from 10-35 μm in diameter), with *Cyclotella* being by far the more numerous. *Prorocentrum scutellum* (between 40 and 50 μm in diameter) completely overwhelmed all other dinoflagellates numerically. The number of diatoms from 10-20 μm in size varied much less over the sampling period than did the dinoflagellates between 40 and 60 μm in size (Fig. 1D). The dinoflagellate population starts falling in the late afternoon and then builds again the following morning.

Discussion

It is apparent that a major constituent of the surface phytoplankton is decimated in the late afternoon. The question arises as to where the plankton go. Possible explanations include vertical advection, sinking migration of the phytoplankton, horizontal advection, and grazing by either zooplankton or fish.

The first of these hypotheses can be examined by determining water density as a function of depth using salinity and temperature data acquired at the same time as the other data. At both 4 o'clock in the afternoon (1600) and 2 o'clock in the morning (0200), (times of maximum and minimum turbidity on July 13-14, 1983), the water column was very stable with sigma-t increasing with depth (Fig. 2). It does not appear that the water is undergoing any natural turnover.

At this same time of maximum surface turbidity (1600 on July 13) there was a very marked decrease in turbidity with depth (Fig. 3). At this time the suspended materials in the water column were greatest in the surface layers. At 0200 the next morning, however, this was reversed with the surface layers having a somewhat smaller amount of light-decreasing material. Data taken at other

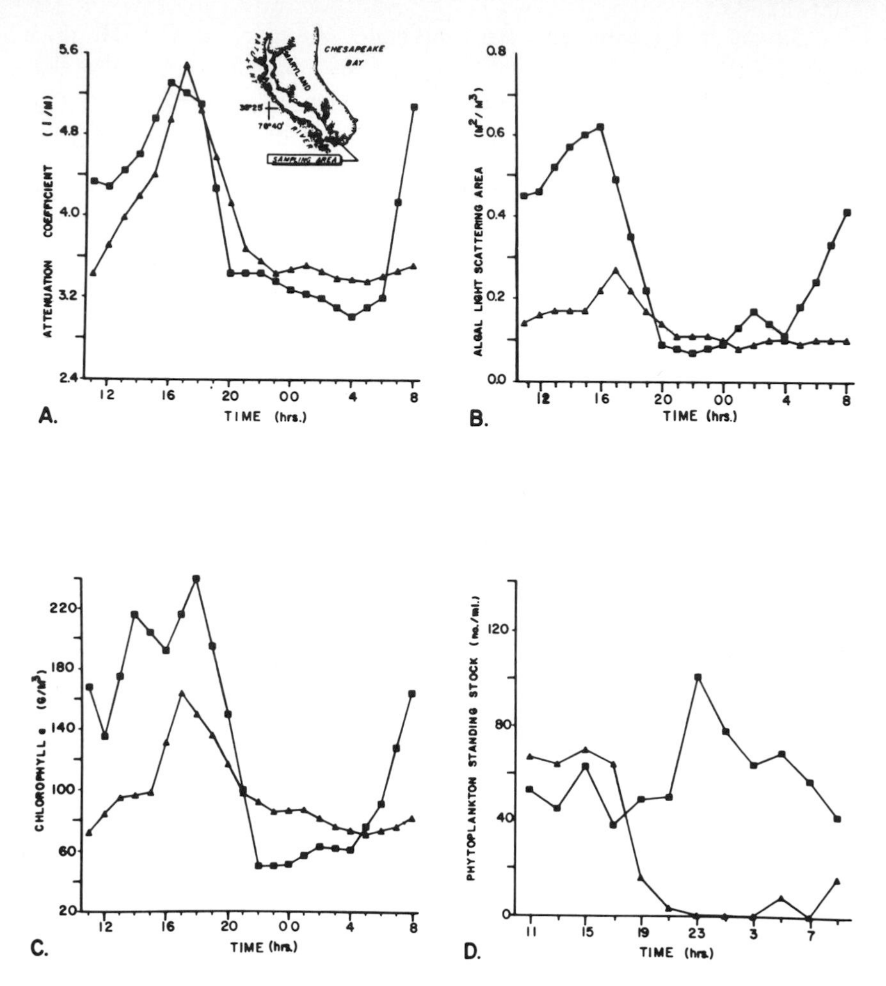

Figure 1. *Hourly variations of four parameters in the Patuxent River estuary during July, 1983 (squares) and June, 1984 (triangles) sampling periods. The beam attenuation coefficient (1A), suspended particle cross-sectional area (1B), chlorophyll* a *(1C), and dinoflagellate standing crop (1D), all show a marked decline at the same time of day. In Figure 1D the squares represent diatom (size between 10 and 20 micrometers) counts while the triangles represent dinoflagellate (size between 40 and 60 micrometers) counts taken in June, 1984.*

times all fall within the envelope represented by the two curves shown in Fig. 3. Apparently, during this diurnal cycle, the surface layer is the one most affected. During this period, however, the material in the surface layers does not fall to a lower layer because the transparency of the lower layers remains about the same over the entire diurnal cycle, while the larger concentration of material in the

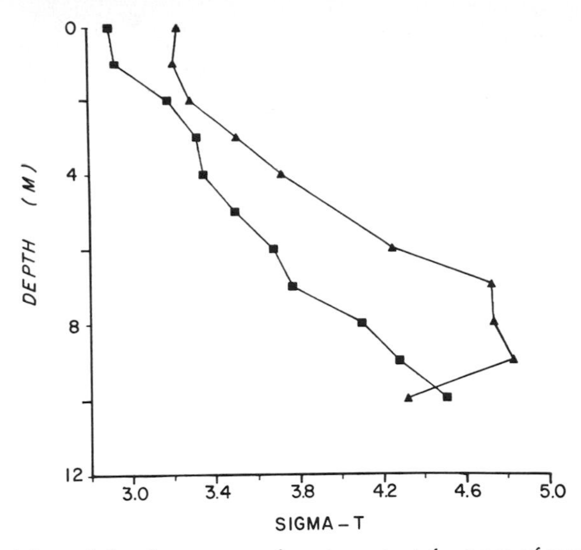

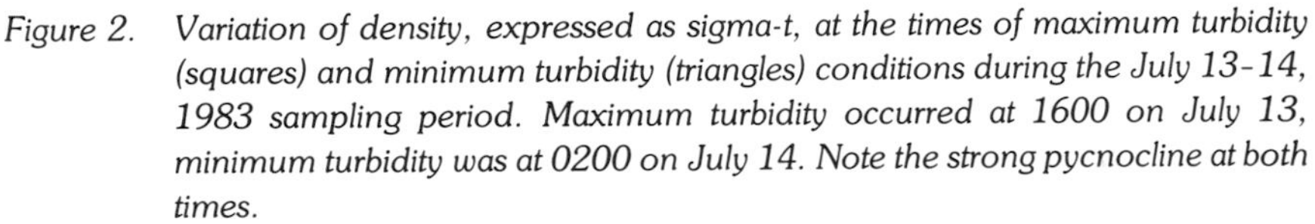
Figure 2. *Variation of density, expressed as sigma-t, at the times of maximum turbidity (squares) and minimum turbidity (triangles) conditions during the July 13-14, 1983 sampling period. Maximum turbidity occurred at 1600 on July 13, minimum turbidity was at 0200 on July 14. Note the strong pycnocline at both times.*

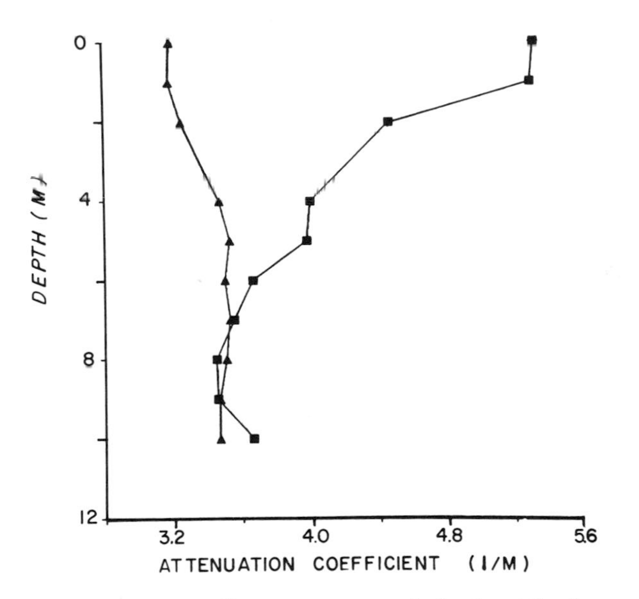

Figure 3. *Beam attenuation coefficient variation with depth, at the times of maximum turbidity (squares) and minimum turbidity (triangles) on July 13-14, 1983. Beam attenuation profiles taken at other times during the sampling period fall between the two extremes illustrated.*

upper layers simply appears to disappear. The phytoplankton apparently are not sinking to a different position in the water column.

The variation in particle size distribution supports the same conclusion. A plot of the surface suspended particulate volume as a function of particle size for the time of maximum turbidity (1600) shows a marked peak at about 35 μm with a somewhat smaller bulge in the vicinity of 10–15 μm (Fig. 4A). These two volume maxima could represent dinoflagellate and diatom populations similar to those found in 1984 (Fig. 1D). The surface suspended particulate volume distribution at the minimum turbidity time (0200) is also shown in Figure 4A. The difference between the two curves in Fig. 4A represents the particle size distribution of the particulates lost between 1600, July 13 and 0200, July 14, 1983. The selective nature of this loss is readily apparent (Fig. 4B) with most of the loss being produced by larger particles.

If the loss in suspended particulates is indeed the result of a decrease in local surface phytoplankton populations, horizontal advection is suspected immediately. But this cause was as unlikely as vertical advection. During the 1983 sampling period the tide began to ebb in the late afternoon as the turbidity started its precipitous change. In 1984, the tide had begun to flood at the outset of the afternoon turbidity change. The two sampling periods had tides almost 180° displaced from each other. Maximum ebb current occurred at 1937 on the afternoon of 13 July, 1983, while maximum flood current occurred at 1954 on the afternoon of 25 June, 1984. One would expect that plankton free water would be arriving from either up- or down-river, but not both.

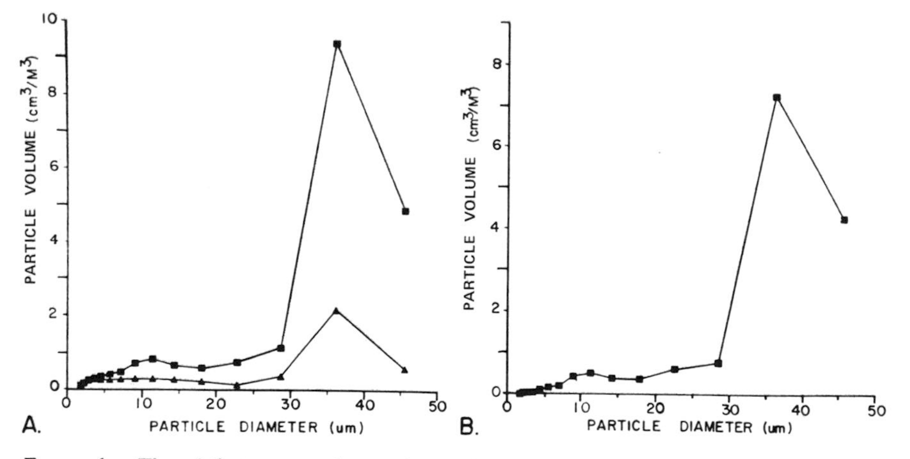

Figure 4. The shift in particulate volume distribution in the Patuxent River estuary. A. Particulate volume distribution of surface samples at maximum turbidity on July 13 (squares) and minimum turbidity on July 14 (triangles). B. Difference between the curves in A illustrating the volume distribution of the suspended particulates lost from the surface waters between 1600, July 13 and 0200, July 14, 1983. Note that most of the lost material falls within the size range of typical dinoflagellates.

Although we have no direct measurements to support the grazing hypothesis, the indirect evidence is very strong. Wilson (1973) has shown that mature and immature stages of *Acartia tonsa,* a common estuarine copepod, ingests only particles larger than 13.8 μm. This value was well in the range of the phytoplankton sampled. The fact that the surface layer essentially is cleared of dinoflagellate-sized particles in late afternoon, and they do not appear in any other portion of the water column makes it difficult to support any hypothesis other than grazing.

A rather simplistic model can be developed to describe the diurnal variation in phytoplankton seen in the surface waters of the Patuxent. We assume that P(t), the phytoplankton population at any time t, is related to P_o, some constant population value; $P_e(t)$, the amount lost by grazing; and $P_g(t)$, the amount gained by growth, by the following:

$$P(t) = P_o + P_g(t) - P_e(t)$$

Since the growth rate is exponential, and photosynthesis cannot proceed without sunlight, $P_g(t)$ can be expressed as:

$$P_g(t) = \int_0^t AS \exp(Bt)\, dt$$

where: S is the relative solar irradiance given by:

$$S = C \sin [D\,(t\text{-}E)]$$

t is the time in hours (noon is 12, midnight is 0), and
A,B,C,D,E are constants of the system.

For example, D is set to make the sine function vanish at sunrise and sunset, and E is the time of sunrise.

The grazing relationship was generated assuming conditions similar to those reported by Duval and Geen (1976). They reported a bimodal feeding activity with peaks occurring at 0200 and 1800, so that:

$$P_e(t) = \int_0^t R_e dt$$

where:

$R_e = F + G \sin [H\,(t - J)]$ from 2200-0600,

$R_e = R + L \sin [H\,(t - J)]$ from 1400-2200,

$R_e = (F - G)$ or $(K - L)$ from 0600-1400, and
F,G,H,J,K,L are constants of the system.

By changing the system constants, the model can be adjusted for differing growth rates, grazing rates, day lengths, feeding patterns, etc. Three variations of the model are illustrated in Fig. 5. In Fig. 5A, the grazing is assumed to be equal during both feeding periods, while in Fig. 5B heavier feeding occurs during the early morning period. Figure 5C illustrates the effect of heavier late after-

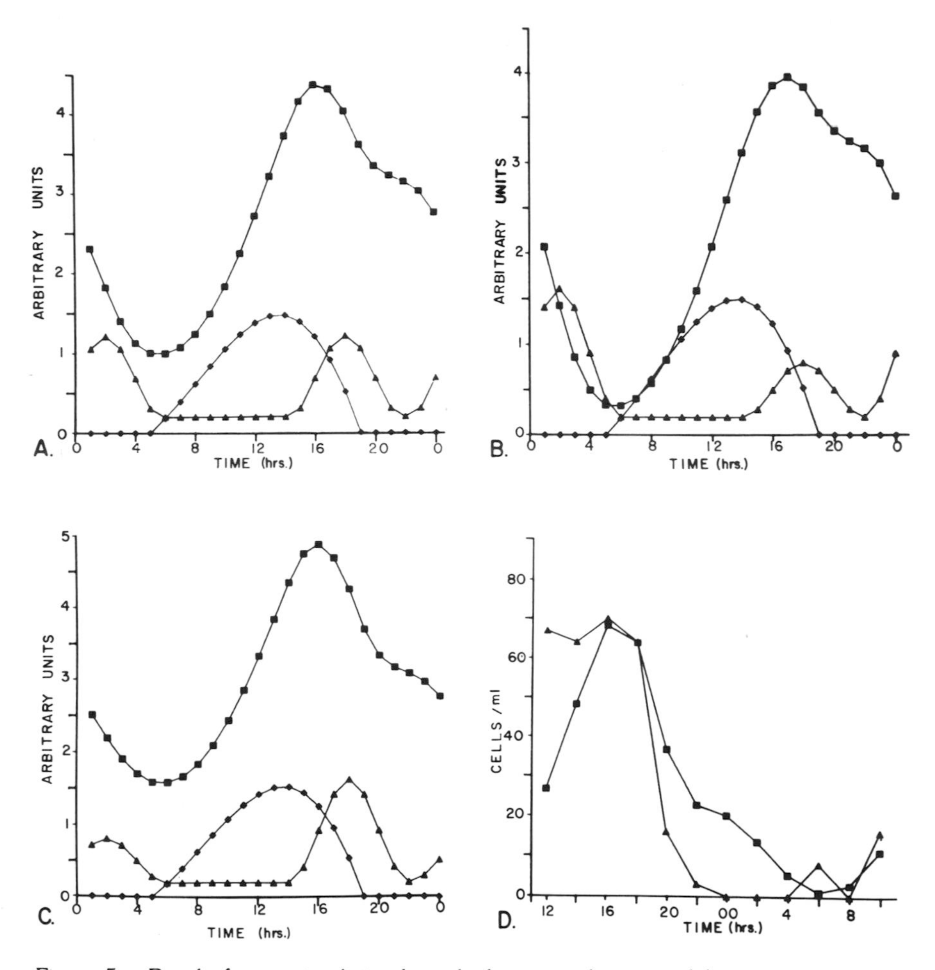

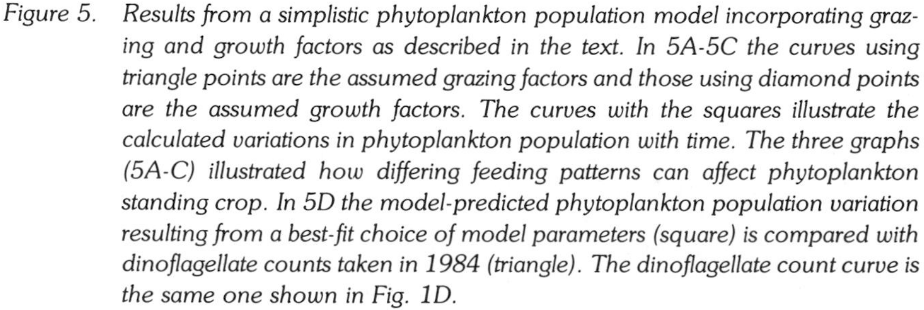

Figure 5. *Results from a simplistic phytoplankton population model incorporating grazing and growth factors as described in the text. In 5A-5C the curves using triangle points are the assumed grazing factors and those using diamond points are the assumed growth factors. The curves with the squares illustrate the calculated variations in phytoplankton population with time. The three graphs (5A-C) illustrated how differing feeding patterns can affect phytoplankton standing crop. In 5D the model-predicted phytoplankton population variation resulting from a best-fit choice of model parameters (square) is compared with dinoflagellate counts taken in 1984 (triangle). The dinoflagellate count curve is the same one shown in Fig. 1D.*

noon feeding. All three of these figures included the specific grazing and growth components of each model along with the associated variation in phytoplankton population resulting from integration over a 24 h period. The constants used in each of the three models are listed in Table 1. Note that the numbers used for the model constants are arbitary and do not represent actual growth or grazing rates. Estimates of the rates necessary to support a grazing hypothesis can be obtained from the 1984 data where the maximum number of dinoflagellates encountered was 65 per cubic meter. Reproduction rates must be large enough to produce this number in about 12 hours, and grazing rates must be large enough to devour this number in about 6 hours.

Table 1. The constants used in the three variants of the model (Figures 5A-5C).

Constant	*Fig. 5A*	*Fig. 5B*	*Fig. 5C*
A	0.5	0.5	0.5
B	0.08440	0.08622	0.08622
C	1	1	1
D	0.2243	0.2243	0.2243
E	5	5	5
F	0.707	0.505	0.909
G	0.505	0.303	0.707
H	0.7854	0.7854	0.7854
J	16	16	16
K	0.707	0.909	0.505
L	0.505	0.707	0.303
P_o	3	3	3

By changing the constants of the model it is possible to account for the effects of a number of different natural parameters. For example, the feeding habits of different species may be easily included. Doubling the magnitude of the late afternoon feeding peak in the model illustrated in Fig. 5C, gives phytoplankton population estimates reasonably consistent with the 1984 plankton data (Fig. 5D).

Summary

The case presented above seems to support our original grazing hypothesis for explaining the decrease in turbidity associated with phytoplankton. Coincidental decreases in chlorophyll *a*, large particles, and dinoflagellates were documented. These occured in a stable water column that had no diurnal change in turbidity below the surface layers. The organisms were disappearing from the surface region and not reappearing anywhere else in the water column and this change occurs during both ebb and flood stages of the tide. The fact that this diurnal variation in abundance can be explained with a simplistic grazing model lends further credence to the grazing hypothesis.

Acknowledgments

This work was carried on with the support of the U.S. Defense Mapping Agency and the U.S. Naval Explosive Ordnance Disposal Technology Center. Thanks are also due to Ralph Wicklund who not only developed a large portion of the data acquisition network utilized but also helped obtain most of our data, and Karen Vorous who performed a large portion of the laboratory analyses.

References Cited

Chisholm, S.W., R.G. Stross and P.A. Nobbs 1975. Environmental and intrinsic control of filtering and feeding rates in arctic *Daphnia*. *J. Fish. Res. Bd. Canada* 32:319-226.

Dagg, M.J. and K.D. Wyman. 1983. Natural ingestion rates of the copepods *Neocalanus plumpchrus* and *N. Cristalus* calculated from gut contents. *Mar Ecol. Prog. Ser.* 13:37-46.

Duval, W.S. and G.H. Geen. 1976. Diel feeding and respiration rhythms in zooplankton. *Limnol. Oceanogr.* 21:823-829.

Flemer, D.A. and J. Olmon. 1971. Daylight incubator estimates of primary production in the mouth of the Patuxent River, Maryland. *Chesapeake Sci.* 12:105-110.

Haney, J.F. and D.J. Hall. 1975. Diel vertical migration and filter-feeding activities of *Daphnia*. *Arch. Hdrobiol.* 75:413-441.

Heinle, D.R. 1966. Production of a calanoid copepod, *Acartia tonsa*, in the Patuxent River Estuary. *Chesapeake Sci.* 7:59-74.

Heinle, D.R. 1974. An alternate grazing hypothesis for the Patuxent Estuary. *Chesapeake Sci.* 15:146-150.

Herman, S.S., J.A. Mihursky and A.J. McErlean. 1968. Zooplankton and environmental characteristics of the Patuxent River Estuary, 1963-1965. *Chesapeake Sci.* 9:67-82.

Jerlov, N.G. 1976. *Marine Optics*. Elsevier, Amsterdam. 232 pp.

MacKas, D. and R. Bohrer. 1976. Fluorescence analysis of zooplankton gut contents and an investigation of diel feeding patterns. *J. Exp. Mar. Biol. Ecol.* 25:77-85.

McNabb, C. 1960. Enumeration of freshwater phytoplankton concentrated on the membrane filter. *Limnol. Oceanogr.* 5:57-61.

Simard, Y., G. Lacroix and L. Legendre. 1985. *In situ* twilight grazing rhythm during diel vertical migrations of a scattering layer of *Calanus finmarchicus*. *Limnol. Oceanogr.* 30:598-606.

Strickland, J.D., and T.R. Parsons. 1972. A practical handbook of seawater analysis, 2nd ed. *Bull. Fish. Res. Bd. Canada* 167:1-311.

Stross, R.G. and J.R. Stottlemeyer. 1965. Primary Production in the Patuxent River. *Chesapeake Sci.* 6:125-140.

Williams, J. 1970. *Optical Properties of the Sea*. U.S. Naval Institute, Annapolis. 123 pp.

Wilson, D.S. 1973. Food size selection among Copepods. *Ecology* 54:907-914.

Yentsch, C.J., and D.W. Menzel. 1963. A method for the determination of phytoplankton chlorophyll and phaeophytin by fluorescence. *Deep Sea Res.* 10:221-231.

SURFACE SEDIMENT STABILIZATION-DESTABILIZATION AND SUSPENDED SEDIMENT CYCLES ON AN INTERTIDAL MUDFLAT

John N. Kraeuter

Crane Aquaculture Facility
Baltimore Gas and Electric
Baltimore, Maryland

and

Richard L. Wetzel

Virginia Institute of Marine Science
College of William and Mary
Gloucester Point, Virginia

Abstract: Sediment trap data were collected for 3.5 years on an intertidal mudflat in Gates Bay on Virginia's Eastern Shore. Order-of-magnitude seasonal cycles of suspended particulate material were found for all three years. The cycles were positively correlated with temperature and negatively correlated with wind speed, sediment shear strength, percent organic matter in the sediment collector and percent sediment surface organics. Sediment water content and vane shear test data indicate that sediment stabilization-destablization occurs on both tidal-cycle and seasonal time scales. These stabilization and destabilization processes allow significant quanities of pore water to exchange with the overlying water. Total quantities of suspended materials collected by sediment traps were similar to those found in other systems, but the strong seasonal pattern in this shallow bay appears to be unique. We propose that these cycles are controlled principally by biological and secondarily by physical-chemical processes.

Introduction

Sources and fates of suspended sediments have been given a great deal of attention in estuarine systems. Sources include rivers and offshore waters (Meade 1972), shoreline erosion (Schubel and Biggs 1969; Schubel 1972; Frostick and McCave 1979) and resuspension of materials from subtidal and intertidal sediments (Oviatt and Nixon 1975; Welsh 1980). The factors controlling sediment dynamics are of interest to geologists in terms of sedimentary budgets and to ecologists because they influence various nutrient cycling processes and the structural and functional characteristics of benthic communities.

Sediment surface stability of intertidal flats is influenced by a complex of factors that include: sediment grain size, porosity and packing (Krumbein and Pettijohn 1938; Sanders 1958; Webb 1959), the presence of certain macroheterotrophs (Rhoads and Young 1970, 1971; Young and Rhoads 1971; Aller and Dodge 1974; Woodin 1976; Yingst and Rhoads 1978), the presence or absence of algal mats, diatom or bacterial films (Webb 1959; Neuman *et al.*

1970; Holland *et al.* 1974; Frostick and McCave 1979; Welsh 1980) and physical factors such as temperature (Anderson 1979), waves (Anderson 1972, 1976), ice (Sasseville and Anderson 1976) and rain (Anderson 1983).

Any factor influencing sediment surface stability affects sediment erodability and thus may cause significant changes in the quantity of suspended matter. The alternating tidal resuspension and deposition of sediments has been shown to be important in establishment of marine benthic communities (Rhoads 1973, 1974), in the exchange of nutrients between sediment and the water column, and in the movement of chlorophyll to and from the benthos (Tenore 1977; Roman 1978; Roman and Tenore 1978; Baillie and Welsh 1980). Fecal pellets of zooplankton and benthic invertebrates are very important in these resuspension/deposition cycles (Haven and Morales-Alamo 1968; Roman 1978; Taguchi and Hargrave 1978; Kraeuter 1976) and may be responsible in part for the observed seasonal changes in sediment stability. Because of the observed sediment pelletization and mixing by benthic organisms, Rhoads *et al.* (1978) proposed that seasonal cycles of resuspension should be present in subtidal areas of Buzzards Bay, but they were unable to demonstrate a predictable cycle. Similar studies by Oviatt and Nixon (1975) did not detect seasonal cycles in the resuspension of subtidal Narragansett Bay sediments.

The principal source for recycled forms of nutrients in shoal benthic habitats has long been postulated to be sediments (Nixon *et al.* 1975; Billen 1978; Nixon 1979; Nixon 1980; Welsh 1980). Nixon *et al.* (1975) and Boynton *et al.* (1980) also proposed that the anomalously low water N/P ratios resulted from denitrification. Analysis of sediment profiles did not indicate selective burial of N, and incomplete degradation of sediment organics could not account for the significant departure from the expected "Redfield" N/P ratios. Seitzinger *et al.* (1980) showed that denitrification was significant in some marine sediments and represented a loss of fixed nitrogen available for recycling.

The mechanisms governing exchange across the sediment water interface are, in part, controlled by concentration gradient, water exchange and turbulent mixing (Aller 1982). Biological activity establishes and maintains the concentration gradient for most nutrient species while tidal currents and climatic conditions drive water exchange and turbulent mixing processes. All are influenced to varying degrees by the stability of the sediment surface. Boynton *et al.* (1980) measured the highest reported subtidal benthic ammonium fluxes in the turbidity maximum zone of the Patuxent River, Maryland. Welsh (1980) suggested that exchange of high interstitial nutrient concentrations in sediments during flood tide resulted in increased water column nutrient concentrations over intertidal flats.

The majority of evidence suggests that interactions between biological processes and physical events control the movement or exchange of dissolved and particulate materials at the sediment-water interface. The present study was undertaken to investigate annual cycles of resuspension over an intertidal mudflat relative to physical and biological processes that might control sediment surface dynamics.

Study Site

The study site is a small circular lagoon known as Gates Bay, midway along the ocean side of the southern portion of the Delmarva Peninsula (Fig. 1), which lies between the entrances to Chesapeake and the Delaware Bays. In this region, barrier islands separate a series of marsh-lagoon complexes from the Atlantic Ocean. These complexes consist of shallow circular or oval bays resembling drowned Carolina Bays. Extensive mudflats are exposed at low tide and are surrounded by salt marshes dominated by *Spartina alterniflora*. Freshwater drainage into the system is minimal and salinities range from 28-32 g l^{-1} throughout most of the region. Near the study site at Wachapreague, Virginia the tides are semidiurnal with a mean range of 1.2 m and a spring range of 1.4 m.

Approximately one km in diameter, Gates Bay has a single outlet for tidal exchange, and at mean low water is about 70% exposed mudflat. It is separated from other bays by surrounding *Spartina alterniflora* salt marsh and is isolated from mainland drainage by Wachapreague Channel. The intertidal flat of the bay is predominantly mud, but extensive oyster reefs (*Crassostrea virginica*) occur near the mouth and small clumps of oysters are scattered throughout the bay.

Materials and Methods

Sampling stations were placed near the shore opposite the inlet and near the center of Gates Bay (Fig. 1). Similar seasonal patterns and magnitudes of sediment resuspension were seen at both stations, and after the first year, most of the studies were conducted at the near shore site. Periodic samples from the center of the bay provided a check on spatial variability throughout the study period. Utilizing the near shore station alleviated much of the disturbance caused by walking on the soft sediments to reach the center of the bay.

Suspended Sediment

Collectors were built to gather three replicate samples of suspended sediment at each site over each collection interval (Fig. 2). When installed, the 10-cm diameter funnel tops were 25 cm above the substrate.

Samples were collected over a three-year period at monthly or greater intervals in winter months and at approximately two-week intervals during times of active deposition. The collector frame and top were cleared of fouling as required. Samples were taken to the laboratory and after the sediment had settled, water was carefully decanted; and any large organisms such as fish, *Fundulus* spp., shrimp, *Palaemonetes* spp. or snails, *Ilyanassa obsoleta,* were removed from the jars. The remaining water was decanted and the sediments were flooded with distilled water to extract salts. The jars were then refrigerated until the sediments had settled. Most of the water was decanted and the sediments were then dried at 50-60°C for 1-3 weeks prior to being weighed on an analytical balance to 0.0001 g. Samples were analyzed for carbon content during the first year with a Leco carbon analyzer. Ash-free-dry-weight (AFDW) (combustion at 500°C for 4-6 h) was used to measure organic matter in the second- and third-year samples.

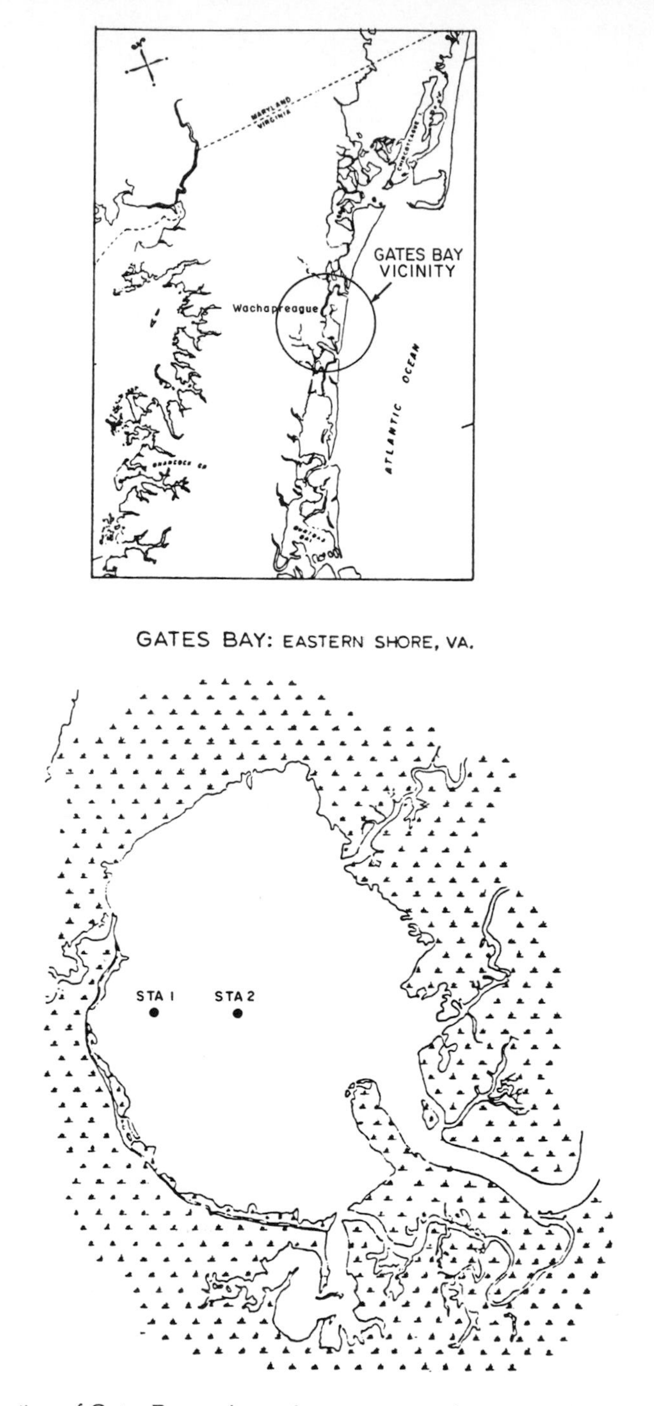

Figure 1. Location of Gates Bay and sampling stations, in the vicinity of Lower Chesapeake Bay, Virginia, USA.

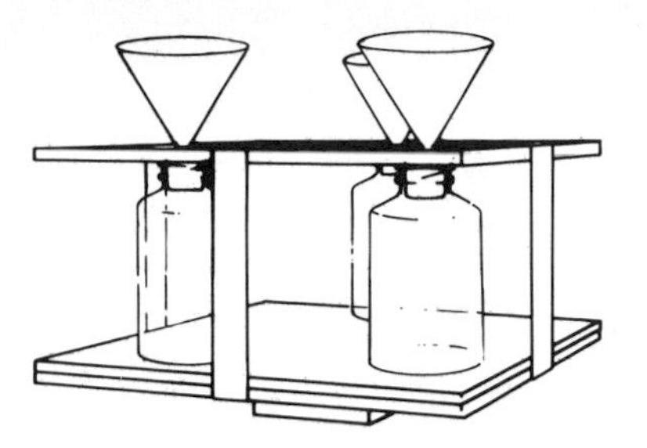

Figure 2. Diagram of sediment collector. The tops of the funnels are 10-cm apart. Sediment surface is below the bottom plate of the collector.

Sediment Water Content and Organic Matter

Cores were taken from the near-shore site during 1977 in conjunction with the sampling of the sediment traps. Some samples were collected at the second site, but this was abandoned because there was little difference in water content, and sampling caused significant disruption of the flat. Three replicate 7.6-cm diameter aluminum tubes 20-cm long were forced nearly full length into the sediment when about 1 cm of water was still present on the sediment surface. The tubes were then capped with rubber stoppers, dug from the substrate, stoppered on the bottom and excess surface water was gently decanted. The cores were returned to the laboratory with as little disturbance as possible and immediately frozen. The frozen cores were extracted from the tube and sliced into sections for analysis (0.5-cm intervals in the top 2 cm and 1-cm intervals in the remainder). To minimize compression effects from the wall of the core tube, only the center-most part of each section was used, and no attempt was made to wash interstitial salt from the sediment. Wet weight, dry weight (50-60°C drying for 1 week) and ash-free-dry-weight (AFDW) (ashing at 500°C for 2 h) were determined to the nearest 0.0001 g using an analytical balance. The relative movement of water to and from the sediment was estimated by calculating differences in sediment water content between sampling periods.

On two occasions when our data indicated resuspension was either high (October) or low (February), a second series of cores was taken to determine the flux of water to and from the sediment over the tidal cycle. Triplicate cores were taken at high tide and again at the end of the ebb, and the top two cm were analyzed as above.

Smaller cores were taken with a 20-cc plastic syringe (luer end cut off) to establish the relationship between the volume change in the sediment and the loss of water. The syringe was placed on the sediment surface and the barrel forced down at the same time the plunger was pulled up. This motion was similar to that of a piston corer so that effects of drag by the corer walls were minimized. In the laboratory these core samples were divided into three 5-cc sections (bottom) and two 2-cc sections (top). The top one cc was discarded because of disturbance by the conical tip of the syringe plunger. All sections were dried and weighed as for other sediment samples. Once dry weights had been obtained, each section was placed in a 50-ml graduated cylinder and water was added from a 25-ml automatic buret up to a known volume on the graduated cylinder. The difference between the cylinder volume and the volume dispensed from the buret gave the dry sediment volume.

Shear Strength

A vane shear test was used to provide a relative measure of sediment surface stability/cohesion and to extend correlations between suspended sediment and water content data. A Torrvane CL-600 (Soiltest, Inc. Evanston, Ill.) vane shear tester was modified for use in the unconsolidated fluid sediments of Gates Bay. We constructed a 15-cm diameter plexiglass disc with eight equally-spaced 6.5-mm deep vanes, extending radially 6.5-cm from the edges toward the center. A 2.0-cm space separated the ends of adjacent blades near the disc center. This modification precluded using the normal calibration supplied with the instrument. Our data are thus relative numbers only, but they are internally consistent and can be utilized to show differences between sampling dates. At each site ten readings were taken each sampling period. Vane shear tests were also taken in conjunction with the studies on tidally-induced changes in sediment water content.

Wind Speed

Daily wind speed and direction were monitored by a recording anemometer located approximately 1.8 km from the sampling site; daily speed and direction were calculated. Mean maximum wind velocity was determined for the periods of time the suspended sediment collectors were set out. Analysis of both absolute wind velocity and vector-averaged wind velocity yielded the same seasonal pattern as maximum wind velocity.

Tidal Exchange Studies

Two 24-h exchange studies were conducted at Gates Bay inlet during periods of maximum (September) and minimum (December) sediment deposition-resuspension. Duplicate water samples were pumped from a depth 1.0 m above the bottom in mid-channel at one-hour intervals. Two water samples were filtered onto 47-mm precombusted, preweighed glass-fiber filters (Gelman GF Type C), which were then weighed, dried at 55°C for 48 h, reweighed, combusted at 500°C for 4 to 6 hours and reweighed to determine the wet weight, dry weight and organic content, respectively, of the suspended sediments. Subsamples (60 ml) were taken from each water sample and syringe-filtered in the field (25-mm cleaned, GF Type C glass-fiber filters) into pre-cleaned, 50-ml polycarbonate centrifuge tubes. These were stored on ice until returned to the laboratory where the samples were frozen (−20°C) until analysis of NH_4, NO_2, NO_3 and PO_4 by standard Technicon Auto Analyzer techniques. Dissolved oxygen concentration and temperature were determined at the depth of water sampling using a Clark-type polarographic oxygen sensor (YSI Model 57) equipped with a thermistor temperature sensor.

Results

Suspended Sediments

The quantity of sediments collected by sediment traps varied according to consistent seasonal patterns for all three years (Fig. 3). Lowest dry weights (3.8-5.1 g m^{-2} d^{-1}) were collected in December and January, with an increase to 10-19 g m^{-2} d^{-1} during February to mid April. The quantity then increased rapidly until August-September when levels reached 140-160 ($\bar{x}$ = 154) g m^{-2} d^{-1} (dry wt.). Quantity of trapped sediment was significantly and positively related to water temperature and (at the 90% level) to water content of the surface sediments (Table 1). Total dry weight of sediment collected over each year was remarkably constant at 20.2-25.8 kg m^{-2} (Table 2).

Suspended Organics

The quantity of organic material varied directly with the total quantity of suspended material collected. The concentration of organic material (percent AFDW) in the collected suspended material showed strong seasonal variation that was, in general, inversely related to the total quantity of sediment collected (Fig. 4), and to temperature and water content of the sediment; and directly correlated with wind speed, shear strength and sediment organic material (Table 1). Percent AFDW was highest from November to March and lowest from July to October. Yearly total AFDW organic materials ranged from 1.43 kg m^{-2} in 1977 to 1.38 kg m^{-2} in 1978 (Table 2).

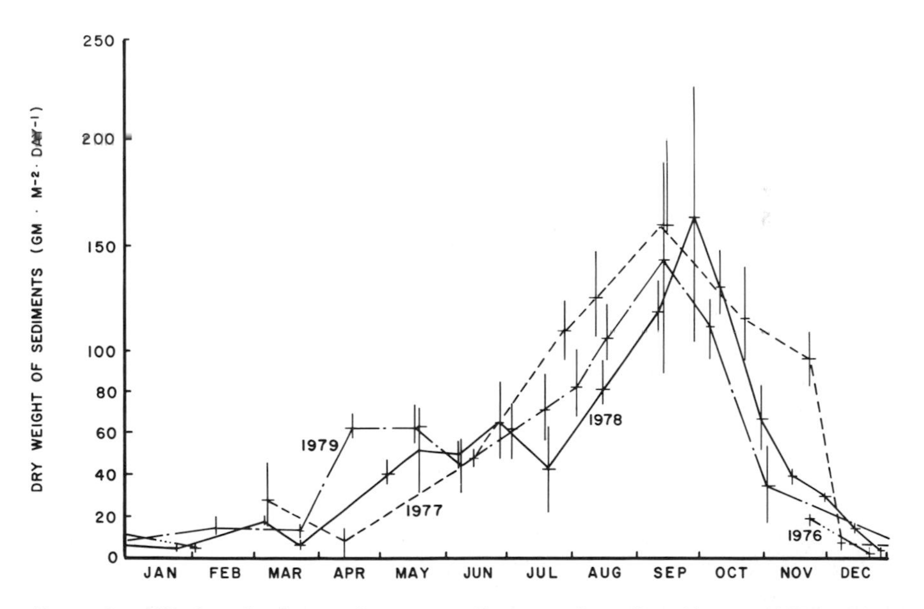

Figure 3. Weights of sediments from trap collectors in Gates Bay, Virginia (1976-1979). Bars represent mean ±1 standard error at t=0.05.

	Suspended Sediment (SS)	Temperature (T)	Wind Speed (W)	Shear Test (ST)	Percent Organics (%0)	Sediment Organics (SO)	Water Content (WC)
SS							
T	0.76 (37)						
W	-0.52 (38)	-0.68 (39)					
ST	-0.45 (37)	-0.54 (36)	0.57 (33)				
%0	-0.58 (42)	-0.68 (41)	0.59 (38)	0.65 (37)			
SO	-0.66 (13)	-0.67 (9)	0.40 (9) NS	-0.41 (5) NS	0.48 (9) NS		
WC	0.59 (9) NS	0.01 (9) NS	-0.67 (9)	0.19 (5) NS	-0.70 (9)	-0.04 (10) NS	

Table 1. Correlations among different parameters related to sediment stability in Gates Bay, Virginia. Suspended sediments (SS) and percent organics (%0) refer to materials collected in sediment traps. Sediment organics (SO) and water content (WC) refer to bottom sediments. Pearson correlation coefficient and number of paired observations ( ) are given for each paired test. NS = not significant. All others significant at the 95% level or better.

Table 2. Yearly total dry weight of suspended sediments and organic material (Ash-Free-Dry-Weight, AFDW) collected by sediment traps above a tidal flat in Gates Bay, Virginia. All values are kg m^2.

Year	1977	1978	1979
Dry weight	25.8	20.2	21.0
AFDW	1.43*	1.38	1.41

*Converted from organic carbon using 2.72 g AFDW = lg organic C.

Figure 4. Mean Ash Free Dry Weight (AFDW) as percent of suspended sediments trapped in Gates Bay, Virginia.

Water Content and Flux

Seasonal trends in surface-sediment (0-0.5 cm) water content were similar to those for the weight of suspended sediments (Fig. 5), but the peak was displaced several months. The amplitude of this seasonal change became less with depth, and below 4 cm the seasonal variation in water content was damped considerably (Fig. 5). Sediment water content was negatively correlated with both wind speed and organic-matter content of suspended sediments. There was no significant correlation with temperature, shear strength, or sediment surface organics, due in part to the limited number of paired samples (Table 1).

Based on the sediment water-content determinations, the seasonal flux of water to and from the sedimentary column was determined (Fig. 6). Seasonal losses of water began in late summer and continued for the majority of the winter. The brief increase in water content above 1.5 cm between December and February was most likely due to ice cover just prior to the February samples. From April through July net water exchange was directed into the sediment, but by late summer the deeper layers appeared to be losing water toward the surface, and this trend continued throughout the fall and into winter.

The cores taken to determine tidal cycle changes in water content indicate dramatic changes (Table 3). Both the water content and the vane shear test data

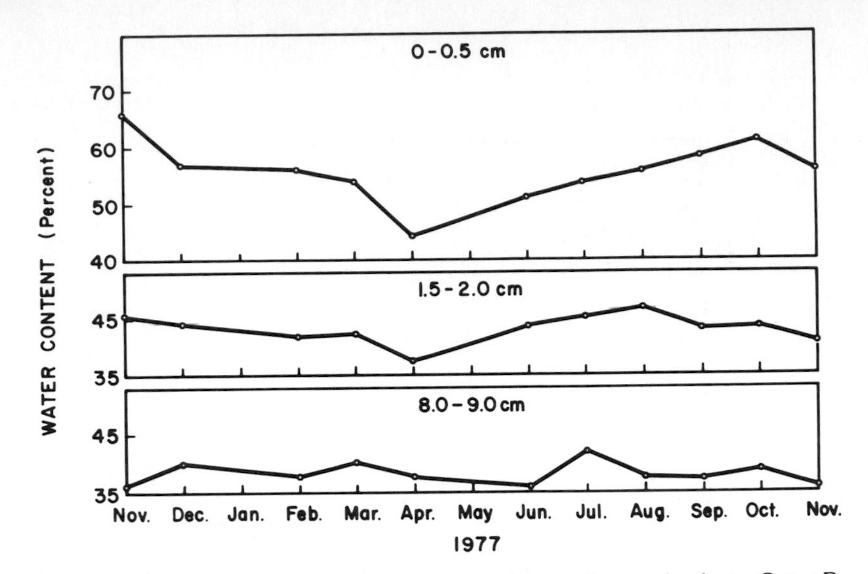

Figure 5. Percent water content of sediments at three sediment depths in Gates Bay, Virginia. The seasonal variability seen near the surface is damped at depth.

indicate least cohesive sediments at high tide during the October samples. Comparison of high and low tide water contents indicated movement of 0.9 l m^{-2} of water from the sediments during the exposed period at low tide. Winter data indicate a loss of 0.38 l m^{-2} water during a similar low tide period. Winter sediments have vane shear tests at high tide comparable to low tide values in October.

Sediment Organic Material

Seasonal cycles in percent AFDW in surface sediments were similar to and positively correlated with percent carbon in suspended sediments (Table 1). Sediment-surface organic material was negatively correlated with both the quantity of suspended sediment and temperature. A decline from high AFDW (5.5%) in winter to lower values in late spring and early summer was followed by a precipitous drop to the August and September minimums of 4.3 and 4.1%, respectively. The rapid increase in sedimentary organic matter during October to 5.6% and subsequent decline to 4.1% by November suggests the dramatic changes taking place during fall cooling and was much different than the gradual constant increase in percent organics in the suspended sediment collectors.

Vane Shear Test

Shear strength data were collected when the substrate was not covered by water during July 1977 through June 1978 (Fig. 7). The remaining data (July 1978 to November 1979) were collected on the ebbing tide while water still covered the sediment surface. Data were taken with and without water cover in July and September 1978, February 1979 and October 1980 for comparison to

Depth (cm)	Nov.- Dec.	Dec.- Feb.	Feb.- Mar.	Mar.- Apr.	Apr.- Jun.	Jun. Jul.	Jul. Aug.	Aug.- Sep.	Sep.- Oct.	Oct. - Nov.
.0 - 0.5	-8	4	-9	-8	8	2	2	2	4	-7
0.5 - 1	-8	4	-5	-1	3	2	4	-6	1	-8
1 - 1.5	-4	1	-3	-2	4	2	4	-4	0	-2
1.5 - 2	-1	-3	0	-4	6	2	2	-4	0	-2
2 - 3	-2	-2	1	-2	1	3	6	-6	-1	-2
3 - 4	-2	-4	0	-2	2	1	1	0	-3	0
4 - 5	6	-4	0	-2	1	-1	2	0	-2	2
5 - 6	4	-3	0	-2	2	5	-2	0	0	-1
6 - 7	5	-4	0	-1	-1	5	-4	-3	5	-1
7 - 8	2	-2	4	-5	-1	6	-3	-1	1	-2
8 - 9	4	-2	2	-3	-2	5	-5	0	2	-3
9 - 10		-2	2	0	-4	6	-5	1	2	-5

Figure 6. *Seasonal depth profile for change in water content in tidal flat sediment, Gates Bay, Virginia. Data are percent change during the period shown at the top of the column.*

Table 3. Vane shear test at High and Low tide in unconsolidated sediments of Gates Bay, Virginia compared with estimated change in water content in top 2 cm of sediments. Water content data are the difference between high and low tide values. Vane shear tests are expressed as Mean ± Standard Error (t = .05).

	Vane Shear Test (Relative units)		
	High tide	Low tide	Change in Water Content ($l\ m^{-2}$)
October	3.1 ± 0.2	4.8 ± 0.2	-0.90
February	4.3 ± 0.2	4.5 ± 0.3	-0.38

the first year's data. The data indicate the effects of water loss during low tides, on sediment-surface stability.

Seasonal trends in shear strength, either on water-covered or drained sediments, showed maximal values in winter and minimal values during summer and fall. The trends were inversely related to temperature and the weight of suspended sediments collected, and directly related to wind speed and percent organic material in the suspended sediment collectors (Table 1). No statistically significant relationship existed between shear strength and sediment-surface organics or water content of the sediment, in part because of the small number of paired tests for sediment surface organics (Table 1).

Wind Velocity

Because Gates Bay is small, shallow, and unobstructed from all sides, maximum wind velocities have a disproportionate effect on mixing when compared with larger and deeper water bodies. Maximum wind velocities averaging 37-45 km h^{-1} were recorded from December through mid-April. Lowest velocities occurred in July and August (27-32 km h^{-1}) (Fig. 8). In general, the highest maximum wind velocity occurred when the least suspended sediment was collected, and the most suspended sediment was collected when average maximum wind velocity was the lowest (Table 1). The same general relationships held for vector-averaged and absolute wind velocities.

Tidal Exchange Studies

During the high suspended-sediment regime in September, suspended particulate concentrations showed a strong ebb-flow-dominated influence that suggests particles were being suspended and exported from the mudflat (Fig. 9). Particulate concentrations were significantly lower and much less variable during flood than during peak ebb. During the low suspended-sediment regime in December, there was no apparent tidal influence. Concentrations remained relatively constant (124 ± 16.8 mg l^{-1} S.D., dry wt.) over the tidal cycle. There was no apparent tidal influence on suspended organic matter for either study period. Mean tidal organic matter concentrations were significantly higher

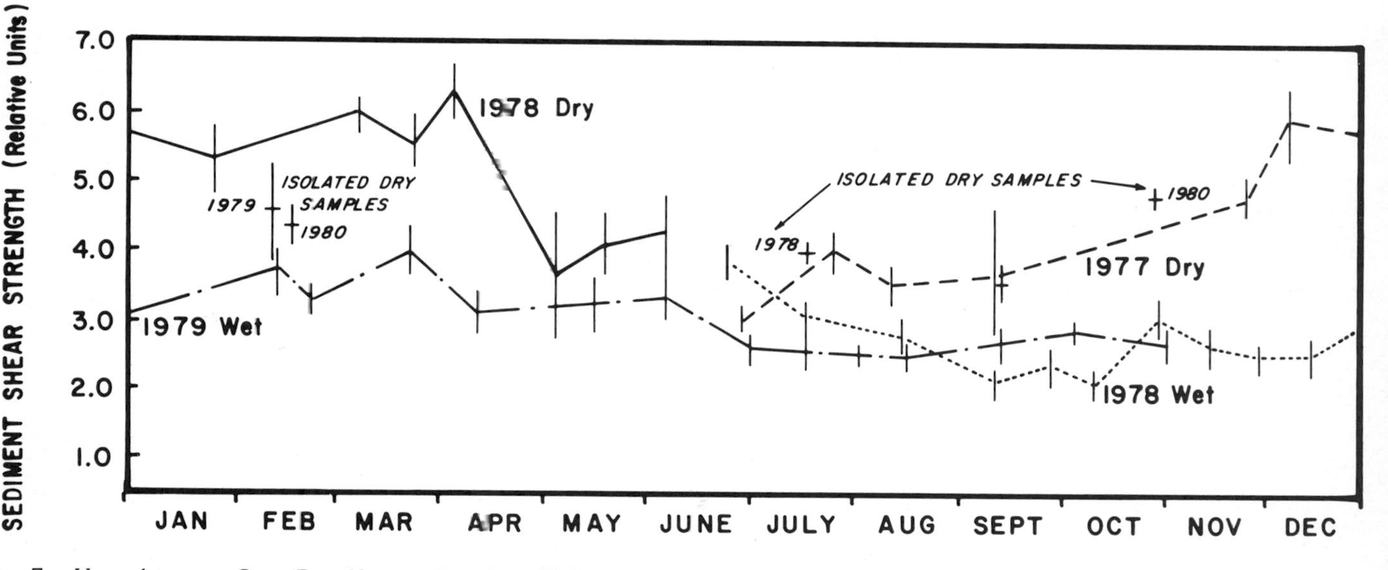

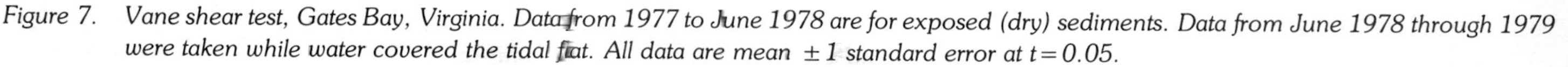
Figure 7. Vane shear test, Gates Bay, Virginia. Data from 1977 to June 1978 are for exposed (dry) sediments. Data from June 1978 through 1979 were taken while water covered the tidal flat. All data are mean ± 1 standard error at $t = 0.05$.

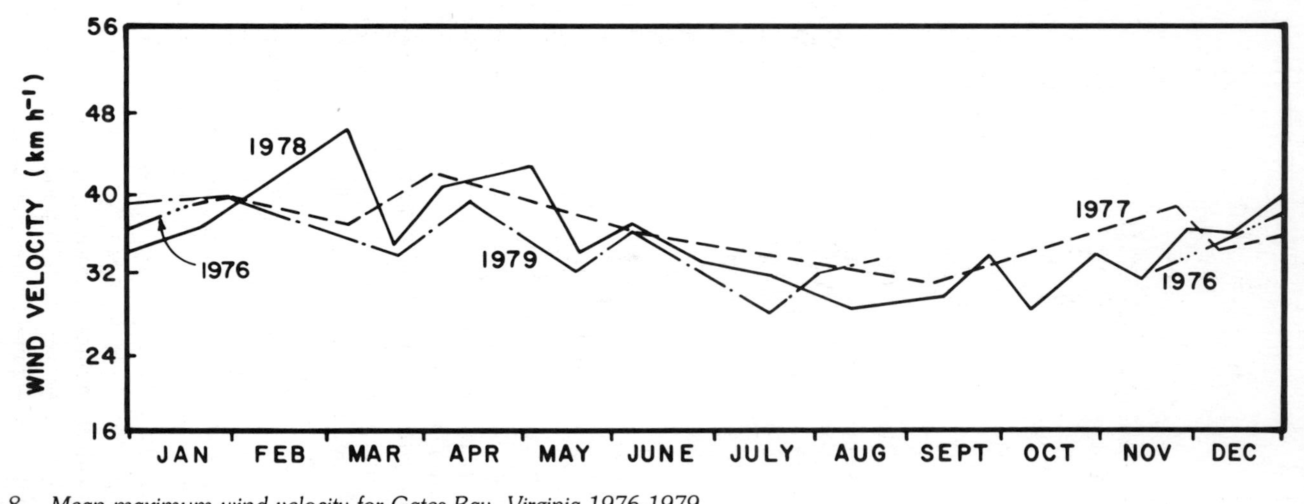

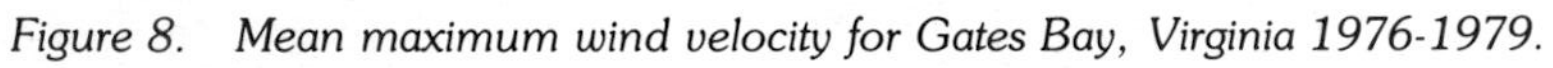
Figure 8. *Mean maximum wind velocity for Gates Bay, Virginia 1976-1979.*

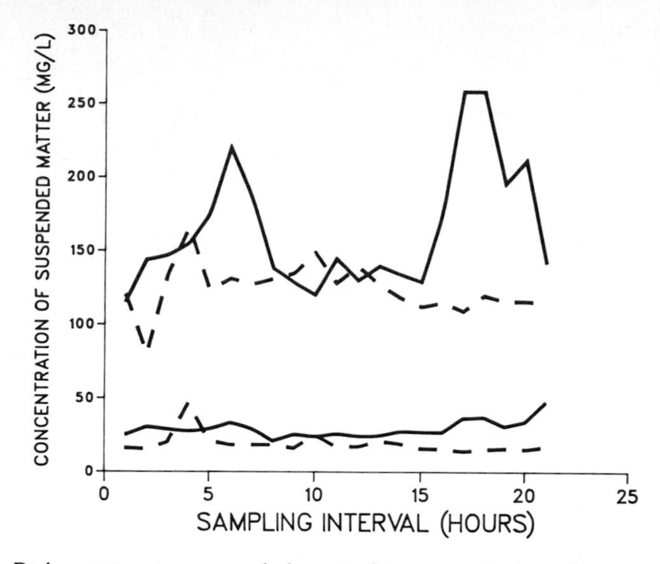

Figure 9. Diel variation in suspended particulate concentrations (upper graphs) and suspended organic matter concentrations (lower graphs) during high (solid lines) and low (dashed lines) suspended sediment regimes in Gates Bay tidal waters.

in September (29.4 ± 5.92 mg l^{-1} S.D., dry wt.) than December (19.0 ± 6.88 mg l^{-1}), but there was no corresponding peak concentration on ebb flow as with suspended particles. The source for the increased ebb-flow particulate concentrations during September was the intertidal mudflat, but the additional material in suspension was inorganic not organic. We suggest that the materials were fine silt-clays whose resuspension by the wind was facilitated by the activity of both resident and tidally-migrating organisms.

Dissolved inorganic nutrient concentrations showed no consistent patterns that could be attributed to either tidal, mudflat, or high-vs-low suspended sediment regimes. During both study periods, dissolved inorganic nitrogen and phosphorus concentrations were low, averaging 2.0, 1.0, 0.35 and 1.0 μg-at l^{-1} for NH_4, NO_2, NO_3 and PO_4, respectively. The data do not indicate effects on nutrient concentrations during extremes in suspended sediment regimes. We conclude that the coupled sediment-water column system acts conservatively, although those factors controlling surface sediment stability obviously interact in the cycling and dynamics of dissolved inorganic nutrients. It is also possible that exchanges and tidal losses of dissolved inorganic nutrients are actually taking place, but are not measurable using the design employed. For example, if interstitial water is exchanged with flood water at a maximum rate of 0.9 l m^{-2} as indicated by the sediment water-content studies, and has an average NH_4 concentration of 100 μg-at l^{-1} (Wetzel, unpublished data), the maximum difference in measured dissolved NH_4 concentration that could be expected at the inlet, based on tidal volumes, would be only 0.9 μg-at l^{-1}. The normal variability in data such as these would obviously mask such a small difference.

Discussion

The present study demonstrates a predictable seasonal cycle of sediment resuspension on an intertidal mudflat. The cycle is correlated with seasonal increases and decreases in sediment surface stability and the changes influence exchange of water with the sediments.

The interpretation of our water content results would be confounded if significant net erosion or deposition (not resuspension) were occurring. We have no direct measurement of these parameters, but either would cause an increase in the sediment in the traps. If sediments were being eroded, thus exposing deeper dryer layers, we should have a negative correlation between suspended sediments and sediment surface water content. We did not find statistically significant correlations between these two parameters (Table 1). The lack of significant correlation is probably due to the small number of paired observations. The coefficient (0.59) is positive, however, and is significant at the 90% probability level. We believe that additional data would confirm that erosion was not a major factor. If significant net deposition were taking place we would expect such a positive correlation between sediment water content and suspended sediment. Three factors, however, suggest net deposition was also not a dominant short-term process in our study area: (1) Gates Bay is surrounded by salt marsh, and this marsh is surrounded in turn by channels and a larger marsh. No rivers enter this system; thus there is no obvious source for depositional material. (2) The strongly seasonal sediment loads were negatively correlated with wind velocity. If deposition were a major factor, then the sediment source would have to be eroded by seasonal factors other than wind. The absence of rivers decreases the probability that rain was a major factor, and rainfall peaks in January and June in the study area, out of phase with sediment peaks. (3) Finally, if sediment in the collectors was due to deposition, the 22.2 kg m^{-2} yr^{-1} would amount to an annual rate of 8 mm (dry sediment) to 22 mm (50% water-sediment mixture). These rates are in excess of the reported 0.1-8 mm yr^{-1} for similar systems (Oviatt and Nixon 1975), and thus seem unlikely.

There also appears to be a tidally-mediated sediment-water flux that follows a seasonal pattern. We propose that tidal pumping is the principle mechanism for driving the exchange across the sediment-water interface of intertidal mudflats similar biogeochemically to Gates Bay. This is similar to the mechanism proposed by Riedl *et al.* (1972) for intertidal sands, but is more strongly controlled by the biota of the mudflat. Our data indicate that temperature cycles interact with variations in sediment surface stability to cause seasonal variability in this tidal pumping mechanism. We suggest that these variations in stability can be explained in turn by biological activity; i.e., seasonal stabilization of the sediment by microbiotic binding and destabilization of the sediment surface by increased macroheterotroph activity. Seasonal changes in accretion and erosion on an intertidal flat in England were mediated by algal binding; but in contrast with Gates Bay, the changes were directly correlated with wind velocity (Frostick and McCave 1979).

The dominant factors controlling stability of the mudflat sediment surface in winter appear to be microbiotic binding, decreased macroheterotroph activity and sediment dewatering. We have no direct independent measures of these factors, but vane shear tests coupled with visual observations of epipelic algal blooms on the sediment surface suggest their importance. Concomittent increases in shear strength and sediment surface organics, a decrease in water content and a decrease in visible activity of benthic invertebrates and fish all imply less disruption and greater sediment cohesiveness in the winter. The decrease in suspended sediments and increase in organic material in the sediment traps coupled with visual observations of increased water clarity during winter all reflect increased stability in spite of increased average maximum wind velocity.

Anderson (1983) stressed the importance of ice as a controlling factor on northern intertidal mudflats. Although ice was much less important for the Gates Bay mudflat than for those described by Anderson (1983), the brief increase in water content exhibited by the surface sediments in 1977 (Fig. 6) may have been due to a two-week period of ice cover prior to the February sampling.

Direct measurement of pigments in intertidal flat sediments have found maximum chlorophyll *a* levels in summer (Pomeroy 1959; Leach 1970) or a decrease in total pigments throughout the summer months (Baillie and Welsh 1980). The data of the latter authors was similar to our sediment surface organic information, but they found no changes in epipelic algal biomass throughout the year. A second factor, benthic activity, may help explain this apparent discrepancy. Benthic activity and fish feeding on the sediment surface peak during mid-to-late summer, and reduction in activity during winter months may allow development of a stable sediment microbial community even though sediment pigment concentrations were not different among seasons.

Anderson (1983) characterizes spring as a period of ice breakup and minimal erosion, because benthic organisms have not yet begun to pelletize the sediment. The mid-Atlantic site of our studies exhibits a relatively stable state from December to March, but there is an abrupt change brought about by the seasonal warming between March and April when average water temperatures exceed 8-10 °C. This temperature range coincides with increased activity of benthic invertebrates in the spring and decreased activity in the fall. The warming trend also coincides with a decrease in sediment shear strength and an increase in suspended sediments. The specific timing of onset of these conditions appears to be the most variable part of the cycle on the Gates Bay mudflat.

Summer conditions in Gates Bay were generally similar to those described by Anderson (1983): increased sediment water-content, biological activity and sediment resuspension. Increased loads of suspended sediment occurred when both sedimentary organic material and suspended organics reached their yearly minimum. This relation suggests that much of the suspended sediment is coming directly from the bottom. This conclusion is supported by Welsh's (1980) observation that incoming waters resuspend both sediments and epipelic algae. Oviatt and Nixon (1975) did not find seasonal differences in suspended organic material in Narragansett Bay, but did find a trend toward an inverse correlation between organic material and total suspended sediments. Anderson and Howell (1984) indicated that pore water content decreased during the summer (July to August) on a New Hampshire tidal flat. This is in direct opposition to our data from Gates Bay where water content increased during the summer. These contrary findings may be due to the timing of the coring activity. Our cores were taken just prior to low tide while water still covered the flat, whereas Anderson and Howell (1984) began their coring after the flat was exposed. This timing is critical during the summer months as can be seen both from Table 3 and from Anderson and Howell's data. Our vane shear test data also indicated significant changes in cohesion as the sediments dewater at low tide.

One factor we did not measure and for which we cannot find literature values is the effect of small fish. We repeatedly observed large numbers of juvenile fish passing across the flat with the incoming tide. These were followed by larger organisms as the water level increased. The numbers of fish seemed to increase from June through early September and then abruptly decrease with the first cold night-time temperatures. The appearance and density of these organisms on the flat may be related to the late summer peak in suspended sediment.

Fall was essentially a period of dewatering, decreased resuspension and decreased biological activity on the Gates Bay flat, accompanied by an increase in sediment cohesion. Anderson (1983) observed similar fall patterns in more northern areas.

Oviatt and Nixon (1975) reported deposition of 18 kg m^{-2} y^{-1} dry weight in subtidal sediment collectors in lower Narragansett Bay, Rhode Island. Our traps approximate those used by Oviatt and Nixon (1975) except ours were fixed to the bottom. Recent studies (Lau 1979; Hargrave and Burns 1979; Gardner 1980 a, b) clearly demonstrate differential collection efficiencies based on trap design. Any trap design will collect consistently but the use of funnels underestimates the quantity of material deposited (Gardner 1980 a,b). The three-year average of 22.2 kg m^{-2} y^{-1} for Gates Bay is similar to Oviatt and Nixon's (1975) estimate for Narragansett Bay. The average organic carbon (510 g m^{-2} y^{-1}) collected in Gates Bay is slightly less than the 640 g m^{-2} y^{-1} reported by Oviatt and Nixon (1975) for lower Narragansett Bay, but is well within the range for the entire Narragansett Bay system. The annual suspended sediment load observed in Gates Bay is well within the range of 0.1-200 kg m^{-2} y^{-1} dry weight, found in temperate-zone high-salinity shallow bays (See Table 7, Oviatt and Nixon 1975), but the predictability and seasonality exhibited by Gates Bay appear to be unique.

Acknowledgments

Contribution No. 1281 from the Virginia Institute of Marine Science. We appreciate the efforts of James Moore, Bob Bisker, Laura Murray, David Ludwig and James Rizzo, who provided invaluable assistance in field collections.

References Cited

Aller, R. C. 1982. The effects of macrobenthos on chemical properties of marine sediment and overlying water. pp. 53-102. *In:* P. McCall and M. J. S. Tevese (eds.), *Animal Sediment Relationships.* Plenum, New York.

Aller, R. C. and R. E. Dodge. 1974. Animal-sediment relationships in a tropical lagoon, Discovery Bay, Jamaica. *J. Mar. Res.* 32:209-232.

Anderson, F. E. 1972. Resuspension of estuarine sediments by small amplitude waves. *J. Sediment Petrol.* 42:602-607.

Anderson, F. E. 1976. Rapid settling rates observed in sediments resuspended by boat waves over a tidal flat. *Neth. J. Sea Res.* 10:44-58.

Anderson, F. E. 1979. How sedimentation pattern may be affected by extreme water temperatures on a notheastern coastal intertidal zone. *Northeastern Geology* 1:122-132.

Anderson, F. E. 1983. The northern muddy intertidal: a seasonally changing source of suspended sediments to estuarine waters-a review. *Canadian J. Fish. Aquat. Sci* 40:143-159.

Anderson, F. E. and B. A. Howell. 1984. Dewatering of an unvegetated muddy tidal flat during exposure—desiccation or drainage? *Estuaries* 7:225-232.

Baillie, P. W. and B. L. Welsh. 1980. The effect of tidal resuspension on the distribution of intertidal epipelic algae in an estuary. *Estuar. Coastal Mar. Sci.* 10:165-180.

Billen, G. 1978. A budget of nitrogen recycling in North Sea sediments off the Belgian coast. *Estuar. Coastal Mar. Sci.* 7:127-146.

Boynton, W. R., W. M. Kemp and C. G. Osborne. 1980. Nutrient flux across the sediment water interface in the turbid zone of a coastal plain estuary. pp. 93-109. *In:* V. S. Kennedy, (ed.), *Estuarine Perspectives.* Academic Press, New York.

Frostick, L. E. and I. N. McCave. 1979. Seasonal shifts of sediment within an estuary mediated by algal growth. *Estuar. Coastal Mar. Sci.* 9:569-576.

Gardner, W. D. 1980a. Sediment trap dynamics and calibration: a laboratory evaluation. *J. Mar. Res.* 38:17-39.

Gardner, W. D. 1980b. Field assessment of sediment traps. *J. Mar. Res.* 38:41-52.

Hargrave, B. T. and N. M. Burns. 1979. Assessment of sediment trap collection efficiency. *Limnol. Oceanogr.* 24:1124-1136.

Haven, D. S. and R. Morales-Alamo. 1966. Aspects of biodeposition of oysters and other invertebrate filter feeders. *Limnol. Oceanogr.* 11:487-498.

Holland, A. F., R. G. Zingmark and J. M. Dean. 1974. Quantitative evidence concerning the stabilization of sediments by marine benthic diatoms. *Mar. Biol.* 27:191-196.

Kraeuter, J. N. 1976. Biodeposition by salt marsh invertebrates. *Mar. Biol.* 35:215:223.

Krumbein, W. C. and F. J. Pettijohn. 1938. *Manual of Sedimentary Petrology.* Appelton Century Crofts, Inc. New York. 549 pp.

Lau, Y. L. 1979. Laboratory study of cylindrical sedimentation traps. *J. Fish. Res. Bd. Canada.* 36:1288-1291.

Leach, J. H. 1970. Epibenthic algal production in an intertidal mudflat. *Limnol. Oceanogr.* 15:514-521.

Meade, R. H. 1972. Transport and deposition of sediments in estuaries. *In:* B. W. Nelson (ed.), *Environmental Framework of Coastal Plain Estuaries.* Geol. Soc. America, Memoirs 133:91-120.

Neuman, A. C., C. P. Gebelein and T. P. Scoffin. 1970. The composition, structure and erodability of subtidal mats, Abaco, Bahama. *J. Sed. Petrol.* 40:274-297.

Nixon, S. W., C. A. Oviatt and S. S. Hale. 1975. Nitrogen regeneration and the metabolism of coastal marine bottom communities. pp. 269-283. *In:* J. M. Anderson and A. McFayden (eds.), *The Role of Terrestrial and Aquatic Organisms in Decomposition Processes.* Blackwell Sci. Publ. Oxford, England.

Nixon, S. W. 1979. Remineralization and nutrient cycling in coastal marine ecosystems. pp. 111-138. *In:* B. Neilson and L. E. Cronin (eds.), *Nutrient Enrichment in Estuaries.* Humana Press, Clifton, New Jersey.

Nixon, S. W. 1980. Between coastal marshes and coastal waters—A review of twenty years of speculation and research on the role of salt marshes in estuarine productivity and water chemistry. pp. 437-535. *In:* P. Hamilton and K. B. MacDonald (eds.), *Estuarine and Wetland Process.* Plenum, New York.

Oviatt, C. A. and S. W. Nixon. 1975. Sediment resuspension and deposition in Narragansett Bay. *Estuar. Coastal Mar. Sci.* 3:201-217.

Pomeroy, L. R. 1959. Algal productivity in the salt marshes of Georgia. *Limnol. Oceanogr.* 4:386-398.

Rhoads, D. C. 1973. The influence of deposit feeding benthos on water turbidity and nutrient recycling. *Am. J. Sci.* 273:1-22.

Rhoads, D. C. 1974. Organism-sediment relations on the muddy sea floor. *Oceanogr. Mar. Biol. Ann. Rev.* 12:263-300.

Rhoads, D. C. and D. L. Young. 1970. The influence of deposit-feeding organisms on sediment stability and community trophic structure. *J. Mar. Res.* 28:150-178.

Rhoads, D. C. and D. L. Young. 1971. Animal-sediment relations in Cape Code Bay, Massachusetts. II. Reworking by *Molpadia oolitica* (Holothuroidea). *Mar. Biol.* 11:255-261.

Rhoads, D. C., J. Y. Yingst and W. J. Ullman. 1978. Sea floor stability in central Long Island Sound Part I. Temporal changes in erodability of fine-grained sediments. pp. 221-244. *In:* M. Wiley (ed.), *Estuarine Interactions.* Academic Press, New York.

Riedl, R. J., N. Huang and R. Machan. 1972. The subtidal pump: a mechanism of interstitial water exchange by wave action. *Mar. Biol.* 13:210-221.

Roman, M. R. 1978. Tidal resuspension in Buzzards Bay, Massachusetts. II. Seasonal changes in the size distribution of chlorophyll, particle concentration, carbon and nitrogen in resuspended particulate matter. *Estuar. Coastal Mar. Sci.* 6:47-53.

Roman, M. R. and K. R. Tenore. 1978. Tidal resuspension in Buzzards Bay, Massachusetts. I. Seasonal changes in resuspension of organic carbon and chlorophyll *a. Estuar. Coastal Mar. Sci.* 6:37-46.

Sanders, H. L. 1958. Benthic studies in Buzzards Bay. I. Animal sediment relationships. *Limnol. Oceanogr.* 3:245-258.

Sasseville, D. R. and F. E. Anderson. 1976. Sedimentological consequences of winter ice cover on a tidal flat environment, Great Bay, New Hampshire. *Rev. Geogr. Montr.* 30:87-93.

Schubel, J. R. 1972. Distribution and transportation of suspended sediment in upper Chesapeake Bay. *In:* B. W. Nelson (ed.), *Environmental Framework of Coastal Plain Estuaries.* Geol. Soc. America, Memoirs 133:151-167.

Schubel, J. R. and R. B. Biggs. 1969. Distribution of seston in upper Chesapeake Bay. *Chesapeake Sci.* 10:18-23.

Seitzinger, S., S. W. Nixon, M. E. Q. Pilson and S. Burke. 1980. Denitrification and N O production in near-shore marine sediments. *Geochim. Cosmochim. Acta.* 44:1853-1860.

Taguchi, S. and B. T. Hargrave. 1978. Loss rates of suspended material sedimented in a marine bay. *J. Fish. Res. Bd. Canada.* 35:1614-1620.

Tenore, K. R. 1977. Food chain pathways in detrital feeding benthic communities: A review, with new observations on sediment resuspension and detrital recycling. pp. 37-53. *In:* B. C. Coull (ed.), *The Ecology of Marine Benthos.* Belle Baruch Library in Mar. Sci. #6. Univ. South Carolina, Columbia.

Webb, J. E. 1969. Biologically significant properties of submerged marine sands. *Proc. Roy. Soc. London B* 174:355-402.

Welsh, B. L. 1980. Comparative nutrient dynamics of a marsh-mudflat ecosystem. *Estuar. Coastal Mar. Sci.* 10:143-164.

Woodin, S. A. 1976. Adult-larval interaction in dense infaunal assemblages: Patterns of abundance. *J. Mar. Res.* 34:25-41.

Yingst, J. Y. and D. C. Rhoads. 1978. Sea floor stability in central Long Island Sound: II. Biological interactions and their potential importance for sea floor erodability. pp. 245-260. *In:* M. L. Wiley (ed.), *Estuarine Interactions.* Academic Press, New York.

Young, D. K. and D. C. Rhoads. 1971. Animal-sediment relations in Cape Code Bay, Massachusetts. I. A transect study. Mar. Biol. 11:242-254.

PROCESS VARIABILITY IN ESTUARIES

Alejandro Yáñez-Arancibia and John W. Day, Convenors

TEMPORAL VARIABILITY IN OXYGEN METABOLISM OF AN ESTUARINE SHOAL SEDIMENT

William M. Rizzo and Richard L. Wetzel

Wetlands Ecology Department
College of William and Mary
Virginia Institute of Marine Science
Gloucester Point, Virginia

Abstract: Benthic microalgal production of a submerged sandy shoal in the York River, Virginia was measured from March to December 1983, with particular emphasis on the variability in metabolic estimates over different time scales. Variation in rates was examined over the photoperiod, between successive sampling days, between tidal condition (mid-day high vs. low tide), and among seasons. The coefficient of variation for 4 to 22 estimates of hourly net production (NP) and respiration (R) over the photoperiod averaged 231% to 75%, respectively. Morning NP was significantly greater than afternoon NP over the study. Mean hourly NP and R were significantly different on successive days in 4 of 6 tests, and 2 of 6 tests, respectively. The coefficients of variation for average NP, R and chlorophyll *a* over these two-day periods ranged from 4 to 43%, 1 to 34% and 0 to 80%. R was significantly higher on days with mid-day low tides (noon ± 2 hours). Mean hourly NP was 49% greater on days with mid-day low tides and R was 70% greater. Hourly NP and R were significantly different among seasons. R peaked in summer and NP in fall. Coefficients of variation for mean hourly NP and R over the seasons were 57% and 59%, respectively. Plots of mean hourly NP and R by month were made using all data for a given month and these were compared to plots made by randomly selecting a single measurement for each month. The latter plots are based on 12 data points, the former on 185 points. The two types of plots produced very similar annual rate estimates but differed radically in their depiction of seasonal changes.

Introduction

Estuaries are productive ecosystems whose functioning is heavily influenced by the metabolism of microscopic organisms. Temporal variability in the metabolism of microautotrophic and heterotrophic communities has immediate consequences for materials cycling and energy flow since these communities are tightly coupled. Knowledge of both the magnitude and the temporal variability of production and consumption by microorganisms is critical to a better understanding of the functioning of estuaries. Unfortunately, few investigations have adequately addressed temporal variability in the estimation of rate processes associated with microorganisms or the consequences of subsequent extrapolation of measurements to different spatial or temporal scales.

Continuous measurements of primary productivity or community metabolism are seldom made for periods longer than one day. More commonly the measurement intervals are a few hours. However, estimates derived from such brief measurements are often extrapolated to daily rates, the daily rates subsequently extrapolated to monthly estimates, and the monthly estimates summed to calculate annual rates. Less commonly, measurements over the photoperiod are made and extrapolations proceed from this point. In addition, changes in either hourly or daily rate estimates between sampling intervals are often interpreted as changes associated with season. If the variation in hourly estimates over a day and/or day-to-day variation over the month is high, extrapolation to longer time periods produces estimates with large and often unknown confidence limits. Similarly, interpreting changes between single monthly rate estimates as the consequence of changes on large scales, i.e. as seasonal changes, may be invalid since the difference may result from a change in the process over periods of hours or days rather than weeks or months. Nevertheless, extrapolated rate measurements are typically the estimates used to compare habitats within estuaries as well as different estuarine systems. The question remains: How reliable are conclusions based on this type of comparison?

Shallow water sediments supporting photosynthesis by benthic microflora are ideal study areas for addressing the temporal variability of community productivity. Over the past 25 years many measurements of the standing crop and productivity of benthic microflora have been made (Pomeroy 1959; Pamatmat 1968; Steele and Baird 1968; Gargas 1970; Leach 1970; Marshall *et al.* 1971; Riznyk and Phinney 1972; Cadee and Hegeman 1974, 1977; Gallagher and Daiber 1974; van Raalte *et al.* 1976; Joint 1978; Riznyk *et al.* 1978; Admiraal and Peletier 1980; Ballie and Welsh 1980; Zedler 1980; van Es 1982; Murray 1983; Shaffer and Onuf 1983; Colijn and de Jonge 1984; Rizzo and Wetzel 1985). The extent of the practice of extrapolating hourly rates to longer periods is illustrated by the fact that only three of these studies actually measure production over the entire photoperiod (Steele and Baird 1968; Cadee and Hegeman 1974; 1977), while ten report daily rates and 18 report annual rates. Since extrapolation of short-term experimental measurements to longer periods of time is virtually unavoidable due to various limitations, estimates of the variability of such measurements over short time scales are needed in order to assess the effects of extrapolations on estimates of the magnitude of processes and the changes in those processes over longer time intervals, i.e. annual scales.

We initiated this study to investigate the variation in measurements of hourly community metabolism of a benthic community over time scales of hours (within the photoperiod), days (day-to-day), tidal conditions (week-to-week) and seasons. We also examined the variability in standing crop of chlorophyll *a* over 2- to 12-day periods.

Study Site

The study site was a permanently submerged sand shoal in the York River, located at the Virginia Institute of Marine Science at Gloucester Point, Virginia (37 14′N, 76 31′W; Fig. 1). The York River is a temperate estuary typical of the tributaries of the Chesapeake Bay. It is approximately 50-km long and contains about 130 km^2 of permanently submerged bottom, based on the mean low water datum plane. Shoal areas less than 2-m deep are within the

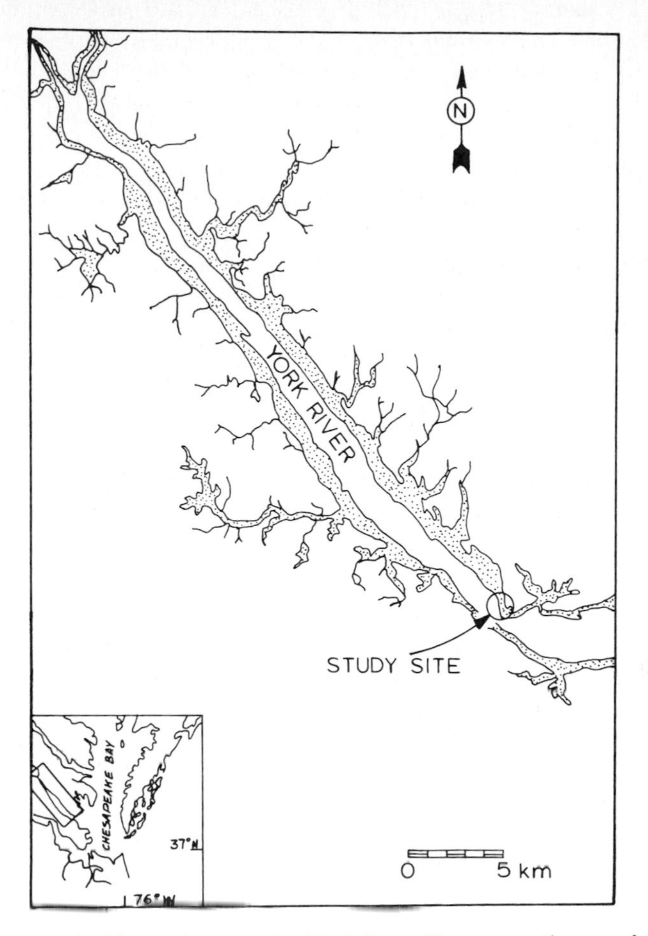

Figure 1. Location of the study site in the York River, Virginia, a tributary of the Chesapeake Bay. Stippled area indicates depths less than 2 meters.

photic zone, support photosynthesis by benthic microalgae and comprise 38% of the bottom surface area. The area has semi-diurnal tides with a range of about 0.8 m.

Materials and Methods

Studies of sediment oxygen metabolism were made each month from March through December 1983. Change in dissolved oxygen concentration was used to calculate net production (NP) and respiration (R) in two transparent and two opaque plexiglass domes, each covering 0.16 m^2 of sediment. Gross production (GP) was calculated as the sum of NP + R. The domes were internally partitioned to provide water column and sediment + water column chambers within the same dome in order to correct for water column metabolism.

Oxygen probes (Model 2710 Oxygen Monitor, Orbisphere Corp., York, ME.) were placed in the bottom chamber of each dome and stirring rods attached to the probe assemblies provided continuous stirring. Oxygen concentrations were determined before and after incubation intervals that ranged from one-half to two hours. Incubations were conducted sequentially over the photoperiod. Experiments were begun about 2.5 h after sunrise and concluded about 1.0 h before sunset. To prevent supersaturation of the transparent domes and oxygen depletion in the opaque domes the domes were flushed at 3-4 h intervals until equilibrium with ambient water column oxygen concentrations was achieved, usually 30–45 minutes. The sediment surface was not visibly disturbed by the flushing process.

Chlorophyll *a* (Chl. *a*) concentrations were determined by extracting the top 1 cm of sediment in 100% acetone overnight at 5 °C (Jacobsen 1978). A second extraction was necessary to achieve complete extraction. Following extraction, the volume was adjusted to 20 ml with 90% acetone. Chl. *a* was estimated spectrophotometrically using a single wavelength (665 nm) and a pheophytin correction (Riemann 1978). Concentrations were calculated using the equation given by Lorenzen (1967). Chl. *a* concentrations were determined on each of five core samples (2.10 cm I.D.) collected for 2-12 consecutive days each month except March, when collections were made only on 5 and 15 March.

Three core samples for organic matter analyses were collected on the chl. *a* sampling dates with the same corer used to sample chl. *a*. The surface centimeter of the sediment was dried to a constant weight at 60 °C, and then ashed for 12 h at 500 °C. The carbon concentrations were determined following procedures described in Rizzo and Wetzel (1985).

Photosynthetically active radiation (PAR) reaching the sediment surface was measured with a LiCor 185A quantum radiometer (LiCor, Inc. Lincoln, NE). PAR was continuously recorded. Temperatures were measured with the thermistors within the oxygen probe assemblies. Salinity was measured with a temperature-compensated salinity refractometer (Model AO 10419 American Scientific Products, Washington, D.C.).

The average hourly morning and afternoon NP and R for equal periods preceding and following solar noon on each sampling day were compared over the study using paired t-tests (n = 19 days). Day-to-day variation in average hourly NP and R for the same time periods on successive days were also compared using t-tests (n varied from 5 to 14 measurements per day). Short-term changes in sediment chlorophyll concentrations during the 2-12 day series each month were examined using one-way analysis of variance. Variation of mean hourly metabolic rates as a function of tidal condition was analyzed using t-tests by comparing hourly NP and R between dates with noon + 2 hour high tides (n = 10) and days with mid-day low tides (n = 7). Seasonal differences in average hourly NP and R were examined using analysis of variance and Student-Newman-Keuls (SNK) multiple range tests (n = 185). All statistical procedures are described in Sokal and Rohlf (1969). When coefficients of variation (CV) are reported for two samples, e.g. variation of mean NP on successive days, an estimate of the standard deviation was made from the range (Wilcoxon and Wilcox 1964).

The effect of extrapolating single monthly measurements of hourly NP and R to an annual estimate was examined by randomly selecting a single measurement from the total measurements made in a given month. The annual rates were calculated by extrapolating the hourly rate to a monthly rate based on the number of hours in the photoperiod and the number of days per month. Monthly estimates were then summed to produce an annual estimate. Monthly rates for the January-February period were linearly interpolated using the March and December sampling days, and the estimate for September was interpolated using the August and October samples. Two curves, derived in this way from randomly selected monthly measurements, were plotted and compared with a third curve derived from the mean of all available monthly data. The resulting seasonal plot and annual production estimate represented by the latter curve was based on 185 data points while each of the other two curves and production estimates was based on 12 data points.

Results and Discussion

Temperature and salinity ranged from 7.3-27.3 °C and 13.5-21.5 ppt, respectively, during the study. Minimum and maximum water depths during the study were 0.22 and 1.32 m. The average hourly rates of GP, NP and R by sampling date are shown in table 1. Over the study, coefficients of variation for mean hourly GP, NP and R over the photoperiod ranged from 23 to 296%, 41 to 1281% and 31 to 127%, and averaged 90%, 231% and 75%, respectively. High variability in metabolic rates occurred in all seasons and often differed markedly from sampling date to sampling date. For example, the coefficient of variation for the respiration measurements made on 8 November was 75%, while the coefficient of variation for the next day was only 15%.

Average morning and afternoon NP and R are shown in fig. 2A. NP was greater in the morning on 16 of 19 sampling dates and R was higher in the afternoon on 11 of 19 sampling days. In data pooled over the study, NP was significantly higher in the morning, but there was no statistically significant difference in R. Over the study, there was no significant difference in the amount of light received by the benthic community between morning and afternoon, although considerable variation was occasionally observed between morning and afternoon light. During sampling dates when light levels were lowest (March, April, and November) NP was higher in the morning even though light conditions were more favorable in the afternoon.

The relatively consistent temporal separation observed between peak NP and R within the photoperiod suggest that the two processes are coupled over a short time frame and endogenous factors rather than exogenous factors such as light or temperature, limit their magnitude. For example, NP may become limited during the afternoon by lack of nutrients. While the upper York River has been characterized as moderately nutrient-enriched (Heinle *et al.* 1980), the lower reaches of the river, including this study site are relatively unenriched.

Table 1. Hourly rates of metabolism of the sand shoal. Mean ± standard deviation ($mg\ O_2\ m^{-2}\ h^{-1}$), (n) = sample size.

DATE	GROSS PRODUCTION		NET PRODUCTION		RESPIRATION	
4 MAR	61.2 ±	49.57 (9)	44.1 ±	43.22 (9)	13.3 ±	11.17 (11)
5 MAR	54.6	19.66 (14)	42.9	20.59 (14)	10.2	8.67 (16)
14 MAR	55.8	33.48 (15)	28.2	27.07 (16)	26.6	25.00 (16)
15 MAR	70.2	45.63 (12)	48.1	38.48 (13)	18.1	21.72 (12)
13 APR	48.6	77.27 (16)	46.1	76.53 (16)	2.5	3.75 (17)
23 MAY	54.7	38.29 (12)	22.9	47.63 (22)	24.7	20.01 (11)
26 MAY	51.8	69.41 (9)	27.4	55.35 (11)	28.3	23.77 (9)
31 MAY	20.7	61.69 (13)	−5.5	70.46 (13)	26.2	20.70 (13)
22 JUN	139.7	118.75 (10)	52.9	87.29 (11)	80.2	72.98 (10)
23 JUN	121.6	51.07 (6)	29.3	37.50 (6)	74.9	38.95 (7)
30 JUN	47.1	87.14 (9)	23.7	69.44 (9)	23.3	25.16 (9)
28 JUL	91.3	108.65 (9)	27.2	74.26 (9)	64.1	53.20 (9)
4 AUG	213.0	83.07 (4)	151.9	92.66 (4)	61.1	56.21 (4)
29 AUG	40.7	32.15 (6)	−2.9	20.24 (6)	43.6	20.49 (6)
30 AUG	72.8	77.17 (5)	38.7	92.49 (6)	61.3	19.00 (5)
3 OCT	137.1	57.58 (6)	83.8	46.93 (6)	53.3	35.71 (6)
4 OCT	114.7	84.88 (4)	81.0	64.80 (5)	30.4	23.10 (4)
8 NOV	168.2	70.64 (5)	141.6	58.06 (5)	26.6	19.95 (5)
9 NOV	89.5	113.67 (4)	57.5	109.25 (4)	31.9	4.79 (4)
15 NOV	74.5	38.00 (2)	33.2	26.56 (2)	42.3	12.27 (2)
8 DEC	99.9	4.00 (2)	91.1	0.91 (2)	8.8	3.08 (2)

Water column concentrations of ammonium were <10 ug-at N l^{-1} on all but one sampling date, and phosphate concentrations were always <2 ug-at P l^{-1} (Rizzo, unpub.). In any case, the availability of water column nutrients to the benthic microflora may be diffusion limited. Consequently, microfloral photosynthesis in the surface sediment may depend on *in situ* heterotrophic nutrient regeneration. Concommittant studies indicate that the release of ammonium from these sediments is highly correlated with benthic respiration (Rizzo, unpub.). Benthic respiration on the other hand, may be limited by lack of labile organic matter. Organic matter concentrations in the surface centimeter of these sediments averaged 0.64% and reached a maximum of 1.63% in summer. These concentrations are similar to sediment carbon determinations made for other submerged sandy shoal sediments in the lower York River (assuming % carbon = 0.45 × % ash-free dry weight). Percent sediment carbon varied little with depth. Concentrations in the surface 15 cm, at five cm intervals, were (dry weight basis) 0.29 ± .16% (standard deviation), 0.28 ± .14% and 0.28 ± .14% over ten monthly samples, suggesting that much of the surface organic matter is refractory. The tendency for R to increase in the afternoon and temporally lag peak NP suggests that microheterotrophic metabolism in the surface sediments relies on the release of labile photosynthate. Increased heterotrophic

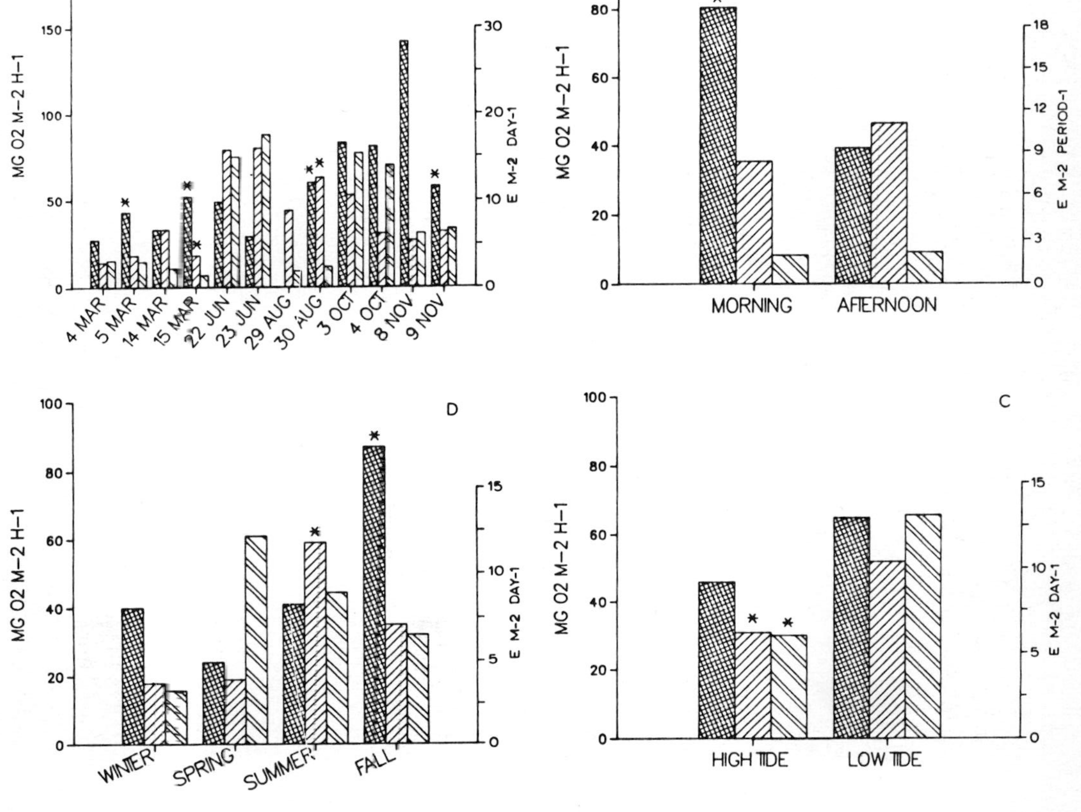

Figure 2.

*Net production and respiration (mg O_2 m^{-2} h^{-1}), and photosynthetically active radiation (E m^{-2} day^{-1}) on an estuarine shoal in the York River. For each time period, the 3 bars represent net production, respiration, and PAR, respectively, from left to right. Statistical results are examined at the 95% significance level. Fig. 2a. * denotes morning rates significantly different from afternoon rates. PAR is in E m^{-2} per morning or afternoon period. Fig. 2b. * denotes a significant difference from the preceding day. Fig. 2c. * denotes high tide rates significantly different from low tide rates. Fig. 2d. * denotes a seasonal rate significantly different from the other three seasons.*

activity in turn would increase nutrient regeneration stimulating autotrophic productivity the following morning. Further investigation into this apparent daily cycle would be highly desirable, and may provide greater insights into day-to-day variability in rate estimates.

The average NP and R for the same sampling periods on consecutive days are shown in fig. 2B. The average coefficient of variation for the six two-day periods was 51% (2 – 164%) for NP and 21% (5 – 39%) for R. Mean hourly NP was significantly different ($P = .05$) over two-day periods in four of the 6 tests, and R was significantly different in 2 of 6 tests. However, average hourly metabolic rates compared on successive days varied less than measurements compared over the photoperiod. Day-to-day sampling was limited to successive days that had similar environmental regimes (i.e. similar tidal, photoperiod, temperature, salinity and climatic conditions) which further suggests that endogenous cycles account for a significant source of variation within a daily period.

The causes of the daily variability in metabolism are not readily apparent. As shown in fig. 2B, daily light reaching the sediment surface was nearly identical between days for each two-day period. The maximum temperature difference between days was 0.9 °C, a difference unlikely to result in significant changes in metabolic rates. The ANOVA'S of the sediment chl. concentrations over 2-12 day periods (table 2) show that significant changes typically occur abruptly. Changes over periods of days were significant in all but one month. Preliminary analyses indicate that these changes are related primarily to changes in wind speed and wind direction (Rizzo, in prep.). However, changes in chl.-*a* biomass do not offer compelling evidence for the changes shown in fig. 2B. Complete chl.-*a* data are unavailable for the March samples. In June, chl. *a* concentrations changed little (115 to 125 mg m^{-2}) and NP was not significantly different between the two days. In August, chl. *a* decreased (128 to 79 mg m^{-2}) and was accompanied by a significant NP increase on the second day. In November a decrease in chl. *a* (299 to 255 mg m^{-2}) was accompanied by a significant decrease in NP, but a similar increase in chl. *a* (151 to 202 mg m^{-2}) in October was not accompanied by any significant change in NP. It also seems unlikely that biomass changes could be responsible for the observed day-to-day significant differences in R, particularly in March, although there are no data on heterotrophic biomass so the possibility cannot be ruled out. Within time periods having relatively constant physical conditions, much of the variability in metabolic rates may result from physiological responses to the proposed endogenous cycles of autotrophy and heterotrophy. Unfortunately, interpretation of such day-to-day changes is hindered by the often rapid changes in the physical environment, by lack of information on time lags in community responses and by lack of information on prior community history.

Tidally-related changes in metabolic rates are shown in fig. 2C. Metabolic rates were higher on days with mid-day low tides, although only R was significantly higher ($P = .05$). Marshall *et al.* (1971) also found that metabolism was higher on days with mid-day low tides in shoal sediments in Rhode Island. Daily PAR at the sediment surface was significantly higher on days with mid-day low tides, and probably accounted for the higher NP. Since temperatures varied only a few degrees between tidal conditions (i.e. between weeks) the significant increase in R is probably a response to increased NP.

Figure 2D shows the changes in average NP, R and PAR among seasons. NP was significantly different among seasons and SNK tests indicated that NP was significantly greater in the fall than during the other three seasons. Similarly, R was significantly different among seasons, with summer R greater than other seasons. Although the peak in R coincided with the seasonal peak in water temperature, a lag is apparent. Spring temperatures increased from about 12 °C to over 20 °C between April and May, but R did not increase significantly until June (temperatures $\leq$ 28 °C). Fall R remained high until mid-November, even though water temperatures were similar to the spring period (ca. 20 °C). NP peaked in fall and lagged spring increases in PAR and water temperature.

Figure 3 shows the results of plotting two curves derived from single, randomly chosen measurements of NP and R from each month compared to a third curve derived from a mean of all the monthly data. Annual estimates of NP calculated from these three curves are 288, 232 and 230 g O_2 m^{-2} y^{-1}. respectively, and for R are 233, 230 and 287. The agreement of the annual estimates from curves 1 and 2 (n = 12) with curve 3 (n = 185) is striking. However, random sampling of a normally distributed population will tend to produce similar estimates if the sample size is large enough. Apparently a sample size of 12, evenly distributed over the year, will give similar estimates of annual metabolism as a much larger sample size. This is supported by the similarity among these three estimates and their general agreement with published values (Pomeroy 1959; Pamatmat 1968; Marshall *et al.* 1971; Riznyk and Phinney 1972; Cadee and Hegeman 1974, 1977; Gallagher and Daiber 1974; van Raalte *et al.* 1976; Joint 1978; Riznyk *et al.* 1978; Baillie and Welsh 1980; Zedler 1980; van Es 1982; Murray 1983; Colijn and de Jonge 1984; Rizzo and Wetzel 1985).

Table 2. Results of analyses of variance of the chlorophyll series. Mean ± standard deviation (mg m^{-2}). Sample size, N = no. of sampling days per month. * denotes significant difference among days (P = .05).

MONTH	CHLOROPHYLL A	N	PROBABILITY OF F
MARCH	33 ± 30	2	.025 *
APRIL	50 35	8	.002 *
MAY	86 40	11	.000 *
JUNE	121 34	12	.002 *
JULY	131 37	9	.000 *
AUGUST	136 53	7	.000 *
SEPTEMBER	208 33	3	.044 *
OCTOBER	176 46	2	.250
NOVEMBER	236 94	4	.001 *
DECEMBER	235 78	3	.002 *

While short-term variability appears to have little impact on estimates of annual production it has major impact on our perception of the seasonal changes in metabolism. In the plot of NP (fig. 3a) the monthly changes suggested by curves 1 and 2 based on 12 data points are much different from curve 3, each differing by an order of magnitude during one month (November and August). Curve 2 indicates net heterotrophy for the benthic community during most of the spring.

The respiration plots show similar differences. Each curve shows a different peak month for R, in September, March and July, for curves 1 to 3 respectively. Again, the differences among the points in most months are substantial, approaching an order of magnitude in September for curve 1.

Summary

While estimates of annual metabolic rates are very important, consideration of temporal distribution of metabolic rates is equally important in the overall functioning of estuaries and also for insight into controls on metabolic processes. The type of sampling design typically used in past studies (single, brief measurements, on a monthly schedule) appears to be adequate for estimation of annual rates, but is inadequate for assessing seasonal changes in metabolism or for providing insights into controls on metabolic processes on shorter time scales. As shown in this study, there is high variability in metabolic rates over the photoperiod, between similar days each month, and between tidal conditions, all imbedded in the changes in metabolism found on the seasonal scale. Variation on the shorter time scales can markedly affect the shape of seasonal curves based on limited data. Adequate description of seasonal changes requires enough sampling to encompass variability occurring on these shorter time scales. At a minimum, multiple measurements over the photoperiod, and an estimate of daily and tidally-related variability within the month is required to improve descriptions of the seasonal changes in metabolism within estuarine communities.

Intensive sampling over the photoperiod and between successive days also revealed temporal patterns in autotrophic and heterotrophic metabolism which are not wholly explained by variation in exogenous variables such as light and temperature. On these scales metabolic variability most likely varies as a result of changes in the physiological state of the community as a function of the past history of that community. Differences in metabolism over longer time scales are more explainable by variability in exogenous variables. NP increases during days with mid-day low tides as a result of improved light conditions. R increases are probably a response to greater availability of organic matter from in situ benthic primary production. Seasonal changes in NP and R are also reasonably interpretable in terms of the seasonal changes in the physical variables of light and temperature. The variability in metabolism over brief periods of time, and the factors which control it on those scales need further attention, and require sampling over extended periods of time, but perhaps at less frequent intervals.

Adequate estimates of the magnitude of community metabolism, the seasonal changes in community metabolism and the factors controlling community

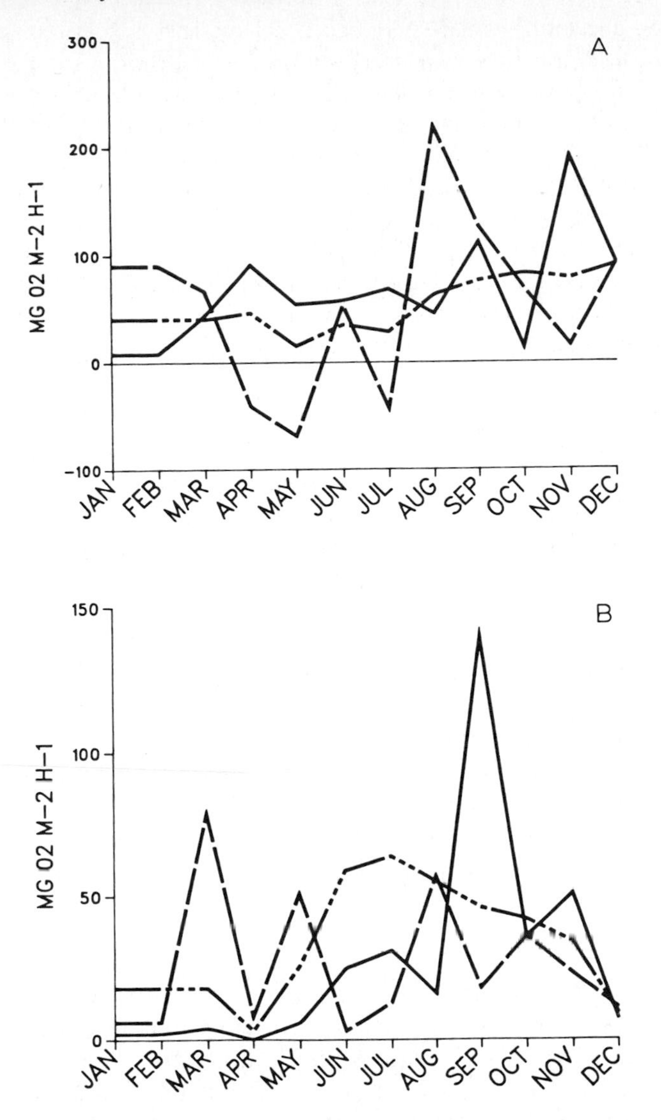

Figure 3. *Plots of net production (A) and respiration (B) in mg O_2 m^{-2} h^{-1}, by month. Curve 1 (———) and curve 2 (- - -) are drawn from a single randomly selected measurement each month (total sample size for each curve = 12). Curve 3 (———- - -) is drawn from a mean of all monthly measurements (total sample size = 185).*

metabolism are essential for a better understanding of benthic community dynamics. Adequate assessments of seasonal change and the factors important in controlling benthic community metabolism must take into account the metabolic changes occurring on shorter time scales.

Acknowledgments

This research was partially funded by a grant-in-aid from Sigma Xi, the scientific research society, and the Department of Wetlands Ecology, Virginia Institute of Marine Science. This is contribution number 1288 from the College of William and Mary, School of Marine Science, and Virginia Institute of Marine Science, Gloucester Point, VA.

References Cited

Admiraal, W. and H. Peletier. 1980. Influence of seasonal variations of temperature and light on the growth rate of cultures and natural populations of intertidal diatoms. *Mar. Ecol. Prog. Ser.* 2:35-43.

Baillie, P.W. and B.L. Welsh. 1980. The effect of tidal resuspension on the distribution of intertidal epipelic algae in an estuary. *Estuar. Coastal Mar. Sci.* 10:165-180.

Cadee, G.C. and J. Hegeman. 1974. Primary production of the benthic microflora living on tidal flats in the Dutch Wadden Sea. *Neth. J. Sea Res.* 8:260-291.

Cadee, G.C. and J. Hegeman. 1977. Distribution of primary production of the benthic microflora and accumulation of organic matter on a tidal flat area, Balzgand, Dutch Wadden Sea. *Neth. J. Sea Res.* 11:24-41.

Colijn, F. and V.N. de Jonge. 1984. Primary production of microphytobenthos in the Ems-Dollard estuary. *Mar. Ecol. Prog. Ser.* 14:185-196.

Gallagher, J.L. and F.C. Diaber. 1974. Primary production of edaphic algal communities in a Delaware salt marsh. *Limnol. Oceanogr.* 19:390-395.

Gargas, E. 1970. Measurements of primary production, dark fixation and vertical distribution of the microbenthic algae in the Oresund. *Ophelia* 8:231-253.

Heinle, D.R., C.,F. D'Elia, J.L. Taft, J.S. Wilson, M. Cole-Jones, A.. Caplins and L.E. Cronin. 1980. Historical Review of Water Quality and Climatic Data from Chesapeake Bay with Emphasis on Effects of Enrichment. U.S. EPA Chesapeake Bay Program Final Report, Grant #R806189010. Chesapeake Research Consortium, Inc. Publication No. 84. Annapolis, MD. 128 pp.

Jacobsen, T.R. 1978. A quantitative method for the separation of chlorophylls *a* and *b* from phytoplankton pigments by high pressure liquid chromatography. *Mar. Sci. Communications* 4:33-47.

Joint, I.R. 1978. Microbial production of an estuarine mudflat. *Estuar. Coastal Mar. Sci.* 7:185-195.

Leach, J.H. 1970. Epibenthic production in an intertidal mudflat. *Limnol. Oceanogr.* 15:514-521.

Lorenzen, C.J. 1967. Determination of chlorophyll and pheo-pigments: Spectrophotometric equations. *Limnol. Oceanogr.* 12:343-346.

Marshall, N., C.A. Oviatt and D.K. Skauen. 1971. Productivity of the benthic microflora of shoal estuarine environments in southern New England. *Int. revue ges. Hydrobiol.* 56:947-956.

Murray, L. 1983. Metabolic and structural studies of several temperature seagrass communities, with emphasis on microalgal components. Ph.D. dissertation. College of William and Mary. Williamsburg, Va. 90 pp.

Pamatmat, M.M. 1968. Ecology and metabolism of an intertidal sandflat. *Int. revue ges. Hydrobiol.* 53:211-298.

Pomeroy, L.R. 1959. Algal productivity in salt marshes of Georgia. *Limnol. Oceanogr.* 4:386-397.

Riemann, B. 1978. Carotenoid interference in the spectrophotometric determination of chlorophyll degradation products from natural populations of phytoplankton. *Limnol. Oceanogr.* 23:1059-1066.

Riznyk, R.Z. and H.K. Phinney. 1972. Manometric assessment of interstitial microalgae production in two estuarine sediments. *Oecologia* 10:193-203.

Riznyk, R.Z., J.I. Edens and R.C. Libby. 1978. Production of epibenthic diatoms in a southern California impounded estuary. *J. Phycol.* 14:273-279.

Rizzo, W.M. and R.L. Wetzel. 1985. Intertidal and shoal benthic community metabolism in a temperate estuary: Studies of spatial and temporal scales of variability. *Estuaries* 8:342-351.

Shaffer, G.P. and C.P. Onuf. 1983. An analysis of factors influencing the primary production of the benthic microflora in a southern California lagoon. *Neth. J. Sea Res.* 17:126-144.

Sokal, R.R. and F.J. Rohlf. 1969. *Biometry.* W.H. Freeman and Co. San Francisco, CA. 776 pp.

Steele, J.H. and I.E. Baird. 1968. Production ecology of a sandy beach. *Limnol. Oceanogr.* 13:14-25.

van Es, F.B. 1982. Community metabolism of intertidal flats in the Ems-Dollard estuary. *Mar. Biol.* 66:95-108.

van Raalte, C.D., I. Valiela and J.M. Teal. 1976. Production of epibenthic salt marsh algae: Light and nutrient limitation. *Limnol. Oceanogr.* 21:862-872.

Wilcoxon, F. and R.A. Wilcox. 1964. *Some rapid approximate statistical procedures.* Lederle Laboratories. Pearl River, NY 60 pp.

Zedler, J.B. 1980. Algal mat productivity: Comparisons in a salt marsh. *Estuaries* 3:122-131.

VERTICAL ACCRETION IN MARSHES WITH VARYING RATES OF SEA LEVEL RISE

J. Court Stevenson and Larry G. Ward

Horn Point Environmental Laboratories
University of Maryland
Cambridge, Maryland

and

Michael S. Kearney

Department of Geography
University of Maryland
College Park, Maryland

Abstract: Marsh systems are often considered sites of major sediment accumulation, where vertical accretion rates are equal to or greater than sea level rise. However, the degree of apparent sea level rise (land subsidence plus change in eustatic sea level) along the coast of the United States is variable, ranging from over 10 mm y^{-1} decline along the coast of southeastern Alaska to almost 10 mm y^{-1} rise along the northeastern Maine and Louisiana coasts. This variability in sea level rise accounts for some of the discrepancy in vertical accretion rates in marshes. We reviewed 15 areas dated with Lead-210 or Cesium-137, however, and at least four are clearly not keeping pace with sea level rise. Among these are Blackwater marsh in Chesapeake Bay, and the backmarsh areas of Barataria Bay, Fourleague Bay and Lake Calcasieu in Louisiana. There is surprisingly strong correlation ($r = 0.83$) between mean tidal range and accretionary balance for 15 marshes where we obtained data. Based on overall sedimentary processes, tidal wetlands can be classified into at least six major categories: emerging coastal, submerging coastal, estuarine, submerged upland, floating, and tidal freshwater marshes. Acceleration of present rates of sea level rise could cause major alterations in each of these six types and reduce their carbon export potential to surrounding coastal waters.

Introduction

Some of the earliest observations on tidal marsh function concerned marsh development processes. Mudge (1862), noting the occurrence of old peat below low tide level in a marsh near Lynn, Massachusetts, postulated that the marsh must have accreted vertically through time. Dawson (1868) concluded from similar observations in Nova Scotia that organic tidal marsh sediments had been deposited on subsiding substrates. In contrast, Shaler (1886) focused on the lateral development of marshes over tidal flats which were often first vegetated by seagrass (*Zostera*) communities. Ganong (1903) surmised that Canadian salt marshes developed on a base of inorganic red sediment over a subsiding shoreline, and posed an important question which remains with us today: "Whence then comes this great store of rich mud?" He theorized that it could only have been derived from resuspension of previously-eroded bottom materials from the Bay of Fundy which were transported subsequently to the marshes. By comparison, Davis (1910) explained marsh accretion as being driven by post-glacial sea level rise. Building upon this very early work in New England, Johnson and York (1915), Chapman (1938), and Miller and Egler (1950) helped establish the concept of marsh accretion as a general paradigm, applicable to many geographic areas (Pethick 1984).

Perhaps the most influential modern synthesis of marsh sediment processes and sea level rise in coastal marshes is Redfield's (1972) study at Cape Cod. Redfield demonstrated from numerous sediment cores from the Barnstable Harbor region that marshes developed after deglaciation behind a prograding spit. These marshes had accreted both vertically and laterally by capturing sediments brought by longshore drift as sea level rose over the last 4,000 years. Redfield's detailed description of sediment accumulation in response to sea level rise reinforced the other studies in New England and led to a conceptual model of marsh function in an environment of rising sea level. Indeed, so persuasive have been Redfield's (1972) inferences on patterns of vertical accretion at several locations in Barnstable Harbor that they, along with McCaffrey and Thomson's (1980) work on Long Island Sound, have often been cited (Frey and Basan 1978; Nixon 1982) as convincing evidence of the ability of coastal marshes to keep abreast of sea level rise.

The view that marshes represent sites of ubiquitous sedimentation (Stumpf 1983) has emerged despite the early work of Hoyt and his colleagues (see Hoyt and Hails 1967) implying that the long-term stability of marshes along transgressive coasts such as southeastern North America was questionable. For example, Newman and Munsart (1968) found that back-barrier marsh systems along the middle Atlantic coast tend to be comparatively young, about 1,000-1,500 years before present (BP) compared to Barnstable marsh (4,000 BP). As barrier islands move landward during periods of rising sea level, the marshes behind them are displaced and are eventually buried or eroded (Shideler *et al.* 1984). Other lines of evidence, such as basal Carbon-14 dates from existing marshes and tidal flux studies, do not support the concept that marshes are able to accrete material during periods of intense sea level rise. Rampino and Sanders (1981) have demonstrated that most present coastal marshes postdated the mid-Holocene rapid rise in sea level. When flux of materials has been measured, most marshes seem to be exporting sediment on an annual basis (Stevenson *et al.* 1985b).

Additional questions regarding the variability of marsh accretion rates with respect to changing sea levels have been raised by increasing evidence of large accretionary deficits in Louisiana marshes associated with locally high rates of coastal submergence (DeLaune *et al.* 1978; Boesch *et al.* 1984; Baumann *et al.* 1984). In recent years, radiometric and other dating methods have provided a long-term perspective on relationship of material fluxes in marshes to sea level changes, a problem inadequately addressed by yearly flux measurements because of their short duration and lack of storm-event sampling (Nixon 1980).

Our purposes here are to review the available accretion rates in coastal marshes and to determine to what extent rising sea level or other physical forces account for variations in these rates. We also attempt an assessment of why certain marshes do not keep up with sea level rise, and offer a conceptual framework for classifying marshes to account for variability of sedimentation and erosion processes in tidal wetlands.

Patterns of Apparent Sea Level Rise

Determination of Apparent Sea Level Rise

The Holocene period has been characterized by a worldwide transgression produced by the melting of the late Pleistocene ice sheets and the thermal expansion in ocean volume as world temperature rose. Most sea level researchers concur that the rate of change in global sea level has decelerated since the mid-Holocene (5,000-8,000 BP) allowing estuaries to reach their present state (Fairbridge 1980). However, the magnitude of sea level fluctuations during the past 3,000-4,000 y still remains controversial (DePratter and Howard 1981). A recent worldwide analysis of tide gauges (Gornitz *et al.* 1982) has provided an estimate of global eustatic trend on the order of 1.2 mm y^{-1} during the last 100 y. This estimate yields a long-term baseline for assessing the contribution of local subsidence or emergence in apparent sea level (ASL) changes as recorded by tide gauges.

Generally, marsh researchers have analyzed accretion versus sea level trends by using ASL rates derived from the nearest comprehensive tide gauge record (McCaffrey and Thomson 1980). Though acceptable for study of a particular marsh, this procedure makes comparison of accretion rates and sea level rise between marsh systems difficult because the time periods covered by individual tide gauge records are highly variable. The overall length of these records ranges from gauges established within the last several decades to the San Francisco Bay data set, which began in 1855 (Fig. 1). Gornitz *et al.*'s (1982) analysis of historic tide gauge records indicates that the rates of eustatic changes in sea level over shorter intervals were not the same as the long-term trend of 1.2 mm y^{-1}.

In order to eliminate the varying lengths of the tide gauge records as a source of possible noise and thus increase resolution of the sometimes small differences in ASL, we have used 1940-1980 as the standard time period because there is wide spatial coverage for this period (Hicks *et al.* 1983). One drawback of this approach is that the rate of global sea level rise during this time has been rather slow (Gornitz *et al.* 1982), producing an underestimate of long-term ASL rise. This situation is reflected in the downturn in the late 1970's in many Atlantic coast tide records and is detectable at the New York City gauge (Fig. 1). In a few cases we calculated 40-y trends by linear regression of annual mean sea level for stations omitted by Hicks *et al.* (1983). Although these records contain gaps, we believe that the relatively small errors incurred by including incomplete data sets are more than balanced by the benefits of having information from the critical

North Carolina and Louisiana coasts, where 1940-1980 trends were otherwise unavailable. The stations where fragmentary data were used include Wilmington, North Carolina; Eugene Is. and Grand Is., Louisiana; and Port Isabel, Texas.

Trends in Apparent Sea Level Changes

The temporal variability of apparent sea level rise along the North American coast is remarkably high (Fig. 1). Detailed spatial analyses carried out by Aubrey and Emery (1983) suggest points of flexure along the east coast at Wilmington and Boston where rates of ASL change direction (Fig. 2). North of Boston, coastal subsidence becomes as high as 9 mm y^{-1} at Eastport, Maine, documented by releveling of land benchmarks (Anderson *et al.* 1984). Trends in ASL also increase south of Boston, possibly because of the subsidence associated with Cameron's Fault or the Connecticut Valley Triassic Graben (Brown 1978). The glacial-outwash coastline of New Jersey has consistently high rates

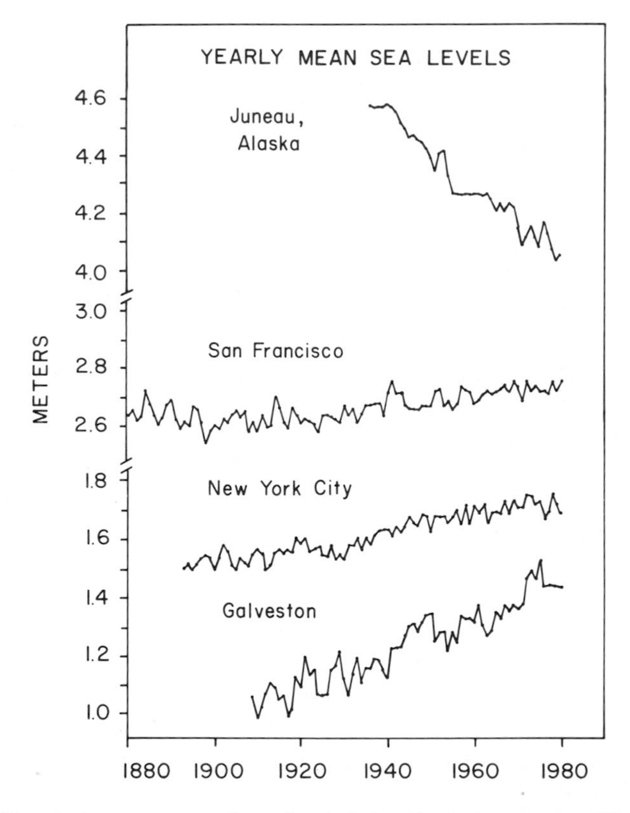

Figure 1. Trends in mean annual sea level during the last century in North America. Vertical scale is not to a common datum, but represents the sea level reported for each locality.

of ASL for both Atlantic City (3.9 mm y^{-1}) and Sandy Hook (4.0 mm y^{-1}), even though the latter may be affected by localized subsidence (Hicks 1972).

In the Chesapeake embayment, rates of sea level rise correspond with the overall pattern of subsidence reported by Holdahl and Morrison (1974), increasing toward the south (Fig. 2) and culminating at about 6 mm y^{-1} in the Great Dismal swamp. Our regression shows only a slight ASL rise (1.9 mm $^{-1}$) for Wilmington, which is in agreement with Brown's (1978) contention of relative uplifting of this portion of the Carolina coast associated with the basement arch seaward of Cape Fear (Stewart 1976). This conflicts with Aubrey and Emery's (1983) analysis, however, which predicts locally high rates of sea level rise along the Carolina coast. More tide gauge data is necessary to resolve sea level trends in this area. Along the coast of South Carolina and Georgia, ASL appears to be rising at similar rates ($\sim$2.5 mm y^{-1}).

Further south, rates of ASL rise decline along the east coast of Florida toward the Cape Canaveral prominence (Fig. 2). In the vicinity of the Everglades, benchmark releveling shows there is a slight uplift of the Florida Gulf coast but apparently no tide gauges are positioned to detect this trend (Holdahl and Morrison 1974). By comparison, Louisiana and Texas are experiencing marked subsidence demonstrated in the long-term ASL record at Galveston (Fig. 1). The high rates of subsidence along this section of the Gulf coast may be attributed to oil and gas drilling and dewatering of aquifers (Poland and Davis 1969) as well as compression due to sediment loading from the Mississippi River (Gosselink 1984).

In contrast to the regional patterns in ASL along the east coast, changes in ASL along the California coast are relatively erratic, reflecting neotectonic activity associated with plate movements (Fig. 2). Although the southern part of this coastline is not an ideal environment for extensive marsh development because of the steep topographic gradients and erratic rainfall (Zedler, this volume), areas farther north in California had extensive tidal wetlands (e.g., San Francisco) prior to development. The ASL rise at the mouth of San Francisco Bay has been about 1.5 mm y^{-1} (Fig. 1), while across the Bay at Alameda, 20 km to the east, sea levels have been remarkably stable (0.1 mm y^{-1}). Although the northern California and Oregon coasts have decreasing ASL, Puget Sound, Washington has increasing ASL, varying from 0.5 mm y^{-1} at Friday Harbor to 2.1 at Seattle (Fig. 2). Sea level trends again reverse sharply further north due to the emerging coastline. At Juneau, Alaska, the decline of ASL has been 12.8 mm y^{-1} (Fig. 1). Earthquakes in this region can catastrophically lift shorelines as much as 10 m! Thus, marshes along coastlines with crustal uplift are precariously balanced with respect to sea level.

Sea Level Rise and Vertical Accretion in Coastal Marshes

Methods for Estimating Vertical Accretion

The first systematic accretion rates suited to analysis of whether marsh surfaces are keeping abreast of recent sea level change were derived from marker

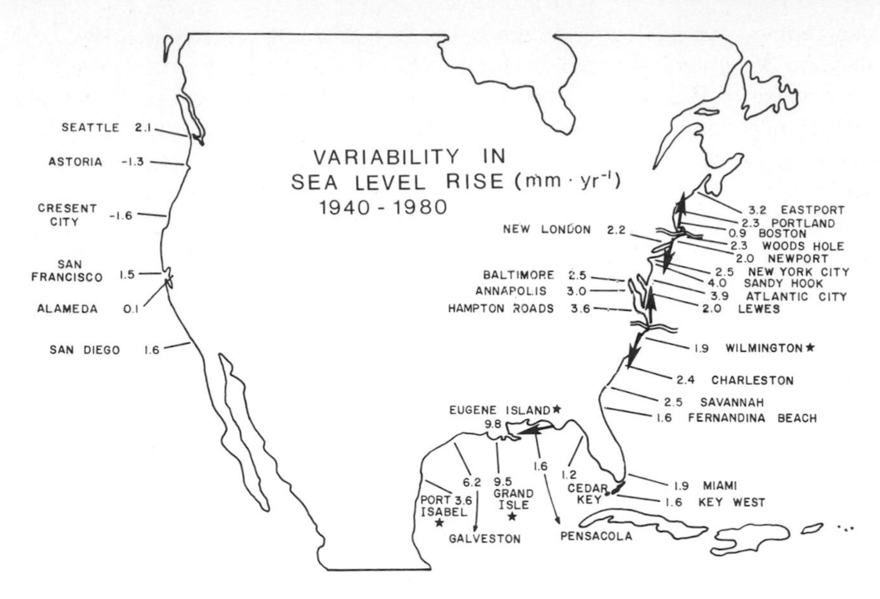

Figure 2. Spatial variability in rates of apparent sea level (ASL) change during the last forty years in the United States. Broken wavy lines are flexures described by Aubrey and Emery (1983) where ASL reverses as indicated by the arrows. Negative values indicate falling ASL. Asterisks designate incomplete data sets.

horizon techniques (Richards 1934). However, near-surface marker horizons and similar methods do not account for the progressive compression of older materials at deeper levels in marsh sediments, brought about by autocompaction of peat (Kaye and Barghoorn 1964). The problem of compaction is evident in Redfield's (1972) data in which surficial horizons have higher "apparent" accretion rates than the deeper measurements integrated over longer time periods. Like other estuarine deposits (Nichols and Biggs 1985), the surface sediments of marshes are often unconsolidated organic debris that becomes compressed with increasing burial (Stevenson *et al.* 1985b). Additional losses in volume may occur from dewatering and from decomposition of organic materials through oxidation, methanogenesis, denitrification and sulfate reduction. Compaction of continental shelf sediments increases as the percentage of organic matter increases (Busch and Kellar 1982). Unfortunately, similar data are unavailable for marsh sediments, where the percent organics is much higher than in most other marine systems. In conclusion, since some unknown amount of compaction will occur in the superficial horizons, it is necessary to obtain rates of mass accumulation from deeper core horizons to assess long-term vertical accretion.

In this review we rely as much as possible on Lead-210 dating (Koide *et al.* 1972; Turekian and Cochran 1978; Schell 1982) for accretion estimates because it accounts for dewatering and decomposition of organics. The available Cesium-137 measurements (Delaune *et al.* 1983) are also included to increase geographic coverage even though we believe they overestimate the long-term (beyond 30 y) vertical accretion due to autocompaction. Lastly, because of the importance of Barnstable marsh as a point of comparison, we use Redfield's (1972) rate derived from historical evidence because it spans the same time period as Lead-210 dating (ca. 150 y at Nauset Harbor). Unfortunately, Redfield's highest accretion estimate (Broadway Island), so often used as a basis for comparison to other marshes, is actually from a subtidal mud flat where accretion rates can exceed those of adjacent marshes (T. Jordan, pers. comm.).

Marsh Accretion Rates

The lowest vertical accretion rate of the marshes we reviewed (Table 1) was a surprising 1.4 mm y^{-1} for Oyster Landing marsh at North Inlet, South Carolina, which was even lower than Blackwater marsh in Chesapeake Bay (1.7 mm y^{-1}). The highest rate was 14 mm y^{-1} for a levee marsh in Barataria Bay of Louisiana. Significantly, both Blackwater (Stevenson *et al.* 1985a) and Barataria Bay (Hatton *et al.* 1983) marshes are adjacent to areas that were once marsh but now are actively submerging. It is possible that the remaining levee marshes in these systems may be trapping material from the more extensive eroding backmarsh areas as well as from the usual riverine sources. Our review showed a much narrower range of accretion rates for most marshes, with the northern marshes having very similar means (4.3-5.5 mm y^{-1}) despite the relatively wide range of ASL rise (Table 1). Barnstable marsh is located in one of the areas of lowest ASL rise, which may account for its ability to keep pace with sea level.

Slightly less than 75% of the marshes for which data were available have a positive accretionary balance when plotted against ASL rise (i.e., they fall above the diagonal in Fig. 3). However, some of the marshes having positive accretion are surprisingly close to the overall ASL rise (such as the Duplin River marshes at Sapelo Island, Georgia). This may be important because the 1940-1980 ASL rise underestimates the long-term average and suggests that marshes close to the diagonal in Fig. 3 could be on the borderline of submergence. Also, the accretion rate of the Sapelo Island marsh was derived from Cesium-137 dating, which usually overestimates long-term accretion since it does not account for autocompaction of sediments. Zarillo (1985) found that the Duplin River system is ebb-dominated and exports sand derived from marsh creek-banks and subtidal areas. This export coupled with the modest accretion rate suggests that this system may be on the verge of submergence. Letzsch and Frey (1980) have observed seasonal erosion at nearby Blackbeard marsh on the northern part of Sapelo Island, although they concluded that those short-term losses appeared to be balanced by depositional processes during their two year study. The dynamics of sediment transport processes are not reported for nearby North Inlet, S.C., although accretion rates (Sharma *et al.* 1986) suggest that marshes there may be even closer to submergence than at Sapelo Island. In comparison, estimates of accretion in the deltaic marshes at the mouth of the Savannah River are as high as 12 mm y^{-1} (Goldberg *et al.* 1979), no doubt because of riverine

Table 1. Summary of rates (mm y^{-1}) reported for accretion and apparent sea level (ASL) rise (1940-1980) for tidal marshes dated with Lead-210 (L) or Cesium-137 (C) compared to Barnstable marsh dated with historical (H) techniques.

Location	Salinity ‰	Technique	Accretion Range	Accretion Mean	ASL Rise	Mean Tidal Range (m)	Reference
Barnstable, MA	20-30	H	3-8	5.5	0.9	2.9	Redfield 1972
Prudence Is., RI	28-32	L	2.8-5.8	4.3	1.9	1.1	Urso and Nixon 1984
Farm River, CN	—(1)	L	—	5.0	1.9	1.8	McCaffrey & Thomson 1980
Fresh P., NY	26	L	—	4.3	2.2	2.0	Clark and Patterson 1985
Flax P., NY	26	L	4.7-6.3	5.5	2.2	2.0	Armentano and Woodwell 1975
Lewes, DE	25-30	L	—	4.7	2.0	1.3	Church *et al.* 1981
Lewes, DE	25-30	C	—	>10.0	2.0	1.3	Brickman 1978
Nanticoke, MD	2-6	L	4.9-7.2	6.1	3.2	0.7	Kearney and Ward 1986
Blackwater, MD	1-5	L	1.7-3.6	2.6	3.9	0.3	Stevenson *et al.* 1985a
North R., NC	—(1)	L	2-4	3.0	1.9	0.9	Benninger and Chanton 1985
North Inlet, SC	30-35	L	1.4-4.5	2.5	2.2	1.6	Sharma *et al.* 1986
Savannah R., GA	—(1)	L	—	11.0	2.5	3.0	Goldberg *et al.* 1979
Sapelo Is., GA	30-35	C	3-5	4.0	2.5	2.1	Hopkinson (pers. comm.)
Barataria, LA							
Fresh	<1%	C	6.5-10.6	6.9			Hatton *et al.* 1983
Brack.	5-10	C	6.4-13.5	7.1			Hatton *et al.* 1983
Inter.	10-15	C	5.9-14.0	6.7			Hatton *et al.* 1983
Salt	>15	C	7.5-13.5	8.1			Hatton *et al.* 1983
Mean				7.2	9.5	0.5	Hatton *et al.* 1983
Fourleague, LA	10-20	C	—	6.6	8.5	0.3	Baumann *et al.* 1984
L. Calcasieu, LA	15	C	6.7-10.2	7.8	9.5	0.6	Delaune *et al.* 1983

(1) Not given in reference

sediment input not present at Sapelo Island or North Inlet and increased tidal energies in this region (Blanton and Atkinson 1978).

When the accretionary balance (vertical accretion rate minus local ASL rise) was plotted against mean tidal range for each marsh (Fig. 4), a moderate correlation was obtained ($r = 0.83$). This improves ($r = 0.86$) if the riverine marshes are not included, but the regression lines have almost identical slopes (2.4 vs. 2.7 mm per meter). This suggests the amount of tidal energy is equally important under high and low sediment inputs in determining rates of marsh accretion. Figure 4 also illustrates that most submergence occurs in marshes of microtidal environments especially where the mean tides are in the range of 0.5 m or less. More accretion rates are needed from areas of low tidal energy to confirm this relationship because other factors that may influence accretion processes (such as hydraulic alterations, grazing by muskrats and other animals, marsh burning and rising salinities) are common features of submerging marshes in Louisiana and the Chesapeake. However, we postulate that brackish and salt marshes (where tides are often weak and irregular) have particular problems

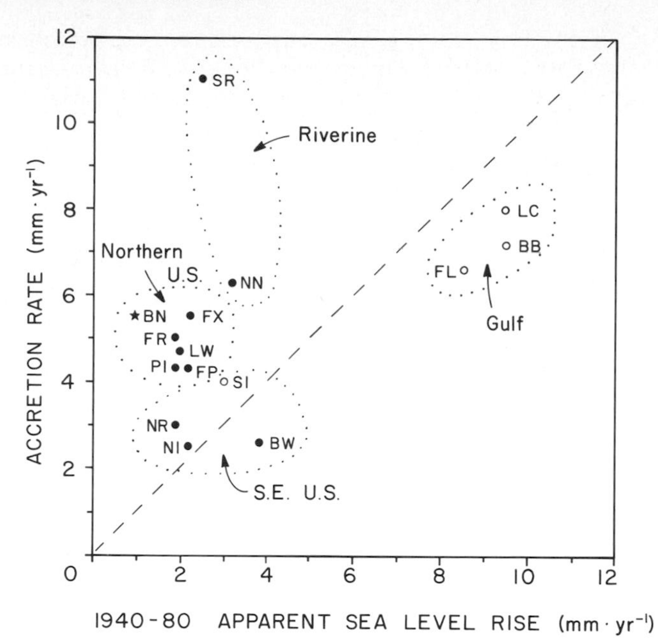

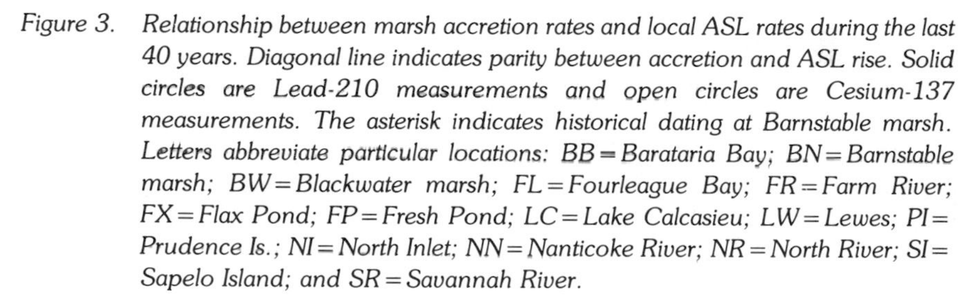

Figure 3. *Relationship between marsh accretion rates and local ASL rates during the last 40 years. Diagonal line indicates parity between accretion and ASL rise. Solid circles are Lead-210 measurements and open circles are Cesium-137 measurements. The asterisk indicates historical dating at Barnstable marsh. Letters abbreviate particular locations: BB = Barataria Bay; BN = Barnstable marsh; BW = Blackwater marsh; FL = Fourleague Bay; FR = Farm River; FX = Flax Pond; FP = Fresh Pond; LC = Lake Calcasieu; LW = Lewes; PI = Prudence Is.; NI = North Inlet; NN = Nanticoke River; NR = North River; SI = Sapelo Island; and SR = Savannah River.*

with rising ASL when sediment inputs are low or reduced from previous levels. Low tidal action also contributes to waterlogging of sediments which causes physiological stress in *Spartina* (Mendelssohn *et al.* 1981) and low overall standing crops of marsh grass (Steever *et al.* 1976). This reduction in productivity limits the accretionary potential of *Spartina* marshes from the standpoint of *in situ* organic inputs. Although not all species seem to be as affected by low tidal amplitudes (e.g., *Juncus roemerianus*), low tidal energy also limits the amount of regular inorganic sediment input available from seaward sources. Thus sporadic sedimentation during major storm events such as hurricanes may be a critical factor in the sediment budget of marshes having microtidal environments (Stumpf 1983; Baumann *et al.* 1984).

The New England marshes have tidal ranges of 1-3 meters which renders them less vulnerable to sea level change because sediment transport potential is greater. Although the length of time that marshes are inundated (influenced by tidal range and storm activity) and the biological productivity (via organic deposi-

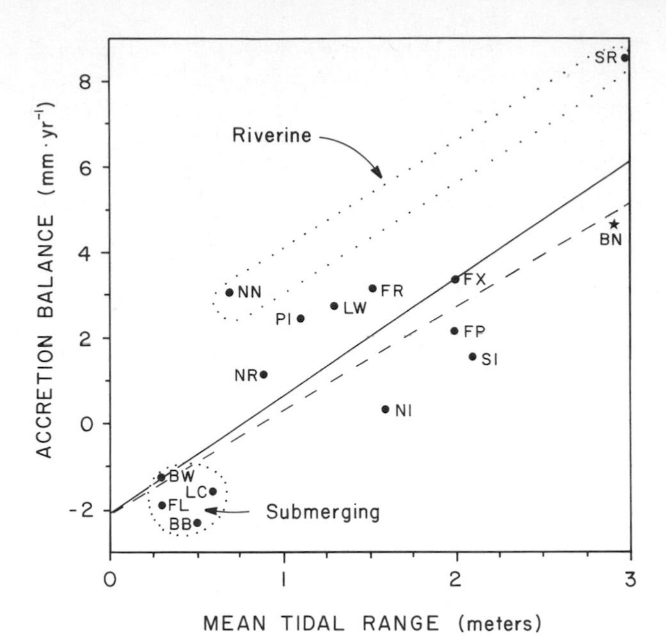

Figure 4. Relationship between accretionary deficit (vertical accretion rate minus local ASL rise) and mean tidal range of marshes (r = 0.83) in Figure 3. Solid regression line (Y = 2.7X − 2.1) is for all 15 marshes, while dashed line (Y = 2.4X − 2.1) is for 13 non-riverine marshes (r = 0.86).

tion), are important in controlling accretion rates, other factors such as proximity to riverine or diffuse source inputs are also crucial (Stevenson *et al.* 1985b). In the estuarine meander marshes of the Nanticoke River, for example, the higher accretion rates were associated with the riverine sediment inputs, and occurred upstream rather than downstream nearer the Chesapeake Bay (Kearney and Ward 1986). In contrast, a Choptank River *Hibiscus* marsh appears to have extremely high accretion rates, exceeding 10 mm y^{-1} (Stevenson and Ward, unpublished). Although this *Hibiscus* marsh has reduced tidal amplitudes due to a culvert at its mouth (Stevenson *et al.* 1977), it also has comparatively high primary and secondary productivities (Cahoon and Stevenson 1986) and receives a large input of terriginous sediment from surrounding agricultural fields. These studies suggest that there are a variety of sedimentary processes that are important to determining rates of vertical accretion in marsh systems.

Marsh Classification Based on Accretionary Relationships

A classification scheme based on broad functional patterns would help to conceptualize the large differences in sedimentary processes and tidal energies associated with marsh ecosystems. Chapman's (1974) regional approach to marsh classification, with an emphasis on floristic considerations and an awkward European system of nomenclature (Braun-Blanquet 1928), is not appropriate. Frey and Basan (1978) adapted Chapman's general scheme for dividing the salt marshes of the United States into four regional types: New England, Southeastern Coastal Plain, Central Pacific and Alaskan-sub arctic. Although this classification has utility, it does not effectively contrast the basic functional differences among marsh systems. We have taken an alternate approach by categorizing tidal wetlands into six types based on marsh response to extrinsic forcing functions (Fig. 5). Since we are introducing a simple classification scheme, most of our categories are highly aggregated and could logically be split into subgroups, but that level of detail is beyond the scope of this review.

Emerging Coastal

This marsh type is common along tectonically active coastlines such as Alaska, Newfoundland and Scandinavia as well as in scattered localities in the southern hemisphere in Australia and Southern Africa (Clark 1977). Where emergence is significant, marsh development along these coastlines is restricted. The marsh peat in these areas becomes part of the upland soil matrix and although halophytic salt-marsh plants will persist for some time, the area is eventually invaded by glycophytes from landward grasslands (Chapman 1974). Therefore, these marshes are clearly transitional and could be considered classic examples of Shaler's (1886) concept of succession.

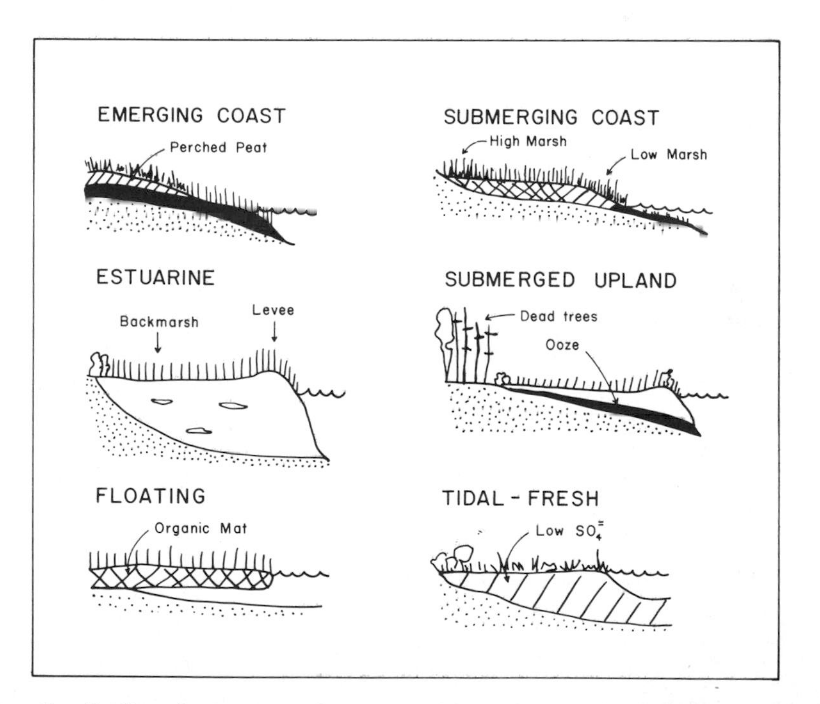

Figure 5. Profiles of six major sedimentary tidal marsh types showing the main distinguishing features of each.

Submerging Coastal

This category includes those marshes which develop on submerging coastlines with characteristically high salinities, such as the coasts of the United States north of Florida, northern Europe (excluding Scandinavia) and the majority of the coastlines of Asia, Africa and South America (Clark 1977). One characteristic of these systems, especially in the northern latitudes where total suspended sediment (TSS) is low, is the presence of "high and low marsh" zones (Redfield 1972) with the latter seaward zone being the most recent (Pethick 1981). When wave energy is lower and TSS is higher, a well-developed levee replaces low marsh. Although these marshes seem to be keeping up with sea level rise, sediment flux estimates in the middle Atlantic and southeastern coastal plain of the United States suggest net export of sediment (Settlemyre and Gardner 1977; Boon 1978; Roman 1981; Ward 1981). A combination of high *in situ* carbon fixation and high deposition during storm events might explain the discrepancies between accretion and flux measurements, but incipient submergence may be occurring where sediment inputs are less than ASL rise.

Estuarine

Estuarine marshes are typically situated in localities with abundant sediment such as river deltas or turbidity maxima of estuaries (Ahnert 1960). In contrast to the "high-low marsh" zones of high-salinity coastal marshes of New England (Nixon 1982), these marshes are characterized by streamside levees with lower elevation backmarshes. Because of allochthonous inputs, estuarine marshes often have extremely high rates of accretion and consequently very deep sedimentary profiles. However, if the sediment supply is reduced, accretionary deficits can quickly develop because of the continuing process of compaction in the deeper peats (Kaye and Barghoorn 1964). As these areas become increasingly waterlogged, primary productivity drops and carbon deposition is reduced to low levels.

Submerged Upland

This category consists of the brackish marshes that develop on recently submerged coastal terraces under low tidal ranges and often intergrade into the previous two types. Originally described for major estuaries such as Chesapeake Bay (Darmody and Foss 1979), this marsh type is also found scattered throughout flat coastal plain terrain and appears well-developed in many areas near the extensive sounds in North Carolina. These marshes differ from the estuarine type by having a relatively thin veneer of sediment over a recently submerged upland soil profile. They can often be identified superficially by the presence of dead trees at the landward edge indicating landward encroachment. Since these wetlands often erode at the seaward edge, they represent the reverse of Shaler's classic successional scheme for New England marshes. They are prone to submergence and possibly erosion because sediment supplies both from the sea and from upland sources are limited. An unconsolidated, unstable organic layer consequently forms beneath the rhizomatous surface mat (Stevenson *et al.* 1985a).

Floating Mat

The unique floating mat marshes constitute a large proportion of coastal wetlands on the coast of the Gulf of Mexico and occur sporadically in Europe, Africa, and South America (Sasser and Gosselink 1984). This marsh type may be related to degraded submerged upland marshes in which the rhizomatous peat mat has broken loose from the subsiding substrate (O'Neil 1949). Because the surface of this marsh is not inundated by tidal action but floats with the tides, it has little access to an inorganic sediment supply. Instead these marshes grow almost entirely from organic material that is deposited by high *in situ* primary productivity (Sasser and Gosselink 1984). This marsh type may be the least affected by sea level rise since it can simply float upward in response in increasing ASL. However, whether floating marshes could withstand the elevated salinities associated with rising sea levels is questionable.

Tidal Freshwater

The floristically-rich tidal freshwater marshes are located at the heads of estuaries in water with salinity less than 0.5 g l^{-1} and where tidal action is often still very strong (Odum *et al.* 1984). These wetlands often accrete rapidly when they receive substantial quantities of riverine inputs from large upstream watersheds and form deltaic marshes. Also, they may deposit more peat than brackish and salt marshes because decomposition rates may be slow due to limited electron acceptors. Howarth and Hobbie (1982) speculated that methanogenesis in freshwater sediments is less efficient at peat decomposition than is sulfate reduction, which occurs when seawater is present. Hence, a large quantity of organic materials can be stored in the sediment pool of these marshes, accounting for their importance in coal formation (Galloway and Hobday 1983). Rapidly rising sea levels could result in massive degradation of freshwater peats since their preservation depends on a pH less than 4 (Cecil *et al.* 1979) and seawater is usually circumneutral. Thus, marine intrusions result in eventual destruction of this marsh type if sediment inputs are inadequate.

Overview

The wide range of variation in vertical marsh accretion which emerges from our literature review and classification scheme demonstrates that concepts concerning marsh development need reformulation. It might be argued that part of the variability is due to methodological problems such as differential deposition, scavenging, remobilization (T. Church, pers. comm.) and bioturbation (R. Gardner, pres. comm.) of Lead-210 and/or Cesium-137. However, the marshes we reviewed where accretionary balances are negative have been submerging in recent years as evidenced by aerial photography (Gagliano *et al.* 1981, Stevenson *et al.* 1985a). While this relationship has been reported only for Louisiana and the Chesapeake Bay, these regions comprise almost 8,500 km^2 or approximately half of the tidal marsh acreage in the United States (Gosselink 1984, Stevenson *et al.* 1985b). Therefore the prospect of submergence cannot be dismissed as inconsequential. We believe that other areas may be facing similar problems and encourage more research in the marsh types identified above where no modern accretion studies have yet been carried out, and especially in areas such as northeastern Maine where ASL rates have been high.

Our greatest concern, however, is the prediction that as CO_2 increases in the atmosphere, eustatic sea level will rise even more dramatically than in the past (Hansen *et al.* 1985). Although scrutiny of past tide gauge records (Fig. 1) makes Hoffman *et al.*'s (1983) prediction of a magnitude increase in ASL (10-20 mm y^{-1}) over the next century questionable, Hansen *et al.* (1985) have concluded that vertical mixing in the oceans may act as a delayed feedback loop. Thus, most of the effects of past CO_2 increases may not yet be realized. Our analysis suggests that any acceleration in ASL on coastal marshes (except perhaps the floating type) could be catastrophic. Some of the prevailing concepts concerning marsh function could become obsolete.

One pertinent example is E.P. Odum's (1968) controversial carbon outwelling hypothesis, which was first formulated from early work at Sapelo Island, Georgia. Marshes such as this, which appear on the borderline of submergence now, may not be able to sustain carbon exports to surrounding systems under high ASL rise conditions. Contrary to indications in previous budgets (Pomeroy and Wiegert 1981; Hopkinson and Hoffman 1984), considerable carbon burial is needed to keep up with sea level rise if the inorganic sediment flux is in balance (Letzsch and Frey 1980). A 10-mm rise in ASL would require the deposition of 2-4 kg m^{-2} y^{-1}—virtually the entire annual productivity of most marshes—to maintain a dry bulk density of 0.2-0.4 g cm^{-3} (Fig. 6). Any carbon export might represent the final degradation and erosion of old refractory materials from previously deposited peat mats, with no nourishment potential for coastal food chains.

The impacts of submergence of marshes, as they fail to keep pace with ASL, have obvious ramifications for dependent food chains. In addition, the effects of rising sea levels on estuarine circulation, bathymetry and sediment budgets need to be more carefully assessed. Schubel (this volume) comments on the aging of estuaries due to high sediment inputs. However, if apparent sea level rise exceeds the rate of overall estuarine sedimentation which, for example, averages 7.6 mm y^{-1} in the Chesapeake Bay (Officer *et al.* 1984), there should be a reversal of the geological aging process. The consequences of this rejuvenation may be profound not only for marsh systems, but also for the entire coastal environment. Changes of marsh acreage to open water could in turn lead to reconfiguration of shorelines and eventually change prevailing currents which influence sediment transport patterns. The extent to which submergence of marshes leads to massive erosion in systems with accretionary deficits should be a focus of future research. We could then, perhaps, slightly rephrase Ganong's (1903) query: Whence then *goes* this great store of rich mud?

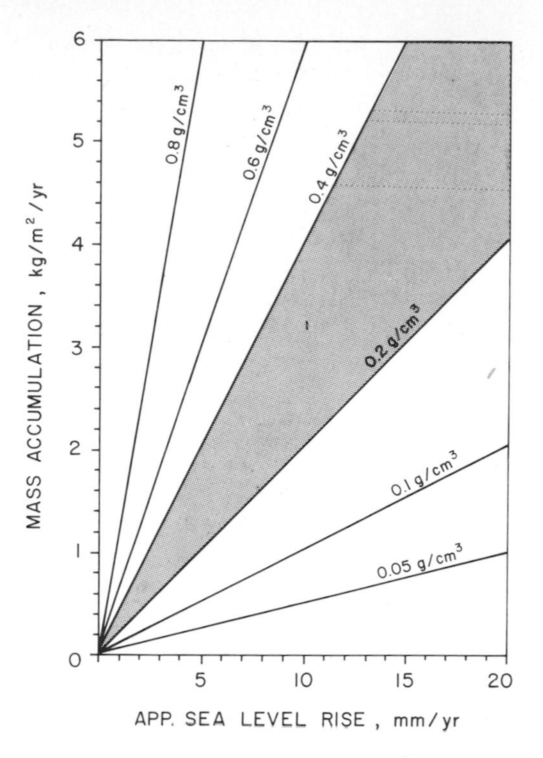

Figure 6. The mass accumulation of marsh sediments necessary to maintain different dry bulk densities (g cm^{-3}) under increasing sea level rise. Shaded area suggests the range needed to minimize autocompaction.

Acknowledgments

This is contribution No. 1672 from the University of Maryland Center for Environmental and Estuarine Studies. We are especially indebted to John Day Jr. for his encouragement to attempt a synthesis of marsh accretion and sea level relationships. A number of investigators supplied critical information during our review efforts. These include W. Boicourt, J. Chanton, T. Church, J. Day, C. Hackney, C. Hopkinson, T. Jordan, V. Lee, S. Nixon, R. Twilley and C. B. Urso. Also we thank L. R. Gardner, J. Gosselink, D. Nummedal, C. P. Stevenson and D. Wolfe for their comments and constructive criticisms of the manuscript. This work was partially supported by grants from the U.S. Fish and Wildlife Service; U.S. Dept. of Interior, Office of Water Resources Research, and the Maryland Department of Natural Resources.

References Cited

Ahnert, F. 1960. Estuarine meanders in the Chesapeake Bay. *Geogr. Rev.* 50:390-401.

Anderson, W. A., J. T. Kelley, W. B. Thompson, H. W. Borns, Jr., D. Sanger, D. C. Smith, D. Tyler, R. S. Anderson, A. E. Bridges, K. J. Crossen, J. W. Ladd, B. G. Anderson, and F. T. Lee. 1984. Crustal warping in coastal Maine. *Geol.* 12:677-680.

Armentano, T. V. and G. M. Woodwell. 1975. Sedimentation rates in a Long Island marsh determined by ^{210}Pb dating. *Limnol. Oceanogr.* 20:452-456.

Aubrey, D. G. and K. O. Emery. 1983. Eigenanalysis of recent United States sea levels. *Contin. Shelf Res.* 2:21-33.

Baumann, R. H., J. W. Day and C. A. Miller. 1984. Mississippi deltaic wetland survival: sedimentation versus coastal submergence. *Science* 224:1093-1095.

Benninger, L. K. and J. P. Chanton. 1985. Fallout ^{239}, ^{240}Pu and natural ^{238}U and ^{210}Pb in sediments of the North River Marsh, North Carolina. *EOS (Transac. Amer. Geophysic. Union)* 66:276 (Abstract).

Blanton, J. O. and L. Atkinson. 1978. Physical transfer processes between Georgia tidal inlets and nearshore waters, pp. 515-532. *In:* M. L. Wiley (ed.), *Estuarine Interactions.* Academic Press, New York.

Boesch, D. F., J. W. Day, Jr. and R. E. Turner. 1984. Deterioration of coastal environments in the Mississippi deltaic plain: options for management, pp. 447-466. *In:* V.S. Kennedy (ed.), *The Estuary as a Filter.* Academic Press, New York.

Boon, J. D. III. 1978. Suspended Solids in a saltmarsh creek—an analysis of errors, pp. 147-161. *In:* B. Kjerfve (ed.), *Estuarine Transport Processes.* University of South Carolina Press, Columbia.

Braun-Blanquet, J. 1928. *Pflanzenoziologie.* Grunadzuge der Vegetationskunde, Berlin. 439 pp.

Brickman, E. 1978. *^{137}Cs Chronology in Marsh and Lake Samples from Delaware.* M.S. Thesis. University of Delaware. Newark. 93 pp.

Brown, L. D. 1978. Recent vertical crustal movement along the east coast of the United States. *Tectonophysics* 44:205-231.

Busch, W. H. and G. H. Keller. 1982. Consolidation characteristics of sediments from the Peru-Chile continental margin and implications for past sediment instability. *Mar. Geol.* 45:17-39.

Cahoon, D. R. and J. C. Stevenson. 1986. Production, predation, and decomposition in a low salinity shrub marsh. *Ecology* 67:(in press).

Cecil, C. B., R. W. Stanton, F. T. Dulong and A. D. Cohen. 1979. Geologic factors that control mineral matter on coal, pp. 43-56. *In:* A. C. Donaldson, M. W. Presley, J. J. Renton (eds.), *Carboniferous Coal Guidebook,* Vol. 3. West Virginia Geol. and Econ. Surv., Wheeling.

Chapman, V. J. 1938. Coastal movement and the development of some New England salt marshes. *Proc. Geol. Assn.* 49:373-384.

Chapman, V. J. 1974. *Salt Marshes and Salt Deserts of the World.* 2nd edition. Verlag von J. Cramer, Lehre Germany. 392 pp.

Church, T. M., C. J. Lord III and B. L. K. Somayajulu. 1981. Uranium, thorium, and lead nuclides in a Delaware Salt Marsh Sediment. *Estuar. Coastal Shelf Sci.* 13:267-275.

Clark, J. A. 1977. A numerical model of worldwide sea level changes on a viscoelastic earth, pp. 525-534. *In:* N. A. Morner (ed.), *Earth Rheology, Isostasy and Eustasy.* John Wiley, New York.

Clark, J. S. and W. A. Patterson III. 1985. The development of a tidal marsh: upland and oceanic influences. *Ecol. Monog.* 55:189-217.

Darmody, R. G. and J. E. Foss. 1979. Soil-landscape relationships of the tidal marshes of Maryland. *Soil Sci. Soc. Amer. Journ.* 43:534-541.

Davis, C. A. 1910. Salt marsh formation near Boston and its geological significance. *Econ. Geol.* 5:623-639.

Dawson, J. W. 1868. *Acadian Geology* (2nd ed.) Macmillan, New York. 694 pp.

DePratter, C. B. and J. Howard. 1981. Evidence for a sea level lowstand between 4500 and 2400 years BP on the southeast coast of the United States. *J. Sed. Petrol.* 51:1287-1295.

DeLaune, R. D., W. H. Patrick Jr. and R. J. Buresh. 1978. Sedimentation rates determined by ^{137}Cs dating in a rapidly accreting salt marsh. *Nature* 275:532-533.

Delaune, R. D., R. H. Baumann and J. G. Gosselink. 1983. Relationships among vertical accretion, coastal submergence, and erosion in a Louisiana gulf coast marsh. *Jour. Sed. Pet.* 53:147-157.

Fairbridge, R. W. 1980. The estuary: its definition and geodynamic cycle, pp. 1-36. *In:* E. Olausson and I. Cato (eds.), *Chemistry and Biogeochemistry of Estuaries.* John Wiley & Sons, New York.

Frey, R. W. and P. B. Basan. 1978. Coastal salt marshes. pp. 101-168. *In:* R. A. Davis (ed.), *Coastal Sedimentary Environments.* Springer-Verlag, New York.

Gagliano, S. M., K. J. Meyer-Arendt and K. M. Wicker. 1981. Land loss in the Mississippi River deltaic plain. *Trans.-Gulf Coast Assoc. Geol. Soc.* 31:295-300.

Galloway, W. E. and D. K. Hobday. 1983. *Terrigenous Clastic Depositional Systems.* Springer-Verlag, New York. 423 pp.

Ganong, W. F. 1903. The vegetation of the Bay of Fundy salt and dyked marshes: An ecological study. *Bot. Gaz.* 36:161-186.

Goldberg, E. D., J. J. Griffin, V. Hodge and M. Koide. 1979. Pollution history of the Savannah River estuary. *Environ. Sci. Tech.* 13:588-594.

Gornitz, V., S. Lebedeff and J. Hansen. 1982. Global sea level trend in the past century. *Science:* 1611-1614.

Gosselink, J. G. 1984. *The Ecology of Delta Marshes of Coastal Louisiana: A Community Profile.* FWS/OBS-84/09. U.S. Fish Wildlife Service. Slidell, Louisiana. 134 pp.

Hansen, J., G. Russell, A. Lacis, I. Fung and D. Rind. 1985. Climate response times: dependence on climate sensitivity and ocean mixing. *Science* 229:857-859.

Hatton, R. S., R. D. Delaune and W. H. Patrick, Jr. 1983. Sedimentation, accretion, and subsidence in marshes of Barataria Basin, Louisiana. *Limnol. Oceanogr.* 28:494-502.

Hicks, S. D. 1972. Vertical crustal movements from sea-level measurements along the east coast of the United States. *J. Geophys. Res.* 77:5930-5934.

Hicks, S. D., H. A. Debaugh Jr., and L. E. Hickman. 1983. *Sea Level Variations for the United States 1855-1980.* U.S. Dept. Commerce, National Oceanic and Atmospheric Administration, Rockville, Maryland. 170 pp.

Hoffman, J. S., D. Keyes and J. G. Titus. 1983. *Projecting Future Sea Level Rise: Methodology, Estimates to the Year 2100, and Research Needs.* U.S. Environmental Protection Agency, Office of Policy and Resources Management, Strategic Studies Staff. Washington, D.C. 121 pp.

Holdahl, S. R. and N. L. Morrison. 1974. Regional investigations of vertical coastal movements in the U.S., using precise relevelings and mareograph data. *Tectonophysics* 23:373-390.

Hopkinson, C. S. and F. A. Hoffman. 1984. The estuary extended—a recipient-system study of estuarine outwelling in Georgia, pp. 313-330. *In:* V.S. Kennedy (ed.), *The Estuary as a Filter.* Academic Press, New York.

Howarth, R. A. and J. E. Hobbie. 1982. Regulation of decomposition and microbial activity in salt marsh soils: a review, pp. 183-208. *In:* V.S. Kennedy (ed.), *Estuarine Comparisons.* Academic Press, New York.

Hoyt, J. H. and J. R. Hails. 1967. Pleistocene shoreline sediments in coastal Georgia: deposition and modification. *Science* 155:1541-1543.

Johnson, D. S. and H. H. York. 1915. *The Relation of Plants to Tide Levels.* Publ. 206. Carnegie Inst., Washington, D.C. 162 pp.

Kaye, C. A. and E. S. Barghoorn. 1964. Late quarternary sea level change and crustal rise at Boston, Massachusetts, with notes on the autocompaction of peat. *Geol. Soc. of Am. Bull.* 75:63-80.

Kearney, M. S. and L. G. Ward. 1986. Vertical accretion rates in brackish marshes of a Chesapeake Bay tributary. *Geo-Mar. Lett.* (in press).

Koide, M., A. Soutar and E. D. Goldberg. 1972. Marine geochronology with ^{210}Pb. *Earth Planet. Sci. Lett.* 11:407-414.

Letzsch, W. S. and R. Frey. 1980. Deposition and erosion in a Holocene salt marsh, Sapelo Island, Georgia. *J. Sed. Pet.* 50:529-542.

McCaffrey, R. J. and J. Thomson. 1980. A record of the accumulation of sediment and trace metals in a Connecticut salt marsh. *Adv. Geophys.* 22:165-236.

Mendelssohn, I. A., K. L. McKee and W. H. Patrick, Jr. 1981. Oxygen deficiency in *Spartina alterniflora* roots: metabolic adaptation to anoxia. *Science 214:439-441.*

Miller, W. B. and F. E. Egler. 1950. Vegetation of the Wequetequock-Pawcatuck tidal marshes, Connecticut. Ecol. Monogr. 20:143-172.

Mudge, B. F. 1862. The salt marsh formations of Lynn, pp. 117-119. *Proc. Essex Inst.* II (1856-1860). Essex, Massachusetts.

Newman, W. S. and C. A. Munsart. 1968. Holocene geology of the Wachapreague Lagoon, Eastern Shore Peninsula, Virginia. *Mar. Geol.* 6:81-105.

Nichols, M. M. and R. B. Biggs. 1985. Estuaries, pp. 77-186. *In:* R. A. Davis Jr. (ed.), *Coastal Sedimentary Environments.* Springer-Verlag, New York.

Nixon, S. W. 1980. Between coastal marshes and coastal waters—A review of twenty years of speculation and research on the role of salt marshes in estuarine productivity and water chemistry, pp. 437-525. *In:* P. Hamilton and K. B. MacDonald (eds.), *Estuarine and Wetland Processes.* Plenum Press, New York.

Nixon, S. W. 1982. *The Ecology of New England High Salt Marshes: A Community Profile.* FWS/OB5-81/55. U.S. Fish and Wildlife Service. Slidell, Louisiana. 70 pp.

Odum, E. P. 1968. A research challenge: evaluating the productivity of coastal and estuarine waters, pp. 63-64. *In: Proceeding 2nd Sea Grant Conference.* University of Rhode Island, Kingston, R.I.

Odum, W. E., T. J. Smith III, J. K. Hoover and C. C. McIvor. 1984. *The Ecology of Tidal Freshwater Marshes of the United States East Coast: A Community Profile.* FWS/OB5-83/17. U.S. Fish and Wildlife Service. Slidell, Louisiana. 177 pp.

Officer, C. B., D. R. Lynch, G. H. Setlock and G. R. Helz. 1984. Recent sedimentation rates in Chesapeake Bay, pp. 131-157. *In:* V.S. Kennedy (ed.), *The Estuary as a Filter.* Academic Press, New York.

O'Neil, T. 1949. *The Muskrat in the Louisiana Coastal Marshes.* La. Wild. Fish. Commun., New Orleans, Louisiana. 152 pp.

Pethick, J. 1981. Long-term accretion rates on tidal salt marshes. *J. Sed. Pet.* 51:571-577.

Pethick, J. 1984. *An Introduction to Coastal Geomorphology.* Edward Arnold Publishers, London. 260 pp.

Poland, J. F. and G. H. Davis. 1969. Land subsidence due to withdrawal of fluids, pp. 187-269. *In: Reviews of Engineering Geology II.* (Geol. Soc. Amer.) Boulder, Colorado.

Pomeroy, L. and R. Wiegert. 1981. *The Ecology of a Salt Marsh.* Springer-Verlag, New York. 271 pp.

Rampino, M. E. and J. Sanders. 1981. Episodic growth of Holocene tidal marshes in the northeastern United States: a possible indicator of eustatic sea-level fluctuations. *Geology* 9:63-67.

Redfield, A. C. 1972. Development of a New England Salt Marsh. *Ecol. Monog.* 42:201-237.

Richards, F. S. 1934. The salt marshes of the Dovey Estuary IV, the rates of vertical accretion, horizontal extension and scarp erosion. *Ann. Bot.* 48:225-259.

Roman, C. T. 1981. *Detrital Exchange Processes of a Delaware Salt Marsh.* Ph.D. Dissertation. University of Delaware. 144 pp.

Sasser, C. E. and J. G. Gosselink. 1984. Vegetation and primary production in a floating freshwater marsh in Louisiana. *Aquat. Bot.* 20:245-255.

Schell, W. R. 1982. Dating recent (200 years) events in sediments from lakes, estuaries, and deep ocean environments using lead-210. *Amer. Chem. Soc. Sympos. Ser.* 176:331-361.

Settlemyre, J. L. and L. R. Gardner. 1977. Suspended sediment flux through a salt marsh drainage basin. *Estuar. Coastal Shelf Sci.* 5:633-663.

Shaler, N. S. 1886. Sea-coast swamps of the eastern United States, pp. 359-368. *U.S. Geological Survey, 6th Annual Report.* Washington, D.C.

Shideler, G. L., J. C. Ludwick, G. F. Oertel and K. Finkelstein. 1984. Quaternary stratigraphic evolution of the southern Delmarva Peninsula Coastal Zone, Cape Charles, Virginia. *Geol. Soc. Amer. Bull.* 95:489-502.

Sharma, P., L. R. Gardner, W. S. Moore and M. S. Bollinger. 1986. Sedimentation and Bioturbation in a salt marsh as revealed by ^{219}Pb ^{137}Cs, and ^{7}Be studies. *Limnol. Oceanogr.* (in review).

Steever, E. Z., R. S. Warren and W. A. Niering. 1976. Tidal energy subsidy and standing crop production of *Spartina alterniflora. Estuar. Coastal Mar. Sci.* 4:473-478.

Stevenson, J. C., D. R. Heinle, D. A. Flemer, R. J. Small, R. A. Rowland and J. F. Ustach. 1977. Nutrient exchanges between brackish water marshes and the estuary, pp. 219-240. *In:* M. Wiley (ed.), *Estuarine Processes, Volume II.* Academic Press, New York.

Stevenson, J. C., M. S. Kearney and E. C. Pendleton. 1985a. Sedimentation and erosion in a Chesapeake Bay brackish marsh system. *Mar. Geol.* 67:213-235.

Stevenson, J. C., L. G., Ward, M. S. Kearney and T. E. Jordan. 1985b. Sedimentary processes and sea level rise in tidal marsh systems of Chesapeake Bay, pp. 37-62. *In:* H. A. Groman *et al.* (eds.), *Wetlands of the Chesapeake.* The Environmental Law Institute Washington, D.C.

Stewart, D. M. 1976. Possible precursors of a major earthquake centered near Wilmington-Southport, North Carolina. *Earthquake Notes* 46:3-19.

Stumpf, R. P. 1983. The process of sedimentation on the surface of a salt marsh. *Estuar. Coastal Shelf Sci.* 17:495-508.

Turekian, K. K. and J. K. Cochran. 1978. Determination of marine chronologies using natural radionuclides, pp. 313-360. *In:* J. P. Riley and R. Chester (eds.), *Chemical Oceanography, Volume* 7. Academic Press, New York.

Urso, S. B. and S. W. Nixon. 1984. *The Impact of Human Activities on the Prudence Island Estuarine Sanctuary as Shown by Historical Changes in Heavy Metal Inputs and Vegetation.* Final report to the Narragansett Bay Estuarine Sanctuary Scientific Committee. Graduate School of Oceanography, University of Rhode Island, Kingston, RI.

Ward, L. G. 1981. Suspended-material transport in marsh tidal channels, Kiawah Island, South Carolina. *Mar. Geol.* 40:139-154.

Zarillo, G.A. 1985. Tidal dynamics and substrate response in a salt-marsh estuary. *Mar. Geol.* 67:13-37.

PROCESS VARIABILITY IN THE GULF OF MAINE – A MACROESTUARINE ENVIRONMENT

Daniel E. Campbell

Maine Department of Marine Resources
West Boothbay Harbor, Maine

Abstract: The geomorphological, physical, chemical, and biological characteristics of the Gulf of Maine are reasonably consistent with the current concept of an estuary. A semi-quantitative index was developed to evaluate the relative estuarinity of coastal waters. An historical reconstruction of variability in the Gulf of Maine environment during the Holocene suggests that it was more estuarine than at present during the periods 12,000-10,000 y before present (BP) and 4,000-2,000 y BP, but less estuarine around 7,000 y BP. A model of the plankton ecosystem in the Gulf of Maine was formulated based on an estuarine analogue in which the variability in planktonic production is affected by variation in import-export exchange processes. The model indicated that plankton productivity was most sensitive to a decrease in slope water exchange with a 50% decline in slope water inflow causing a 15% decline in primary production. This response is consistent with the observed ranges of variability for wind conditions which drive the slope water inflow and for primary production in the Gulf of Maine. The perception of the Gulf of Maine as an estuary illuminates the causes of process variability within its boundaries and provides a basis for further understanding of this complex body of water.

Introduction

This paper examines the variability of ecological processes in the Gulf of Maine. The manner in which physical process variability has affected the estuarine character of the Gulf of Maine over geologic time is considered. Also, one physical process (nontidal exchange) is examined as a factor that influences year-to-year variation in plankton production within the Gulf.

The Gulf of Maine is macroestuarine in both a spatial and a temporal sense. It is about 14 times larger than the Chesapeake Bay and has existed as an estuary for at least the past 12,000 years. Even though the Gulf of Maine is a part of the present continental shelf, its similarity to an estuary has long been recognized. Bigelow (1927) observed that the nontidal circulation of the Gulf of Maine was estuarine in nature with lighter fresher waters moving seaward at the surface and heavier more saline water moving landward at depth. He pointed out that this pattern might have been anticipated based on the Gulf's geomorphology and the large amount of freshwater discharged into it.

Two hierarchical levels of process variability are examined for the Gulf of Maine. The first level of variability is defined by the temporal scale (1000-10000 y) of an estuary's life cycle. On a geologic time scale estuaries are known to be ephemeral and probably rare structures (Emery 1967; Russell 1967; Schubel and Hirschberg 1978). By virtue of its longevity, the Gulf of Maine presents a unique opportunity to examine how long-term variability in the processes that create and maintain estuaries affects their relative estuarinity. At the second level of temporal variability (1-10 y), the effect of variation in new nitrogen supplied by net nontidal exchange on plankton production in the Gulf of Maine is examined using a marine ecosystem model.

The Estuarine Character of the Gulf of Maine

Many scientists have recognized that coastal and shelf waters are similar to estuaries in some respects. Ketchum and Keen (1955) considered the coastal waters of the Mid-Atlantic Shelf to behave as an estuary with respect to the mixing of oceanic and river water. Yentsch (1975) characterized these same waters as an infinite estuary in adapting the work of Ketchum and Keen to consider the distribution of phosphorus across the continental shelf. Schubel and Hirschberg (1978) note that the similarity between estuaries and low salinity coastal waters may have been an important factor in the evolutionary development of the present dependence of many fish species on estuarine nursery areas. Hopkinson and Hoffman (1984) proposed that Georgia coastal waters represent an extended estuary by virtue of the large contribution of organic matter from salt marsh estuaries to nearshore waters.

The Gulf of Maine is obviously estuarine in the broad sense of Ketchum (1951) who required only that an estuary be a body of water within which river water mixes with an measurably dilutes seawater. However, the Gulf of Maine is also described by the more restrictive and widely accepted definition of Pritchard (1967) which reads: "An estuary is a semi-enclosed coastal body of water which has a free connection with the open sea and within which seawater is measurably diluted with freshwater derived from land drainage".

Pritchard (1967) developed his original definition so that large bodies of water such as the Gulf of Maine would not be defined as estuaries. Limitation of the definition was necessary so that it would be more useful in describing typical estuaries. To further our understanding of estuarine variability it is useful to consider how the concept of an estuary may be extended to describe other bodies of water that are more or less estuarine in character. In this regard it is helpful to examine how the Gulf of Maine fits the definition point by point.

The present coastline of the Gulf of Maine appears as a slightly enclosed area of the coast (Fig. 1a). However, 12,000 y ago the DeGeer Sea was almost completely enclosed (Fig. 1b, Lougee 1953). Even though the coastal banks are submerged today, they still serve the function of enclosure and control water movements within the Gulf (Bigelow 1927; Bumpus 1973).

The word *coastal* was used by Pritchard (1967) to preclude large bodies of water from being defined as estuaries. This restriction recognized that lateral boundaries have reduced importance in determining the kinematics and dynamics of water movements in large water bodies. Lateral boundaries of the Gulf of Maine are important in determining the observed circulation pattern (Bigelow 1927; Bumpus 1973; Brooks 1985). Pritchard's exclusion of large structures is somewhat artificial in light of recent work on the geometry of coast-

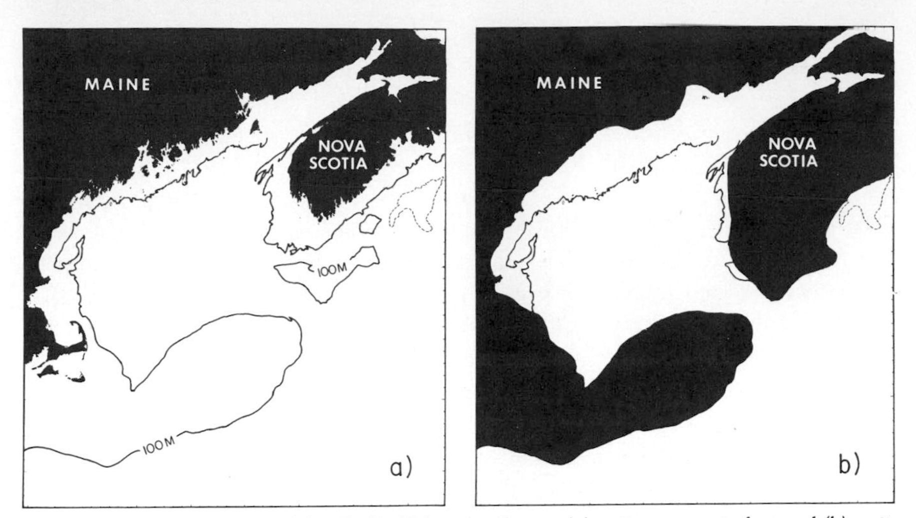

Figure 1. *The coastline of the Gulf of Maine is shown (a) as it appears today and (b) as it probably looked 12,000 y BP. The coastline in (b) was redrawn and simplified from Lougee (1953).*

lines. Mandelbrot (1977) has proposed that coastlines are physical entities of irrational dimension (between Euclidean dimension 1 and 2) and self similar structure. Practically, self similarity means that structures which appear on one spatial scale are also found on larger and smaller scales. Based on this view of coastline geometry, size alone is not a sufficient condition for excluding a body of water from consideration as an estuary.

Pritchard's third criterion is that the body of water have a free connection with the open sea. Communication between the ocean and the estuary must be sufficient to allow continuous exchange of tidal energy and sea salts. The Gulf of Maine satisfied this criterion during the low stand of sea level 12,000 y before the present (BP) when the entrance at the Northeast Channel was 40 km wide (Grant 1970).

Pritchard's final criterion for an estuary is that seawater be measurably diluted by freshwater derived from land drainage so that the resulting density gradients drive an estuarine circulation pattern. There is strong evidence that freshwater is an important factor contributing to the observed circulation patterns in the Gulf of Maine (Bigelow 1927; Brooks 1985).

Historical Variability of the Macroestuarine Environment

The three criteria from Pritchard's definition of an estuary were combined into a semi-quantitative index used to express the relative estuarinity of a coastal body of water:

$$E = F\,(S+T) \qquad (1)$$

The estuarinity, E, is indicated by the product of freshwater runoff, F, and the sum of the degree of enclosure, S, and tidal exchange, T, where F, S and T are expressed on an arbitrary scale from 0 to 1. When $S = 0$ this expression reduces to the case of a straight coast and estuarinity depends only on the freshwater input and the tidal exchange. By definition when $T = 0$, S must be equal to 1. This condition represents complete enclosure and estuarinity is determined by freshwater inflow; e.g., a coastal lagoon cut off from the sea. If freshwater inflow is zero, estuarinity is also zero. This index was used to estimate the past estuarinity of the Gulf of Maine relative to present conditions. A normalized version of this index might prove useful in comparing the estuarinity of different coastal bodies of water, but that exercise is not attempted here.

Figure 2 shows a plot of F, S, T, and E in the Gulf of Maine over the past 12,000 y. The plotted values are relative to present conditions in the Gulf of Maine. The degree of enclosure is expressed as a fraction of the Gulf's total perimeter (Grant 1970). Freshwater inflow was estimated relative to present conditions (arbitrarily set equal to 0.5) from the historical vegetation record (T. Webb pers. comm.). The variation in tidal exchange is reconstructed from changes in tidal amplitude (Grant 1970; Scott and Greenberg 1983). Present conditions were arbitrarily set equal to 0.5, and the tidal exchange 12,000 y BP was assumed to be 10% of the present value.

Figure 2 emphasizes the point that estuaries are not static but evolve in time becoming more or less estuarine as physical processes and characteristics vary. The relative estuarinity index shows that the Gulf of Maine was slightly more estuarine from 4,000 to 2,000 y BP and from 12,000-10,000 y BP than it is today. However, the character of the estuary was undoubtedly very different during these two periods. The Gulf of Maine was probably similar to a lagoonal estuary from 12,000-10,000 y BP because tidal exchange was low and the basin was more than 90% enclosed. The estuarinity index is determined largely by freshwater inflow during these two time periods. About 7,000 y BP estuarinity of

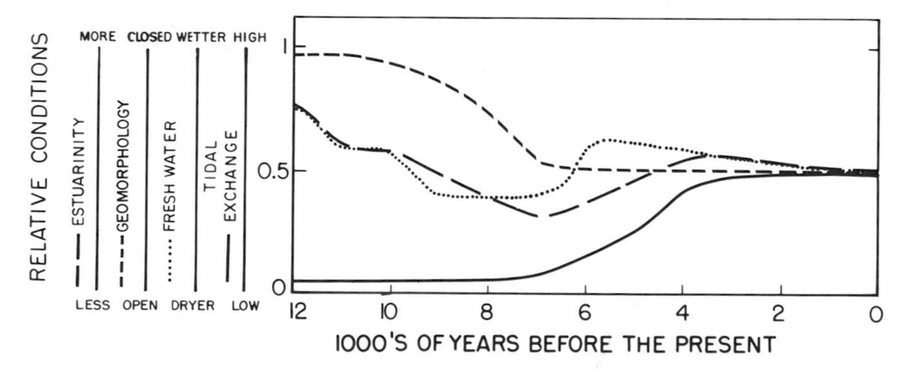

Figure 2. *A reconstruction of physical conditions (relative to present values) related to the estuarinity of the waters in the Gulf of Maine from 12,000 y BP to the present.*

the Gulf reached a minimum due to a dryer climate and a more open coastline caused by eustatic sea level rise. From 7,000-4,000 y BP the Gulf of Maine became increasingly more estuarine because of greater fresh water inflow and the onset of tidal resonance as a result of continued eustatic sea level rise.

Spatial variability in the physical processes that create estuarine conditions produces a well-recognized continuum of estuarine types. Temporal variability of these same processes affects the estuarine character of a body of water through time, as is well illustrated by the Gulf of Maine—a large and long-lived macroestuary.

Plankton Production and the Variability of Non-Tidal Exchange

Productive processes in estuaries are affected by many physical variables, i.e., sunlight, tidal exchange, nutrient inflows, freshwater runoff, wind, etc. A prominent characteristic of estuaries is that the import and export of materials and organisms play important roles in controlling biological production within the system (Margalef 1967). If the Gulf of Maine functions as a macroestuary, the variability of net nontidal exchange should in part determine the levels of plankton production observed within the system. Riley (1967a) first modeled the effects of shoreward nutrient transport on the productivity of coastal waters off southern New England. He concluded that nutrient transport was an important factor explaining the distribution of biological productivity across the continental shelf. In the present study an estuarine ecosystem model was used to examine the sensitivity of plankton productivity in the Gulf of Maine to variability in the physical transport of water into and out of the system.

A Model of the Gulf of Maine Plankton Ecosystem

The physical factors that control plankton productivity were incorporated into an energy circuit model (Odum 1971, 1983) of the plankton ecosystem in the Gulf of Maine including Georges Bank (Fig. 3). The plankton system is simplified to three components phytoplankton, P; zooplankton, Z; and a limiting nutrient, nitrogen, N. The remineralizing functions of bacteria are included at the hexagon marked B. The forcing functions, storage compartments, and pathway fluxes in Fig. 3 are defined in Table 1, along with references and assumptions used in assigning values to them. The differential equations for P, Z, and N follow (see Fig. 3, Table 1):

$$dP/dt = k_1NPJ_R - k_2P - k_4P - k_3ZP + k_{14}J_CP_C - k_6P(J_C + J_N + J_{SW} - J_E) \quad (2)$$

$$dZ/dt = k_7ZP - k_8Z - k_9Z + k_{13}J_CZ_C - k_{10}Z(J_C + J_N + J_{SW} - J_E) \quad (3)$$

$$dN/dt = k_{11}J_NN_R + k_{12}J_CN_C + k_{17}Z + k_{18}P + k_{19}P + k_{15}J_{SW}N_{SW} - k_{16}N(J_C + J_N + J_{SW} - J_E) - k_5NPJ_R \quad (4)$$

The model represents an homogeneously mixed layer from 0-75 m, which implies that annual vertical mixing is sufficient to mix this depth. Seventy-

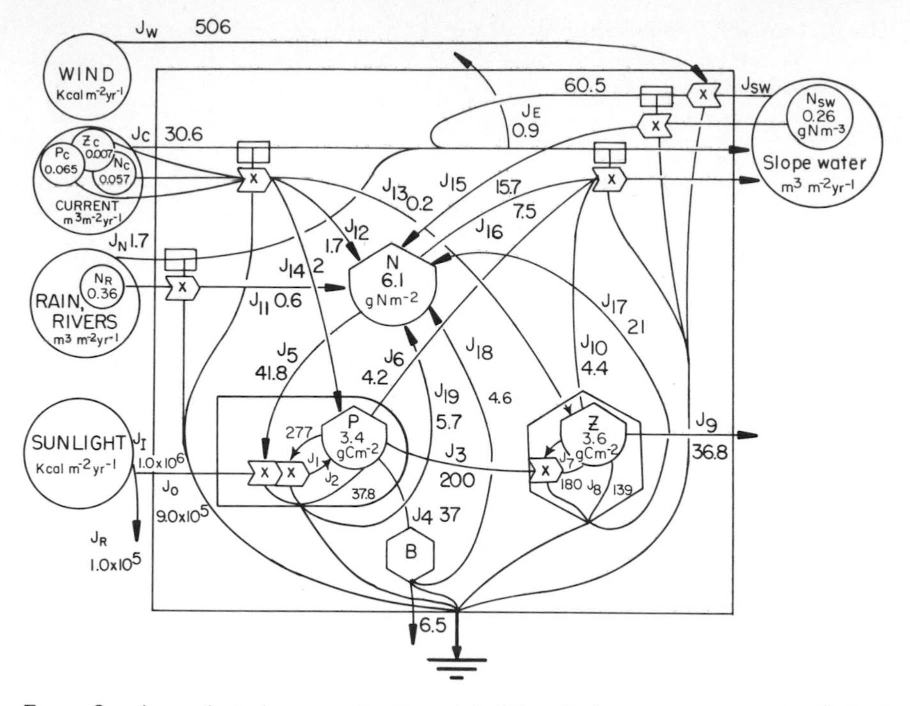

Figure 3. An evaluated energy circuit model of the plankton ecosystem in the Gulf of Maine. The J's are annual fluxes of energy (kcal), carbon or nitrogen (g), or water (m^3), expressed per m^2 of surface area. Storages have units of gN m^{-2} or gC m^{-2}, and values inside small circles (P_C, Z_C, N_C, N_R) are concentrations with units gC m^{-3} or gN m^{-3}.

five meters corresponds approximately to the sill depth in the Great South Channel. Wind is assumed to be the major factor responsible for the pressure field which drives water transport through the Northeast Channel (Ramp *et al.* 1982). Water from three sources contributes to the physical transport of materials into and out of the Gulf of Maine. The largest amount of water enters through the Northeast Channel below a depth of 75 m. All of the nitrogen contained in the net inflow of slope water is assumed to become available to phytoplankton via upwelling within the system. A second source of water enters through rainfall and runoff, but this inflow is less than 2% of the total. The Scotian Current brings a third flow of water into the Gulf which is about half of the amount that enters through the Northeast Channel. All three of these transports contribute to the export of organisms and materials from the Gulf.

Redfield (1941) hypothesized that year-to-year variations in zooplankton production in the Gulf of Maine could be explained by variations in surface water exchange. He reasoned that Scotian Current water enters the Gulf with an impoverished fauna while the waters exported from the Gulf carry a rich fauna. In years of high Scotian Current flow there would be a net loss of zooplankton

Table 1. Definition and evaluation of the forcing functions, storage compartments, and pathway fluxes used in the energy circuit model of the Gulf of Maine plankton ecosystem (Fig. 3).

Variable	Definition	Source of Value
Forcing Functions		
J_I	Solar isolation at water surface (43 deg. N lat.)	List (1971)
$J_R = \frac{J_I}{1 + k_0 NP}$	albedo for 43 deg. N lat.	Von Arx (1962)
J_N	runoff and rainfall	Schlitz and Cohen (1984)
N_R	nitrogen conc. in river water and rainfall	Schlitz and Cohen (1984)
J_W	wind energy transfer to water surface	Campbell and Wroblewski (1986)
J_C	Scotian Current inflow	Smith (1983)
N_C	Scot. Current nitrogen concentration	O'Boyle *et al.* (1984)
P_C	Scot. Current phytoplk. concentration	O'Boyle *et al.* (1984), Strickland (1960)
Z_C	Scot. Current zooplk. concentration	O'Boyle *et al.* (1984), Wiebe *et al.* (1975)
J_{SW}	net transport of slope water	Ramp *et al.* (1982)
N_{SW}	nitrogen in slope water	Ramp *et al.* (1982)
J_E	evaporation	Schlitz and Cohen (1984), Brown and Beardsley (1978)
Storages		
P	phytoplankton biomass	Apollonio and Applin (1972)
Z	zooplankton biomass	Schlitz and Cohen (1984)
N	nitrogen storage	Apollonio and Applin (1972) Pastuzak *et al.* (1982)
Pathway Fluxes		
$J_0 = k_0 NPJ_R$	absorbed solar radiation	$(J_I - J_R)$
$J_1 = k_1 NPJ_R$	gross primary production	Cohen and Grosslien (1986)
$J_2 = k_2 P$	phytoplankton respiration and excretion	Steeman Nielson and Hansen (1959), Hellebust (1965)
$J_3 = k_3 NP$	zooplankton ingestion	90% assimilation

Table 1. (Cont'd)

Variable	Definition	Source of Value
$J_4 = k_4P$	phytoplankton detritus	by difference on P
$J_5 = k_5NPJ_R$	nitrogen required by phytoplankton	Schlitz and Cohen (1984)
$J_6 = k_6P(J_C + J_N + J_{SW} - J_E)$	phytoplankton exported	calculated from model
$J_7 = k_7ZP$	zooplankton assimilation	equal to losses from Z
$J_8 = k_8Z$	zooplankton respiration and excretion	from J_{17} using Redfield (1934), Cohen and Grosslein (1986)
$J_9 = k_9Z$	zooplankton production	Cohen and Grosslein (1986)
$J_{10} = k_{10}Z(J_C + J_N + J_{SW} - J_E)$	zooplankton exported	calculated from model
$J_{11} = k_{11}J_NN_R$	nitrogen input in runoff and rainfall	calculated from model
$J_{12} = k_{12}J_CN_C$	nitrogen import by current	calculated from model
$J_{13} = k_{13}J_CZ_C$	zooplankton import by current	calculated from model
$J_{14} = k_{14}J_CP_C$	phytoplankton import by current	calculated from model
$J_{15} = k_{15}J_{SW}N_{SW}$	nitrogen import in slope water	calculated from model
$J_{16} = k_{16}N(J_C + J_N + J_{SW} - J_E)$	nitrogen exported	calculated from model
$J_{17} = k_{17}Z$	zooplankton remineralization	Schlitz and Cohen (1984)
$J_{18} = k_{18}P$	protozoa and bacteria remineralization	by difference on N
$J_{19} = k_{19}P$	phytoplankton remineralization	from J_2 using Redfield (1934), Cohen and Grosslein (1986)

from the system and productivity would decline. However, in years of high slope water exchange, additional nitrogen entering through the Northeast Channel would have a positive effect on plankton producitivity. Since increased transport can have both positive and negative effects on productivity, there is no way to determine *a priori* which effect will dominate. The model was used to test these hypothesized relationships.

Simulation Methods

The model was simulated to determine the sensitivity of plankton productivity to variations in nontidal exchange. First, storage and flow values from Fig.

3 were used to determine values for the coefficients (k's), which were assumed to be constants, and a steady state solution was established by running the model. One forcing function at a time was then varied by a fraction of its steady state value and the new steady state values of model storages and flows were recorded after a run time of 10 (simulated) years. The results of this sensitivity analysis are presented as the fraction of the original steady state value assumed by each variable under post-perturbation steady state conditions (Figs. 4, 5). Each curve in Figs. 4 and 5 is the result of 21 simulation runs.

Simulation Results and Discussion

Planktonic productivity was sensitive to changes in the inflow of slope water (Fig. 4a). Slope water inflow supplies 87% of the new nitrogen (Dugdale and Goering 1967) to the system (Fig. 3). New nitrogen supports new production that is critical to maintaining production levels against losses from export and

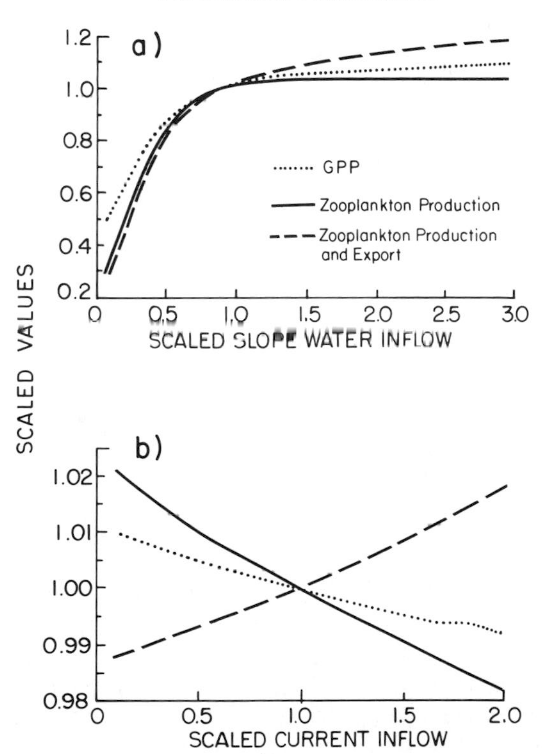

Figure 4. The sensitivity of gross primary production, zooplankton production, and zooplankton production plus export to variability in (a) slope water and (b) Scotian Current inflow.

sinking (Eppley and Petersen 1969). Gulf of Maine sediments are a sink for nitrogen (Christiansen 1985), so new nitrogen import must supply sediment denitrification as well as water column primary production. Gross primary production (GPP) increased slightly as slope water inflow increased, but it fell off sharply as slope water inflow declined (Fig. 4a). A 50% increase in slope water inflow caused a 3.3% increase in GPP, but an equal decrease in slope water influx resulted in a 15% decrease in GPP. This result indicates that a certain amount of new nitrogen is necessary to maintain high productivities on Georges Bank and in the Gulf of Maine, but beyond this point additional nutrients have little stimulating effect. Low slope water inflow may be responsible for years of low primary production, but highly productive years are not accounted for by the small enhancement of productivity in years of large slope water influx.

There are a few field measurements that indicate the model response is reasonable. Ramp *et al.* (1982) used a current meter array to measure transport through the Northeast Channel for two years. They showed that although the yearly mean transports were about the same, there was considerable seasonal variability within each year and between the two years. Average slope water inflow from November to February exhibited a 10% variation between years, and the inflow occurred in pulsed flows which were strongly correlated with winter storm events and northwest winds (Ramp *et al.* 1982). The extreme variation in average monthly wind velocity taken from Portland, Maine climatological data from 1970-1981 was only −8% to +10% from the 12-year mean. However, the range of variability in the number of consecutive (2 or more) days with northwest winds exceeding 16 km h^{-1} was −49% to +36% of the mean. If slope water transport is closely coupled with winter storm events a variability of plus or minus 50% may not be unreasonable. The year-to-year seasonal variation of primary production in waters of Georges Bank and the Gulf of Maine is about 10-15% (O'Rielly and Busch 1984; J. E. O'Rielly pers. comm.). Thus, the range of sensitivity shown by the model is consistent with present information on the variability of wind and primary production in these waters.

Figure 4b shows plankton production as a function of changes in the Scotian Current inflow. This current brings little new nitrogen and low concentrations of phytoplankton and zooplankton. Therefore, the negative effects of biomass export dominate any stimulation of production by nitrogen or organism influx. The changes in productivity generated by variations in the Scotian Current are an order of magnitude smaller than those caused by similar percentage variations in slope water inflow (scale difference Fig. 4a, b). Zooplankton production plus export increases with increased Scotian Current flow despite the fact that GPP and zooplankton production decrease. Zooplankton export increases by a much larger percentage than that by which zooplankton production decreases (Fig. 5b), thereby eroding the stock of organisms in the Gulf of Maine, just as Redfield (1941) hypothesized. The magnitude of this flushing effect was small, but the model did not account for export produced by the combined effects of spatial heterogeneity in the plankton and temporal pulsing of nontidal exchange.

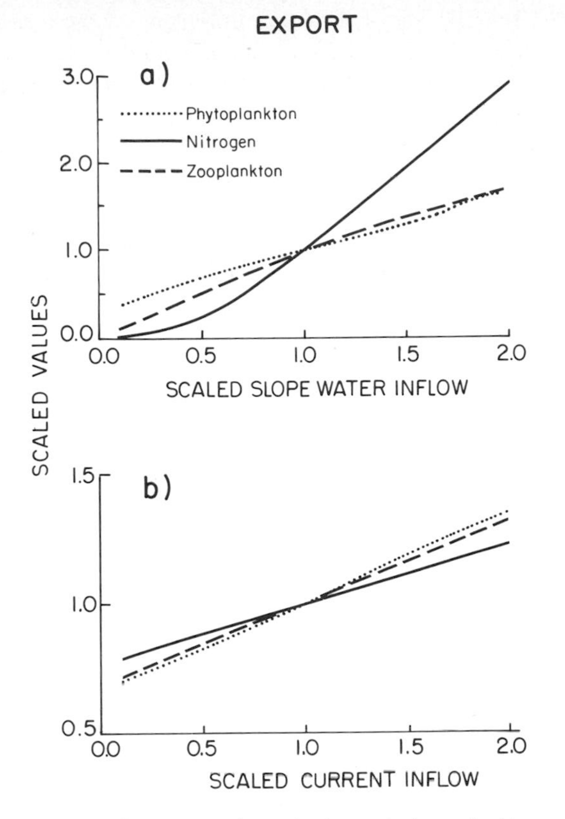

Figure 5. The sensitivity of nitrogen, phytoplankton, and zooplankton export to variability in (a) slope water and (b) Scotian Current inflow.

Nitrogen export increases linearly with large values of slope water inflow, but is highly nonlinear at low inflows (Fig. 5a). High nitrogen fluxes are largely controlled by physical processes, but for lower nitrogen supplies biological processes play the major role. Export of phytoplankton, zooplankton and nitrogen caused by changes in the Scotian Current (Fig. 5b) is about half of that caused by similar percentage changes in the slope water inflow, reflecting the relative proportions of the two waters in the export function. Since increased export of all ecosystem components is the result of increased flow from either source one would expect areas down the shelf from the Gulf of Maine to benefit in years of increased exchange.

Nutrients are supplied to estuaries principally in river runoff, in sewage outfalls, and through the subsurface counter current that is characteristic of many estuaries (Ketchum 1967). The relative importance of these three sources of nutrient supply varies widely among estuaries. Boynton *et al.* (1982) compared the physical and chemical factors affecting plankton production in 63 different estuarine systems. The Gulf of Maine appears most similar to embayment estuaries such as the Sheepscot River in Maine, where the major source of new nitrogen is the inflow of deep water (Garside *et al.* 1978). Estuaries in more urban areas (such as Narragansett Bay), are supplied with new nitrogen largely by river runoff and sewage (Nixon and Pilson 1984). In the Gulf of Maine the ratio of recycled nitrogen to new nitrogen derived from Slope Water inflow is 2:1. A similar ratio ranged from 2:1 to 8:1 for eight estuaries (Kemp *et al.* 1982). Therefore, new nitrogen is more important in supplying the nutrient requirements of primary production in the Gulf of Maine than in many estuaries. Peak and annual levels of primary production in estuaries are apparently enhanced by increased new nitrogen supply (Kemp *et al.* 1982 and Boynton *et al.* 1982), which is the central premise of Riley's model and of the model examined here.

Although vertical mixing, solar insolation, and other factors were held constant in this sensitivity analysis, they are recognized as important in determining levels of primary production in the Gulf of Maine. For example, Yentsch and Garfield (1981) and Campbell and Wroblewski (1986) have examined the importance of vertical mixing in determining primary production in the Gulf of Maine.

Conclusions

The variability of the Gulf of Maine macroestuarine environment throughout the Holocene was reconstructed from estimations of the variability of freshwater runoff, degree of enclosure, and tidal exchange. The estuarinity of this water body has varied markedly over this period, ranging from a lagoonal system 12,000-10,000 y BP when sea level was much lower than present and the Gulf was 90% enclosed, to a more open coastal system about 7,000 y BP when sea level was higher but the climate was somewhat dryer than now. The present Gulf of Maine is characterized by large tidal exchange, intermediate fresh water inflow, and a partly closed perimeter.

A model of phytoplankton production in the Gulf of Maine was constructed and the variability of plankton production with changes in nontidal exchange was examined. The model indicated that low Slope Water inflow could be responsible for years of low primary production. The Gulf of Maine functions like a large estuary with regard to the export of materials and organisms. Since zooplankton growth rates are fast compared to the turnover time of the Gulf, the model provides evidence that the Gulf retains a native population as expected for a partially enclosed system (Riley 1967b). Bigelow (1924), Fish and Johnson (1937), and Redfield (1941) provide empirical evidence that the Gulf maintains native zooplankton populations. The range of variability in nontidal transport and the consequent changes in primary productivity were consistent with the currently existing observations from the Gulf of Maine. Comparative analysis of the nitrogen budget in the Gulf of Maine places it within the spectrum of estuarine bodies of water.

As Ketchum (1967) and Schubel and Hirschberg (1978) point out, the greatest importance of estuaries is in their unique relationship to man. Mankind's exceedingly intimate relationship with estuaries is contradictory in nature. On the one hand we expect to harvest bountiful fisheries and use clear waters for recreation, while on the other hand we expect estuaries to receive and process human and industrial wastes, and provide transportation for cargoes that are deadly to life if an accident releases them. The New England states depend on the Gulf of Maine for fisheries, recreation, transportation, potential mineral resources, and waste disposal. In this socio-economic context, the Gulf of Maine is like an estuary, and like our smaller estuaries there is already evidence of its vulnerability to the waste products and exploitation of an industrial society (Larsen *et al.* 1985).

Acknowledgments

I thank J. W. Day and H. T. Odum for the opportunity to present this paper; R. W. Langton, C. S. Yentsch, and my colleagues at the Maine Department of Marine Resources, and the Bigelow Laboratory for Ocean Sciences for valuable information and support; D. A. Wolfe and three reviewers for constructive criticism and suggestions, and L. A. Codispoti who provided helpful insights and discussion during the initial stages of this work. This work was supported by contract #NA-83-FA-C-00047 from the National Marine Fisheries Service Woods Hole, MA.

References Cited

Apollonio S. and H. H. Applin Jr. 1972. An atlas of the distribution of inorganic nutrients in coastal waters of the Gulf of Maine. Maine Dept. Sea and Shore Fish.: 122 pp.

Bigelow, H.B. 1924. Plankton of the offshore waters of the Gulf of Maine, U.S. Dept. Commerce, Bur. of Fish. 968:1-509.

Bigelow, H. B. 1927. Physical oceanography of the Gulf of Maine. U.S. Dept. Commerce, Bur. of Fish. 969:511-1027.

Boynton, W. R., W. M. Kemp and C. W. Keefe. 1982. A comparative analysis of nutrients and other factors influencing estuarine phytoplankton production, pp. 69-91. *In:* V. S. Kennedy (ed.), *Estuarine Comparisons.* Academic Press, New York.

Brooks, D. A. 1985. Vernal circulation in the Gulf of Maine. *J. Geophys. Res.* 90 (C3):4687-4705.

Brown, W. S. and R. C. Beardsley. 1978. Winter circulation in the Gulf of Maine; Part I. Cooling and water mass formation. *J. Phys. Oceanogr.* 8:265-277.

Bumpus, D. F. 1973. A description of the circulation on the continental shelf of the east coast of the United States. *Prog. Oceanogr.* 6:111-157.

Campbell, D. E. and J. S. Wroblewski. 1986. Fundy tidal power development and potential fish production in the Gulf of Maine. *Can. J. Fish. Aquat. Sci.* 43(1):78-89.

Christensen, J. P. 1985. Nutrient regeneration and denitrification in Gulf of Maine sediments. *In: Abstracts Gulf of Maine Workshop.* ARGOM, University of Maine, Orono.

Cohen, E. B. and M. D. Grosslein. 1986. Production on Georges Bank compared with other shelf ecosystems. *In:* R. Backus (ed.), *Georges Bank.* MIT Press, Boston. In Press.

Dugdale, R. C. and J. J. Goering. 1967. Uptade of new and regenerated forms of nitrogen in primary production. *Limnol. Oceanogr.* 12:196-206.

Emery, K. O. 1967. Estuaries and lagoons in relation to continental shelves pp. 9-11. *In:* G. H. Lauff (ed.), *Estuaries.* Amer. Assoc. Advan. Sci. Pub. No. 83, Washington, D.C.

Eppley, R. W. and B. J. Peterson. 1979. Particulate organic matter flux and planktonic new production in the deep ocean. *Nature* 282:677-680.

Fish, C. J. and M. W. Johnson. 1937. The biology of the zooplankton population in the Bay of Fundy and the Gulf of Maine with special reference to production and distribution. *J. Biol. Board Can.* 3:189-322.

Garside, C., G. Hull and C. S. Yentsch. 1978. Coastal source waters and their role as a nitrogen source for primary production in an estuary in Maine, pp. 565-575. *In:* M. L. Wiley (ed.), *Estuarine Interactions.* Academic Press, New York.

Grant, D. R., 1970. Recent coastal submergence of the Maritime Provinces, Canada. *Can. J. Earth Sci.* 7:676-689.

Hellebust, J.A. 1965. Excretion of some organic compounds by marine phytoplankton. *Limnol. Oceanogr.* 10:192-206.

Hopkinson, C. S. and F. A. Hoffman, 1984. The estuary extended—A recipient-system study of estuarine outwelling in Georgia, pp. 313-330. *In:* V. S. Kennedy (ed.), *The Estuary as a Filter.* Academic Press, New York.

Kemp, W. M., R. L. Wetzel, W. R. Boynton, C. F. D'Elia and J. C. Stevenson. 1982. Nitrogen cycling and estuarine interfaces: Some current concepts and research directions, pp. 209-230. *In:* V. S. Kennedy (ed.), *Estuarine Comparisons.* Academic Press, New York.

Ketchum, B. H. 1951. The exchanges of fresh and salt waters in tidal estuaries. *J. Mar. Res.* 10:18-38.

Ketchum, B. H. 1967. Phytoplankton nutrients in estuaries. pp. 329-335. *In:* G. H. Lauff (ed.), *Estuaries.* Amer. Assoc. Advan. Sci. Pub. No. 83, Washington, D.C.

Ketchum, B. H., and D. J. Keen. 1955. The accumulation of river water over the continental shelf between Cape Cod and Chesapeake Bay. *Deep-Sea Res.* 3(suppl.):346-357.

Larsen, P. F., D. F. Gadbois and A. C. Johnson. 1985. Observations on the distribution on PCB's in deepwater sediments of Gulf of Maine. *Mar. Pollut. Bull.* 16(11):439-442.

List R. J. 1971. *Smithsonian Meteorological Tables.* Smithsonian Institution Press, Washington, D.C. 527 pp.

Lougee, R. C. 1953. A chronology of post-glacial time in Eastern North America. *Sci. Month.* 76:259-276.

Mandelbrot, B. B. 1977. *Fractals.* S. H. Freeman and Co., San Francisco. 365 pp.

Margalef, R. 1967. Laboratory analogues of estuarine plankton systems, pp. 515-521. *In:* G. H. Lauff (ed.), *Estuaries.* Amer. Assoc. Advan. Sci. Pub. No. 83, Washington, D.C.

Nixon, S. W. and M. E. Q. Pilson. 1984. Estuarine total system metabolism and organic exchange calculated from nutrient ratios: An example from Narragansett Bay, pp. 261-290. *In:* V. S. Kennedy (ed.), *The Estuary as a Filter.* Academic Press, New York.

O'Boyle, R. N., M. Sinclar, R. J. Conover, K. H. Mann and A. C. Kohler. 1984. Temporal and spatial distribution of ichthyoplankton communities of the Scotian Shelf in relation to biological, hydrological, and physiographic features. *Rapp. P-v. Reun. Cons. int. Explor. Mer.* 183:27-40.

Odum H. T. 1971. An energy circuit language for ecological and social systems, its physical basis, p. 139-211. *In:* B. Patten (ed.), *Systems Analysis and Simulation in Ecology,* Vol. 2. Academic Press, New York.

Odum, H. T. 1983. *Systems Ecology.* John Wiley and Sons, New York. 644 p.

O'Reilly, J. E. and D. A. Busch. 1984. Phytoplankton primary production on the northwestern Atlantic shelf. *Rapp. P.-v. Reun. Cons. int. Explor. Mer.* 183:255-268.

O'Reilly, J. E. 1985. Personal communication. NOAA, NMFS, NEFC, Sandy Hook Laboratory, Highlands, NJ 07732.

Pastuzak, M., W. R. Wright and D. Patanjo. 1982. One year of nutrient distribution in the Georges Bank region in relation to hydrography, 1975-1976. *J. Mar. Res.* 40 (suppl.): 525-542.

Pritchard, D. W. 1967. What is an estuary: Physical viewpoint, pp. 3-5. *In:* G. H. Lauff (ed.), *Estuaries.* Amer. Assoc. Advan. Sci. Pub. No. 83, Washington, D.C.

Ramp, S. R., R. J. Schlitz and W. R. Wright. 1982. On the deep transport of mass, heat, and nutrients through the Northeast Channel, Gulf of Maine, Natl. Mar. Fish. Serv., MARMAP Contribution No. MED/NEFC 82-7 Woods Hole, Mass. 47 pp.

Redfield, A. C. 1934. On the proportions of organic derivatives in sea water and their relation to the composition of the plankton. *James Johnstone Mem. Vol.,* Liverpool, UK. 176 pp.

Redfield, A. C. 1941. The effect of the circulation of the water on the Calanoid community in the Gulf of Maine. *Biol. Bull.* 80:86-110.

Riley, G. A. 1967a. Mathematical model of nutrient conditions in coastal waters. *Bull. Bingham Oceanogr. Coll.* 19:72-80.

Riley, G. A. 1967b. The plankton of estuaries, pp. 316-325. *In:* G. H. Lauff (ed.), *Estuaries.* Amer. Assoc. Advan. Sci. Pub. No. 83, Washington, D.C.

Russell, R. J. 1967. The origins of estuaries, pp. 93-99. *In:* G. H. Lauff (ed.), *Estuaries,* Amer. Assoc. Advan. Sci. Pub. No. 83, Washington, D.C.

Schlitz, R. J. and E. B. Cohen. 1984. A nitrogen budget for the Gulf of Maine and Georges Bank. *Biol. Oceanogr.* 3(2):203-221.

Schubel, J. R. and D. J. Hirschberg. 1978. Estuarine graveyards, climatic change and the importance of the estuarine environment, pp. 285-303. *In:* M. L. Wiley (ed.), *Estuarine Interactions.* Academic Press, New York.

Scott, D. B. and D. A. Greenberg. 1983. Relative sea-level rise and tidal development in the Fundy tidal system. *Can. J. Earth Sci.* 20:1554-1564.

Smith, P. C. 1983. The mean and seasonal circulation off Southwest Nova Scotia. *J. Phys. Oceanogr.* 13:1034-1054.

Steemann Nielson, E. and V. Kr. Hansen. 1959. Measurements with the carbon-14 technique of the respiration rates in natural populations of phytoplankton. *Deep-Sea Res.* 5:222-233.

Strickland, J. D. H. 1960. Measuring the production of marine phytoplankton. *Fish. Res. Board Can. Bull.* 122:1-172.

Von Arx, W. S. 1962. *An Introduction to Physical Oceanography.* Addison-Wesley, Reading, Massachusetts. 422 pp.

Webb, T. 1985. Personal communication. Dept. of Geol. Sci., Brown Univ. Providence, Rhode Island.

Wiebe, P. H., S. Boyd, and J. L. Cox. 1975. Relationships between zooplankton displacement volume, wet weight, dry weight, and carbon. *Fish. Bull.* 73(4):777-786.

Yentsch, C.S. 1975. New England coastal waters—An infinite estuary, pp. 608-617. *In:* T. M. Church (ed.), *Marine Chemistry in the Coastal Environment.* ACS Symposium Series No. 18. American Chemical Society, Washington, D.C.

Yentsch, C. S. and N. Garfield. 1981. Principal areas of vertical mixing in the Gulf of Maine, with reference to the total productivity of the area, pp. 303-312. *In:* J. F. R. Gower (ed.), *Oceanography from Space.* Plenum, New York.

THE RELATION OF DENITRIFICATION POTENTIALS TO SELECTED PHYSICAL AND CHEMICAL FACTORS IN SEDIMENTS OF CHESAPEAKE BAY

Robert R. Twilley and W. Michael Kemp

Horn Point Laboratory
Center for Environmental and Estuarine Studies
University of Maryland
Cambridge, MD

Abstract: The variability of denitrification potentials in estuarine sediments was measured during 15-18 October 1984 at ten stations in Chesapeake Bay. Selected physical and chemical characteristics of the sediments were also measured to investigate factors that may regulate denitrification in this estuarine ecosystem. Denitrification was measured in slurries of surface sediments (0-15 mm depth) amended with a range of nitrate concentrations (0-250 μg-at l^{-1}) using the acetylene blockage technique. These "denitrification potentials" increased at higher concentrations of NO_3^-, and except for two stations, linear transformations of rectangular hyperbolic formulations had correlation coefficients (r^2) > 0.73. Maximum nitrate-saturated denitrification rates (D_{max}) ranged from 8 to 556 ng-at N gdw^{-1} h^{-1}. Half-saturation constants (K_s) ranged from 5 to 90 μg-at l^{-1} as NO_3^- for sediments where significant kinetic relations were obtained. Multiple stepwise regression analysis revealed a significant model for D_{max} (Y) in which the atomic ratio (X) of total sediment nutrients (N:P) was the only significant independent variable [$Y = -8.39(X) + 120.1$, $r^2 = 0.71$]. K_s (Y) was associated with salinity (X1) and organic carbon content of sediments (X2) [$Y = -34.22(X1) + 88.06(X2) + 348.27$, $r^2 = 0.79$]. The landward stations of the 3 tributaries studied had a lower affinity for NO_3^- but higher D_{max}, while the seaward stations had higher K_s but lower D_{max}. The kinetic constants K_s and D_{max}, as well as the denitrification rates based on ambient NO_3^- concentrations, were generally higher in the tributaries (particularly Choptank and Patuxent River estuaries) than in the mainstem Bay. Ambient denitrification rates appear to vary spatially (due to the kinetic nature of denitrification capacity) to about the same degree as they vary seasonally (due to differences in NO_3^- concentration).

Introduction

Denitrification may contribute to the general tendency for nitrogen to be limiting for phytoplankton growth in shallow estuaries (Seitzinger *et al.* 1980; Nixon 1981). Indirect evidence based on the nonconservative behavior of NO_3^- in oligohaline waters during the spring (Boynton and Kemp 1985; Ward and Twilley 1986) along with measurements of NO_3^- uptake by sediments (Boynton *et al.* 1980) suggests that denitrification rates may indeed contribute to the loss of NO_3^- from two tributaries of the Chesapeake Bay. Direct measurements using N_2 evolution and acetylene inhibition techniques indicate that denitrification may account for 30-50% loss of nitrogen from estuarine systems (Seitzinger *et al.* 1980; Smith *et al.* 1985). Coupled with global loss of gaseous nitrogen from the oceans (Hattori 1983), denitrification is considered an important process in nitrogen cycles of estuarine and coastal sediments.

Estimates for total N loss from estuarine systems must take into account significant spatial and temporal variation in sediment denitrification rates. Most of this observed variance in denitrification kinetics for sediments may be due to sediment factors such as NO_3^- concentrations, organic content, size composition, and redox. For example, Kohl *et al.* (1976) demonstrated that the kinetic relations of NO_3^- reduction in soils depend on availability of organic matter. Such a relationship, however, has not been established for estuarine sediments (Hattori 1983). Information on this fundamental relationship would improve our understanding of the spatial heterogeneity of denitrification in marine ecosystems.

To extend our knowledge on spatial heterogeneity of this process in Chesapeake Bay, we investigated the relationships between denitrification and NO_3^- concentration for different types of estuarine sediments. By using a common method on a diverse group of sediments, we overcame many of the problems encountered when results from dissimilar types of experimental designs are interpreted. In addition, contemporaneous measurements of sediment characteristics at ten sites allowed us to evaluate selected physical and chemical factors regulating denitrifier activity.

Methods

Field Sampling

Denitrification potentials were measured at ten stations in the upper portion of Chesapeake Bay during a cruise aboard the R/V Aquarius from 15 to 18 October 1984 (Fig. 1). The stations represented two salinity regimes in the Potomac, Patuxent and Choptank River tributaries, along with four sites along the axis of the mainstem Bay. Hydrocasts were made at each station to sample bottom waters at ~0.5 m above the sediment-water interface. Dissolved oxygen was measured with a Model 57 polarographic electrode and meter (Yellow Springs Instrument, Yellow Springs, Ohio), and temperature and salinity were measured with a Model RS5-3 Salinometer (Beckman, Cedar Grove, Calif.). Water samples were collected with a submersible pump and one-liter subsamples were filtered through precombusted GF/C glass fiber filters (1.1 μm). The filtrate was analyzed for dissolved inorganic nitrogen (NO_3^-, NO_2^-, NH_4^+) and phosphorus (reactive with molybdate) with an Auto-Analyzer II system (Technicon Corp., Tarreytown, N.Y.) using standard colorimetric techniques (U.S. Environmental Protection Agency 1979).

Sediments were collected at each station with a modified Bouma box corer (Boynton *et al.* 1985). Subsamples taken with a 5-cm² acrylic core to a depth of 15 cm were analyzed for total carbon (TC), nitrogen (TN) and phosphorus (TP), and dissolved inorganic nitrogen and phosphorus in the pore waters (Boynton *et al.* 1985). The top 2 cm of a box core was homogenized and analyzed for size using methods described by Folk (1974), and water content and bulk density by drying 3 cm³ fresh sediment at 85°C to constant weight (Blake 1965). Redox was measured on intact box cores at 1-cm intervals to a depth of 15 cm using platinum wire and a standard calomel electrode connected

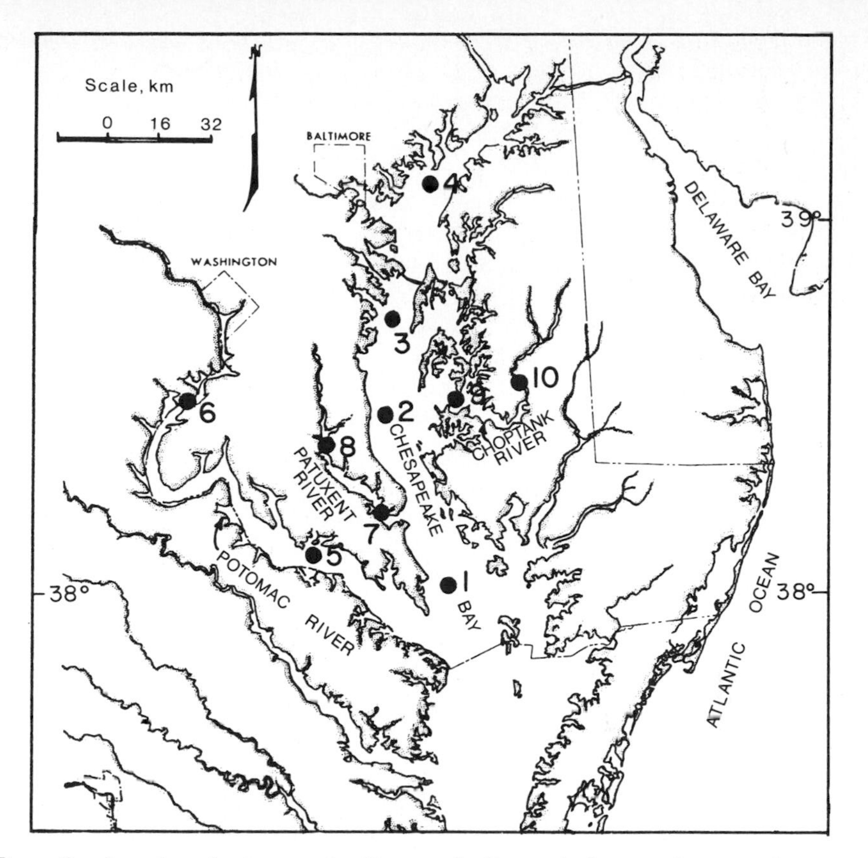

Figure 1. Location of stations in the Chesapeake Bay and adjacent tributaries where surface sediments were sampled to determine denitrification potentials.

to an Altex Monitor II meter (Altex). Eh was standardized with 0.001 M ferricyanide in 0.01 M KCl (Whitfield 1969).

Denitrification Potentials

The potential for denitrification in slurries of estuarine sediments was measured using an acetylene technique which inhibits the reduction of N_2O to N_2 (Balderston *et al.* 1976; Sørensen 1978; Oremland *et al.* 1984). Surface (1.5 cm deep) subsamples were taken from box cores at each station, mixed thoroughly and refrigerated (4°C) in the dark until initiation of incubations (within 48-96 h). Duplicate sediment plugs (10 cm^3) were placed in individual, sterilized 250-ml serum bottles along with a solution (150 ml) of artificial seawater (Instant Ocean, Aquarium Systems, Mentor) at ambient salinity amended with NO_3^- at concentrations of 5, 20, 50, 100, 175 and 250 μg-at l^{-1}. Bottles were capped with a rubber serum stopper, purged with He for 15 min and amended with acetylene from a cylinder (Air Products Inc., Trenton, N.J.) to a partial pressure of 15.2 kPa. This level of C_2H_2 is sufficient to block N_2 production in sediments (Sørensen 1978; Kaspar 1982; Oremland *et al.* 1984). Bottles were incubated with continuous gentle shaking in a temperature-controlled incubator at 17°C. Headspace samples were taken 1.5 h and 5.5 h after the addition of C_2H_2. Previous C_2H_2 inhibition experiments have shown N_2O production to be linear over 6-8 h for various Chesapeake Bay sediments (Twilley and Kemp, unpublished). Thus we have avoided problems caused by substrate depletion and N_2O consumption that have been observed in samples incubated for several days (Knowles 1979; Van Raalte and Patriquin 1979; Klingensmith and Alexander 1983). Headspace was sampled with 4-ml Becton-Dickinson vacutainers and assayed for N_2O by injecting one-ml samples with a gas tight syringe into a Packard-Becker Model 417 gas chromatograph fitted with a Nickel-63 electron capture detector at 350°C. Nitrous oxide was assumed in equilibrium between the gas and liquid phases, and aqueous concentrations were calculated from Bunsen coefficients calculated from Weiss and Price (1980) using appropriate temperature and salinity values. Nitrous oxide was separated on a 15 mm (ID) × 25 cm column packed with carbosieve S and heated at 100°C. Concentrations were determined by integrating peak areas with a Spectra-Physics Minigrator relative to standard curves produced with dilutions of bottled N_2O (Air Products Inc., Trenton, N.J.). At the termination of each experiment slurries were centrifuged at 2000 rpm, and the supernatant was filtered through GF/C filters (1.1 μm) and assayed for concentrations of NO_3^- and NH_4^+ using the techniques described above.

Data Analysis

The measured denitrification potentials, expressed as N_2O produced per g dry mass of sediment, and NO_3^- concentration were transformed by the method of Dowd and Riggs (1965). Equation 1 describes the rectangular hyperbolic relations between denitrification (D) and nitrate concentration $[NO_3^-]$ based on maximum denitrification rate (D_{max}) and the substrate concentration at which denitrification is one-half the maximum rate (K_s):

$$D = (D_{max} \cdot [NO_3^-]) / (K_s + [NO_3^-]) \quad (1)$$

The independent variable $[NO_3^-]$ and the dependent variable D are curvilinear and estimations of the two parameters are made by the following linear transformation:

$$[NO_3^-]/D = \frac{1}{D_{max}}[NO_3^-] + K_s/D_{max} \quad (2)$$

After solving for this linear model ($y = ax + b$), D_{max} is calculated as the reciprocal of the slope and K_s can then be determined from the y intercept.

Relationships of K_s and D_{max} to selected sediment characteristics were determined with multiple stepwise regressions (Statistical Analysis System 1982). Sediment variables considered in this study include: salinity, bulk density, water content, redox potential (Eh at 2-cm depth), sand content (arcsin of % composition), TC, TP, TN, and atom ratios C:N and N:P.

Results

Salinities were higher than normal, particularly in the headwaters of the Choptank and Patuxent River estuaries, reflecting extreme meteorological conditions. For example, salinities at Windy Hill in the Choptank during October are normally <3 g l^{-1} (Ward and Twilley 1986), yet during this survey salinity was 8.9 g l^{-1}. In the mainstem Bay, salinities ranged from 8.4 to 19.1 g l^{-1} and the lowest value at all 10 stations was 6.0 g l^{-1} (Table 1). Water temperature was fairly constant—ranging from 17.0 to 19.9°C. Nitrate concentrations ranged from 1.37 to 29.8 μg-at l^{-1} and were inversely related to salinity and somewhat lower than average (Ward and Twilley 1986) (Table 1). Lowest dissolved oxygen concentrations were measured in the mainstem Bay at Sta. 2 and 3 (Table 1); bottom waters at these two stations were anoxic during the summer of 1984 (Boynton *et al.* 1985).

Surface sediments were dominated by silts and clays (ca. 50:50) at all stations except Sta. 1 where sediments contained 81% sand (Table 2). TC concentrations for most sediments ranged from 2.2% to 4.6% dry wt except for Sta. 3 which had a concentration of 9.8% at 1 cm and 18.0% at 2 cm depth (Table 2). Total nitrogen in sediment did not vary substantially, ranging from 0.24% to 0.48% dry wt. Atomic ratios of carbon:nitrogen (C:N) ranged from 8.4 to 24.8, with the highest value occurring at Sta. 3. Total phosphorus was generally higher in the more oligohaline stations representing areas of higher organic matter deposition (Boynton and Kemp 1985). Redox at 2-cm depth ranged from +14 to +334 mV, and the only two stations with values $< +100$ were Sta. 3 and 5 (Fig. 2). At greater depths, Eh values were generally lower in the mainstem Bay and Potomac River estuary than in the Choptank or Patuxent River estuaries (Fig. 2).

In general, there was a first-order relation between denitrification potentials and NO_3^- concentration up to some higher NO_3^- concentration above which rates exhibited zero-order kinetics (Fig. 3). The rectangular hyperbolic model fit by the linear transformation of Dowd and Riggs (1965) adequately described these curvilinear relationships for all but two stations (Table 3). Correlation coefficients (r^2) for this Michaelis-Menten type model ranged from 0.73 to 0.98 for eight of the stations, and only Sta. 3 and 4 had nonsignificant relationships ($r^2 = 0.04$ and 0.30, respectively; Table 3).

Nitrate-saturated denitrification rates (D_{max}) ranged from 8.0 to 555.6 ng-at N gdw^{-1} h^{-1}, with the highest rates in sediments from the oligohaline regions of the Potomac and Choptank River estuaries (Table 3). D_{max} was higher in sediments from the Choptank than in those from the Patuxent River estuary, and rates in tributary stations tended to be higher than those in the mainstem Bay. Where kinetics conformed to the Michaelis-Menten model, half-saturation constants (K_s) ranged from 2.0 to 92.2 μg-at l^{-1} NO_3^- (Table 3). The tributary stations had K_s values >60 μg-at l^{-1} in the oligohaline regions and <25 μg-at l^{-1} in the mesohaline regions. In the mainstem Bay stations that had significant linear models (Sta. 1 and 2), K_s values were similar at 8.5 μg-at l^{-1} (Table 3).

Table 1. Physical and chemical characteristics of bottom water at the ten stations where sediment denitrification was measured in Chesapeake Bay.

Station	Latitude (N)	Longitude (W)	Depth (m)	Temp. (°C)	Salinity (g l^{-1})	Dissolved Oxygen (mg l^{-1})	Nutrient Concentrations (*μg-at* l^{-1})			
							NH_4	NO_3+NO_2	PO_4	SiO_3
CHESAPEAKE BAY										
1	38°07.95	76°15.04	13.4	18.2	19.1	6.6	8.2	1.37	0.27	13.7
2	38°33.49	76°25.65	19.0	18.7	18.4	4.7	12.6	2.99	0.40	19.0
3	38°57.62	76°26.31	16.4	19.9	18.9	4.0	14.5	2.02	0.41	24.3
4	39°20.86	76°10.89	10.0	17.1	8.4	7.6	5.3	29.40	0.27	25.8
POTOMAC TRIBUTARY										
5	38°09.79	76°11.53	15.5	18.2	15.5	6.2	8.9	2.18	0.25	12.9
6	38°21.37	76°11.53	9.5	18.7	6.0	6.6	8.7	29.80	1.09	36.2
PATUXENT TRIBUTARY										
7	38°22.76	76°30.09	7.0	18.4	14.2	7.5	6.7	3.72	0.47	24.3
8	38°20.98	76°29.83	4.0	18.4	11.2	6.2	8.5	2.87	1.38	47.2
CHOPTANK TRIBUTARY										
9	38°37.08	76°07.83	8.0	17.1	13.5	9.0	1.4	2.10	0.31	23.9
10	38°41.41	75°58.42	4.6	17.0	8.9	8.5	1.8	5.94	0.94	38.7

Table 2. Chemical and physical characteristics of sediments at the ten stations where denitrification potentials were measured in Chesapeake Bay.

Station	Depth (cm)	Water Content (%)	Bulk Density (gdw cm^{-3})	Size Analysis (%)			Total Nutrients (%)			Atom Ratios	
				Sand	Silt	Clay	C	N	P	C:N	N:P
CHESAPEAKE BAY											
1	0-1	50.8	0.71	81	6	13	3.47	0.48	0.08	8.4	13.1
	1-2						3.44	0.46	0.07	8.7	15.2
2	0-1	67.7	0.40	6	43	51	2.30	0.36	0.06	7.5	15.0
	1-2						3.83	0.30	0.05	14.9	12.7
3	0-1	71.6	0.34	20	40	40	9.78	0.46	0.09	24.8	11.6
	1-2						17.98	0.48	0.07	43.7	15.9
4	0-1	46.0	0.88	9	58	33	4.69	0.24	0.07	22.8	8.3
	1-2						4.83	0.25	0.07	22.5	8.2
POTOMAC TRIBUTARY											
5	0-1	74.8	0.31	9	43	48	3.47	0.48	0.08	8.4	13.1
	1-2						3.44	0.46	0.07	8.7	15.2
6	0-1	61.2	0.33	3	55	42	2.79	0.30	0.12	10.9	5.9
PATUXENT TRIBUTARY											
7	0-1	68.5	0.39	11	32	57	2.42	0.28	0.06	10.1	10.5
	1-2						2.29	0.27	0.06	9.9	10.8
8	0-1	64.4	0.46	27	22	51	2.19	0.27	0.13	9.5	4.6
	1-2						2.34	0.29	0.15	9.4	4.6
CHOPTANK TRIBUTARY											
9	0-1	61.3	0.51	12	54	34	4.03	0.25	0.06	18.8	9.7
	1-2						2.19	0.26	0.06	9.8	10.2
10	0-1	67.7	0.40	4	55	41	4.52	0.36	0.10	14.6	8.1
	1-2						7.05	0.52	0.13	15.8	9.4

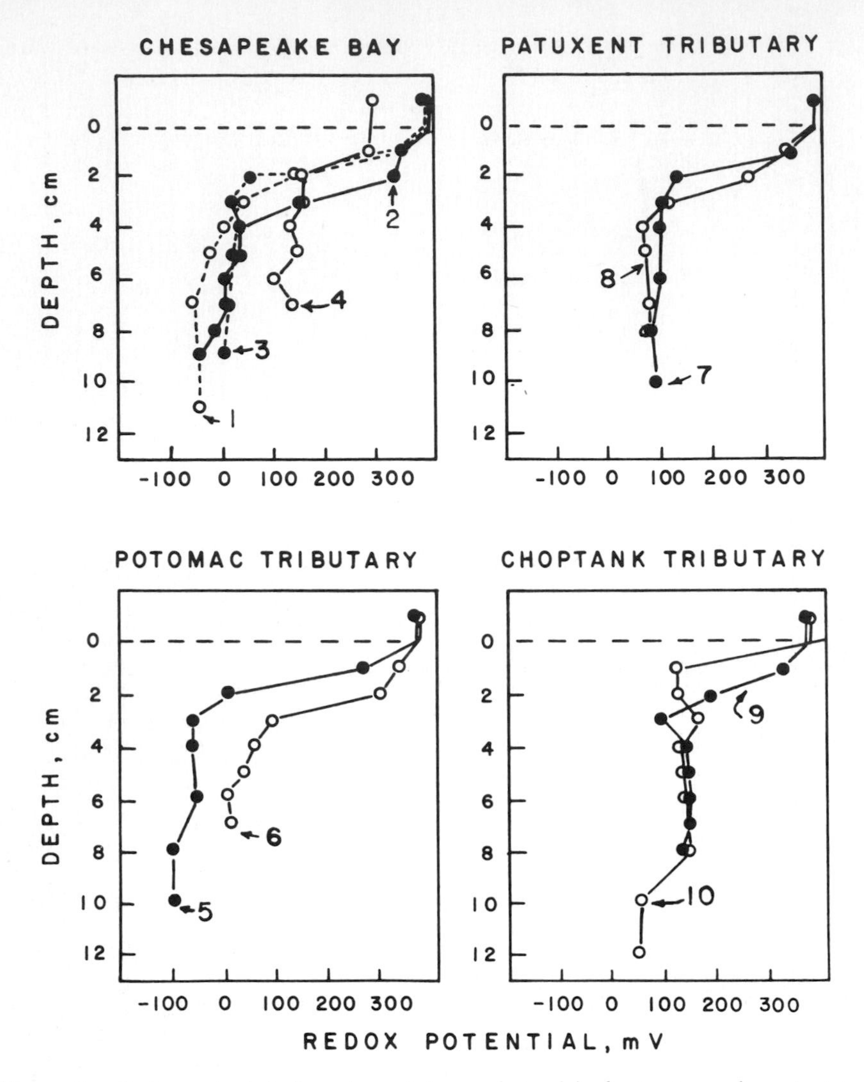

Figure 2. Redox potentials (based on platinum electrode) of estuarine sediments at ten stations in Chesapeake Bay.

Stepwise regression analysis with various sediment characteristics revealed significant regression models for both independent variables K_s and D_{max} (Table 4). Only those observations from stations that had significant kinetic constants were analyzed (i.e., Sta. 3 and 4 were excluded). K_s was significantly related to the N:P ratio of surface sediments, and D_{max} was correlated with salinity of bottom water and TC of surface sediments (Table 4). Both models were significant at $P<0.025$.

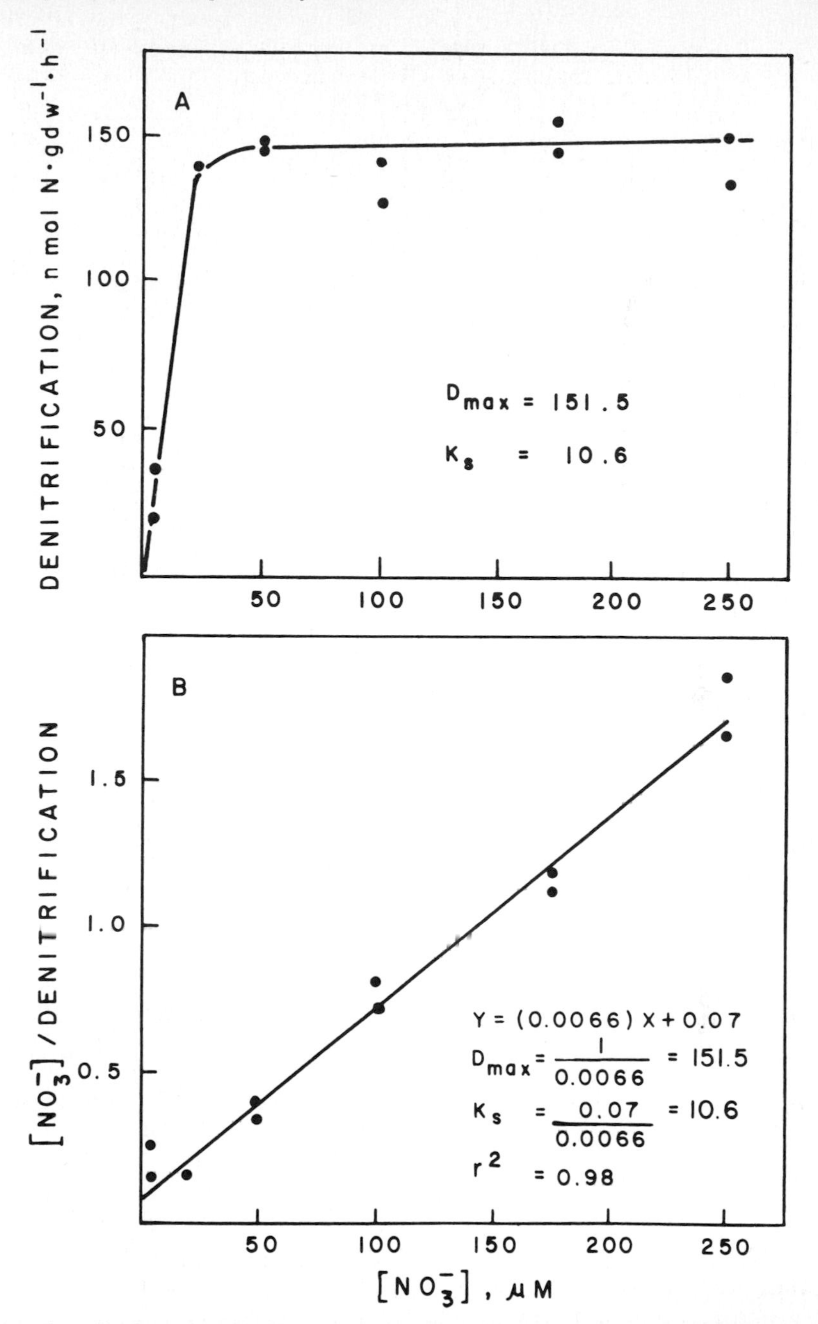

Figure 3. *A) Measured denitrification potentials expressed as ng-at N $g^{-1}h^{-1}$ (sediment dry weight basis), at different aqueous concentrations (µg-at l^{-1}) of nitrate for estuarine sediments from Choptank River estuary (Horn Point), Maryland. B) Linear transformation of data from (A), as used to calculate Michaelis-Menten kinetic constants of denitrification rates.*

Table 3. Kinetic constants based on linear transformation of rectangular hyperbolic functions including maximum denitrification rates and K_s (half-saturation constant) for estuarine sediments in Chesapeake Bay.

Station	Maximum Denitrification (D_{max})		K_s	r^2
	ng-at N $cm^{-3}h^{-1}$	ng-at N $gdw^{-1}h^{-1}$	(μg-at l^{-1})	
CHESAPEAKE BAY				
1	6.99	9.8	8.4	0.85
2	6.67	24.2	8.5	0.87
3	*	*	*	0.04
4	*	*	*	0.30
POTOMAC TRIBUTARY				
5	2.65	8.0	2.0	0.92
6	124.35	401.3	67.2	0.89
PATUXENT TRIBUTARY				
7	57.14	146.4	22.5	0.94
8	24.37	52.9	78.2	0.73
CHOPTANK TRIBUTARY				
9	77.33	151.5	10.6	0.98
10	222.30	555.6	92.2	0.76

*values not given because relation was nonsignificant

Table 4. Significant correlations between constants D_{max} (maximum denitrification potentials) or K_s (half-saturation constants) and selected chemical and physical factors, based on stepwise regression analysis.

Dependent Variable	Independent Variables	Slope (Intercept)	r^2	P>F
K_s	N:P Ratio	−8.394 (120.136)	0.705	0.009
D_{max}	Salinity Carbon	−34.22 88.06 (348.27)	0.790	0.021

Discussion

The response of denitrification potentials to NO_3^- concentrations followed rectangular hyperbolic type kinetics in 8 of the 10 stations of Chesapeake Bay sediments investigated. These kinetic relations could be characterized by the constants D_{max} and K_s that are the maximum denitrification rate and half-saturation constant, respectively. In reporting zero-order response of denitrification to NO_3^- concentration, Kaspar (1982) argued that the kinetic nature of denitrification observed by Oren and Blackburn (1979) was possibly due to diffusion. We used well-mixed slurries with high water: sediment ratios in our experiments to exclude diffusion as a rate-limiting process. Our results demonstrate that in most cases denitrification potentials in Chesapeake Bay sediments respond to increasing NO_3^- concentrations in a kinetic fashion (Twilley and Kemp, unpublished).

This kinetic response and the constants used to describe it resemble those used to interpret first-order enzyme kinetics (Dowd and Riggs 1965). However, even though the curves may be similar, the application of these enzyme-system constants may not be appropriate for complex heterogenous populations (Williams 1973; Goldman and Gilbert 1983). In our studies it may be assumed that K_s represents a measure of the physiological affinity of the denitrifier community for NO_3^- substrate. Denitrification potentials were not normalized to microbial biomass, so that D_{max} may be taken as an index of the density of denitrifying bacteria in a constant volume of sediment (McLaren 1976). However, interpretations of K_s values should not be affected by bacterial abundance as long as no physiological adaptation of the population occurs during the course of the experiment (Nedwell 1975).

The range in K_s values measured in Chesapeake Bay sediments is similar to that observed in other estuarine systems (Table 5). K_s values <25 μg-at l^{-1}, which were common in mainstem Chesapeake Bay, have been reported for Japanese estuarine systems (Koike *et al.* 1972; Koike *et al.* 1978; Koike and Hattori 1979) and for sediments in the Bering Sea (Koike and Hattori 1979). Average values observed for Patuxent and Choptank River sediments (about 50 μg-at l^{-1}) are similar to those measured for the Scheldt estuary in Belgium (Billen 1978) and Izembek Lagoon, USA (Iizumi *et al.* 1980). One exceptionally high K_s value (344 μg-at l^{-1}) has been reported for Kysing Fjord (Oren and Blackburn 1979). Other sediments with much higher nitrogen loading characteristics such as those adjacent to sewage effluents or ditches draining agricultural fields have correspondingly higher K_s values (300-2500 μg-at l^{-1}) and higher denitrification rates (Nedwell 1975; Van Kessel 1977; Kohl *et al.* 1976). However, these extreme K_s values are based on NO_3^- uptake by intact sediments, and may thus represent effects of diffusion rather than enzyme kinetics. These methods may also overestimate actual denitrification potentials since a portion of the nitrate may be reduced to end products other than those associated with denitrification (e.g., NH_4^+; Stanford *et al.* 1975). In Chesapeake Bay, higher K_s values occurred in the landward stations of the tributaries where springtime NO_3^- concentrations are much higher than the downstream stations (Kemp and Boynton 1984; Ward and Twilley 1986). This pattern suggests higher K_s values may represent enzymatic adaptations to higher substrate concentrations (Fenchel and Blackburn 1979). Yet Yoshinari *et al.* (1977) observed that amendments of soils with glucose caused an increase in K_s for denitrification, suggesting that the generally higher organic deposition rates in these estuarine regions (Kemp and Boynton 1984; Boynton and Kemp 1985) may partially account for this pattern as well.

Table 5. Half-saturation constants (K_s) describing the kinetic response of denitrification (or nitrate reduction) rates to nitrate concentrations for marine and freshwater sediments, and soils.

Site	K_s (μg-at l^{-1} NO_3^-)	Process	Reference
Chesapeake Bay, USA	2.0-92.2	Denitrification	This study
Bering Sea	3.5-9.1	Denitrification	Koike and Hattori 1979
Tokyo Bay, Japan	24	Denitrification	Koike *et al.* 1978
Odawa Estuary	25	Denitrification	Koike *et al.* 1972
Mangoku-Ura, Japan	27-42	Denitrification	Koike *et al.* 1978
Soldier Key, USA	30	Denitrification	Capone and Taylor 1980
Scheldt Estuary, Belgium	50	Denitrification	Billen 1978
Izembek Lagoon, USA (eelgrass sediments)	53	Denitrification	Iizumi *et al.* 1980
Kysing Fjord, Denmark	344	Denitrification	Oren and Blackburn 1979
Vatuwaga River, Fiji (mangrove sediments)	180-600	Nitrate Uptake	Nedwell 1975
Danish Lakes			
Aerobic Sites	376-6027	Nitrate Uptake	Anderson 1977
Anaerobic Sites	7-893		
Ditch Bank Sediments	14,696	Nitrate Uptake	Van Kessel 1977
Arable Soil	4,759	Nitrate Uptake	Van Kessel 1977
Agriculture Soil	290-3,479	Nitrate Uptake	Kohl *et al.* 1976

Nitrate-saturated denitrification potentials were generally higher for sediments in the Choptank than for those in the Patuxent, and rates were generally higher in the tributary stations than in the mainstem Bay. Similar spatial patterns among Choptank, Patuxent and mainstem Bay sediments have also been observed for ambient and potential denitrification rates using $^{15}N\text{-}NO_3^-$ Twilley and Kemp, unpublished). Springtime nitrate concentrations in the oligohaline zone are generally >100 μg-at l^{-1} in the Choptank (Ward and Twilley 1986), compared to 50-75 μg-at l^{-1} in the Patuxent (Kemp and Boynton 1984), and 25-50 μg-at l^{-1} in the mainstem Bay (Boynton and Kemp 1985). Both Koike *et al.* (1972) and Sørensen (1982) have suggested that higher concentrations of NO_3^- may stimulate the enzyme activation level of nitrate-reducing bacteria. Thus, while the affinity for NO_3^- (K_s) by denitrifiers may be lower in regions with higher NO_3^- concentration, increased denitrification capacity (D_{max}), which reflects differences in population density, may also result from higher ambient NO_3^- concentrations.

Ambient denitrification rates, estimated from the constants for the rectangular hyperbolic response described above and ambient NO_3^- concentrations in

Table 6. Denitrification rates of marine and freshwater sediments.

Site	Denitrification (ng-at N $gdw^{-1}h^{-1}$)	Temp (°C)	Depth (cm)	$[NO_3+NO_2]$ (μg-at l^{-1})	Reference
Chesapeake Bay, USA					This study
Sta. 1	1.4	17	0-1.5	1.4 (OW)*	
Sta. 2	6.3	17	0-1.5	2.9 (OW)	
Sta. 5	4.2	17	0-1.5	2.2 (OW)	
Sta. 6	123.3	17	0-1.5	29.8 (OW)	
Sta. 7	20.8	17	0-1.5	3.7 (OW)	
Sta. 8	1.4	17	0-1.5	2.9 (OW)	
Sta. 9	25.1	17	0-1.5	2.1 (OW)	
Sta. 10	33.6	17	0-1.5	5.9 (OW)	
Limfjord, Denmark	5.0-39.9	5			Sørensen 1978
Izembek Lagoon, USA					Iizumi *et al.* 1980
Sta. 1	0.13	11-15	0-2	5.0 (PW)*	
Sta. 4	0.05	11-15	0-2	2.0 (PW)	
Sta. 10	1.2	11-15	0-2	15.0 (PW)	
Bering Sea					Koike and Hattori 1979
Sta. 12	1.4	2.5	0-2	7.9 (PW)	
Sta. 14	0.91	2.5	0-2	2.9 (PW)	
Sta. 19	1.3	2.5	0-2	3.9 (PW)	
Tokyo Bay, Japan					
Sta. 1	2.0	8.0	0-2	1.6 (PW)	Koike and Hattori 1979
Sta. 2	4.2	8.0	0-2	5.9 (PW)	
Sta. 2	8.4	20.0	0-3	15 (Exp)	Koike and Hattori 1978a
Mangoku-Ura, Japan					
Sta. 7	11	21	0-2	62.6 (PW)	Koike and Hattori 1978b
Sta. 10	7.6	21	0-2	2.3 (PW)	Koike and Hattori 1978b
Sta. 3	0.15	15	0-2	34.2 (PW)	Koike and Hattori 1978b
Sta. 3	8.9	3.5	0-2	22.2 (PW)	Koike and Hattori 1979
	293	21	0-3	30.0 (Exp)*	Koike and Hattori 1978a
Tokyo Bay, Japan	8.4	20	0-3	15.0 (Exp)	Koike and Hattori 1978a
Simoda Bay, Japan	17.8	26	0-3	15.0 (Exp)	Koike and Hattori 1978a
Delaware Inlet	3.9-39.5	22	0-7.5	1,000 (Exp)	Kaspar 1982
New Zealand	0.12-1.40	22	0-7.5	2.0 (PW)	
Louisiana Lakes, USA					Smith and DeLaune 1983
Freshwater	4.5	25			
Saline	1.8	25			
Toolik Lake	0-0.12	4	0-12	<5 (PW)	Klingensmith & Alexander 1983
Lake Naivasha, Kenya	1.4-82.9	27	0-2	13,000 (Exp)	Viner 1982

*OW = Ambient concentration of overlying water
PW = Ambient concentration of pore water
Exp = Experimental concentration

overlying water, ranged from 1.4 to 123.3 ng-at N gdw^{-1} h^{-1} in Chesapeake Bay in October. These rates are applicable only for the top 1-2 mm of sediment and may decrease with depth given that NO_3^- levels in pore waters also generally decrease with depth. Most denitrification rates for other estuarine sediments are <10 ng-at N gdw^{-1} h^{-1} (Table 6), except for those amended with NO_3^- (Koike and Hattori 1978a). Hattori (1983) calculated denitrification rates for eutrophic Tama Estuary (150 ng-at N gdw^{-1} h^{-1}) similar to the upper range of values for Chesapeake Bay. Combining the kinetic constants from our study with higher NO_3^- concentrations (ca. 100 μg-at l^{-1}) that occur in the Choptank and Patuxent River estuaries during the spring, we estimate that denitrification rates at our stations range from 26 to 289 ng-at N gdw^{-1} h^{-1} in Chesapeake Bay. These estimated ranges suggest that actual denitrification rates may vary as much spatially due to the kinetic nature of denitrification capacity to about the same extent as the seasonal variation due to differences in NO_3^- concentration.

A significant portion of the variability observed for D_{max} and K_s in Chesapeake Bay sediments can be explained statistically in terms of gross sediment characteristics. Nitrate-saturated denitrification rates (D_{max}) were inversely correlated with salinity ($r^2=0.65$). Elsewhere, it has been shown that mean NO_3^- concentration in overlying water is also inversely related to salinity in Chesapeake Bay (Kemp and Boynton 1984; Ward and Twilley 1986). Thus, salinity serves as a good predictor of long-term mean NO_3^- levels. As we have discussed above, higher ambient NO_3^- concentrations presumably lead to higher growth rates and more abundant denitrifier populations. D_{max} was also directly correlated with TC in sediments, which is consistent with the concept that increased availability of electron donors can also stimulate denitrifier growth (Hattori 1983). Other investigators have demonstrated higher denitrification capacities with higher organic matter concentrations in sediments (Terry and Nelson 1975; Koike and Hattori 1978b) and soils (Bowman and Focht 1974; Yoshinari *et al.* 1977).

Similarly, K_s was inversely correlated to salinity (although the relation is less significant that for D_{max}); again this suggests a lower affinity for NO_3^- for those denitrifier populations generally exposed to higher substrate concentrations. K_s was also significantly related to N:P ratio of sediment particulates. This unexpected result might reflect the effect of higher denitrification rates decreasing the residual particulate nitrogen levels in these sediments. Thus, it appears that sediment characteristics influence the kinetic relations for denitrifer bacteria in estuarine sediments in various ways. The correlative information presented in this paper suggests that physical and chemical features of the sediment and overlying water are largely responsible for the observed ranges in kinetic parameters for sediment denitrification in this estuarine system.

Acknowledgments

We thank W. Boynton, L. Lubbers, K. Wood, J. Barnes and W. Keefe, all of Chesapeake Biological Laboratory for sampling sediments around the Chesapeake Bay and for analytical services. M. Shenton-Leonard, J. Caffrey and L. Lane added valuable laboratory assistance at Horn

Point Laboratory with the slurry experiments. D. Kennedy and J. Metz drafted the illustrations and A. R. McGuinn typed the manuscript. This project was funded by the U.S. Environmental Protection Agency/Chesapeake Bay Program under contract X-003310-01-0. This article is contribution No. 1688 from the University of Maryland Center for Environmental and Estuarine Studies. The present address of R.R. Twilley is Department of Biology, University of Southwestern Louisiana, Lafayette, LA.

References Cited

Anderson, J. M. 1977. Rates of denitrification of undisturbed sediment from six lakes as a function of nitrate concentration, oxygen and temperature. *Arch. Hydrobiol.* 80:147-159.

Balderston, W. L., B. Sherr and W. J. Payne. 1976. Blockage by acetylene of nitrous oxide reduction in *Pseudomonas perfectomarinus*. *Appl. Environ. Microbiol.* 31:504-508.

Billen, G. 1978. A budget of nitrogen recycling in North Sea sediments off the Belgian Coast. *Estuar. Coastal Mar. Sci.* 7:127-146.

Blake, G. R. 1965. Bulk density, pp. 374-390. *In:* C. A. Black *et al.* (eds.), *Methods of Soil Analysis.* American Society of Agronomy, Madison, WI.

Bowman, R. A. and D. D. Focht. 1974. The influence of glucose and nitrate concentrations upon denitrification rates in a sandy soil. *Soil Biol. Biochem.* 6:297-301.

Boynton, W. R., W. M. Kemp, and C. G. Osborne. 1980. Nutrient fluxes across the sediment-water interface in the turbid zone of a coastal plain estuary, pp. 93-109. *In:* V. Kennedy (ed.), *Estuarine Perspectives.* Academic Press, New York.

Boynton, W. R. and W. M. Kemp. 1985. Nutrient regeneration and oxygen consumption by sediments along an estuarine salinity gradient. *Mar. Ecol. Prog. Ser.* 23:45-55.

Boynton, W. R., W. M. Kemp, L. L. Lubbers, K. V. Wood and C. Keefe. 1985. *Maryland Office of Environmental Programs, Maryland Chesapeake Bay Water Monitoring Program, Ecosystem Processes Component Data Report No. 1.* University of Maryland Center for Environmental and Estuarine Studies Reference No. 84-109.

Capone, D. G. and B. F. Taylor. 1980. Microbial nitrogen cycling in seagrass community, pp. 153-162. *In:* V. S. Kennedy (ed.), *Estuarine Perspectives.* Academic Press, New York.

Dowd, J. E. and D. S. Riggs. 1965. A comparison of estimates of Michaelis-Menten kinetic constants from various linear transformations. *J. Biol. Chem.* 24:863-869.

Fenchel, T. and T. H. Blackburn. 1979. *Bacteria and mineral cycling.* Academic Press, New York.

Folk, R. L. 1974. *Petrology of sedimentary rocks.* Hemphill, Austin, Texas, 182 pp.

Goldman, J. C. and P. M. Glibert. 1983. Kinetics of inorganic nitrogen uptake by phytoplankton, pp. 233-274. *In:* E. J. Carpenter and D. G. Capone (eds.), *Nitrogen in the Marine Environment.* Academic Press, New York.

Hattori, A. 1983. Denitrification and dissimilatory nitrate reduction, pp. 191-232. *In:* E. J. Carpenter and D. G. Capone (eds.), *Nitrogen in the Marine Environment.* Academic Press, New York.

Iizumi, H., A. Hattori, and C. P. McRoy. 1980. Nitrate and nitrite in interstitial waters of eelgrass beds in relation to the rhizosphere. *J. Exp. Mar. Biol. Ecol.* 47:191-201.

Kaspar, H. F. 1982. Denitrification in marine sediment: Measurement of capacity and estimate of *in situ* rate. *Appl. Environ. Microbiol.* 43:522-527.

Kemp, W. M. and W. R. Boynton. 1984. Spatial and temporal coupling of nutrient inputs to estuarine primary production: the role of particulate transport and decomposition. *Bull. Mar. Sci.* 35:522-535.

Klingensmith, K. M. and V. Alexander. 1983. Sediment nitrification, denitrification, and nitrous oxide production in a deep Arctic Lake. *Appl. Environ. Microbiol.* 46:1084-1092.

Knowles, R. 1979. Denitrification, acetylene reduction, and methane metabolism in lake sediment exposed to acetylene. *Appl. Environ. Microbiol.* 38:486-493.

Kohl, D. H., F. Vitayathil, P. Whitlow, G. Shearer, and S. H. Chien. 1976. Denitrification kinetics in soil systems: The significance of good fits to mathematical forms. *Soil Sci. Soc. Am. Proc.* 40:249-253.

Koike, I. and A. Hattori. 1978a. Simultaneous determinations of nitrification and nitrate reduction in coastal sediments by a ^{15}N dilution technique. *Appl. Environ. Microbiol.* 35:853-857.

Koike, I. and A. Hattori. 1978b. Denitrification and ammonia formation in anaerobic coastal sediments. *Appl. Environ. Microbiol.* 35:278-282.

Koike, I. and A. Hattori. 1979. Estimates of denitrification in sediments of the Bering Sea shelf. *Deep-Sea Res.* 26:409-415.

Koike, I., A. Hattori, and J. J. Goering. 1978. Controlled ecosystem pollution experiment: Effect of mercury on enclosed water columns. VI. Denitrification by marine bacteria. *Mar. Sci. Commun.* 4:1-12.

Koike, I., E. Wada, T. Tsuji, and A. Hattori. 1972. Studies on denitrification in a brackish lake. *Arch. Hydrobiologia* 69:508-520.

McLaren, A. D. 1976. Rate constants for nitrification and denitrification in soils. *Rad. and Environm. Biophys.* 13:43-48.

Nedwell, D. B. 1975. Inorganic nitrogen metabolism in a eutrophicated tropical mangrove estuary. *Wat. Res.* 9:221-231.

Nixon, S. W. 1981. Remineralization and nutrient cycling in coastal marine ecosystems, pp. 111-138. *In:* B. J. Nielson and L. E. Cronin (eds.), *Estuaries and Nutrients.* Humana Press, Clifton.

Oremland, R. S., C. Umberger, C. W. Culbertson and R. L. Smith. 1984. Denitrification in San Francisco Bay intertidal sediments. *Appl. Environ. Microbiol.* 47:1106-1112.

Oren, A. and T. H. Blackburn. 1979. Estimation of sediment denitrification rates at *in situ* nitrate concentrations. *Appl. Environ. Microbiol.* 37:174-176.

Seitzinger, S., S. Nixon, M. E. Q. Pilson, and S. Burke. 1980. Denitrification and N_2O production in near-shore marine sediments. *Geochim. Cosmochim. Acta.* 44:1853-1860.

Smith, C. J. and R. D. DeLaune. 1983. Nitrogen loss from freshwater and saline sediments. *J. Environ. Qual.* 12:514-518.

Smith, C. J., R. D. DeLaune, and W. H. Patrick, Jr. 1985. Fate of riverine nitrate entering an estuary: I. Denitrification and nitrogen burial. *Estuaries* 8:15-21.

Sørensen, J. 1978. Denitrification rates in a marine sediment as measured by the acetylene inhibition technique. *Appl. Environ. Microbiol.* 36:139-143.

Sørensen, J. 1982. Reduction of ferric iron in anaerobic marine sediment: an interaction with reduction of nitrate and sulfate. *Appl. Environ. Microbiol.* 43:319-324.

Stanford, G., J. O. Legg, S. Dzienia and E. C. Simpson, Jr. 1975. Denitrification and associated nitrogen transformation in soils. *Soil Sci.* 120:147-152.

Statistical Analysis System. 1982. *SAS User's Guide: Statistics.* SAS Institute Inc., Cary, North Carolina.

Terry, R. E. and D. W. Nelson. 1975. Factors influencing nitrate transformation in sediments. *J. Environ. Qual.* 4:549-554.

U.S. Environmental Protection Agency. 1979. Methods for chemical analysis of water and wastes. USEPA-600/4-79-020, Cincinnati, Ohio.

Van Kessel, J. F. 1977. Factors affecting the denitrification rate in two sediment-water systems. *Wat. Res.* 11:259-267.

Van Raalte, C. D. and D. G. Patriquin. 1979. Use of the "acetylene blockage" technique for assaying denitrification in a salt marsh. *Mar. Biol.* 52:315-320.

Viner, A. B. 1982. Nitrogen fixation and denitrification in sediments of two Kenyan lakes. *Biotrop.* 14:91-98.

Ward, L. G. and R. R. Twilley. 1986. Distribution of dissolved and particulate material in a coastal plain tributary of the Chesapeake Bay. *Estuar.* (in press).

Weiss, R. F. and B. A. Price. 1980. Nitrous oxide solubility in water and seawater. *Mar. Chem.* 8:347-359.

Whitfield, M. 1969. Eh as an operational parameter in estuarine studies. *Limnol. Oceanogr.* 14:547-558.

Williams, P. J. LeB. 1973. The validity of the application of simple kinetic analysis to heterogeneous microbial populations. *Limnol. Oceanogr.* 18:159-165.

Yoshinari, T., R. Hynes and R. Knowles. 1977. Acetylene inhibition of nitrous oxide reduction and measurement of denitrification and nitrogen fixation in soil. *Soil. Biol. Biochem.* 9:177-183.

TEMPORAL VARIABILITY OF SALT MARSH VEGETATION: THE ROLE OF LOW-SALINITY GAPS AND ENVIRONMENTAL STRESS

Joy B. Zedler and Pamela A. Beare

San Diego State University
San Diego, California

Abstract: Extreme variability (including floods and drought) characterizes estuaries in semi-arid regions. At Tijuana Estuary, California, the winter floods of 1980 reduced channel salinities to 0 g l^{-1} and salt marsh soil salinities to 15 g l^{-1}. During the drought of 1984, the estuarine mouth was blocked; channel salinities reached 70 g l^{-1}; and marsh soils exceeded 100 g l^{-1}. Changes in the growth and distribution of salt marsh plants were documented by a monitoring program that began in 1979. The expansion of *Spartina foliosa*, initiated in 1980, was reversed in 1984; the response of succulents was reciprocal. Populations of two short-lived marsh plants crashed during the hypersaline drought. At the San Diego River salt marsh, prolonged flooding due to reservoir drawdown eliminated halophytes and allowed fresh-to-brackish species to invade in 1980. Reestablishment of salt marsh species has been slow under the recent, more saline conditions.

The causes of these vegetation dynamics are summarized in a conceptual model of community composition: Invasion and expansion of species are controlled by the annual "low-salinity gap," which varies both in duration and degree of salinity reduction. Population declines and local extinctions are caused by hypersaline drought and prolonged inundation, to which species have differential tolerance.

Introduction

Southern California coastal salt marshes differ from those of the Atlantic and Gulf coasts of the United States in that they are more often dominated by succulent halophytes than by *Spartina*, and a mixture of species, rather than a monoculture, is commonly found (Zedler 1982). A major cause of these regional differences appears to be climate, which controls the regional hydrology. The Mediterranean-type rainfall pattern (wet winters, dry summers) in Southern California provides only brief periods of freshwater influx to coastal wetlands. This restriction of freshwater influence to the winter season means that the salinity of coastal waters is only intermittently estuarine. Substantial year-to-year differences in streamflow and occasional extreme events add to the normal temporal variability.

Rare events are very important to overall wetland structure and functioning in southern California (Zedler and Onuf 1984). Wherever catastrophic floods or droughts occur, wetland communities are affected; however, documentations of how such events influence establishment and persistence are few. This paper focuses on the role of extreme conditions in controlling the most basic structural feature of a plant community: its species composition. We hypothesize that germination and establishment are limited to the "low-salinity gap" that follows winter rainfall, and that two environmental stresses, hypersaline drought and excessive inundation, limit the expansion and persistence of many plant populations. Extreme stresses of these types cause local extinctions.

Methods

Forty-four years of streamflow records (U.S. Geological Survey 1937-1981) were examined for the most downsteam gaging stations of Tijuana River (32°34′N, 117°7′W) and San Diego River (32°N,46′N, 117°14′W), San Diego County, California. Unpublished data for the 1982-83 San Diego River flows were obtained from USGS. Mean annual discharges and coefficients of variation were obtained. Records of rainfall at San Diego State Univ. and soil salinity data for coastal marshes of both rivers were used to characterize conditions in recent years for which streamflow records were unavailable.

Baseline studies of salt marsh composition were carried out at Tijuana Estuary in 1974 (Zedler 1977), using 357 0.25-m^2 quadrats along three transects that spanned the lower and upper marsh elevations. Species presence was recorded, and estimates of cover were made. *Salicornia bigelovii* was sampled monthly in 1975 across mid-marsh elevations (0.25-m^2 quadrats, $n = 10$) to document establishment and mortality of seedlings. In 1979, a monitoring program was developed for the lower intertidal salt marsh of Tijuana Estuary. Salinities of interstitial soil water were measured (American Optical salinity refractometer) in April and September at approximately 100 permanent stations distributed at 5-m intervals along eight transects (Zedler 1983). Vegetation was assessed each September. The data used here include counts of living *Spartina foliosa* plants in 0.25-m^2 circular quadrats at each station and percent cover of other species in the same quadrats. Beginning in April 1984, the monitoring program was expanded to include 115 stations along six transects in the upper salt marsh. All species present in each 0.25-m^2 circular quadrat were recorded and their percent cover estimated (Zedler and Covin 1984).

Observations have been made periodically since 1975 at the San Diego River marsh, where salt marsh species declined in 1980 and *Typha domingensis* and other fresh-to-brackish marsh species invaded (Zedler 1981). A monitoring program was established in 1982 to test the prediction that *Typha domingensis* would decline and halophytes reinvade following the return of hypersaline conditions. Sampling was carried out each summer along ten 50-m transects and 0.25-m^2 circular quadrats at 5-m intervals ($n = 250$) to record species presence and estimate percent cover.

To evaluate the relative ability of different marsh plants to establish across a broad range of salinities, seed-germination experiments were performed in 1984-85 using the 14 species for which sufficient seeds could be collected. Sea salt solutions of 0, 5, 10, and 20 g l^{-1} were used as treatments. For each species, 100 seeds were distributed equally among 4 petri dishes for each salinity treatment. Seeds were placed on filter paper that floated on a styrofoam disk in each salt solution (Beare 1984). A second set of germination experiments utilized species mixtures to assess collective salinity responses; seed-rich debris from the high-tide line at Tijuana Estuary was supplemented with seeds from the above 14 species. Equal volumes (50 ml) of this mixture were sown into 15 × 25-cm bins containing 8 cm of sand. A perforated test-tube inserted in the center of each bin allowed maintenance of saturated soil conditions and salinity treatments of 0, 16, and 34 g l^{-1}. There were three replicate bins for each treatment.

In both types of seed-germination experiments, treatments were maintained in the laboratory under fluorescent light banks. Germination was assessed weekly by counting seedlings in the single-species experiments and by determining frequency of occurrence in the mixed-species experiments. For the latter, each bin was sampled with twenty 6.25-cm^2 quadrats.

Results and Discussion

Streamflow records from 1937-77 document high variability in both monthly and annual flows for Tijuana and San Diego Rivers. For Tijuana R., mean discharge was 20.8×10^6 m^3 (16,882 ac-ft) per year, with a coefficient of variability of 325%. Several years had zero flow, while flooding in 1941 had a maximum of over 40×10^6 m^3. For San Diego R., the mean discharge was 11.7×10^6 m^3 (9,481 ac-ft) per year, and the coefficient of variability was lower, 245%, possibly because four large dams ameliorate floods and irrigation runoff augments summer flows. Since 1977, variability has been even greater, with major flooding in 1978 (4x previous mean flows) and "100-year" flooding in 1980 (15x the mean at San Diego R., and 28x the mean at Tijuana R.; Fig. 1).

The 1979-84 soil and vegetation sampling included years of near-average rainfall, unusually high rainfall with flooding, and unusually low rainfall with drought conditions at the Tijuana Estuary (Table 1). The San Diego River marsh

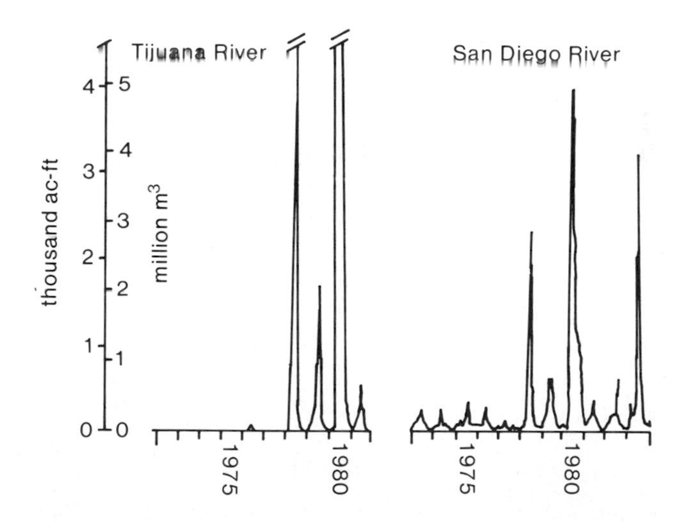

Figure 1. Recent streamflows in Tijuana River (at Nestor) and San Diego River (at Santee).

did not undergo drought, because the river remained connected to the ocean and irrigation runoff maintained streamflow during 1984. Substantial year-to-year variability in species composition and plant growth was documented for both marshes; in the following section we summarize first the plant population expansion and then the declines in species abundance.

Seedling Establishment

The species that establishes seedlings most successfully at Tijuana Estuary is the annual plant, *Salicornia bigelovii.* It was common in the 1930's (Purer 1942); in 1974 it exhibited 64% frequency (Table 2) and 35% mean cover at 60-70 cm MLLW (Zedler 1977). In 1975, densities of *Salicornia bigelovii* ranged from a maximum average of 3,072 seedlings/m² in May, declining to an average of 892 plants/m² in October (Zedler 1975). The population remained abundant through 1983, due to annual establishment from seed.

Suaeda esteroa is a common short-lived (2-3 yr) plant that does not reproduce vegetatively (Purer 1942). Seedlings are frequently encountered, but they are seldom dense. Good recruitment ability is assumed from the widespread occurrence of the species at Tijuana Estuary, both in the 1930's (Purer 1942) and in 1974, when it had 37% frequency (Table 2) and 18% mean cover at 90-100 cm MLLW (Zedler 1977).

Table 1. Recent environmental conditions. Rainfall data are from San Diego State Univ., 16 km inland on the San Diego River (long-term average rainfall = 24 cm).

Year	San Diego Rainfall	Tijuana Estuary	San Diego R.
1979	49 cm; above average	open to tidal flow	open to tidal flow
1980	54 cm; above average	100-yr flood event; soil became brackish in spring	flooding prolonged artificially by reservoir discharge; soils 0-10 g l^{-1} from Mar.-July
1981	23 cm; near average	open to tidal flow	open to tidal flow
1982	38 cm; above average	open to tidal flow	open to tidal flow
1983	57 cm; winter rains through April	open to tidal flow except for 2 weeks of seawater impoundment in winter following extreme storm; summer reservoir discharge reduced salinities	open to tidal flow
1984	18 cm; below average with 4 cm during the 1984 growing season	closed to tidal flow on April 6; estuary dredged open on Dec. 18, 1984; soils averaged 104 g l^{-1} in Sept.	open to tidal flow

Table 2. Frequency of occurrence of species in two salt marshes. The Tijuana Estuary marsh (lower and upper marsh combined) was sampled in 1974 (n = 357) and in 1984 (n = 217). The San Diego River marsh was sampled in 1978 (n = 20), 1982 and 1984 (n = 250). Taxonomic nomenclature follows Munz (1974).

	Tijuana Estuary		**San Diego River**		
Species	**1974**	**1984**	**1978**	**1982**	**1984**
Salt marsh species					
Salicornia virginica L.	69	75	100	18	20
S. bigelovii Torr.	64	0			
Jaumea carnosa Less. (Gray)	55	24	0	59	49
Batis maritima L.	51	38	0	7	6
Frankenia grandifolia Cham. & Schlecht	49	26	20	19	11
Monathochloe littoralis Engelm.	41	16			
Suaeda esteroa Ferren & Whitmore	37	<1			
Salicornia subterminalis Parish	18	10			
Spartina foliosa Trin.	16	46	0	3	0
Distichlis spicata (L.) Greene	13	4	5	0	0
Limonium californicum (Boiss.) Heller	10	4			
Triglochin concinnum Davy	9	16			
Cressa truxillensis HBK	7	3			
Cuscuta salina Engelm.	6	4	0	5	3
Juncus acutus L.			0	1	1
Brackish marsh species					
Typha domingensis Pers.			0	66	35
Scirpus robustus Pursh.			0	46	1
Cotula coronopifolia L.			0	35	6
Atriplex patula L.			0	12	0

The *Spartina foliosa* population grew substantially in 1980. Densities increased vegetatively within the long-term monitoring area (Table 3), and approximately 70 new clones became established in areas dominated by *Salicornia virginica* (Zedler 1983). The relatively low soil salinities (15 $g\ l^{-1}$ in April 1980) may have promoted seedling establishment; seedlings of *S. foliosa* were not seen in other census years.

Table 3. Lower marsh comparisons at Tijuana Estuary. Data are from permanent sampling stations (n = 102).

	1979	**1980**	**1981**	**1982**	**1983**	**1984**	**1985**
Salicornia virginica							
% occurrence	65	79	76	no data	76	75	88
mean % cover	50	38	50	no data	38	47	64
Spartina foliosa							
% occurrence	90	92	88	85	94	78	38
density (no./sq.m.)	49	60	50	70	113	42	7

The most dramatic species invasions occurred in the San Diego River marsh in 1980 (Zedler 1981) following a prolonged period of reservoir discharge (Table 2). Reservoir drawdown extended the normal period of low salinities well beyond the wet season, and *Typha domingensis, Scirpus robustus, Scirpus californicus,* and *Cotula coronopifolia* rapidly invaded the salt marsh from seed (Table 2). The very low soil salinities (0-5 g l^{-1}) allowed *Typha* seeds to germinate, and the prolonged period (2-3 months) of low soil salinities allowed seedlings to establish. *Typha* seedlings could not survive high salinities until they had produced rhizomes (Beare 1984). Thus, 2-3 months of low salinity are required for successful *Typha* invasion. Reinvasion by the pre-flood community has been slow in the San Diego R. marsh (Table 2). The previous dominant, *Salicornia virginica* had not recovered by 1984, and *Jaumea carnosa* persisted as the most successful colonist.

Seed germination data (Fig. 2) help to explain the observed cases of seedling establishment. A broad spectrum of responses to salinity occurs for common species of coastal wetlands. Of the species tested, only half would germinate at 20 g l^{-1}, five were restricted to 10 g l^{-1}, and two would not germinate in salinities beyond 5 g l^{-1}. The three *Salicornia* species are most likely to germinate without freshwater influxes; of all the species tested, their germination rates were least reduced by 20 g l^{-1} sea salt. In other experiments, where mixtures of seeds were treated with sea salt solutions of 35 g l^{-1}, only the *Salicornia* spp. germinated after one month, whereas 8 species germinated at 16 g l^{-1}, and 11 species germinated at 0 g l^{-1} (Table 4).

More detailed examination of the data revealed how salinity both delays and reduces germination rates. For the single-species tests (Fig. 2), an analysis of covariance on 1- and 4-week rates at 0 g l^{-1} and 10 g l^{-1} salinities indicated that some species responded to 10 g l^{-1} with reduced germination rates (e.g., *Typha domingensis, Scirpus robustus, Scirpus californicus, Monanthochloe littoralis,* and *Jaumea carnosa*). Germination of other species was merely delayed by 10 g l^{-1} salt (*Cotula coronopifolia, Triglochin concinnum, Juncus acutus,* and *Eclipta alba*). Thus, either 2-3 months of brackish salinities (e.g., 10 to 20 g l^{-1} salt) or 2-3 weeks of fresh water influx appear to be necessary for most species to invade saline marshes. Neither condition is common in the region's intertidal marshes. This explains why so many species invaded the San Diego R. marsh only in 1980, when a substantial gap in hypersaline conditions occurred and soil salinities were below 10 g l^{-1} for several months.

Population Expansions

For the longer-lived perennials, vegetative reproduction is more common than seedling establishment. Seedlings are rarely encountered. *Spartina foliosa* expanded vegetatively in 1980, 1982, and 1983 at stations monitored each September. In each case, expansions followed periods of above-average rainfall and streamflow (Table 3, Fig. 1). In 1983, an overall 60% increase in stem densities was related to the summer influx of fresh water from reservoir drawdown and sewage spills from Mexico (Zedler and Covin 1984; Zedler *et al.* In press). River flows at the Mexico border were the highest on record for April through

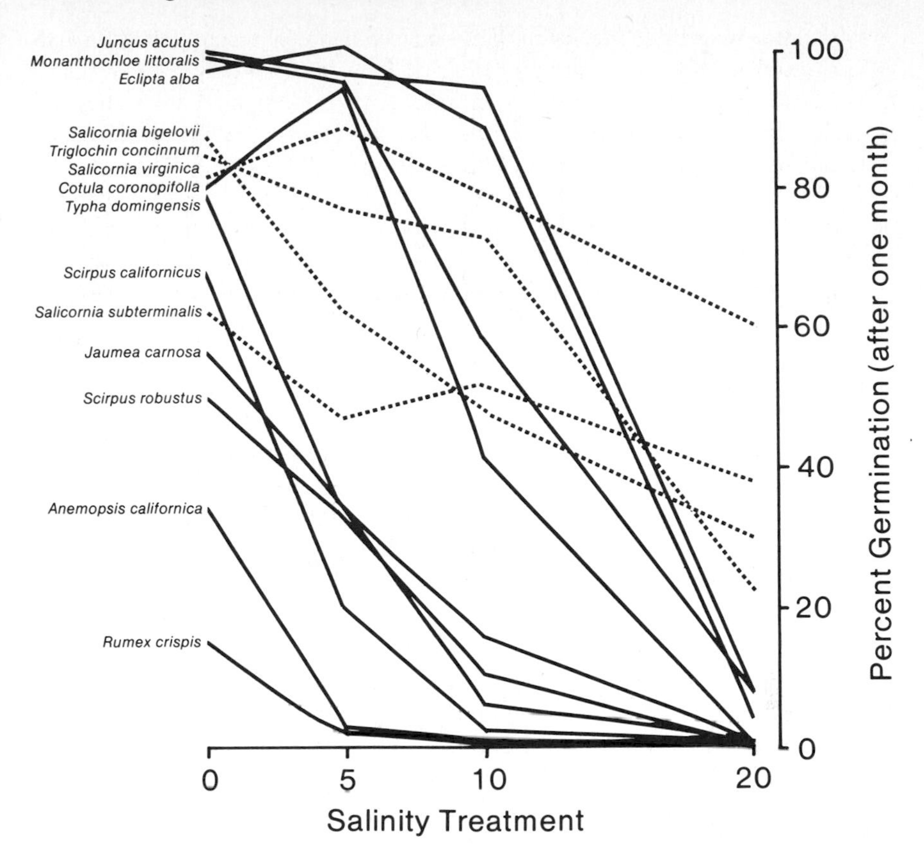

Figure 2. *Germination data for 14 species that occur in coastal marshes (fresh-to-hypersaline) of southern California. Dotted lines indicate high salt-tolerance. Salinity treatments = g l^{-1}.*

Table 4. Results of mixed-species germination experiments after 1, 4, and 9 weeks of growth. Data are percent occurrence (frequency) in 60 6.25-cm² quadrats (20 in each of 3 replicate germination bins).

	Week 1			Week 4			Week 9		
Salinity treatment (g l^{-1})	0	16	35	0	16	35	0	16	35
Number of species germinating	4	4	1	11	8	2	11	8	5
% occurrence of common species									
Salicornia virginica	58	57	8	85	92	57	35	88	67
Salicornia bigelovii	2	2	0	2	13	12	2	15	13
Juncus acutus	0	3	0	27	7	0	8	8	2
Rumex crispus	0	2	0	17	2	0	20	2	0

August, 1983 (International Boundary and Water Commission 1983), and *S. foliosa* growth was also the highest on record.

Salicornia virginica increased during years when *Spartina foliosa* did not. In 1981 and 1984, *S. virginica* cover expanded vegetatively within the low marsh (Table 4). Both were non-flood years, and 1984 was also non-tidal. In 1984, cover of *S. virginica* doubled. A similar response of *S. virginica* occurred at Estero de Punta Banda, Baja California following the diking and restriction of tidal flushing to part of the low marsh (Ibarra-Obando and Escofet, Unpublished ms). The same response at both estuaries suggests a causal relationship with reduced tidal flushing.

Population Declines

In the 10-year comparison for Tijuana Estuary (Table 2), species lists were similar for 1974 and 1984 except for the near extinctions of *Salicornia bigelovii* and *Suaeda esteroa*. Their near extinctions coincided with the hypersaline drought in 1984. Seedlings of *S. bigelovii* were obvious in spring 1984, but they failed to establish; no plants were found in the September census. Seedlings of *S. esteroa* are usually hidden by the vegetation canopy, so it is unclear whether this species failed to germinate or it experienced massive seedling mortality. Only one plant occurred in the September 1984 census. *Spartina foliosa* also declined in 1984, as seen in the September census data (Table 3). By November 1984, mass mortality was apparent along the edges of tidal creeks, which had dry, cracked sediments. Extensive declines were also documented at Estero de Punta Banda under non-tidal hypersaline conditions (Ibarra-Obando and Escofet, Unpublished ms). The low-marsh and short-lived species were negatively affected by hypersaline drought.

Local extinctions have also followed prolonged inundation, with mortality apparently due to anaerobic soils. *Salicornia virginica* was nearly eliminated from the San Diego River marsh after prolonged flooding in 1984 (Table 2; Zedler 1981). *S. virginica* also declined at Tijuana Estuary in 1983 (Table 2), when tidal waters were impounded for two weeks in winter.

Contrary to expectation, the *Typha domingensis* population that invaded the San Diego R. marsh was not eliminated by the return of saline to hypersaline conditions (Table 2). Once salt-tolerant rhizomes developed, persistence was possible. Some mature plants even survived hypersaline soil (45 g l^{-1}) in a year-long experimental treatment (Beare 1984).

Conceptual Model for Community Composition

A conceptual model suggests how interruptions or gaps in hypersaline conditions control invasions and extreme stresses control extinctions (Fig. 3). With average winter rainfall and streamflow (Fig. 3A), 2-3 weeks of salinity reduction stimulates seed germination and allows seedling establishment. Brackish marsh species cannot invade during a narrow low-salinity gap. The full complement of salt marsh species persists in the absence of extreme environmental stresses.

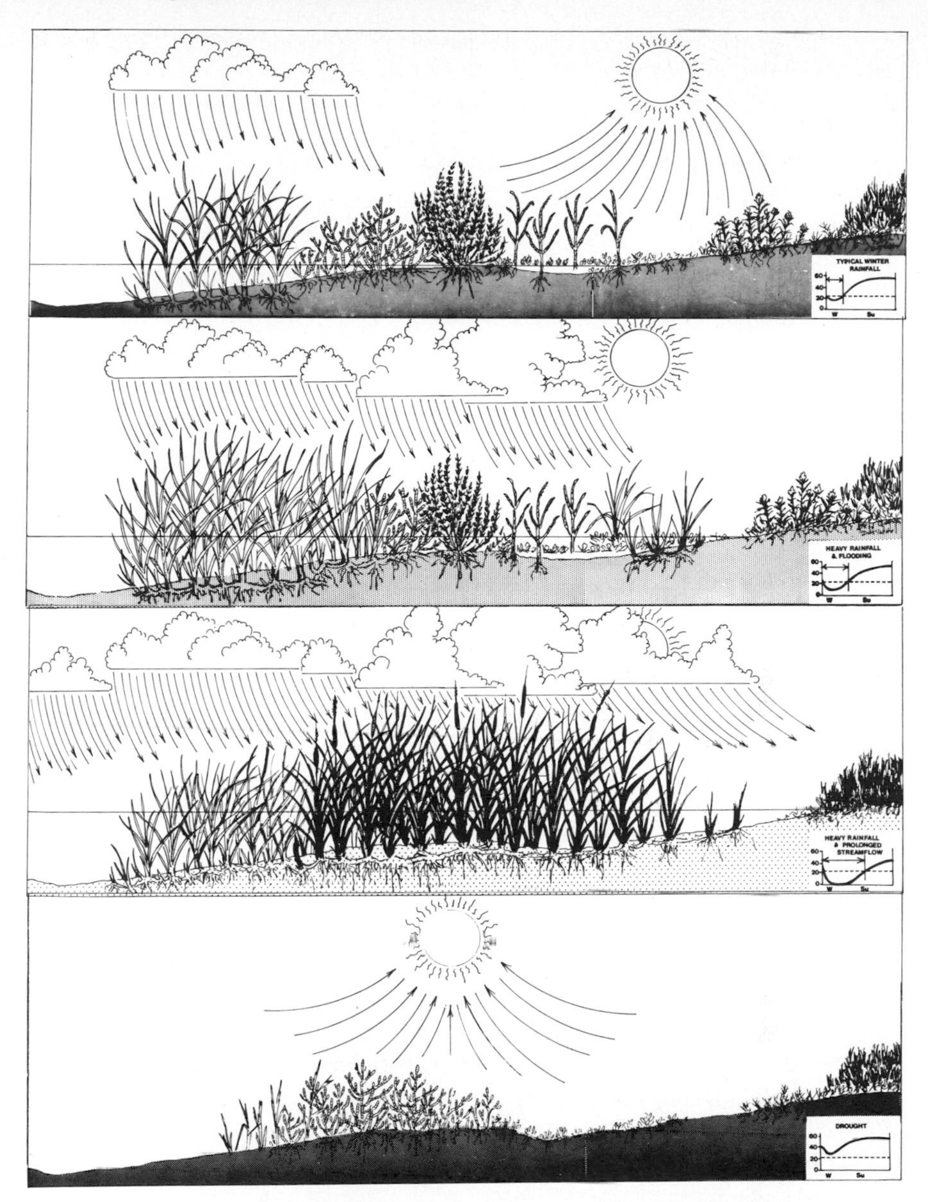

Figure 3. Conceptual model of invasion of and extinction of salt marsh vegetation. At the top are typical conditions: brief low-salinity gap allows salt marsh species to germinate and reestablish; none go extinct. Upper middle: floods decrease salinity sufficient for expansion of Spartina foliosa *both vegetatively and from seed. Lower middle: prolonged flooding (unlikely without reservoir discharges) eliminates most salt marsh vegetation and allows brackish marsh species to invade. Bottom: years without rainfall or tidal flushing prevent reestablishment of most short-lived species; however,* Salicornia virginica *expands in response to reduced inundation (drawn by D. McIntire).*

Extreme flooding (Fig. 3B) reduces salinities substantially; *Spartina foliosa* and other halophytic species can become established from seed. However, because the low-salinity gap is still brief (3-6 weeks), seedling establishment is limited to the normal complement of salt marsh species.

If the low-salinity gap is artificially prolonged (Fig. 3C), a variety of fresh and brackish marsh species can germinate and become established. If they grow to rhizome stage within the gap, they may persist after hypersaline conditions return. Species that cannot tolerate prolonged inundation (i.e., species with little aerenchyma) will undergo heavy mortality or go extinct.

Without winter rainfall (Fig. 3D), seeds of only a few species may germinate, but seedling establishment can be limited by drought and hypersalinity, especially when an estuary closes to tidal flushing. Short-lived species that rely on seedling recruitment undergo local extinction; only drought-tolerant perennials persist.

Salinities determine germination, and persistence of the required salinities and soil moisture conditions determine seedling survival. This is consistent with the regeneration niche concept of Grubb (1977), who hypothesized that multiple characteristics of both species and environment influence establishment. The species characteristics of southern California coastal wetlands do not conform to the classical separation of halophytes and glycophytes on the basis of tolerance to 0.5 g l^{-1} salinities (Waisel 1972). There is a spectrum of tolerances, and establishment appears to be determined by the degree and duration of freshwater influence: i.e., the low-salinity gap.

Once established, salt marsh plant populations may persist or go extinct, depending on environmental conditions that may differ greatly from those controlling establishment. Drought affects short-lived species such as *Salicornia bigelovii* and *Suaeda esteroa*. Excessive inundation limits *Salicornia virginica*. This proposed cause-effect relationship is consistent with the experimental findings of Mahall and Park (1976) for *S. virginica* in San Francisco Bay. In the Netherlands, experimental studies of several halophytes indicated that warm temperatures reduce inundation tolerance (Groenendijk 1984). In southern California, root respiration rates should be high even in winter, and inundation may be highly stressful to species with little root/rhizome aerenchyma.

Conditions that limit *Spartina foliosa* (hypersaline drought) differ from those that limit *Salicornia virginica* (inundation). In addition, *S. virginica* outcompetes *S. foliosa* for space and nutrients (Zedler 1983, Covin 1984), which helps to explain their reciprocal responses to recent environmental conditions. With prolonged flooding, *S. virginica* appears to lose its competitive advantage. The diverse plant communities found in this region may well result from the highly variable environment, if no one species has optimal environmental conditions, or a competitive advantage, indefinitely.

Mediterranean-type climate seems responsible for the control that hypersalinity has on the region's salt marshes. Along the Atlantic and Gulf of Mexico coasts, where rainfall and streamflow are substantial all year, salt marshes are more constant in composition, and vascular plant growth is under much less salinity stress. On the Atlantic coast, salinity plays a role in the spatial, rather than the temporal patterns of *Spartina alterniflora* growth (Chalmers 1982). Annual variations in species distributions are not large, although disturbances can induce changes, e.g., experimental sewage applications in Sippewissett Marsh (Valiela 1984). There has been some interest in characterizing temporal changes in European coastal marshes, especially long-term successional patterns and expansion of introduced *Spartina* (Ranwell 1972, Long and Mason 1983). However, studies of population variation have focused on spatial rather than annual differences, e.g., studies of *Salicornia europea* by Jefferies *et al.* (1981). Although soil salinity is important in wetter climates, it does not have the strong temporal variation seen in semi-arid California.

Extreme events alter wetland structure dramatically in southern California. The wetland vegetation is resilient as a whole, but individual species undergo dynamic change. The resulting population fluctuations have helped to reveal cause-effect relationships.

Acknowledgments

We thank Jordan Covin, Chris Nordby, Phil Williams, John Boland, Stuart Perry, and others for help with field monitoring and data analysis. Donovan McIntire developed and drew Fig. 3. The research was funded in part by Department of Commerce, NOAA Office of Coastal Resource Management, Sanctuary Programs Division, Grant No. NA85AA-D-CZ-30 and NOAA, National Sea Grant College Program, under grant number NA80AA-D-00120, project R/CZ-73, and California State Resources Agency. The U.S. Government is authorized to reproduce and distribute for governmental purposes.

References Cited

Beare, P. A. 1984. Salinity tolerance in cattails (*Typha domingensis* Pers.): Explanations for invasion and persistence in a coastal salt marsh. M. S. Thesis, San Diego State Univ. 57 pp.

Chalmers, A. 1982. Soil dynamics and the productivity of *Spartina alterniflora*, pp. 231-242. *In:* V. C. Kennedy (ed.,) *Estuarine Comparisons.* Academic Press, New York.

Covin, J. 1984. The role of inorganic nitrogen in the growth and distribution of *Spartina foliosa* at Tijuana Estuary, California. M.S. Thesis, San Diego State Univ., 60 pp.

Groenendijk, A. M. 1984. Tidal management: Consequences for the salt-marsh vegetation. *Water Sci. Technol.* 16:79-86.

Grubb, P. J. 1977. The maintenance of species richness in plant communities: The importance of regeneration niche. *Biol. Rev.* 52:107-145.

Ibarra-Obando, S., and A. Escofet. Unpublished Ms. Coastal Oil-Industry: A case study in Baja California. Centro de Investigacion y Educacion Superior Baja California, Ensenada, Mexico.

International Boundary and Water Commission. 1983. Flow of the Colorado River and Other Western Boundary Streams and Related Data. *Western Water Bull. 1983.* El Paso, Texas. 93 pp.

Jefferies, R. L., A. J. Davy, and T. Rudnik. 1981. Population biology of the salt marsh annual *Salicornia europaea. J. Ecol.* 69:17-32.

Long, S. P., and C. F. Mason. 1983. *Saltmarsh Ecology.* Blackie & Son, Ltd., Bishopbriggs, Glasgow. 160 pp.

Mahall, B., and R. Park. 1976. The ecotone between *Spartina foliosa* Trin. and *Salicornia virginica* L. in salt marshes of northern San Francisco Bay. III. Soil aeration and tidal immersion. *J. Ecol.* 64:811-820.

Munz, P. A. 1974. *A Flora of Southern California.* Univ. of California Press, Berkeley. 1086 pp.

Purer, E. 1942. Plant ecology of the coastal salt marshlands of San Diego County, California. *Ecol. Monogr.* 12:81-111.

Ranwell, D. S. 1972. *Ecology of Salt Marshes and Sand Dunes.* Chapman and Hall, London. 258 pp.

United States Gological Survey, 1937-81. *Water Supply Papers,* 1937-70; and *Water Resources Data for California, Water Years 1971-1980. Separate annual volumes, Washington, D.C.*

Valiela, I. 1984. Marine Ecological Processes. Springer Verlag, New York. 546 pp.

Waisel, Y. 1972. *Biology of Halophytes.* Academic Press, New York. 395 pp.

Zedler, J. B. 1975. Salt marsh community structure along an elevation gradient. *Bull. Ecol. Soc. Am.* 56:47.

Zedler, J. B. 1977. Salt marsh community structure in Tijuana Estuary, California. *Estuar. Coastal Mar. Sci.* 5:39-53.

Zedler, J. B. 1981. The San Diego River Marsh before and after the 1980 flood. *Environment Southwest* 495:20-22.

Zedler, J. B. 1982. *The Ecology of Southern California Salt Marshes: A Community Profile.* FWS/OBS-81/54. U.S. Fish and Wildlife Service, Biological Services Program, Washington, D.C. 110 pp.

Zedler, J. B. 1983. Freshwater impacts in normally hypersaline marshes. *Estuaries* 6:346-355.

Zedler, J. B., and J. Covin. 1984. Ecosystem monitoring at the Tijuana River National Estuarine Sanctuary. Final Report to National Oceanic and Atmospheric Administration Sanctuary Programs Division. Washington, D.C. 16 pp.

Zedler, J. B., J. Covin, C. Nordby, P. Williams, and J. Boland. In press. Catastrophic events reveal the dynamic nature of salt marsh vegetation in southern California. *Estuaries.*

Zedler, J. B., and C. P. Onuf. 1984. Biological and physical filtering in arid-region estuaries: Seasonality, extreme events and effects of watershed modification, pp. 415-432. *In:* V. C. Kennedy (ed.), *The Estuary as a Filter.* Academic Press, New York.

SPACE AND TIME VARIABILITY OF NUTRIENTS IN THE VENICE LAGOON

Stefania Facco

Department of Environmental Sciences
University of Venice, Venice, Italy

Danilo Degobbis

"Rudjer Bŏsković" Institute
Center for Marine Research
Rovinj, Jugoslavia

and

Adriano Sfriso and Angelo A. Orio

Department of Environmental Sciences
University of Venice, Venice, Italy

Abstract: We examined nutrient variability in the sediments, interstitial and lagoon waters at 15 stations in the central, most-polluted part of the Venice Lagoon, with 7 cruises between February 1984 and June 1985. Strong concentration gradients were observed, decreasing from the internal border of the lagoon towards the sea. Total phosphorus and total nitrogen in the sediments ranged from 261 to 599 μg g^{-1} and from 200 to 1800 μg g^{-1}, respectively. The total inorganic nitrogen (TIN), mainly ammonium, and reactive phosphorus (RP) in the interstitial waters varied from 4 to 1800 and from 0.1 to 38 μg-at l^{-1}, respectively. The most marked changes, both in time and space, were found in the overlying waters (TIN from 0.7 to 106 and RP from 0.03 to 15.7 μg-at l^{-1}). The higher concentrations of TIN were observed in October and February, whereas the higher RP concentrations appeared in June and October. The maximum phytoplankton biomass occurred in April and June in correspondence with lower nitrogen concentrations. External nutrient contributions, biological recycling and tidal mixing determine high nutrient concentration variability in the Venice Lagoon waters. The results suggest that nutrient release from the sediments can additionally influence this variability, particularly during warm seasons, when maximum rates were measured (up to 23.5 and 1.58 μg-at m^{-2} d^{-1} for TIN and RP, respectively).

Introduction

Coastal lagoons are highly productive and ecologically complex systems frequently characterized in temperate and cold latitudes by strong seasonal and spatial variations of nutrient concentrations (UNESCO 1981). Within these ecosystems sediments may play an important role as a nutrient source for primary production in the water column, as shown for several shallow coastal regions (Harrison 1980).

Oceanic sediments act as an ultimate sink for nutrients incorporated in the refractory fraction of the sedimented particulate matter (Emery *et al.* 1955). In shallow enclosed marine systems characterized by low water-exchange with the open sea, however, a significant portion of the organic matter produced by photosynthesis is deposited on the biologically very active sediment surface, where it is regenerated and released to the overlying water (Degens and Mopper 1976). Previous studies on the Venice Lagoon (Degobbis *et al.* 1986) correlated high alkaline phosphatase activity on the sediment surface with nutrient concentrations in the water column. Furthermore, significant nutrient-release rates were measured in laboratory experiments using undisturbed lagoon sediment cores (Cossu *et al.* 1983a; Amendola 1984). We undertook the present study to understand the time-and-space variability of nutrient concentrations within the lagoon and its relationship with processes of nutrient release from the sediments. The data also provided information on the role of sediments in the eutrophication processes occurring in the Venice Lagoon. Quantitative knowledge of nutrient exchange at the sediment-water interface is also relevant to the mass balance of nutrients in this ecosystem.

In this communication we report on the preliminary results obtained during seven cruises performed during the period February 1984-June 1985 at 15 stations in the central part of the Venice Lagoon, where we measured oceanographic and phytoplankton parameters, nutrient concentrations and release rates from the sediments.

The Study Area

The Venice Lagoon is an enclosed embayment in the northern Adriatic with an average depth of about 1 m and a total surface of 549 km^2 (Fig. 1). The average tidal difference is 0.6 m (Pirazzoli 1974). During an ebb or flood period $1.6\text{-}5.2 \times 10^8$ m^3 of water is exchanged through the port entrances of Lido, Malamocco and Chioggia (Ministry of Public Works 1979). The extreme values of exchange correspond to syzygy and quadrature tides, respectively. During the last century 32% of the surface area of the lagoon has been reclaimed for fishing ponds or filled for industrial, commercial and touristic development (Avanzi *et al.* 1980). The lagoon receives sewage effluents from about 300,000 inhabitants of Venice, Mestre Marghera and Chioggia; wastes from chemical and steel plants located in the 30-km^2 Porto Marghera industrial area; and runoff from an intensively-cultivated, 2000-km^2 drainage basin. High concentrations of heavy metals and nutrients have been measured in the surface sediments of the central part of the lagoon, where the historical center of Venice is located (Donazzolo *et al.* 1984). We focused our attention on this area where the impact due to urban and industrial pollutants is very marked.

In order to cover the eutrophication gradients within the lagoon, measurements were made in different areas (Fig. 1). Area A, located between the Porto Marghera industrial zone and the Venice historical center, receives significant amounts of industrial and urban waste waters. Area B is influenced by sewage

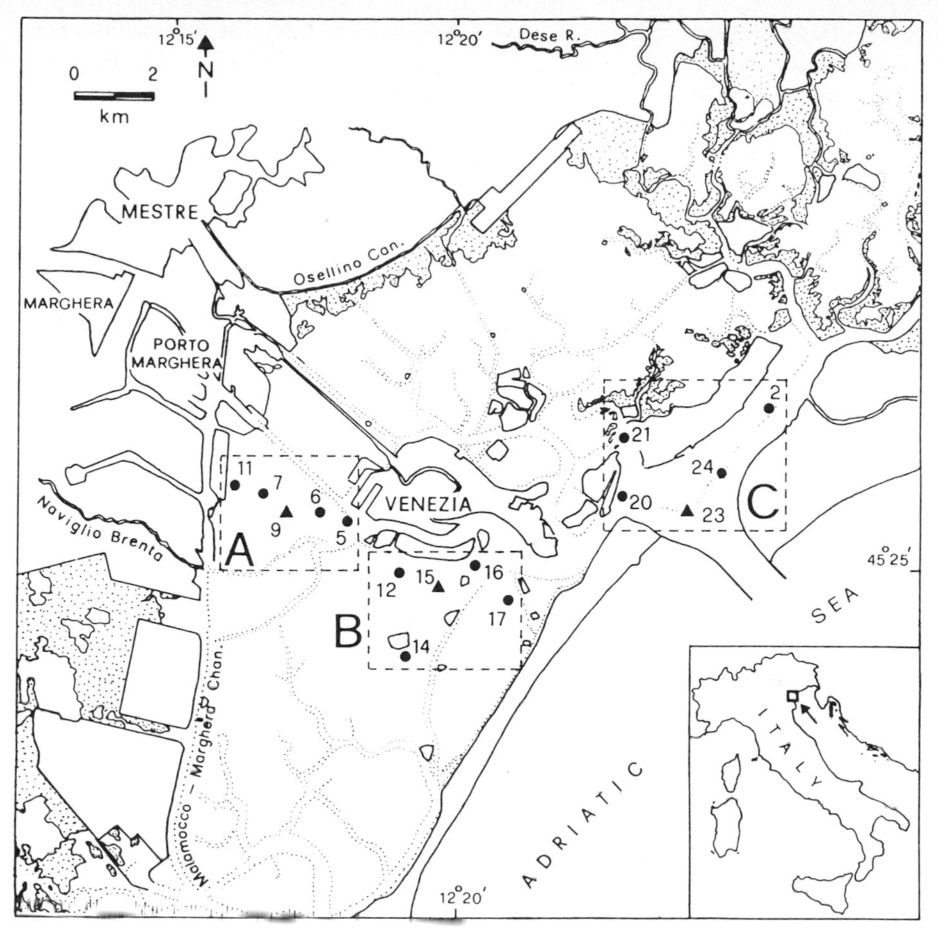

Figure 1. The study areas in the Venice Lagoon. (▲) main stations. (●) oceanographic stations.

discharges from the town of Venice, while area C is mainly characterized by water exchange with the open northern Adriatic Sea.

Sampling and Analytical Procedures

The field measurements and sampling were performed in February, April, June and October 1984, and February, April and June 1985. In each month we sampled during late morning of three consecutive days starting from area A (Fig. 1) about half an hour after the beginning of the ebb. On each day 5 stations of an area were visited within 1-2 hours. The depth of the stations varied from 0.5 to 1.5 m. Lagoon water was sampled with 5-l Van Dorn bottles. Sediment was sampled using 10-cm diameter plexiglass corers. The interstitial waters were extracted aboard the boat immediately after sample collection using stainless steel presses.

Basic meteorological parameters, air and water temperature, water transparency (Secchi disk) and redox potential (Eh) of the sediment samples (Pt electrode and mV-meter; Orsenigo & Co, Milan, Italy) were determined in the field. Water samples were refrigerated and transported to the shore laboratory as soon as possible. The analyses (except salinity) started within two hours from sample collection. Salinity was determined argentometrically by a modified Knudsen method (Oxner 1962) and dissolved oxygen by a Winkler titration technique. A pH-meter (Metrohm Herisau E-500; Switzerland) with glass/silver chloride electrodes was used for pH measurements. Duplicate analyses of ammonium, nitrite, nitrate and reactive phosphorus (RP) in the water samples were performed spectrophotometrically by standard oceanographic methods (Strickland and Parsons 1972), using Perkin-Elmer (Norwalk, Connecticut) spectrophotometers (Mod. 320 or 156). Chlorophyll-*a* was measured in acetone extracts of suspended matter (filtered through Whatman GFC glass filters) using a Perkin-Elmer fluorescence spectrophotometer MPG-44B (Strickland and Parsons 1972).

Sediment samples (5-cm surface layer) for total phosphorus and nitrogen determinations were tightly wrapped in polyethylene foils, stored at -30°C, and analyzed within a few weeks. Total and inorganic phosphorus were extracted from the sediments by shaking the samples 16 hours with 1 M HCl before and after combustion at 550°C and determined by the phosphomolybdenum blue method (Aspila *et al.* 1976). Organic phosphorus was calculated by difference. Total nitrogen was measured on freeze-dried sediments with a Perkin-Elmer Elemental Analyzer 240 B. The pelitic sediment fraction (<63 μm) was determined by wet sieving. All results were recalculated to sediment dry weight (105°C).

The rates of nutrient release from the sediments were measured *in situ* using three PVC and three plexiglass chambers of the type described by Nowicki and Nixon (1985) but modified to allow direct sampling from the boat through PVC tubing. The chambers were inserted in the sediments to a depth of 10 cm at the main stations (9, 15, 23; Fig. 1) enclosing 2.5 l of overlying water. Water samples (100 ml) were withdrawn by syringes from the chambers 0.5-1 and 4-6 h after insertion for subsequent analysis of nutrients and dissolved oxygen. Prior to sampling, the water in the chambers was mixed gently by pumping with the syringe. Three dark glass bottles, filled with overlying water, were also incubated near the chambers to detect possible changes in nutrient concentration in the water.

Results and Discussion

In general the sediments in the inner parts of the Venice Lagoon consist of fine material (<63 μm). The grain size increases seaward to almost pure sand at the port entrances (Barillari 1978). The granulometric analyses provided information on the gradients present in the lagoon, as well as within each of the investigated areas (Fig. 1). The average content of fine fractions in the central part of area A (stations 6, 7 and 9) was about 90%. However, at the borders of this area (stations 5 and 11) the percentage of fine material was much lower (60% and 42%, respectively). The fraction of fine sediment was around 85% at all stations in Area B. In Area C the gradients of textural composition were very marked. Almost pure sand (around 96.5%) was found at the stations (20, 23 and 24) that were directly influenced by strong tidal currents (up to 1.5 m s^{-1}) flowing through the Lido port entrance. At stations 21 and 25, however, the average percentage of sand was only 34.5%.

At the surface of the sediments with a high content of fine material, Eh values (+30 to −300 mV) were characteristic of reduced or highly reduced sediments (Patrick and Delaune 1977). Higher Eh values (+80 to +440 mV), typical of moderately reduced to oxidized conditions, were measured in the sandy sediments. The highest values were generally measured in February.

While the sediment contents of total phosphorus and total nitrogen did not change markedly with time at any station, considerable differences were observed among the three areas. Average concentrations and standard deviations of total phosphorus concentrations in areas A, B and C were 599 ± 171, 391 ± 31 and 261 ± 79 $\mu g\ g^{-1}$ (dry weight), respectively. The higher concentration values and variability in the surface sediments of area A were ascribed to recent pollution as shown from analyses of three radiodated cores (Facco 1983). In fact, the core layers deposited earlier than 50 years ago contained on average only 413 $\mu g\ g^{-1}$ of total phosphorus. Organic phosphorus accounted for 24.3, 24.7 and 12.0% of total phosphorus for areas A, B and C, respectively. The values of total nitrogen at the main stations (9, 15, 23; Fig. 1) were 1800, 1200 and 200 $\mu g\ g^{-1}$ respectively. While the water content of area A sediments was in the range 19-50%, a smaller range (22-35%) was observed for sediments of the other two areas.

Significant concentration gradients existed from area A to area C for nutrients in the interstitial waters of the surface sediments (Table I). The reactive phosphorus concentrations were usually slightly higher in interstitial waters than in the overlying lagoon waters, and generally varied from station to station with patterns similar to those in the overlying water. This confirms the previous assumption that the sediments play an important role in controlling the distribution of this nutrient in the water column when phytoplankton standing crop is low (Degobbis *et al.* 1986). Ammonium concentration in the interstitial waters was highly variable seasonally and generally decreased from area A to area C.

The hypothesis that significant interaction occurs in the Venice Lagoon between sediments and overlying water is also supported by previous results on vertical nutrient distribution in interstitial waters. Strong vertical gradients were observed in sediment cores (40-60 cm) collected in area A (Facco 1983; Amendola 1984) indicating high capability of sediments to release nutrients. Ammonium concentrations ranged from 150 μg-at l^{-1} near the surface to 1000 μg-at l^{-1} in the deeper layers, while RP ranged between 4 and 60 μg-at l^{-1} in surface and deep sediments, respectively.

We observed high spatial variability for most parameters measured in lagoon water (Figs. 2 and 3), as is typical for shallow aquatic environments

Table 1. Concentration ranges (μg-at l^{-1}) of ammonium and reactive phosphorus (RP) in interstitial water from the investigated areas.

Date	Nutrient	Areas		
		A	B	C
Feb 84	NH_4^+	230 - 1840	67 - 1680	980 - 1470
	RP	2.4 - 8.6	4.6 - 30	0.6 - 1.1
Apr 84	NH_4^+	14 - 580	21 - 71	14 - 49
	RP	2.6 - 85	2.1 - 10	1.0 - 3.4
Jun 84	NH_4^+	30 - 76	50 - 300	20 - 58
	RP	4.7 - 30	1.0 - 37	1.1 - 1.4
Oct 84	NH_4^+	47 - 208	42 - 201	28 - 62
	RP	6.4 - 38	1.5 - 12	0.4 - 1.0
Feb 85	NH_4^+	45-80	27-54	15-57
	RP	1.6-9.4	0.5-1.9	0.1-2.5
Apr 85	NH_4^+	4 - 126	27 - 45	42 - 100
	RP	4.1 - 10	1.3 - 15	0.5 - 3.3
Jun 85	NH_4^+	39 - 79	32 - 138	77 - 104
	RP	10 - 24	1.5 - 13	0.9 - 2.0

strongly influenced by anthropogenic impact. Higher concentrations of inorganic nitrogen and phosphorus were generally present at the stations of area A, usually associated with lower salinity values. This observation indicates that the combined influence of land runoff, and sewage and industrial waste water discharges is quite strong, in spite of the waste-water treatment plants recently constructed. Land runoff seems to contribute significantly amounts of nutrients, particularly after heavy rainfall in spring and summer, when marked algal blooms are induced under favorable light intensities (Anonymous 1979).

Strong horizontal nutrient gradients were also observed in area A. Nutrient concentrations increased 2- to 10-fold between station 5 (near the town of Venice) and station 11 of the industrial area (Fig. 1). For example, in April 1985 the concentrations of total inorganic nitrogen (ammonium, nitrite and nitrate) and reactive phosphorus at stations 5, 6, 9, 7, 11 were: 3.6, 7.1, 11.8, 17.3, 68.4, and 0.85, 1.43, 1.96, 1.77, 5.95 μg-at l^{-1}, respectively. In spite of these nutrient gradients no corresponding changes of chlorophyll *a* (19, 27, 33, 6.4 and 7.9 μg l^{-1}) were observed at the same stations. Thus the algal response to eutrophication forcing in a polluted lagoon ecosystem may not depend predominantly on nutrient concentrations. The nutrient gradients were less marked between areas B and C, where differences of about 2-3 times were generally observed. During phytoplankton blooms the chlorophyll *a* concentration increased

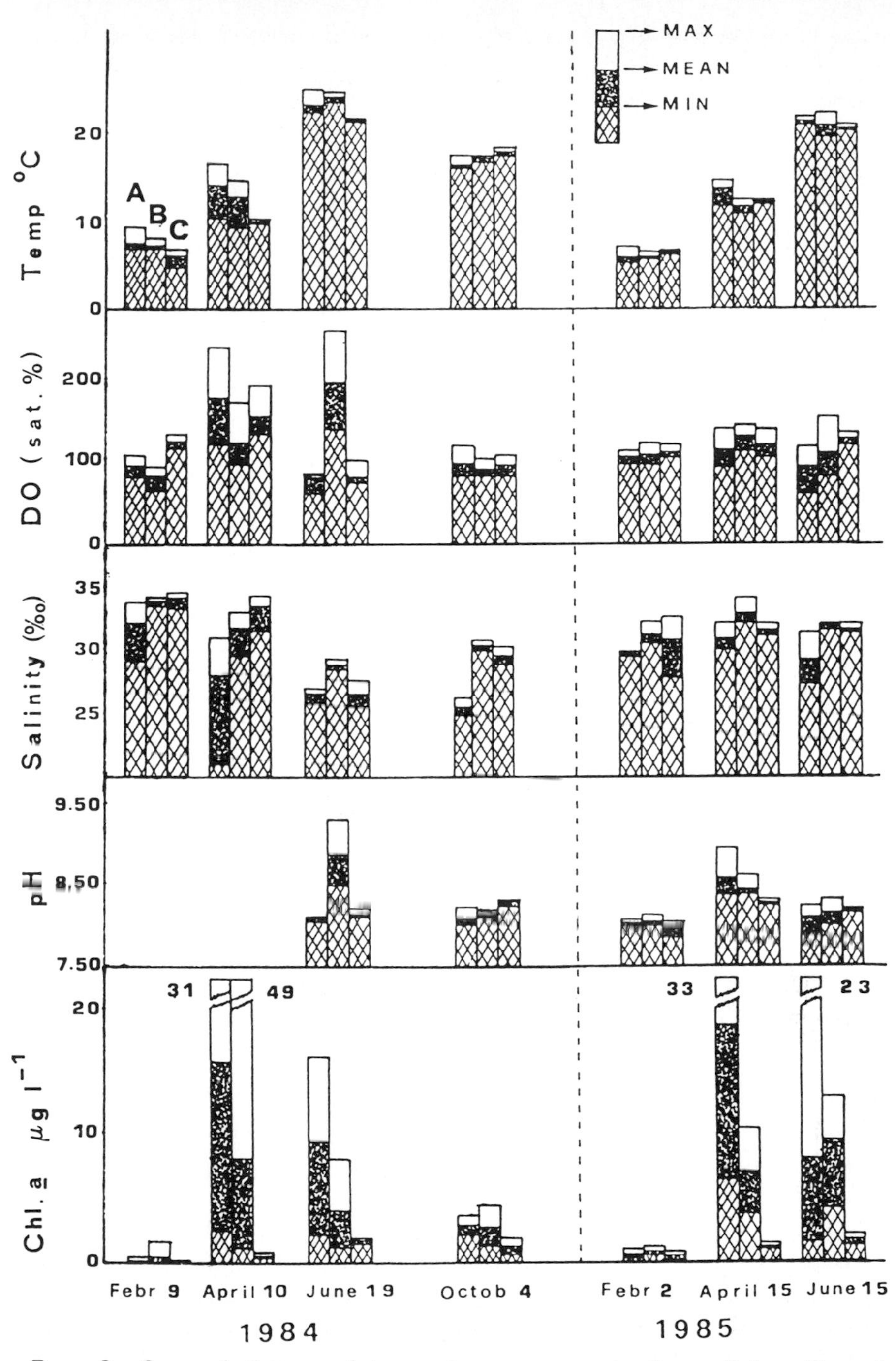

Figure 2. *Seasonal changes of temperature, oxygen saturation, salinity, pH, and chlorophyll* a *in the study areas (A, B, C).*

at least an order of magnitude in the areas A and B relative to that observed in the area C.

Many of the parameters determined in lagoon water underwent seasonal variations (figs. 2 and 3). The observed changes were related to biological processes. Higher pH values and dissolved oxygen saturations coincided with increased chlorophyll *a* contents during April and June phytoplankton blooms (Fig. 2). Nitrogen concentrations were higher in the fall and winter months and lower in the spring-summer period, the opposite of the pattern for chlorophyll *a* (Fig. 3). Quite surprisingly the photoplankton biomass was low in October even though the nutrient availability was very high. Similar results were obtained during three cruises performed in the period October-November 1982 (Degobbis *et al.* 1986). At this time of year the light level and temperature conditions are still favorable and phytoplankton blooms normally occur in the northern Adriatic Sea (Revelante and Gilmartin 1976).

The highest nitrate concentrations were measured in February (Fig. 3), when nitrification prevailed over regeneration processes (which are slowed due to the lower production of organic matter). Higher values of ammonium were observed in October as a result of the predominance of regeneration over assimilation processes. Quite differently, the highest concentrations of reactive phosphorus were observed in June and October. In summer regeneration and release of phosphorus from the sediments are faster than assimilation, which is not the case for nitrogen. In area A we found the same relationship between the reactive phosphorus concentrations in lagoon water vs. interstitial water in October and February when the phytoplankton crop was at a minimum as in April during a phytoplankton bloom (Cossu *et al.* 1984). By contrast, ammonium was almost completely exhausted during the bloom.

The nutrient release rates from the sediments measured *in situ* show a high variability (Fig. 4), and the data so far available are not enough for a quantitative explanation of all mechanisms in the nutrient cycles involving the sediments. However, some conclusions can be drawn on the basis of these preliminary results.

Measurements in dark bottles showed that no significant net nutrient regeneration occurred in the incubation water, even when very high release rates from the sediments were measured. This is in agreement with the hypothesis that the sediments play an important role in the regeneration processes occurring in the Venice Lagoon. The release was very significant in June and October, but was practically zero in February and April. Thus the seasonal deposition of organic matter may be more important than the geochemical processes in the sediment column. The temperature effect might also explain these differences at least in part, since at low temperatures the metabolism of heterotrophic organisms is minimal. In October, when regeneration processes prevail, the release rate measured in each chamber was very reproducible. The high variability observed in June, when assimilation and regeneration processes occur at comparable rates, could reflect a more complex situation. Furthermore, marked heterogeneity in macroalgal biomass and distribution was observed in

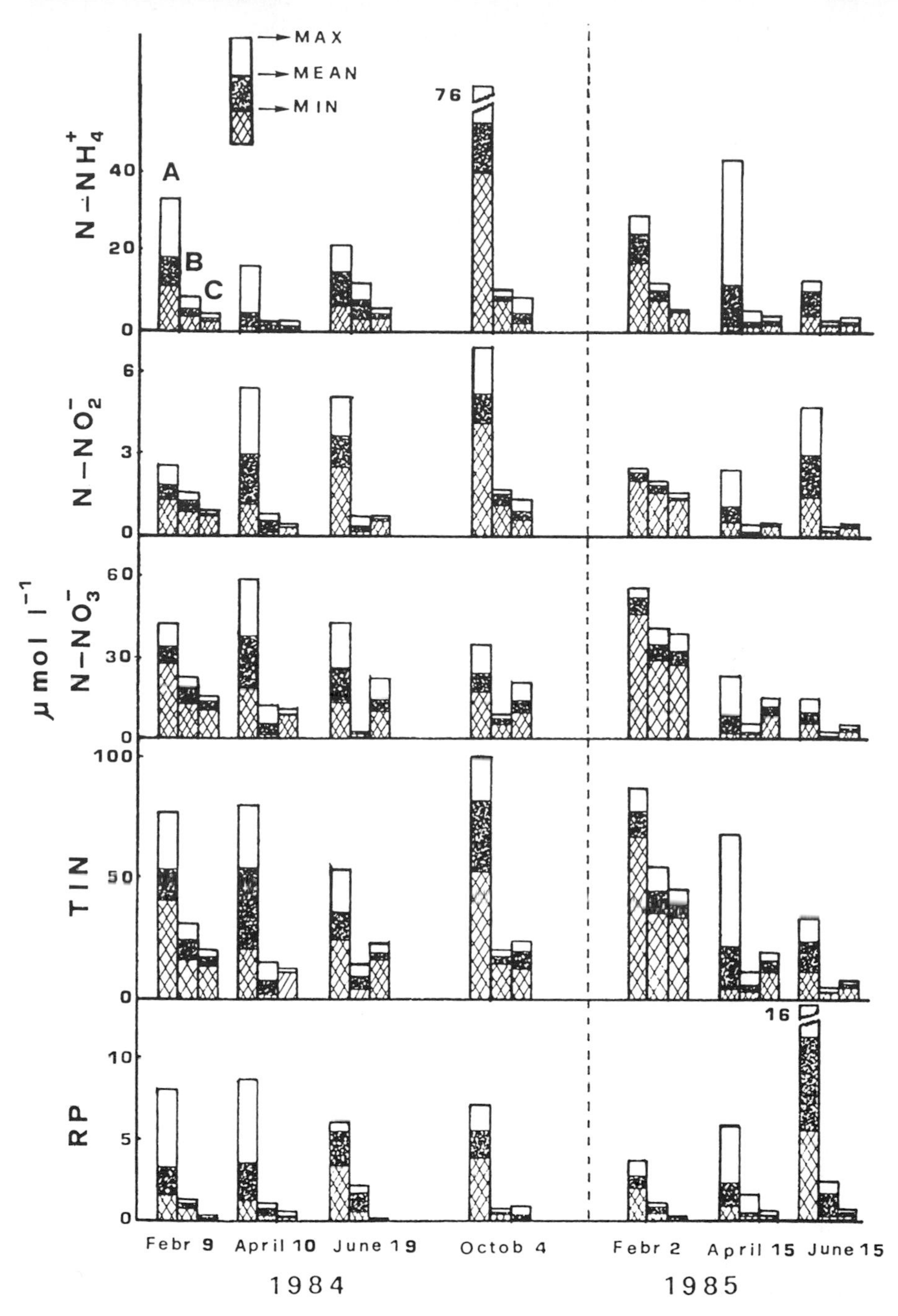

Figure 3. Seasonal changes of nutrient concentrations in the water of the study areas (A, B, C).

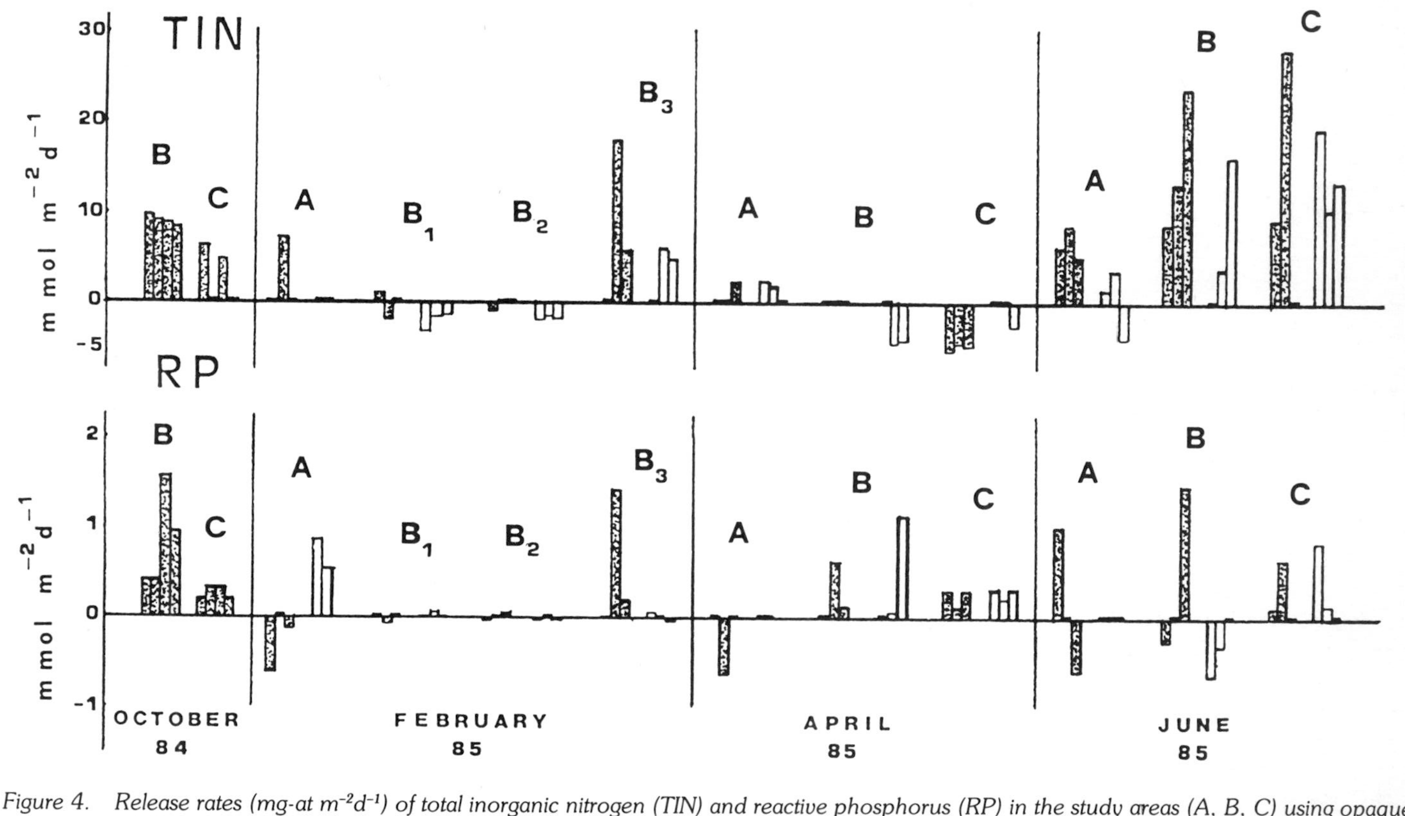

Figure 4. *Release rates (mg-at $m^{-2}d^{-1}$) of total inorganic nitrogen (TIN) and reactive phosphorus (RP) in the study areas (A, B, C) using opaque PVC chambers (dark histograms) and transparent plexiglass chambers (empty histograms). B1, B2 and B3 refer to experiments carried out on three consecutive days in the same area.*

June, which certainly affected the organic matter content of the surface sediments.

Dissolved oxygen concentration in the chambers decreased significantly during incubation when nutrients were released rapidly from the sediment, but generally remained higher than 50% of saturation. Only a few values as low as 30% were measured in dark chambers in area A in June and October.

Among nitrogen species, only ammonium was apparently released from the sediment. In only a few cases was a slight decrease of nitrate concentration observed in dark chambers, perhaps due to bacterial reduction activity. The results obtained in June in areas A and B showed an apparent decrease in the nutrient release (or a reversal of the flux direction) in the transparent chambers which was probably due to algal assimilation.

Conclusions

The Venice Lagoon is a shallow eutrophied marine system that receives high quantities of nutrients from several (agricultural runoff, sewage and industrial waste-water discharges, atmospheric fallout) external sources (Cossu *et al.* 1983b), and in which very marked algal blooms occur (Anonymous 1979). In the inner parts of the lagoon the water mass exchange is considerably lower than in the area of the port entrances (Nyffeler 1976). Thus, the dispersion of nutrients discharged in those inner areas is expected to be relatively low. The complex interactions between nutrient input, physical dispersion and biological recycling induce high variability of nutrient concentrations in the Venice Lagoon water and sediments both in space and in time.

Moreover, the results obtained to date seem to support the assumption that sediments may contribute significantly to the variability of nutrient concentrations observed in the Venice Lagoon water. For instance, in the warm seasons, the sediments may become a source of nutrients so significant that it is comparable to the external anthropogenic contributions to the lagoon. In fact, the average release rates in June and October (10.5 and 0.26 mg-at m^{-2} d^{-1} for total inorganic nitrogen and reactive phosphorus, respectively) are the same order of magnitude as external inputs (3.7 and 0.38 mg-at m^{-2} d^{-1} for TIN and RP, respectively) (Cossu *et al.* 1983b).

Acknowledgments

We wish to thank the following scientists for their valuable contribution to the research presented in this paper: Romano Donazzolo, Antonio Marcomini and Bruno Pavoni from the University of Venice, Nenad Smodlaka and Robert Precali from the Center for Marine Research at Rovinj and Malvern Gilmartin from the University of Maine at Orono (USA). The research was supported by EEC grant No. ENV-751-I SB, CNR (Rome) grant 83.02434.03, and Self-Management Community of Interest for Scientific Work of SR Croatia (SIZ III).

References Cited

Amendola, P. 1984. *Release of Nutrients from the Venice Lagoon Sediments.* Doctoral Thesis. Department of Environmental Sciences, University of Venice, Venice. 71 pp. (in Italian).

Anonymous. 1979. *Environmental Study in the Venice Lagoon for the Definition of a Restoration Plan and Control of Lagoon Water Quality Conditions.* Venice Municipality, Venice. 63 pp. (in Italian).

Aspila, K. I., H. Agemian and A. S. J. Chau. 1976. A Semi-automated method for the determination of inorganic, organic and total phosphorus in sediments. *Analyst* 101:187-197.

Avanzi, G., V. Fossato, P. Gatto, R. Rabagliati, P. Rosa Salva and A. Zitelli. 1980. *Restoration, Conservation and Use of the Venice Lagoon Ecosystem.* Municipality of Venice, Venice, 199 pp. (in Italian).

Barillari, A. 1978. First information on the distribution of surface sediments in the central basin of the Venice Lagoon. Atti Istituto Veneto Scienze Lettere Arti 136:125-34 (in Italian).

Cossu, R., D. Degobbis, R. Donazzolo, E. Maslowska, A. A. Orio and B. Pavoni. 1983a. Nutrient release from the sediments of the Venice Lagoon. *Ingegneria Sanitaria* (5/6):16-23.

Cossu, R., E. de Fraja Frangipane, R. Donazzolo and A. A. Orio. 1983b. Elements for a nutrient budget of the Venice Lagoon. *Ingegneria Ambientale* 13:80-90 (in Italian).

Cossu, R., D. Degobbis, R. Donazzolo, E. Homme Maslowska, A. A. Orio and B. Pavoni. 1984. The role of sediments in the eutrophication of the Venice Lagoon. *Ingegneria Sanitaria* (4):1-9 (in Italian).

Degens, E. T. and K. Mopper. 1976. Factors controlling the distribution and early diagenesis of organic material in marine sediments, pp. 59-113. *In:* J. P. Riley and R. Chester (eds.), *Chemical Oceanography,* Vol. 6, Academic Press, New York.

Degobbis, D., E. Maslowska, A. A. Orio, R. Donazzolo, and B. Pavoni. 1986. The role of alkaline phosphatase activity in the sediments of the Venice Lagoon on nutrient regeneration. *Estuar. Coastal Shelf Sci.* (in press).

Donazzolo, R., A. A. Orio, B. Pavoni and G. Perin. 1984. Heavy metals in the sediments of the Venice Lagoon. *Oceanol. Acta* 7:25-31.

Emery, K. O., W. L. Orr and S. C. Rittemberg. 1955. Nutrient budgets in the oceans. pp. 299-310. *In: Essays in the Natural Sciences in Honor of Captain Allan Hancock.* University of Southern California, Los Angeles.

Facco, S. 1983. *Vertical Distribution of Nutrients and Pollutants in the Venice Lagoon Sediments.* Doctoral Thesis. Department of Environmental Sciences, University of Venice, Venice. 91 pp. (in Italian).

Harrison, W. G. 1980. Nutrient regeneration and primary productivity in the sea. *Environ. Sci. Res.* 19:433-460.

Ministry of Public Works. 1979. *The tidal Currents in the Venice Lagoon.* Institute of Hydraulics, Padua University, Padua. 95 pp. (in Italian).

Nowicki, B. L. and S. W. Nixon. 1985. Benthic community metabolism in a coastal lagoon ecosystem. *Mar. Ecol. Progress Ser.* 22:21-30.

Nyffeler, F. 1976. *The Hydrodynamic Regime of the Venice Lagoon. Influence on transport phenomena.* Fonds National Suisse de la Recherche Scientifique, Geneve. 87 pp. (in French).

Oxner, M. 1962. *The Determination of Chlorinity by the Knudsen Method and Hydrographical Tables.* G. M. Manufacturing Co., New York. 63 pp.

Patrick W.H., Jr., and R. D. Delaune. 1977. Chemical and biological redox system affecting nutrient availability in the coastal wetlands. *Geosci. Man* 18:131-137.

Pirazzoli, P. 1974. Historical data on the mean tide level in Venice. *Atti Accademia Scienze Istituto Bologna* 13:124-148 (in Italian).

Revelante, N. and M. Gilmartin. 1976. The effect of the Po River discharge on phytoplankton dynamics in the Northern Adriatic Sea. *Mar. Biol.* 34:259-271.

Strickland, J. D. H. and T. R. Parsons. 1972. *A Practical Handbook of Seawater Analysis.* Bull. Fish. Res. Bd. Canada (167), 2nd ed., Ottawa. 310 pp.

UNESCO. 1981. *Coastal Lagoon Research, Present and Future.* UNESCO Technical Papers in Marine Science (32), Paris. 95 pp.

SPATIAL VARIABILITY IN ESTUARIES

W. C. Boicourt, Convenor

INTERMITTENCY IN ESTUARINE MIXING

K. R. Dyer and A.L. New

Institute of Oceanographic Sciences
Crossway, Taunton
Somerset, United Kingdom

Abstract: A simplified mixing criterion is proposed for partially mixed estuaries, based on the layer Richardson Number (Ri_L). For $Ri_L > 20$, bottom-generated turbulence appears ineffective in decreasing the stratification; for $20 > Ri_L > 2$, mixing is increasingly active; and for $Ri_L < 2$, fully developed mixing occurs. These critical values are used to illustrate that mixing occurs preferentially in the shallow parts of the estuarine cross-section, and increases towards spring tides. We also examine cases of mixing produced by bridge supports, and by the topography of the estuary bottom, as further examples of the intermittency of estuarine mixing in time and in space.

Introduction

Quantification of mixing processes and rates of turbulent exchange of salt are among the largest outstanding problems of research in estuaries. Mixing may result from a combination of bottom friction, internal shear instabilities, and breaking of lee waves, with wind mixing occasionally being significant. The tidal oscillation of the whole water mass over the bottom roughness elements and around topographic irregularities produces turbulence on various scales creating a vertical flux of momentum which works against the buoyancy forces to progressively break down the stratification. Internal mixing occurs only when the water mass is layered—with fresher, less-dense water on the surface overlying denser saline water at the bottom. When the velocity shear across the interface between the layers exceeds a critical value, small disturbances form and grow into waves or billows and eventually break. This thickens the interface, decreases the gradients of density and velocity, and restabilizes the stratification.

The balance between the mixing produced by internal shear and that produced by bottom shear varies with estuarine type (Abrahams 1980); internal shear predominates in salt wedge estuaries with little tidal motion, and bottom shear predominates in well-mixed estuaries. In highly stratified estuaries with a significant tidal oscillation, the flow over topographic features causes lee waves that break when the velocity approaches critical conditions. This situation causes an increase in the surface-layer salinity and thickness, and decreases the degree of stratification. However, quantification of mixing requires either measuring the salt fluxes directly, or considering the salt budget in a water mass and deriving the fluxes from the changes in salt content with time. In the field it is very difficult to separate the changes caused by local mixing from those caused by advection.

Apparent changes in thickness of an upper layer can be produced, for example, by the tilting of an interface either in a longitudinal or lateral direction, by horizontal advection of a tilted surface, or by mixing.

The estuarine flow continually changes in magnitude during the tidal cycle, and maximum bottom-induced mixing is likely to occur at maximum flow velocities. The mixing produced at any particular point in the estuary is advected along the estuary, appearing elsewhere later in the tidal cycle. A water mass may be sufficiently modified by the mixing to undergo further mixing later as the tidal cycle progresses. On the other hand, the mixed water may be transported along the estuary without further mixing only to return almost to the same position at the same stage of the next tide, where it may then undergo further mixing. On successive tides, the mean flow through the estuary will gradually push the water further seaward, so that mixing produced at maximum current velocity at one position in the estuary will be gradually advected into other parts of the estuary where mixing may not have been as vigorous. Thus, bottom-induced mixing is intermittent both in space and in time and the mixing produced locally in only a few parts of the estuary may dominate in creating the overall structure of the estuary when averaged over a sufficiently long time. Additionally, the maximum internal shear may not necessarily occur at the time of maximum velocity, but may occur near the turn of the tide. In estuaries where both internal shear and bottom mixing are important, these two mechanisms may interact and the internal shear early in the tide may reduce the stratification to the extent that bottom-induced mixing becomes more effective at maximum current. Hence intermittency in mixing would appear to be a common feature among estuaries, but the topographical complexity of estuaries makes it difficult to describe without extensive Lagrangian measurements following the evolving structure of the water mass as it is moved around in the estuary.

In this paper we apply a simplified criterion for mixing to the results of conventional Eulerian measurements so that qualitative estimates can be made of the intensity of mixing occurring at different times and at different positions in the estuary. We also examine some other situations where various features produce either spatially or temporally variable mixing.

Parameterization of mixing

At present there is no simple way of knowing from conventional measurements whether or not mixing is taking place. In this section we present theoretical and practical approaches on which a simplified criterion might be based.

In a uniform steady flow far from boundaries, the effect of the density gradient and shear-induced turbulence are compared by the gradient Richardson Number:

$$\mathrm{Ri} = -\frac{g}{\varrho}\frac{\partial\varrho/\partial z}{(\partial u/\partial z)^2} \qquad (1)$$

where ϱ is the water density, and u the velocity at height z. Theoretically, above a value of Ri = 1/4 the flow is laminar and though the interface may move up and down, negligible mixing occurs. Below that value the flow becomes unstable, perturbations on the interface grow and break, and mixing takes place. In the field, however, the transition is far from distinct at any single value of Ri, but occurs over a wide range between about Ri = 0.03 where the flow is near neutrally stable, and Ri = 1 where it is stable. Various laboratory and field measurements have provided a number of different relationships within this range (Dyer 1986) though the relationships give an average critical value of Ri = ¼. Because salinity and velocity are usually measured at discrete depths, the calculated Richardson number is affected by the measurement interval, resulting generally in an overestimate. Consequently the gradient Richardson number may not be a good practical criterion for mixing in estuaries.

In a flow where the mixing is dominated by the velocity shear produced at the bed, an alternative parameterization may be suitable. This is the bulk or layer Richardson number:

$$Ri_L = \frac{gh}{U^2} \frac{\Delta\varrho}{\varrho} \qquad (2)$$

where g is gravitational acceleration, h is the water depth, U the depth-mean velocity and $\Delta\varrho$ the surface-to-bottom density difference.

Inspection of the salinity fluctuations at a point in the flow may show whether they are dominated by internal waves or by turbulence. Comparison can then be carried out with the Richardson number. For an equilibrium spectrum obtained from Fourier analysis of a fluctuating variable such as salinity, there is a range of frequencies known as the inertial subrange, over which the decay in the energy of turbulence is proportional to f^{-p} (where f is the frequency). In this range the turbulence is generally isotropic and p = 5/3. For stratified conditions there is additionally a buoyancy subrange at somewhat lower frequencies, where the loss of energy due to working against gravity can be comparable with that transferred to higher frequencies. In the buoyancy range, p = 3 (Weinstock 1985). For an oceanic internal wave spectrum Garrett and Munk (1975) predicted a value of p = 2.5 and Muller *et al.* (1978) obtained p = 2.4 ± 0.4 from observations. On the other hand, Weinstock (1978) considered the exponent p in the buoyancy subrange to be variable, and at $p>5/3$ he postulated an energy sink on short scales with non-stationary and decaying turbulence. Thus with a high Ri and low turbulence intensity the exponent p is likely to be in the range 2.5-3.0, gradually decreasing with increasing degree of turbulent mixing to a value of about 5/3.

Examples of observed spectra of salinity fluctuations at two different values of Ri_L are shown in Fig. 1. These were calculated using a fast Fourier-transform routine, from 1000-s sections of continuous salinity measurements obtained at mid-depth, just below the halocline in the Tees Estuary. The estimates of salinity variance (in units of $‰^2$) were averaged into 15 frequency bands spaced on a logarithmic scale. The spectrum at high Ri_L shows a distinct internal wave peak at a period of about 62 s, and a decay exponent p of 3.0.

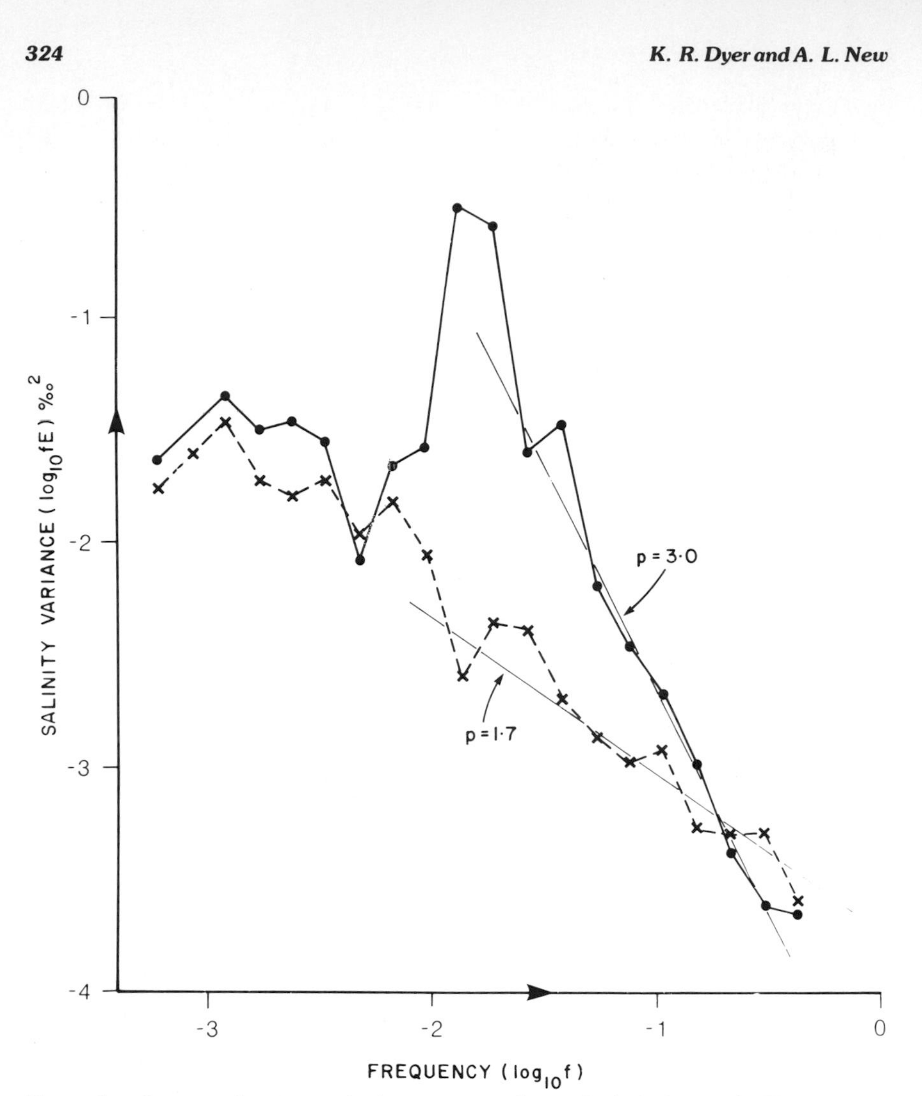

Figure 1. *Spectra of variance of salinity measured near the halocline in the Tees estuary. Straight lines were fitted to the high frequency parts of the plots to estimate the values of the spectral decay exponent, p.*
(•) 24 May 1984, HW + 20 min. Internal waves present. $Ri_L = 500$, $p = 3.0$, $U = 3.8$ *cm* s^{-1}.
(x) 23 May 1984, HW + 2 h 30 min. Internal waves absent. $Ri_L = 2.5$, $p = 1.7$, $U = 50.0$ *cm* s^{-1}.

At low Richardson number the spectrum shows no such peak, and the decay range can be represented by an exponent of 1.7 throughout.

At the same time that these spectra were recorded, a 200 KHz echo-sounder was traversed past the measuring vessel. The record taken at the high Ri_L confirmed the presence of internal waves, but these were absent at the lower Ri_L (Fig. 2), when the structure of the flow suggested that mixing was occurring.

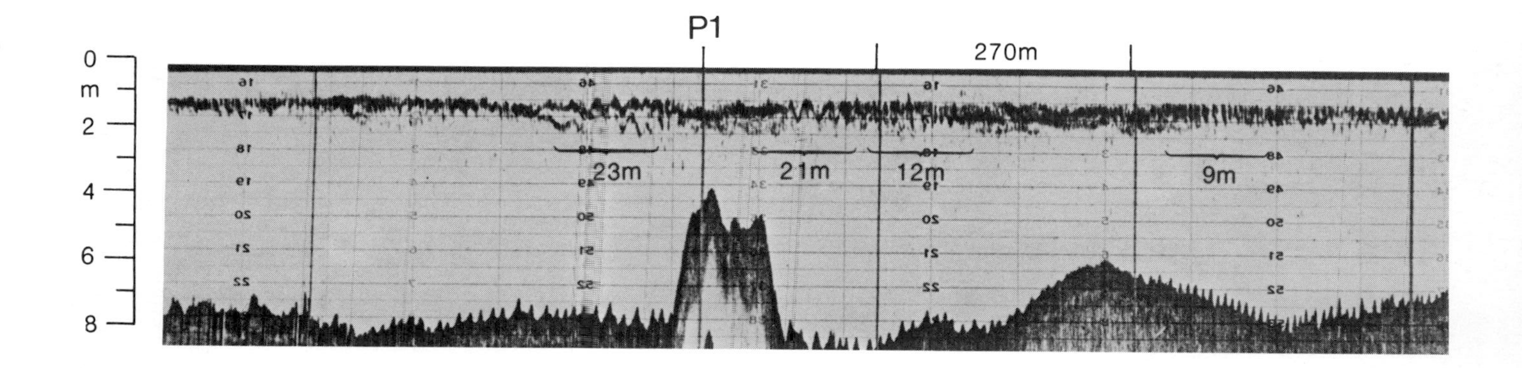

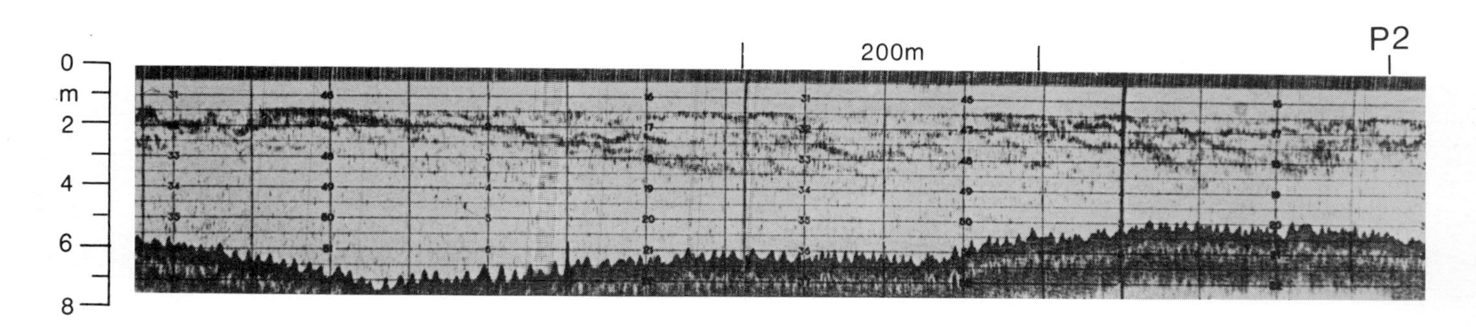

Figure 2. *Echo-sounding recordings during the times at which the spectra in Fig. 1 were obtained. P1 and P2 indicate the positions at which the salinity fluctuations were measured. (a) 24 May 1984. Wavelengths of the internal waves are indicated. (b) 23 May 1984. Internal waves absent.*

An interesting result is apparent when the decay exponent of observed spectra from two different estuaries is plotted against Ri_L (Fig. 3). At Ri_L above about 20, internal wave features dominate the spectrum and it is presumed that there is little mixing. Dyer (1980) proposed that this value (Ri_L = 20) was the limit below which turbulent mixing started near the halocline in Southampton Water. Below that value the exponent gradually approaches the value of $-5/3$ with decreasing Ri_L. We interpret this to result from the increased effectiveness of bottom turbulence in reducing the intensity of the stratification. Below a value of $Ri_L \sim 2$ the turbulence is presumably effectively isotropic and mixing is fully developed. Given enough time, the halocline would be broken down and the flow would become completely mixed. Though Ri_L characterizes the entire mixing in terms of the boundary shear, there was internal shearing at the time of the measurements. Consequently the critical values derived above must empirically include the effects of any internal mixing produced by this shear. Despite the possible differences between the two estuaries in the relative importances of the internal and boundary shear, the results compare reasonably, suggesting that they could apply more widely. These results were obtained at about mid-depth and below the halocline, and it is unclear how the p versus Ri_L relationship would apply for salinity fluctuations measured at other heights, particularly above the halocline. Consequently care must be used in applying the critical values, except in a very general way, until the relationships are better defined.

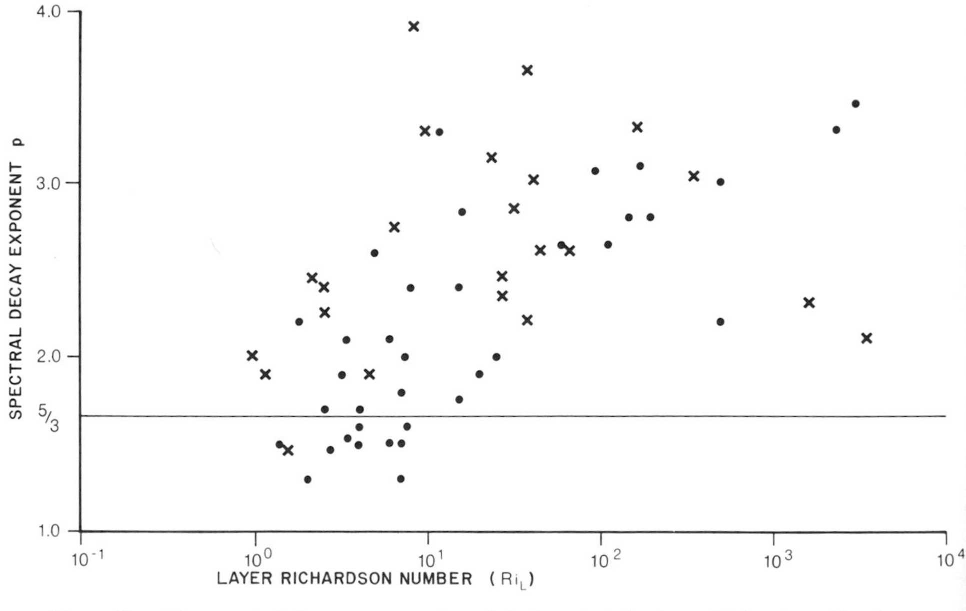

Figure 3. *The spectral decay exponent p plotted against the layer Richardson Number. (x) Observations in Southampton Water. (•) Observations in the Tees Estuary.*

Mixing Diagrams

The above criteria can be used to illustrate the temporal occurrence of mixing at various positions within an estuary by considering the relative contribution, throughout the tide, of the density gradient and the velocity shear to the value of Ri_L. From this point of view it is instructive to plot the surface-to-bottom salinity difference (ΔS) against $U|U|$, where U is the depth-mean current. This modulus form retains the directional character of the flow. A hysteresis loop is plotted to represent the tidal variation in stratification and bed shear stress. The limits of $Ri_L = 2$ and 20 can be shown for the appropriate water depth, assuming the salinity stratification is directly related to the density stratification.

Since the stratification in a partially-mixed estuary normally increases towards the head, the hysteresis loop at a downstream station would be an open anti-clockwise one, in the absence of any mixing (Fig. 4a). With mixing as well as advection more complicated loops occur, depending on the degree of mixing and its extent upstream and downstream of the measurement position. Figure 4b shows an illustrative diagram with both advection and mixing. Starting at high water, advection down the estuary occurs early in the ebb tide of increasingly stratified water. The velocity at A is then strong enough to create mixing at the measuring station, and this intensifies to B at maximum velocity. As the velocity decreases there is continued mixing because of the reduced stratification and the mixing presumably reaches a maximum intensity at C, with a slight phase lag after the maximum current. As the current diminishes, advection of more stratified water from upstream causes increased stratification until low water. On the flood tide less stratified water would be advected initially from downstream, and then ΔS would decrease further as the water mixed on the ebb came past again, undergoing renewed mixing. The decrease in ΔS between the two high water times is a measure of the total amount of mixing. Diagrams such as these should be interpreted with care, however, as different water masses are being considered during the tide, and no account is made of changes in mean salinity with time. On the ebb tide, mixing and advection cause opposing changes in the degree of stratification, and the curves are easier to interpret than on the flood, where they act in concert. Sharp deviations from a smooth curve will occur when fronts or patches of water of anomalous stratification pass by.

Let us now apply the mixing diagram to some actual observations taken on an estuarine cross section in Southampton Water. (Fig. 5). Hourly observations were taken over a tidal cycle at four stations by profiling a salinometer and a current meter at intervals through the water column. Stations B1 and B2 were on opposite sides of a deep channel in about 10 m mean depth, and Stations B3 and B4 were on a wide shallower (5 m) shelf on one side of the channel. The limits $Ri_L = 20$ and 2 are shown for the appropriate water depths, assuming that the tidal variation of water depth was small and that the salinity and density differences were related by $\Delta\varrho = 7.8 \times 10^{-4}\ \Delta S$. For illustrative purposes these assumptions are reasonable.

Stratification was reasonably high at the stations in the deep channel, but some mixing occurred throughout the tidal cycle except for short times around

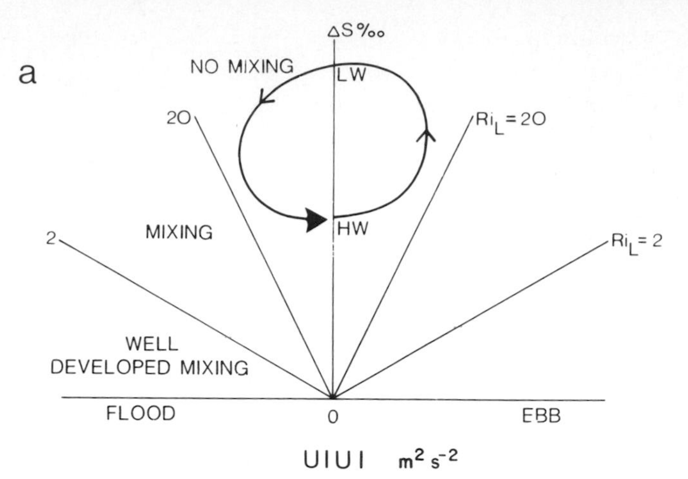

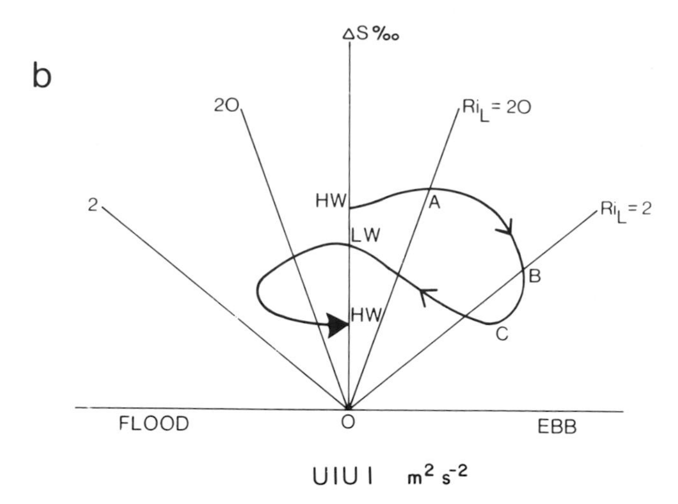

Figure 4. Idealized mixing diagrams: (a) for tidal advection without mixing; and (b) for tidal advection with mixing. See text for explanation.

slack water. Fully developed mixing was attained only at Station B1 on the flood tide. Mixing occurred in the shallower water most of the time, and mixing was fully developed over considerable periods during both flood and ebb tides. At the two central stations (B2 and B3), stratification increased during ebb tide, which suggests that the bottom shear failed to mix the more stratified water advecting from upstream. Near slack water, however, stratification was markedly reduced, indicating advection of better-mixed water from upstream. In contrast, the marginal stations (B1 and B4) exhibited decreasing stratification during

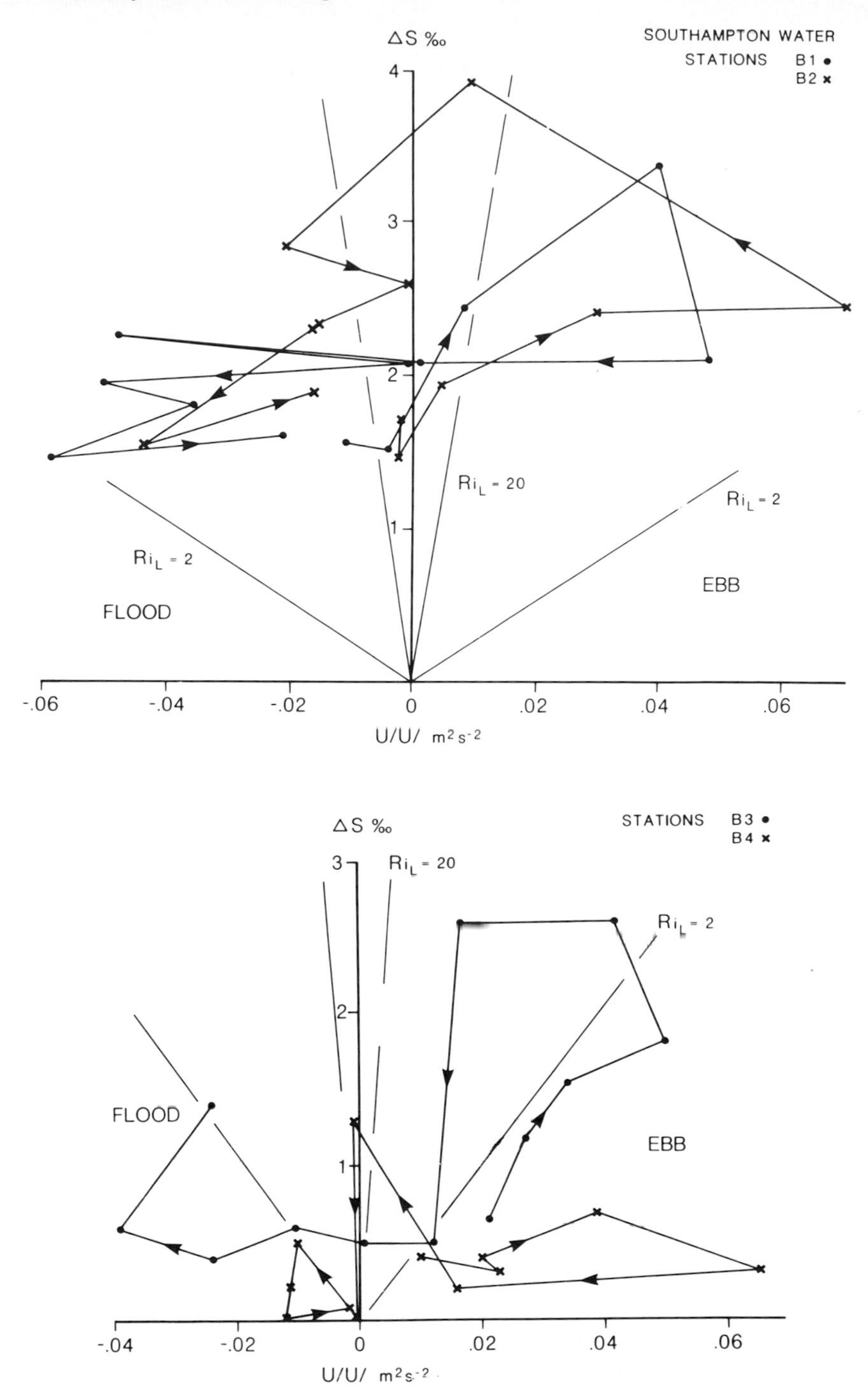

Figure 5. Mixing diagrams for four stations on a cross-section of Southampton Water with observations at hourly intervals over a single tidal cyle.

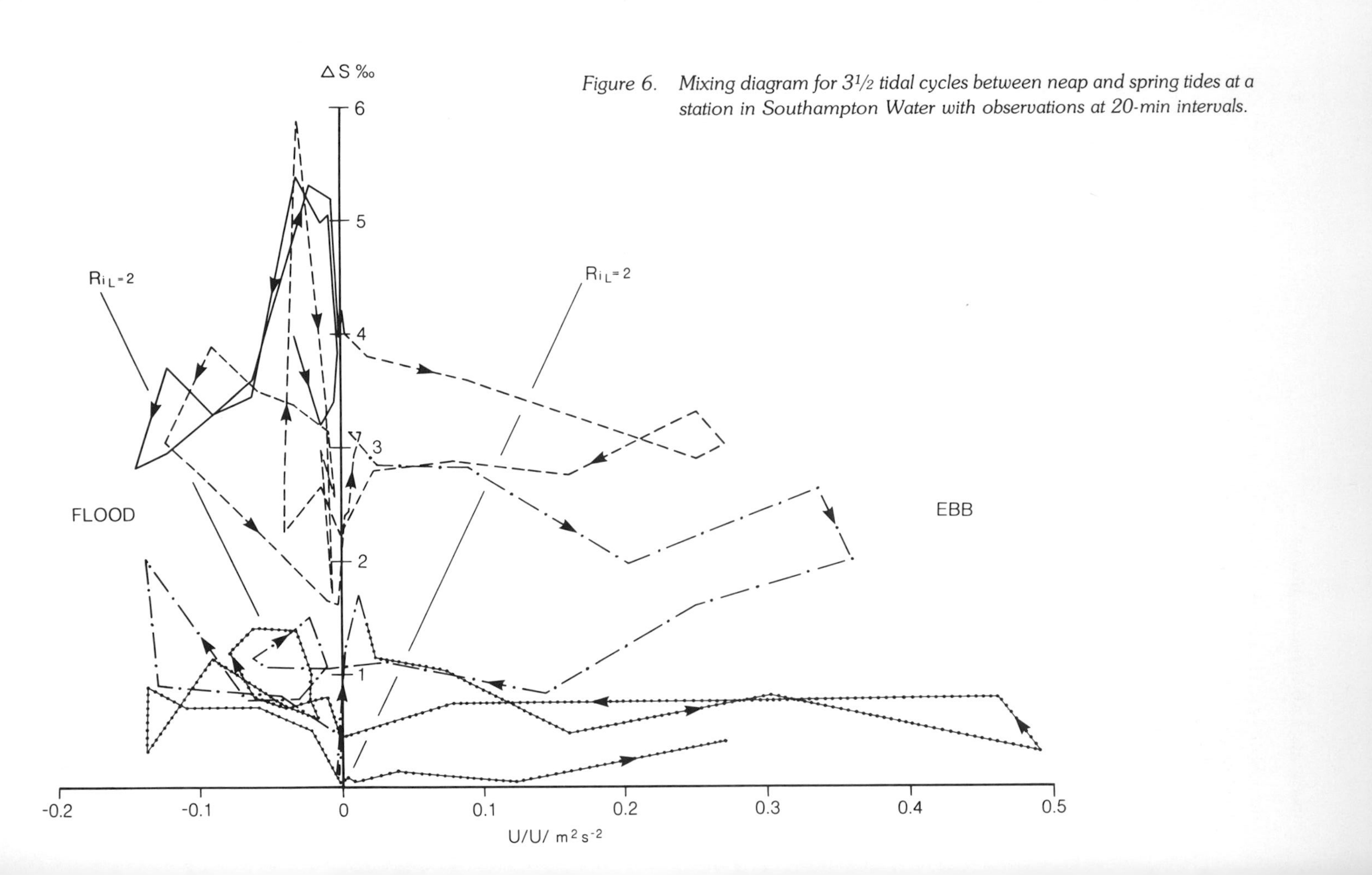

Figure 6. *Mixing diagram for 3½ tidal cycles between neap and spring tides at a station in Southampton Water with observations at 20-min intervals.*

the ebb tide. Consequently, mixing occurred preferentially for that tide in the shallow parts of the cross section. Many estuaries exhibit this behavior.

Figure 6 shows a similar plot for 3½ tidal cycles with observations taken every 20 minutes near Station B2. This was the same time that the spectral results in Fig. 3 were obtained. On this occasion (near neap tides), well-developed mixing occurred only on the ebb tide, and as the peak velocities increased towards spring tides, stratification diminished progressively, largely as a result of mixing that occurred on the ebb tide. On the flood tide the velocities were considerably less than on the ebb tide, but Ri_L was often less than 2 because of the decreased stratification created during the ebb tide. The progressive reduction of stratification towards spring tides is also a common feature of partially-mixed estuaries. As neap tides approach, restratification would follow from reduced peak-current velocities and reduced mixing intensity late in the ebb tide near the head of the estuary.

Figure 7 shows a plot of ½-hourly measurements made over two tidal cycles on consecutive days, and taken at two positions about 1.5 km apart in the upper Tees estuary. These measurements were obtained at the same time as the spectral data in Fig. 3. Significant mixing occurred on the ebb tide with little on the flood. Stratification was markedly decreased during both ebb tides for a period of about 4 hours. These periods coincided with the occurrence of an intense mixing period (IMP). This feature is fairly well known in the estuary (Lewis and Lewis 1983; New *et al.* 1986a) and is characterized by an abrupt increase of about 3-5‰ in surface salinity and an increase in the surface-layer thickness. As will be discussed below, this feature is initiated by other aspects of the estuarine flow. If the IMP did not occur, however, the amount of mixing that occurs on the ebb tide would be significantly reduced, as the stratification would be considerably greater. The occurrence of the IMP implies a localized patch of high salinity water on the surface with fresher water both upstream and downstream.

Intense Mixing Periods

Intense mixing periods have been observed in several estuaries and may result from several causes. They occur on the ebb and are manifest as an abrupt increase in salinity and thickness on the surface layer. First observed in the Duwamish Estuary (Partch and Smith 1978) they were considered to be caused by breaking internal waves traveling upstream until the current was equal and opposite to their propagation velocity. Gardner and Smith (1978) concluded that the mixing was caused by a hydraulic jump at an abrupt shallowing of the estuary, and that the mixed water-mass then advected downstream. Gardner and Smith (1978, 1980) also observed intense mixing further downstream in the same estuary near a bridge. This mixing seemed to occur only during high-velocity ebbs as a result of flow constriction caused by the bridge piers.

The appearance of the IMP in the Tees estuary is shown by the variation of salinity at various depths during a tidal cycle (Fig. 8). Mean water depth at this station was 6 m. High water occurred at about 1040 h, when the main halocline was between 1 and 2 m, and salinity was almost uniform below 2 m. At about

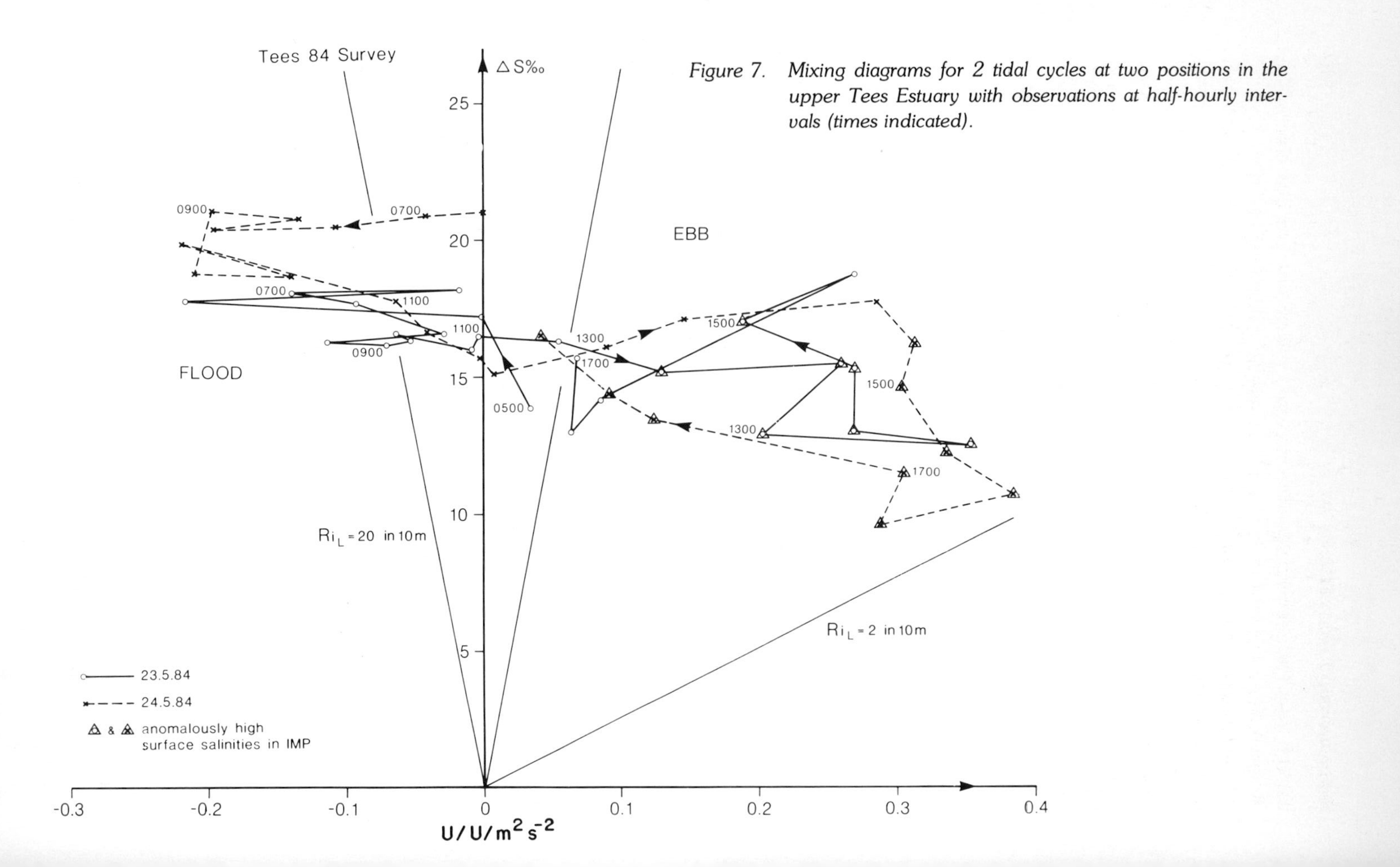

Figure 7. *Mixing diagrams for 2 tidal cycles at two positions in the upper Tees Estuary with observations at half-hourly intervals (times indicated).*

1230 h the surface salinity started to increase, while the surface layer attained an almost uniform salinity and thickened appreciably. The timing of the onset of the IMP varies along the estuary, but the appearance of the IMP coincides well with the time of advection from a series of closely-spaced bridges near the head of the estuary (Fig. 9). Figure 9 shows the time-distance curves for advection of surface water originating at the bridges soon after high water on typical neap and spring tides. Visual observation and dye studies of the flow through the bridges suggest that the turbulent wakes of the supporting piers may create the mixing in this case, rather than a constriction of the flow leading to an internal hydraulic jump.

Internal Waves

In an estuary internal waves (oscillations of the water column localized near the density interface) may result from the interaction of the tide with various topographic features. These internal waves may break near the site of their generation, or at some distance away after a considerable time delay. For example, when the ebb tide flows over a topographic depth increase a depression of the pycnocline is likely to form downstream from the feature (Maxworthy 1979). If the maximum ebb flow is sufficiently strong, this lee wave may break and cause a region of mixed water. The lee wave must have a large upstream phase velocity to remain stationary behind the topographic feature, and the wave may travel upstream as the ebb decreases, evolving as a series of solitary internal waves. This mechanism was substantiated by Farmer and Smith (1980) in a field survey of Knight Inlet. We have obtained echo-soundings from a number of estuaries to further clarify the situation, and these records indicate the potential importance of internal waves as an estuarine mixing process.

Figure 10 presents a sequence of five successive recordings obtained with a 2-MHz echo-sounder in the upper reaches of the Test estuary, Southampton, during spring tides. The ebb flows from right to left over a sharp increase in depth which corresponds to the upstream end of a turning basin for ships. The topographic break is the thin dark line at the bottom right of Fig. 10a, and above this step a train of internal waves is evident on the pycnocline at about 2-m depth. The waves remain stationary behind the topographic feature, and as the ebb increases, they achieve the necessary increase in upstream phase velocity by increasing their wave length (Fig. 10b). As the flow increases still further, however, the waves seem to break (Fig. 10c), and form a large-scale overturning rotor (Fig. 10d) which appears to suspend sediment from the bed. This suspension forms a plume, at 2-3 m depth, which advects downstream (Fig. 10e). Note the changes in horizontal scale and the change of the gain on the echo-sounder. On spring tides, then, significant internal-wave mixing seems to occur near maximum ebb and close to the topographic step.

The survey was repeated at neap tides, and the response of the estuary was markedly different (Fig. 11). Figure 11a is a recording from a 200-KHz echo-sounder close to the time of the maximum ebb, and shows the topography more completely. Two large lee waves are apparent downstream from the step, but these were not observed to break. The structure of the flow near the top of

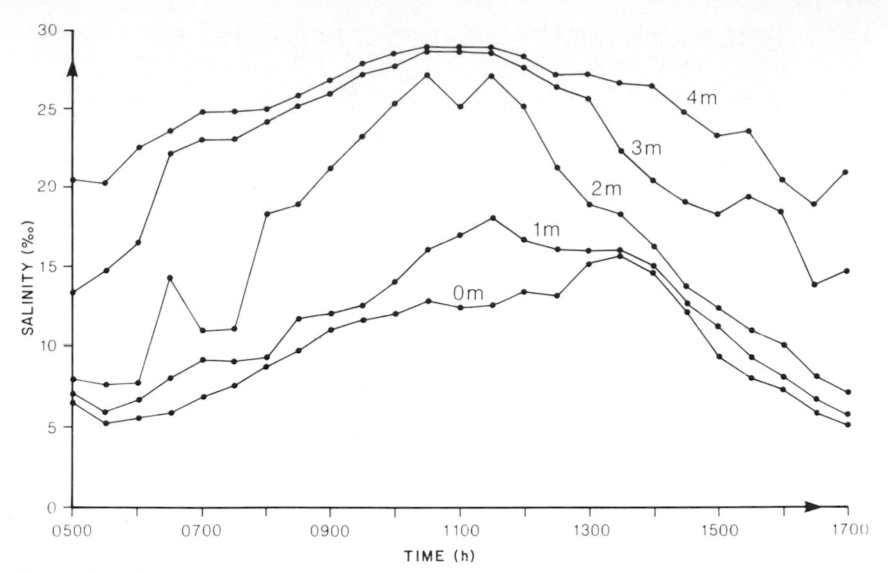

Figure 8. *Salinities at various depths during a tidal cycle at a position in the upper Tees Estuary. High Water 1040. An Intense Mixing Period (IMP) started at 1230.*

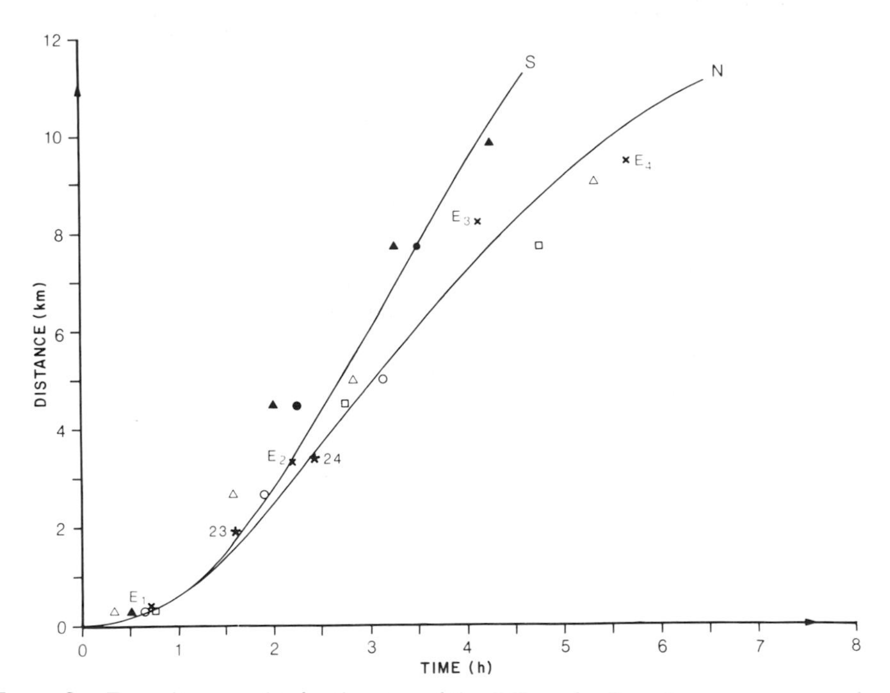

Figure 9. *Time-distance plot for the start of the IMP in the Tees Estuary, in terms of hours after HW and distance downstream from the bridge obstructing the flow. The observations were taken at various times over several years. N = calculated curve for average neap tidal velocities, S = calculated for spring tidal velocities.*

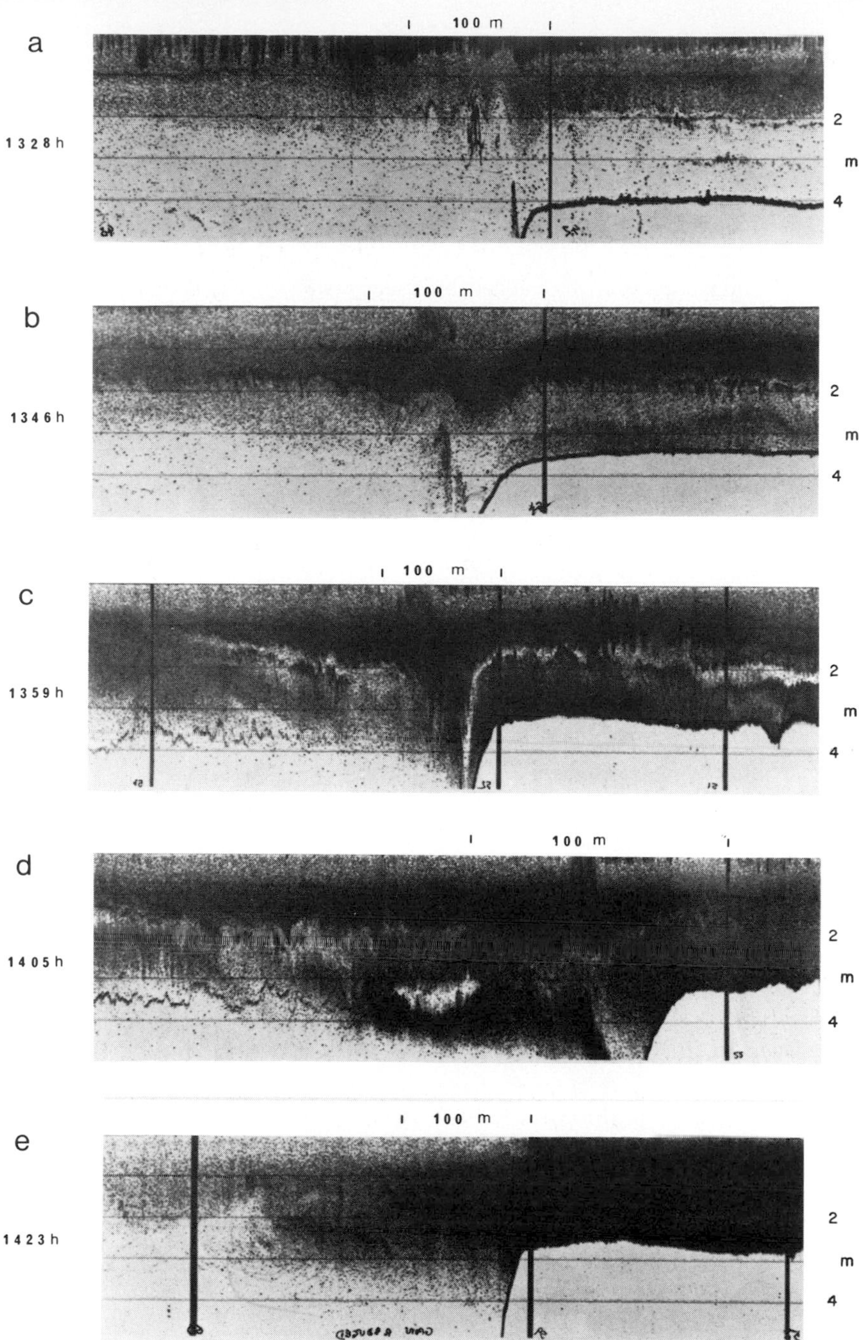

Figure 10. Sequence of 2-MHz echo-sounding traverses over a topographic step in Southampton Water during part of an ebb tide, showing the generation and breaking of waves. Upstream is on the right-hand side of the recordings. The variable horizontal scale is indicated above each record.

the step is shown more clearly in a simultaneous recording obtained with the 2-MHz echo-sounder (Fig. 11b). The upper layer thickens and waves develop on the downstream side of the step. This structure may be interpreted as an upstream-travelling undular bore (Baines 1984). If viscosity is neglected, the bore would be expected to evolve as a series of solitary internal waves. This appears to be the case (Fig. 11c), as the waves traveled about 60 m upstream on the decreasing ebb. These waves appear to break a short time later (Fig. 11d). Since the waves are now nearer the bed and probably in a region of higher shear, the breaking could be caused by shear instability when $Ri < 1/4$, as discussed earlier. The internal-wave mixing in this case occurred about 120 m upstream from the step. A similar mixing region would presumably be produced on other neap ebb tides.

Finally, a patch of waves observed in the Tees estuary near the time of maximum flood indicates the importance of internal waves as an estuarine mixing process (Fig. 12). These were produced as lee waves over the ripples in the bottom topography, and had a similar length scale (about 12 m). The bottom ripples (and the internal waves at this stage in the tidal cycle) extended over several km in the upper reaches of the Tees. As the flood decreased, the internal waves appeared to travel downstream for approximately 500-600 m before disappearing (New *et al.* 1986b). The energy in the waves must be dissipated and converted by mixing either into a potential energy increase or eventually into heat. If 20 percent of the wave energy goes into a potential energy increase (Osborn 1980), then calculations using a two-layer approximation show that internal waves of a 0.3-m amplitude would produce an increase in surface salinities equivalent to approximately one-tenth of that seen in the IMP. The mixing produced towards the end of the flood tide by the breaking of internal waves was therefore somewhat less than that causing the IMP below the bridge on the ebb tide, but probably extended nonetheless over several km of the estuary. While these two processes produced mixing at different times and locations, advection between these areas and others where internal waves were being produced would probably cause some degree of interactive enhancement of the mixing.

Conclusions

Mixing in estuaries is intermittent in both space and time. This intermittency may originate from: (a) interaction of the flow with lateral bottom topography which results in more intense mixing in shallow water at the sides of the estuaries; (b) interaction of the flow with longitudinal changes in topography (the flow over the bed causes lee waves on the interface which may break near their generation site at spring tides or may travel against the flow on weaker tides and break elsewhere in the estuary); and (c) obstructions to the flow (such as bridges), which may cause mixed patches of water. Three mixing regimes can be characterized in space and in time by a layer Richardson Number:

1. $Ri_L > 20$. A stable interface with internal wave activity, but no significant bottom induced mixing.

2. $20 > Ri_L > 2$. A mixing regime where the interface is being modified by bottom-produced turbulence.

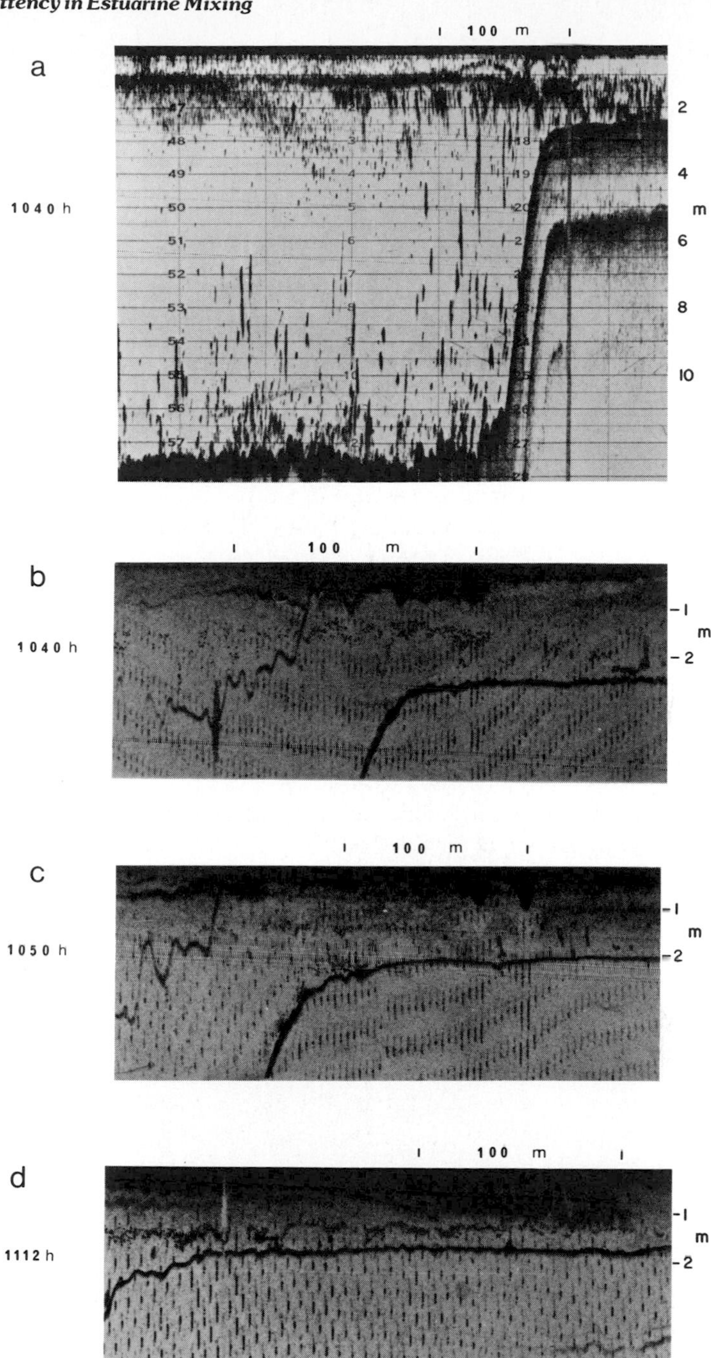

Figure 11. *Sequence of echo-sounding traverses during a neap ebb tide over the same topographic step in Southampton Water as shown in Fig. 10, showing progression of internal waves before breaking.*

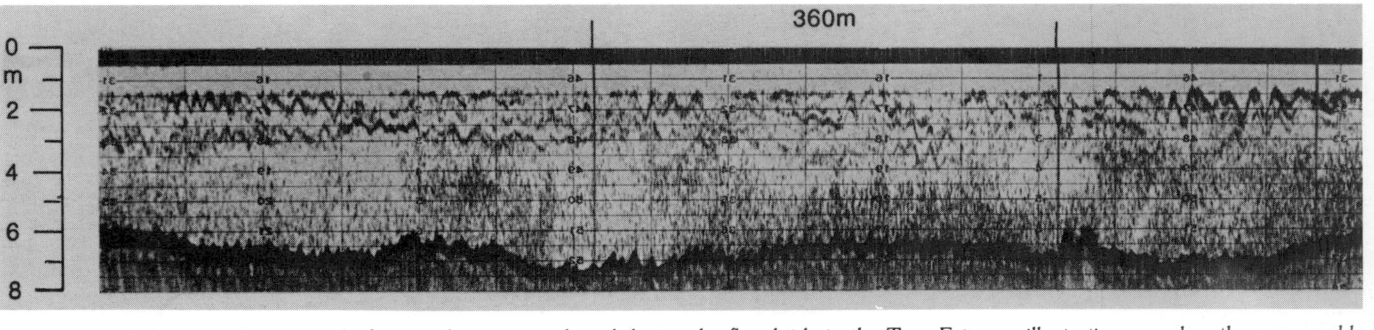

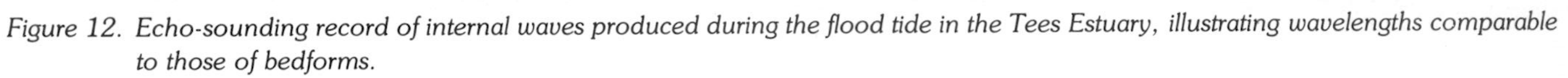

Figure 12. *Echo-sounding record of internal waves produced during the flood tide in the Tees Estuary, illustrating wavelengths comparable to those of bedforms.*

3. $Ri_L < 2$. A fully three-dimensional turbulent modification of the halocline, in which only time and the advection of more stratified water from upstream prevents completely homogenous conditions from becoming established.

The mixing produced in different parts of the estuary will depend on the tidal currents, with more mixing during spring tides than at neap tides. As the water is gradually advected through the estuary by the mean circulation, the mixing at one site will eventually affect the stratification and the mixing at other positions in the estuary. The overall structure of the estuary is likely to be a weighted average of all these events and the introduction or removal of only a few of the mixing sites could have major implications for the overall mixing, the flushing times, and the dispersion rates.

References Cited

Abrahams, G. 1980. On internally generated estuarine turbulence. *Proc. 2nd Int. Symp. Stratified Flows.* Trondheim. 1:344-353.

Baines, P. G. 1984. A unified description of two-layer flow over topography. *J. Fluid Mech.* 146:127-167.

Dyer, K. R. 1982. Mixing caused by lateral internal seiching within a partially mixed estuary. *Estuar. Coastal Shelf Sci.* 15:443-457.

Dyer, K. R. 1986. Tidally generated estuarine mixing processes. *In:* B.J. Kjerfve (ed.), *Hydrodynamics of Estuaries.* CRC Press, Boca Raton, Florida. In Press.

Farmer, D. M. and J. D. Smith. 1980. Tidal interaction of stratified flow with a sill in Knight Inlet. *Deep Sea Res.* 27:239-254.

Gardner, G. B. and J. D. Smith. 1978. Turbulent mixing in a salt wedge estuary. pp. 79-106. *In:* J. C. J. Nihoul (ed.), *Hydrodynamics of Estuaries and Fjords.* Elsevier Oceanography Series Vol. 23. Elsevier, Amsterdam.

Gardner, G. B. and J. D. Smith. 1980. Observations of time-dependent, stratified shear flow in a small salt-wedge estuary. *Proc. 2nd Int. Symp. Stratified Flows.* Trondheim. 2:944-951.

Garrett, C. and W. Munk. 1975. Space-time scales of internal waves: a progress report. *J. Geophys. Res.* 80:291-297.

Lewis, R. E. and J. O. Lewis. 1983. The principal factors contributing to the flux of salt in a narrow partially stratified estuary. *Estuar. Coastal Shelf Sci.* 16:599-626.

Maxworthy, T. 1979. A note on the internal solitary waves produced by tidal flow over a three-dimensional ridge. *J. Geophys. Res.* 84:338-346.

Muller, P., D. J. Olbers and J. Willebrandt. 1978. The IWEX spectrum. *J. Geophys. Res.* 83:479-500.

New, A. L., K. R. Dyer and R. E. Lewis. 1986a. Predictions of the generation and propagation of internal waves and mixing in a partially stratified estuary. *Estuar. Coastal Shelf Sci.* In Press.

New, A. L., K. R. Dyer and R. E. Lewis. 1986b. Internal waves and intense mixing periods in a partially stratified estuary. *Estuar. Coastal Shelf Sci.* In press.

Osborn, T. R. 1980. Estimates of the local rate of vertical diffusion from dissipation measurements. *J. Phys. Oceanogr.* 10:83-89.

Partch, E. N. and J. D. Smith. 1978. Time dependent mixing in a salt wedge estuary. *Estuar. Coastal Mar. Sci.* 6:3-19.

Weinstock, J. 1978. On the theory of turbulence in the buoyancy subrange of stably stratified flows. *J. Atmos. Sci.* 35:634-649.

Weinstock, J. 1985. On the theory of temperature spectra in a stably stratified fluid. *J. Phys. Oceanogr.* 15:475-477.

ABIOTIC FACTORS INFLUENCING THE SPATIAL AND TEMPORAL VARIABILITY OF JUVENILE FISH IN PAMLICO SOUND, NORTH CAROLINA

Leonard J. Pietrafesa, Gerald S. Janowitz

Department of Marine, Earth and Atmospheric Sciences
North Carolina State University
Raleigh, North Carolina

John M. Miller, Elizabeth B. Noble

Department of Zoology
North Carolina State University
Raleigh, North Carolina

and

Steve W. Ross and Sheryan P. Epperly

North Carolina Division of Marine Fisheries
Morehead City, North Carolina

Abstract: We have examined the relationship between the abundance of juvenile spot (*Leiostomas xanthurus*) in the nursery areas along the western boundary of Pamlico Sound, North Carolina, and the circulation patterns in the Sound. Juvenile fish abundance data are related to wind measurements taken two to three weeks prior to the fish sampling program. When the wind has a significant eastward component, i.e., from the nursery areas toward the inlets during the period preceding the sampling program, juvenile spot are relatively abundant compared with periods when the wind is not in that direction. Our numerical model of circulation in Pamlico Sound indicates that this wind field produces surface currents towards the east, and near-bottom currents towards the nursery areas, i.e., towards the west. The near-bottom currents are driven by the pressure-gradient forces associated with the set up of sealevel in the down-wind direction. It is postulated that the bottom-seeking juvenile spot ride these favorable bottom currents across the sound from the inlets to the nursery areas.

Introduction

Pamlico Sound is a barrier island estuary, located in North Carolina along the eastern seaboard of the United States (Fig. 1). The small bays and riverine tributaries along the westernmost periphery of Pamlico Sound comprise the principal nursery areas for fish juveniles in N.C. (Ross and Epperly 1984). These juveniles principally include spot (*Leiostomus xanthurus*), menhaden (*Brevoortia tyrannus*), croaker (*Micropogonias undulatus*) and flounder (both *Paralichthys lethostigma* and *P. dentatus*). The sound has come under scrutiny in recent years by state regulatory and environmental groups because of increased utilization of water and adjacent lands by commercial, municipal and recreational users and because of its ultimate importance to commercial fishing interests.

Pamlico Sound is approximately 140 km long and 25-60 km wide, has a mean depth of 4 m and covers some 4300 km² (Roelofs and Bumpus 1953). The Outer Banks, an island chain consisting of Hatteras, Ocracoke and Portsmouth Islands, provides a barrier between the sound and the coastal ocean. Oregon Inlet at the northern end of Hatteras Island, Hatteras Inlet between Hatteras and Ocracoke Islands, and Ocracoke Inlet between Ocracoke and Portsmouth Islands, connect the sound to the coastal ocean. (Fig. 1).

The sound has an annual, spatially averaged salinity of about 20‰, with virtually no vertical stratification (Schwartz and Chestnut 1973) because of its shallow depth and the influence of mechanical wind mixing. Highly saline (34.5-37‰) Carolina Capes coastal waters enter through Ocracoke and Hatteras inlets, whereas relatively low salinity (31-34‰) Virginia coastal waters enter through Oregon Inlet (Pietrafesa *et al.* 1985). This feature, coupled with exchange of Pamlico Sound waters with the fresher waters of Albemarle and Croatan Sounds to the north, contributes to a general north-to-south positive salinity gradient within the sound. Several large rivers (Neuse, Bay, Tar-Pamlico, Trent and Pungo) feed into the southwest margin of Pamlico Sound creating a net west-to-east positive salinity gradient. Occasionally, however,

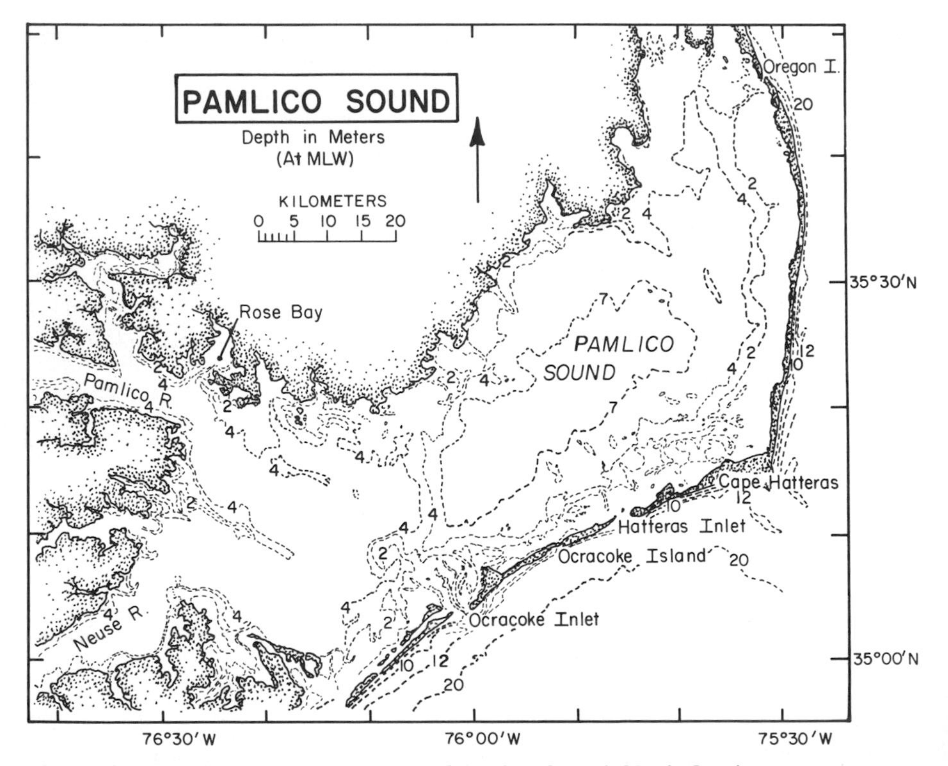

Figure 1. Bathymetry and geography of Pamlico Sound, North Carolina.

these typical salinity patterns may be reversed. While water temperatures in Pamlico Sound are closely coupled to air temperatures, the peripheral nurseries are generally colder in the winter and warmer the remainder of the year than is the sound.

The migratory routes of estuarine-dependent fish larvae along the North Carolina continental shelf have been discussed in the literature (Clark *et al.* 1969; Kjelson *et al.* 1976; Nelson *et al.* 1977; Leggett 1977; Warlen 1981; and Miller 1984a). Five species (spot, menhaden, croaker and the two flounders) constitute only 10 percent of the fish species found in the estuarine nursery areas; yet comprise 90 percent of the total fish biomass in the nurseries. These species all spawn near the Gulf Stream front from late fall to winter, passively migrate 100 km to barrier island inlets and then another 25-100 km to the various estuarine nurseries in the bays and rivers adjoining Pamlico Sound and other similar systems throughout the region. This typical migration scenario for larval transport from the spawning area to a nursery area in Pamlico Sound, is illustrated in Fig. 2 (Miller *et al.* 1984a).

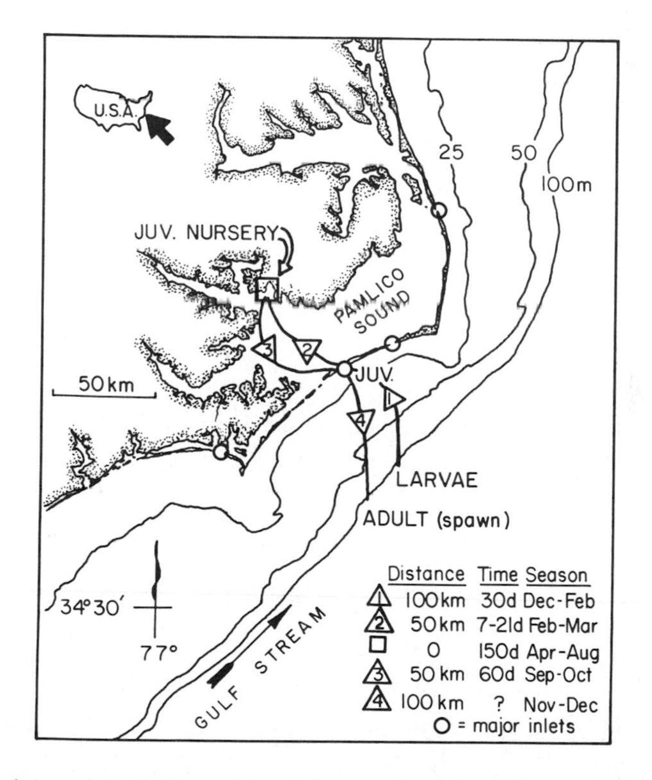

Figure 2. Age and distribution of larval Spot and croaker off N.C. (from Miller et al. *1984a).*

Studies by Ross and Epperly (1984) of the North Carolina Division of Marine Fisheries (NCDMF) and others from the National Marine Fisheries Service (NMFS) (e.g., Lewis and Judy 1983) document large month-to-month and year-to-year variations in overall abundances of juvenile fish and shrimp within the Pamlico Sound nursery areas. Although the ranks of nursery areas change slightly from year to year, Ross and Epperly (1984) were able to group the nursery areas into five categories of productivity. The variations in productivity among the nurseries may be attributable to differences in habitat quality, including such factors as food availability, predator abundance, average salinity (and amplitudes of fluctuations about the mean), and agricultural, municipal, industrial and freshwater drainage. However, Ross and Epperly (1984) hypothesized that the major determinant in the productivity of a particular nursery is its geographic location. In short, they proposed that some nurseries receive more fish recruits because of their strategic locations, and those that receive more produce more adult fish.

If one accepts the geographic location hypothesis of Ross and Epperly (1984), then the obvious questions which follow are: how are the juveniles consistently transported to the preferred nurseries and are they merely passive riders of convenient currents? Miller *et al.* (1984) postulated that the metamorphosed larvae could not swim strongly enough to negotiate the width of the sound, but that they may move vertically in response to hydrographic cues, such that the physics of the sound would dictate the transport mechanism. The vertical migration behavior of the postlarval and juvenile fish would therefore influence their ultimate distribution. The strategic location hypothesis of Ross and Epperly (1984) may be operative, however, only during certain conditions of local marine hydrodynamic climatology. Major climatological differences would be expected to cause markedly different patterns in nursery catch.

We have analyzed several years of sampling data for juvenile fish at selected nursery stations in Pamlico Sound (Fig. 3) and compared these data to atmospheric winds during the sampling period, in an effort to find causal relationships between fish distribution and abundance and wind magnitude and direction. We used a mathematical model of the current patterns in Pamlico Sound to develop relationships between winds, currents and potential migratory pathways across Pamlico Sound from barrier island inlets to nurseries.

Data

The NCDMF has sampled juvenile fish at stations along the western periphery of Pamlico Sound (Fig. 3) for ten of the eleven years during 1974-1984 in an attempt to establish monthly fish statistics within the nurseries. Stations were sampled monthly with a two-seam otter trawl with 3.2-m headrope length, 6.4-mm bar mesh wings and body and a 3.2-mm tail bag mesh. The net was pulled at a speed of 1.1 m s^{-1} over a distance of 68.6 m. All stations were sampled during daylight. Despite occasional sampling gaps, the monthly data sets are fairly complete. More complete details about the sampling technique and sample statistics are provided by Ross and Epperly (1984). In 1985 a

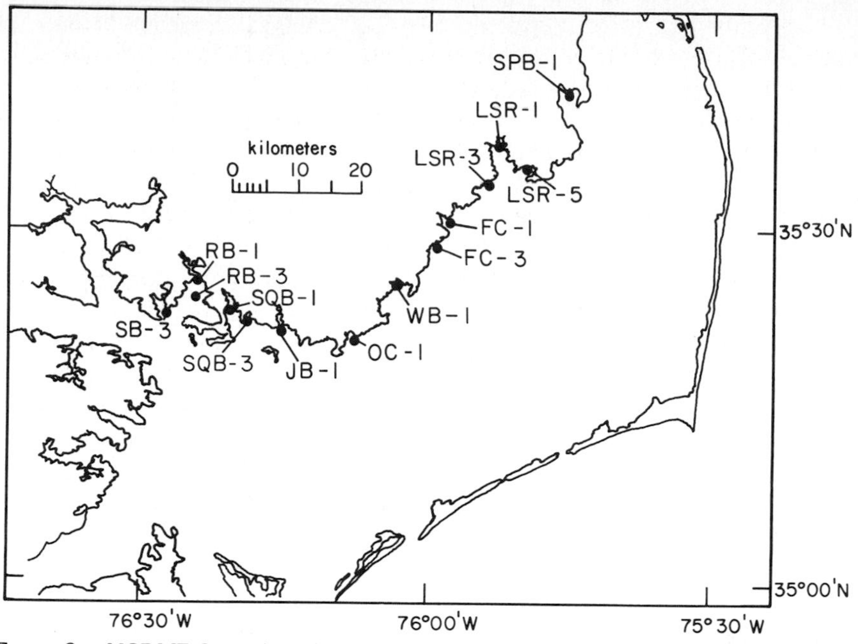

Figure 3. NCDMF Juvenile Fish Sampling Stations during years 1974-1984.

North Carolina Department of Administration juvenile fish sampling program was begun in Pamlico Sound by North Carolina State University (NCSU) scientists, using methods similar to those in the NCDMF studies. Preliminary results from this first year's study are also included here.

Meteorological data measured at the National Weather Service meteorological station located at Cape Hatteras (Fig. 1) were obtained from the NOAA National Climatic Center in Asheville, N.C. The 3-hourly wind data were filtered using Lanczos cosine tapir filters with half-power points of 40 and 400 hours, respectively, to produce 6-hourly and daily time series of the winds. Details of data processing are provided in Pietrafesa and Janowitz (1980).

Juvenile Sampling Results

In the NCDMF samples, juvenile fish size was generally less than 35 mm during the period March-April, less than 55 mm in May and less than 65 mm in June. We have confined our analysis to the months of March and April, the major recruitment months, and to only two species, spot (*Leiostomus xanthurus*) and menhaden (*Brevoortia tyrannus*). Spot are demersal, i.e., benthic or bottom-seeking, whereas menhaden are pelagic, or surface-oriented. The relationship between wind direction during February-April and abundance of juvenile spot sampled in March-April is shown in Figs. 4 and 5. Analysis of the NCDMF larval and juvenile fish data from 1974-1984 reveals several major findings:

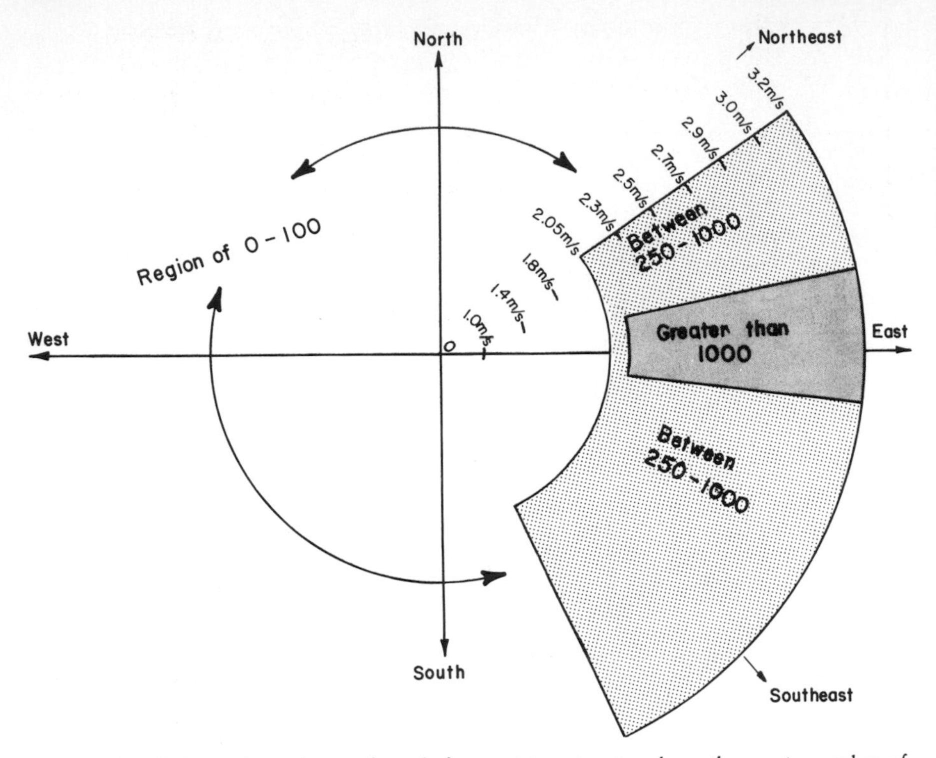

Figure 4. *Relationship of size of catch for spot in estuaries along the western edge of Pamlico Sound, to the windspeed and direction toward which the wind blows. Spot recruit data are from the NCDMF during March and April for the years 1974-1984.*

(1) During the period February through April, post-larval and juvenile spot are found in the nurseries in greater abundances, i.e., factors of three or more, within 2-3 weeks after winds blow predominantly from the southwest to northwest quadrant at magnitudes in excess of 2 m s^{-1}. After winds blew from the WNW to WSW the number of fish sampled increase to 7-12 times the numbers obtained after winds blew from other directions (Fig. 4).

(2) When winds blow predominantly from the northwest at a minimum speed of 2 m s^{-1}, fish yields are higher between stations SB and JB than from stations SPB to FC (Table 1, Fig. 3).

(3) By contrast, winds from the southwest, when dominant for 2-3 weeks at magnitudes in excess of 2 m s^{-1} are followed by greater increases in nursery catch at stations SPB to FC than at stations SB to JB (Table 1, Fig. 3).

(4) Increased numbers of spot recruits are observed at Oregon Inlet 8-20 days before corresponding increases in recruits at stations RB1 and RB3. Wind speeds must exceed 7 m s^{-1} in magnitude in order for this relationship to hold (Fig. 5).

Table 1. Comparison of catch of postlarval and juvenile spot (*Leiostomas xanthurus*) at different stations in Pamlico Sound in relation to wind direction. Stations SPB to FC represent the northern cluster, while SB to JB are the southern stations (Fig. 3).

Wind Direction and Mean Velocity ($m\ s^{-1}$)	Duration of Wind Pattern	Mean Fish Catch per Station		Fish Sampling Dates
		north	south	
from Northwest (4.02)	8-28 March 1975	10	478	25-31 March 1975
from Southwest (5.25)	28 Feb-18 March 1977	419	51	16-21 March 1977

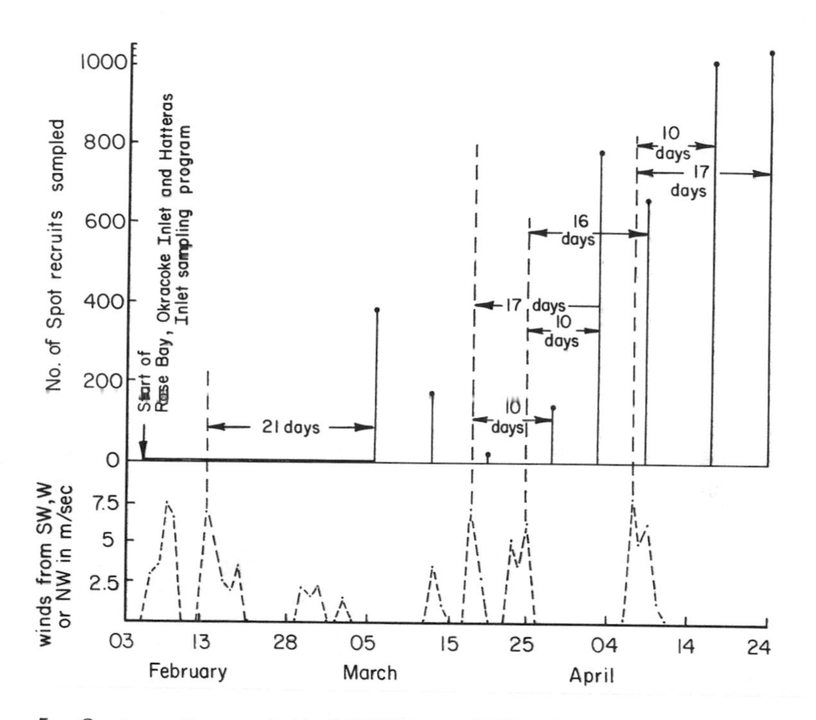

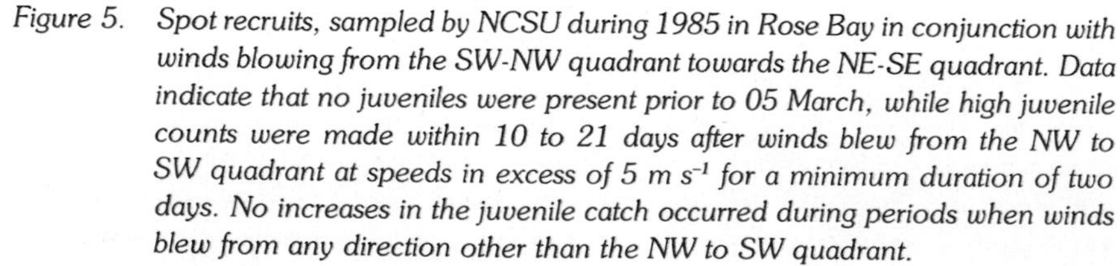

Figure 5. *Spot recruits, sampled by NCSU during 1985 in Rose Bay in conjunction with winds blowing from the SW-NW quadrant towards the NE-SE quadrant. Data indicate that no juveniles were present prior to 05 March, while high juvenile counts were made within 10 to 21 days after winds blew from the NW to SW quadrant at speeds in excess of 5 m* s^{-1} *for a minimum duration of two days. No increases in the juvenile catch occurred during periods when winds blew from any direction other than the NW to SW quadrant.*

Analysis of juvenile menhaden abundance versus wind direction did not reveal any consistent patterns. Possible reasons for this difference are presented in the conclusions section. In the next section, we present a rudimentary model of the wind-induced circulation in Pamlico Sound to identify potential mechanisms for transport of juveniles across the sound.

The Circulation Model

The subtidal frequency circulation in Pamlico Sound is believed to be due largely to the atmospheric windfield which varies both spatially and temporally. However, since the gradients in the windfield are small over distances the size of the sound (Weisberg and Pietrafesa 1983) we assume the wind varies only in time. Also, a uniform density field is assumed since the mean depth of the sound is rather shallow. In fact, the ratio of the baroclinic to the barotropic pressure gradient can be written as $\Delta\varrho H/\varrho_o\eta$, where $\Delta\varrho$ is the scale of density fluctuations, H is the mean depth, ϱ_o is a mean density and η is the scale of sea level fluctuations. For Pamlico Sound, this ratio is <0.05, so a homogenous model is acceptable. Although spatial variations of salinity and temperature do exist within the sound and may be important as cues to the fish, the fronts are not important to the physical dynamics. Also, since the ratio of η to H is generally <0.05, the model is assumed linear. We allow our model to be three-dimensional; as opposed to vertically integrated. In vertically integrated models, the bottom stress is always structured to oppose the mean motion throughout the water column. However, in reality, the bottom stress opposes the mean bottom motion, which is frequently in opposition to the mean or vertically integrated motion. Hence, the bottom stress is incorrectly specified for wind-driven cases by a vertically integrated model. Turbulent eddy stresses are modeled using a spatially and temporally uniform eddy viscosity coefficient. While the Coriolis acceleration terms are retained in the model, the vertical Ekman number is of order unity in this shallow basin, so the rotational effect is slight.

The equations which govern the flow in the sound in a right-handed Cartesian coordinate frame (with $+x$ cross-sound southeast towards the coast, $+y$ along-sound towards the northeast, and z positive up, all rotating at $f_o/2$) are:

$$\frac{\partial \vec{v}\,'}{\partial t'} + f_o \vec{e}_z \times \vec{v}\,' = -g\nabla'\eta' + A\frac{\partial^2 \vec{v}\,'}{\partial z'^2}$$

and

$$\frac{\partial M'x}{\partial x'} + \frac{\partial M'y}{\partial y'} + \frac{\partial \eta'}{\partial t'} = 0$$

where $M'x = \int_{-h}^{0} u'\,dz'$, $\quad M'y = \int_{-h}^{0} v'\,dz'$

and $\vec{v}\,'$ is the horizontal velocity vector, η' is the free surface elevation, h is the local depth, f_o is the Coriolis parameter, A is the constant eddy viscosity coefficient,

$M'x$ and $M'y$ are the respective volume flux vector components in the x and y horizontal directions, and $\vec{e}_z$ is the unit vector in the vertical.

At the surface, the boundary conditions which drive the model are:

$$z = 0, \qquad A \frac{\partial \vec{v}\,'}{\partial z'} = \frac{\vec{\tau}_s\,'}{\varrho}$$

where $\vec{\tau}_s\,'$ is the effective surface wind stress; while at the bottom:

$$z' = -h', \qquad \vec{v}\,' = 0.$$

Additionally, at the lateral boundaries, $\vec{M}\,' \bullet \vec{n} = 0$, where $\vec{n}$ is a horizontal vector normal to the boundaries; except at the inlets and river junctions where a normal flux proportional to $|\eta'|^{1/2}$ is taken. At the coast, the normal gradient of sea level is adjusted to ensure that the normal volume flux vanishes. The resulting system of equations is next nondimensionalized and then written in finite difference form. Central differences are used for spatial derivatives and forward differences for the temporal derivatives.

To substantiate the findings of the fish sampling program summarized previously in the text and visualized in Figs. 4 and 5, we considered two model cases. The first model run was for recruitment-favorable winds while the second was unfavorable to recruitment. The value of 5 m s^{-1} was chosen because the NCDMF data (Fig. 4) suggest a windspeed cutoff of 2 m s^{-1} while the NCSU data (Fig. 5) suggest a cutoff of 7 m s^{-1}. An eddy viscosity coefficient of 100 cm^2 s^{-1} was also used.

We applied a 5 m s^{-1} northwesterly wind (blowing southeastward) to the surface of Pamlico Sound. Ten hours after the onset of the wind the surface and bottom currents assumed the pattern shown in Fig. 6. The surface flow tends to be directed with the wind, at about 5 cm s^{-1} while the bottom flow is directed against the wind at speeds of 1-2 cm s^{-1}. The surface flow field is mechanically driven by the wind while nearbottom currents are forced by a cross-sound pressure gradient caused by the slope of the water surface (Fig. 7), which can be approximated as $\partial\eta/\partial x \sim 10^{-8}\, u_w^2\, h^{-1}$ where u_w is the wind speed and h is the depth.

Circulation induced within the sound by northwestward winds (unfavorable to recruitment) is towards the north-northwest at the surface and south-southeast at the bottom (Fig. 8).

These model results suggest that Pamlico Sound reaches a quasi-steady state condition in a period of less than ten hours after the onset of a steady wind. Recent studies of the coastal meteorology in this region (Weisberg and Pietrafesa 1983) indicate that the monthly-to-seasonal mean winds generally repeat from year to year, but the major portion of the wind variability occurs over time scales of 2 days to 2 weeks. Since spot cross the sound over the period of about 10-15 days (Miller *et al.* 1984b) the fluid physics matches the biology of interest.

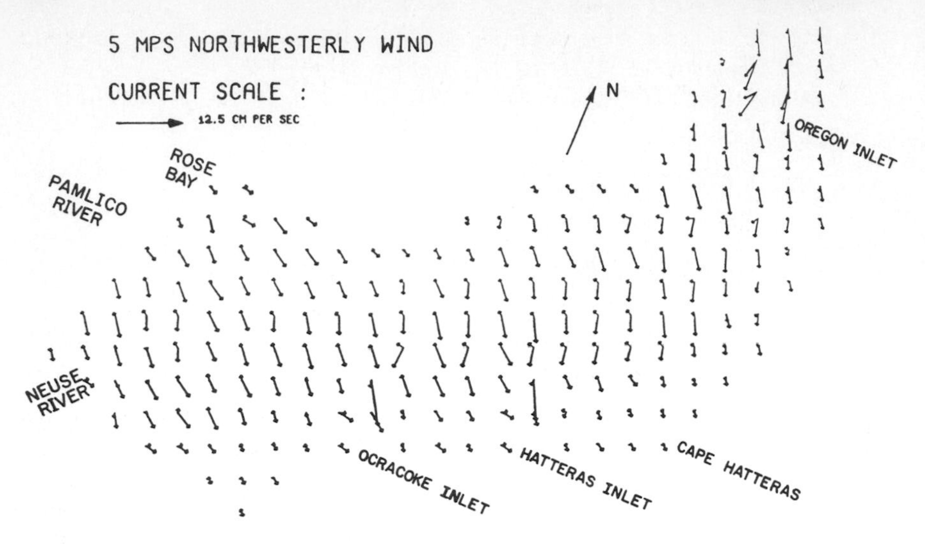

Figure 6. *Surface (long arrow) and near-bottom (short arrow) model current-vector velocities throughout Pamlico Sound in response to a 5 m s⁻¹ northwesterly (southeastward) wind. Bottom velocities are computed 1 meter above the bottom.*

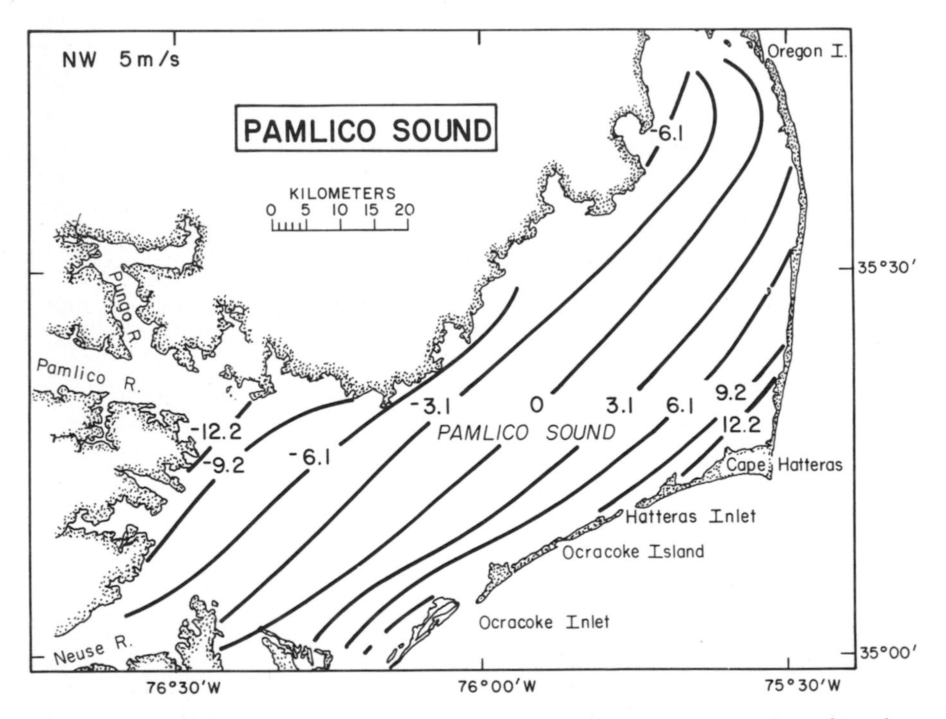

Figure 7. *Model sea level distribution in Pamlico Sound in response to a 5 m s⁻¹ southeastward wind.*

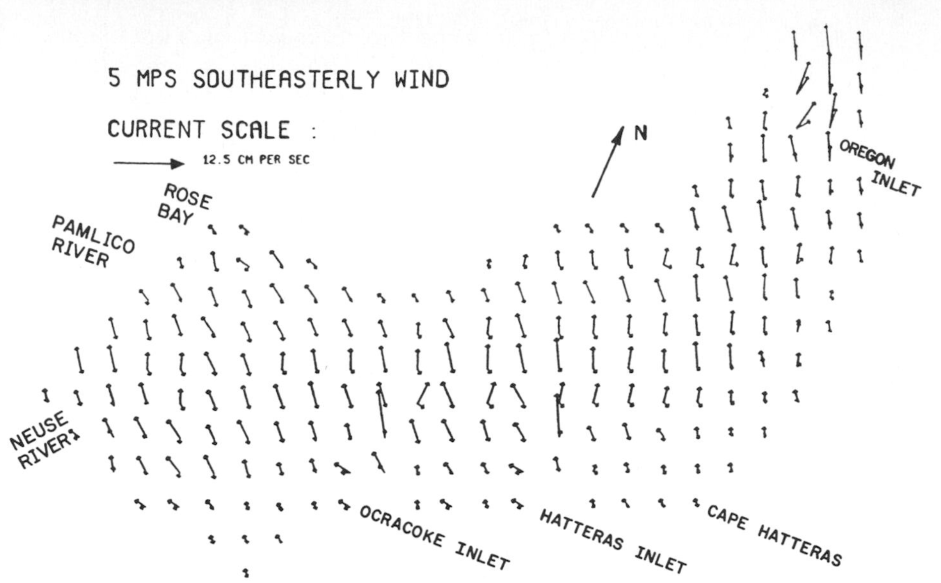

Figure 8. *Surface (long arrow) and near-bottom (short arrow) model current-vector velocities in Pamlico Sound for 5 m s⁻¹ northwestward winds. (Same as Fig. 5, but opposite wind fields). Surface currents are driven northwestward across the sound.*

Conclusions

The rudimentary numerical model of the wind-driven circulation within Pamlico Sound may explain the variations for both juvenile spot and menhaden abundances in nurseries. Winds directed south-southeastward to north-northeastward indirectly create near-bottom currents that are favorable for the transport of the demersal spot from barrier island inlets to the nurseries. The southeastward winds aid recruitment of spot from Ocracoke and Hatteras Inlets primarily towards the more southwesterly nurseries of Pamlico Sound and secondarily, toward those nurseries to the north. Eastward winds also aid recruitment of spot from Oregon Inlet to the more northerly nurseries. Alternatively, west-southwestward to northwestward winds directly drive surface waters from the inlets towards the nurseries, and should create conditions favorable for the recruitment of the pelagic postlarval menhaden.

Although westward winds should aid nursery recruitment of menhaden from all barrier island inlets, our study provides no direct confirmation of this. We assume that the postlarval-juvenile menhaden do not simply swim to the nurseries, but our samples did not show consistent differences in abundance under the different wind conditions. It is likely that our sampling apparatus, (a 0.6 m × 3 m trawl dragged along the bottom) did not properly sample this pelagic species. The fishing gear that has proven most effective in catching larval and juvenile menhaden is designed to fish at or just below the surface (Lewis *et al.* 1970; Hettler 1979; Turner 1973).

Finally, we reiterate that spot (and menhaden) could be cued to nursery location by temperature and or salinity. Miller *et al.* (1984b) suggested that if larval spot vertically positioned themselves in the most saline, warmest outershelf water between December-February, then they would be transported to the barrier island inlets. Once through the inlets, if the juvenile spot continue to track the warmest water, they would be passively transported towards the nurseries under the right wind conditions from February to April. The important factors here are that in the yearly cylces of inner shelf, inlet and estuarine nursery water temperatures, the nursery waters become warmer than inlet and ocean waters in February and bottom waters are warmer than surface waters at this time of year. Stable stratification is maintained by slight salinity compensation.

In summary, temperature and salinity may combine with wind-driven currents and pressure-gradient-driven currents to provide potentially advantageous water mass cues and transport mechanisms that help juvenile spot (and possibly other species) to reach the nurseries of Pamlico Sound. Existing data sets provide inadequate detail, however to understand fish migrations across the sound on the short physical and biological time scales that appear to govern the process.

Acknowledgments

This study was supported by the Office of Sea Grant, NOAA, U.S. Department of Commerce under Grant No. NA81AA-D-00026 and Project No. RES-27, the North Carolina Department of Administration, and the Water Resources Research Institute.

References Cited

Clark, J., W. A. Smith, A. W. Kendall, Jr., and M. P. Fahay. 1969. Studies of estuarine dependence of Atlantic coastal fishes. Data Report 1: Northern Section, Cape Cod to Cape Lookout. R. V. Dolphin cruises 1965-66. Zooplankton volumes, midwater trawl collections, temperatures and salinities. U.S. Fish and Wildlife Service Technical Papers 28:1-132.

Hettler, W. F. 1979. Modified neuston net for collecting live larval and juvenile fish. *Progressive Fish-Culturist* 41(1):32-33.

Kjelson, M. A., G. N. Johnson, R. L. Garner, and J. P. Johnson. 1976. The horizontal-vertical distribution and sample variability of ichthyoplankton populations within the nearshore and offshore ecosystems of Onslow Bay. pp. 287-341. *In:* Atlantic Estuarine Fisheries Center Annual Report to ERDA (Energy Research and Development Administration), U.S. National Marine Fisheries Service, Beaufort, North Carolina.

Leggett, W. C. 1977. The ecology of fish migrations. *Ann. Rev. Ecol. Systematics* 8:285-308.

Lewis, R. M. and M. H. Judy. 1983. The occurrence of spot, flounder and menhaden and Atlantic croaker larvae in Onslow Bay and the Newport River estuary, N.C., U.S. Natl. Mar. Fish. Serv. *Fish. Bull.* 81:405-412.

Lewis, R. M., W. F. Hettler Jr., E. P. H. Wilkens and G. N. Johnson. 1970. A channel net for catching larval fishes. *Chesapeake Sci.* 11:196-197.

Miller, J. M., J. P. Reed and L. J. Pietrafesa. 1984a. Patterns, mechanisms and approaches to the study of estuarine dependent fish larvae and juveniles. pp. 209-225. *In: Mechanisms of Migration in Fishes,* Proceedings of NATO Advanced Research Institute Conference, Dec. 1982. Plenum Publishing Corp., Acquafredda DiMaratea, Italy.

Miller, J. M., S. W. Ross and S. P. Epperly. 1984b. Habitat choices in estuarine fishes: Do they have any? pp. 337-352. *In:* B. J. Copeland, K. Hart, N. Davis, and S. Friday (eds.), *Research for*

Managing the Nation's Estuaries. UNC Sea Grant College Publ. UNC-SG-84-08. UNC Sea Grant College Program. North Carolina State University, Raleigh.

Nelson, W. R., M. C. Ingham, and W. E. Schaaf. 1977. Larval transport and year-class strength of Atlantic menhaden, Brevoortia tyrannus. U.S. Natl. Mar. Fish. Serv. *Fish. Bull.* 75:23-41.

Pietrafesa, L. J. and G. S. Janowitz. 1980. On the dynamics of the Gulf Stream front in the Carolina Capes. pp. 184-197. *In: Proc. of 2nd Int'l Symp. on Stratified Flows.* Tapir Publications, Trondheim, Norway.

Pietrafesa, L. J., G. S. Janowitz and P. A. Wittman. 1985. Physical oceanographic processes in the Carolina Capes. pp. 23-32. *In:* L. P. Atkinson, D. W. Menzel, and K. A. Bush (eds.), *Oceanography of the Southeastern U.S. Continental Shelf,* Coastal and Estuarine Sciences Series 2. American Geophysical Union, Washington, D.C.

Roelofs, E. W. and D. F. Bumpus. 1953. The Hydrography of Pamlico Sound. *Bull. Mar. Sci. Gulf and Caribbean.* 3(3):181-205.

Ross, S. W. and S. P. Epperly. 1984. Utilization of shallow estuarine nursery areas by fishes in Pamlico Sound, North Carolina and adjacent tributaries. *In:* A. Yáñez-Aranciba (ed.), Fish Community Ecology in Estuaries and Coastal Lagoons: Towards an Ecosystem Integration. Publ. Instituto de Ciencias del Mar y Limnologia, Mexico City.

Schwartz, F. J. and A. F. Chestnut. 1973. *Hydrographic Atlas of North Carolina Estuarine and Sound Waters, 1972.* Sea Grant Publication, UNC-SG-73-12. UNC Sea Grant College Program. North Carolina State University, Raleigh.

Turner, W. R. 1973. Estimating year-class strength of juvenile menhaden. pp. 37-43. *In:* A. L. Pacheco (ed.), *Proceedings of a Workshop on Egg, Larval and Juvenile Stages of Fish in Atlantic Coast Estuaries, June 1968.* Tech. Publ. 1, Natl. Mar. Fish. Serv. Middle Atlantic Coastal Fisheries Center, Beaufort, North Carolina.

Warlen, S. M. 1981. Age and growth of larvae and spawning time of Atlantic croaker in North Carolina. *Proc. 34th Ann. Conf., Southeastern Assoc. Fish and Wildlife Agencies* 34:204-214.

Weisberg, R. H., and L. J. Pietrafesa. 1983. Kinematics and correlation of the surface wind field in the South Atlantic Bight. *J. Geophys. Res.,* 88(C8):4593-4610.

LOW-FREQUENCY SHELF-ESTUARINE EXCHANGE PROCESSES IN MOBILE BAY AND OTHER ESTUARINE SYSTEMS ON THE NORTHERN GULF OF MEXICO

William W. Schroeder

Marine Science Program
The University of Alabama
Dauphin Island, Alabama

and

Wm. J. Wiseman, Jr.

Coastal Studies Institute
Louisiana State University
Baton Rouge, Louisiana

Abstract: Two years of data from Mobile Bay show that subtidal shelf-estuarine exchanges at periods of 2 to 4 days are driven by the cross-shelf along-estuary wind stress. At periods of 3 to 20 days, coastal Ekman convergence/divergence, driven by the alongshore wind stress, becomes important in driving the exchanges. Such a situation is common to many estuaries around the northern rim of the Gulf of Mexico. At seasonal scales, river runoff becomes important to both the barotropic and baroclinic exchanges.

Introduction

The past 15 years have witnessed an extensive growth in our understanding of subtidal frequency estuarine dynamics and shelf-estuarine exchange processes. This work has been reviewed by Carter *et al.* (1979) and Wiseman (in press). These low-frequency flows may contain both a barotropic and/or a baroclinic component. Since the long-term records necessary to study these flows are available principally from tide gauge records, attention has tended to focus on net exchange with the coastal ocean. Most of the fundamental work on this problem has been carried out on the east coast of the United States (Weisberg 1976; Elliott and Wang 1978; Wang 1979a,b). The most noteworthy contributions from the Gulf coast arise from Smith's work in Corpus Christi Bay (Smith 1977, 1978). This paper presents new data concerning subtidal exchange processes in Mobile Bay and compares those processes with low-frequency flows in numerous other estuarine situations across the northern rim of the Gulf of Mexico.

Mobile Bay (30.5°N, 88°W) is located on the northern coast of the Gulf of Mexico east of the Mississippi River Delta (Fig. 1). Geomorphologically, it is a combination of the drowned river valley and bar-built estuarine types. The Bay is

50-km long, 14-km wide in the upper, northern half and has a maximum width of 34-km in the lower, southern half (Fig. 2). It has a surface area of 984 km^2 and an average depth of approximately 3 m. Considering its shallow depths, the Bay is bathymetrically complex, with deep holes found throughout the northeastern part and a manmade island in the northwestern area (Figs. 2, 3). A dredged ship channel, 120-m wide and 12-m deep, cuts through the Bay from Main Pass to the Port of Mobile. Midway up the Bay, two shorter channels branch off to the west. There are two openings in the lower part of the Bay: Main Pass, which provides a direct link to the Gulf of Mexico, and Pass-aux-Herons, which connects with the east end of Mississippi Sound. The exchange of waters through these two passes is estimated to be 85% and 15% of the total exchange, respectively (Schroeder 1978).

The principal source of run-off into the Bay is the Mobile River system, which accounts for approximately 95% of the freshwater input and enters through five distributaries in the Mobile River Delta (Schroeder 1979). The average discharge of the system for the period 1929 to 1983 is 1848 m^3 sec^{-1}, but the annual discharge varies considerably from year to year (Fig. 4).

Subtidal Circulation in Mobile Bay

The Data

For the 2-year period 1980-81, we had available: (1) hourly water level records from the NOAA-NOS tide guage at Dauphin Island, (2) hourly water level records from the USACOE tide gauge at the Port of Mobile, (3) hourly wind records from the Dauphin Island Sea Lab, (4) 3-hourly atmospheric pressure records from the Dauphin Island Sea Lab, and (5) daily discharge records for the Mobile River system.

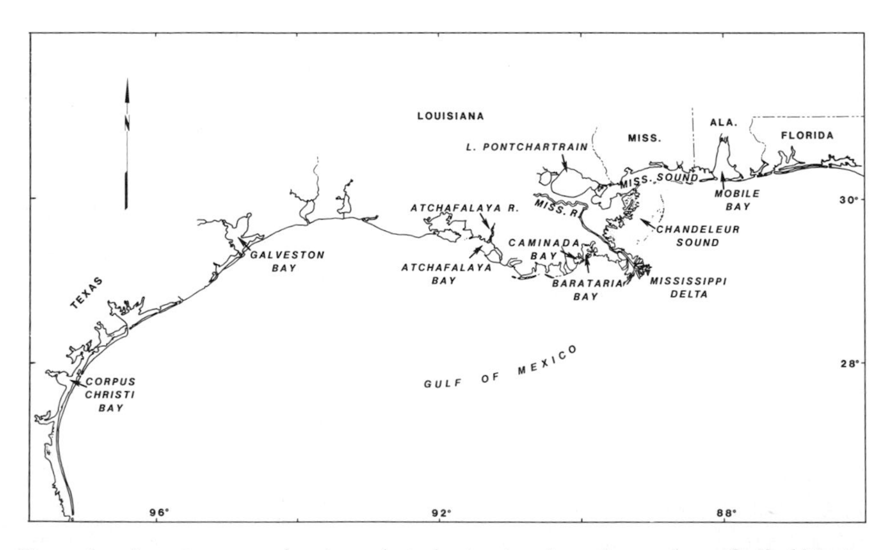

Figure 1. Location map showing selected estuaries along the northern Gulf of Mexico.

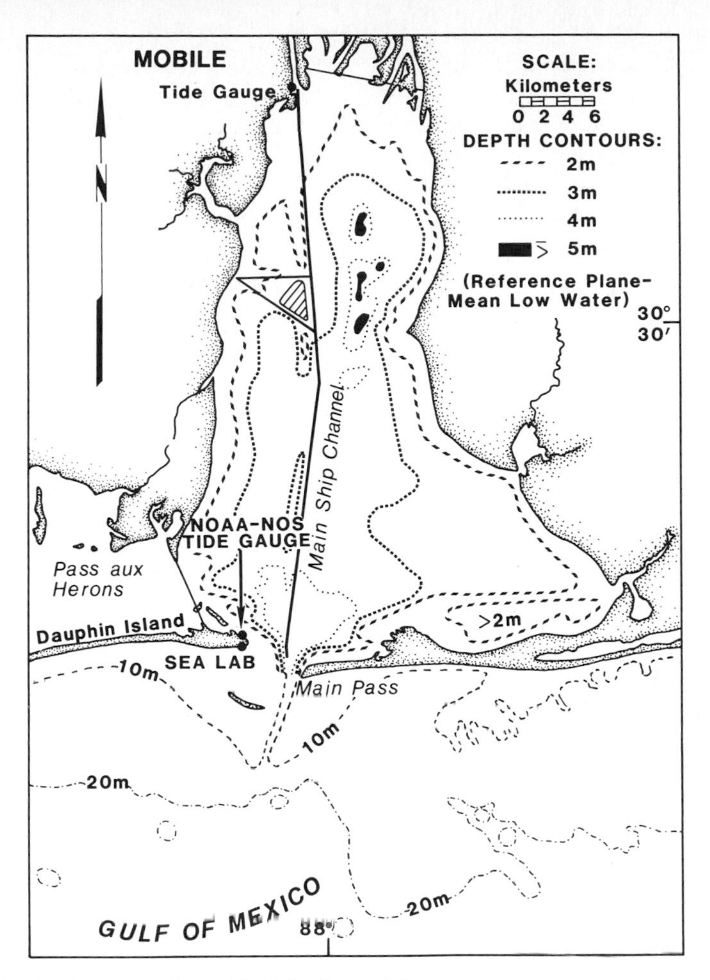

Figure 2. Bathymetric chart of Mobile Bay, Alabama.

The wind velocity, $\vec{v}$, was converted to a quantity proportional to stress, $\vec{v}|\vec{v}|$. This was resolved into north-south and east-west components. The water level, atmospheric pressure, and wind stress were then low-pass filtered with a perfect 40-hour cutoff constructed by setting high-frequency coefficients of the discrete Fourier Transform to zero and then inverse transforming (Cannon 1969). The water level records were then corrected for atmospheric pressure.

Results

Energy spectra for the 1980-81 period were computed from the periodogram (Jones 1965) by averaging over 9 adjacent spectral estimates (Blackman and Tukey 1958), thus giving 18 degrees of freedom (Fig. 5). The north-south wind stress exhibits a peak in the 0.004 to 0.01 cycles-per-hour (cph) frequency band

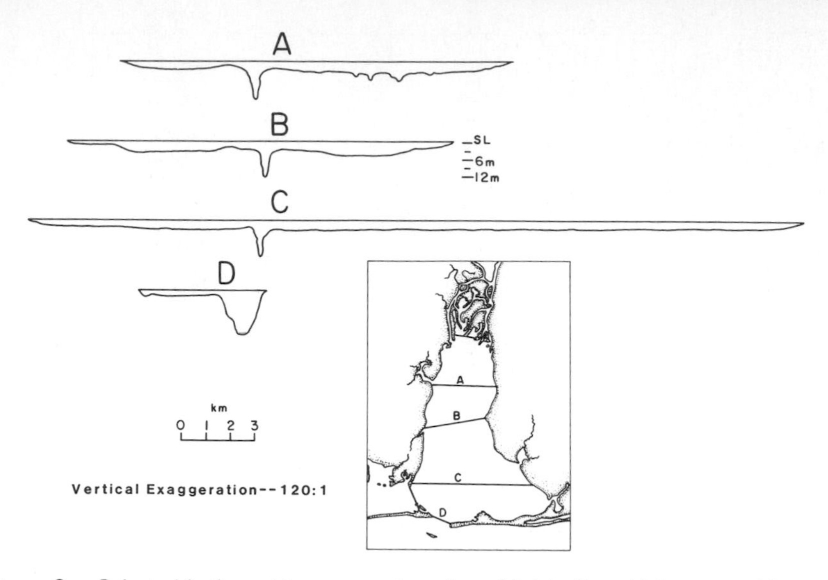

Figure 3. Selected bathymetric cross sections from Mobile Bay, Alabama, and location map (inset). Vertical exaggeration on the cross sections is 120.

(4 to 10 day periods), which is the band associated with winter front passages (Fig. 5A). The east-west wind stress spectrum is less interesting, with a relatively flat spectrum below 0.01 cph (Fig. 5B). The spectra of corrected water level at both Port of Mobile and Dauphin Island are red (Figs. 5C, D), although an energy plateau occurs at the Port of Mobile between 0.003 and 0.01 cph (4 to 13 day periods) (Fig. 5C). Water level slope, estimated as Port of Mobile water level minus Dauphin Island water level, has a generally red spectrum with an energy peak between 0.004 and 0.01 cph (Fig. 5E). Thus, the spectra suggest that strong northerly winds associated with frontal passages are particularly effective in altering water levels in the shallow upper bay and thus changing the barotropic pressure gradient driving the mean flow. The spectrum of the Mobile River system discharge is, as expected, extremely red (Fig. 5F).

Cross-spectral analyses were performed at the same frequency resolution as the spectral estimates (Fig. 6). The cross-spectra are presented as phase and coherence. The coherence is a measure, as a function of frequency, of the fraction of one signal, which is linearly predictable from the other. The phase represents, again as a function of frequency, the relative lag of one signal with respect to the other. At frequencies lower than 0.02 cph (periods longer than 2 days), water levels throughout Mobile Bay are highly coherent (Fig. 6A), with the water level at the Port of Mobile leading that at Dauphin Island except at frequencies between 0.005 and 0.0008 cph (8 to 50 days), where the phase is near zero or slightly positive. Dauphin Island water level is incoherent with north-south wind stress, while the water level at the Port of Mobile (Fig. 6B) is strongly coherent with this stress component between frequencies of 0.02 and 0.005 cph (2 to 80 days) with

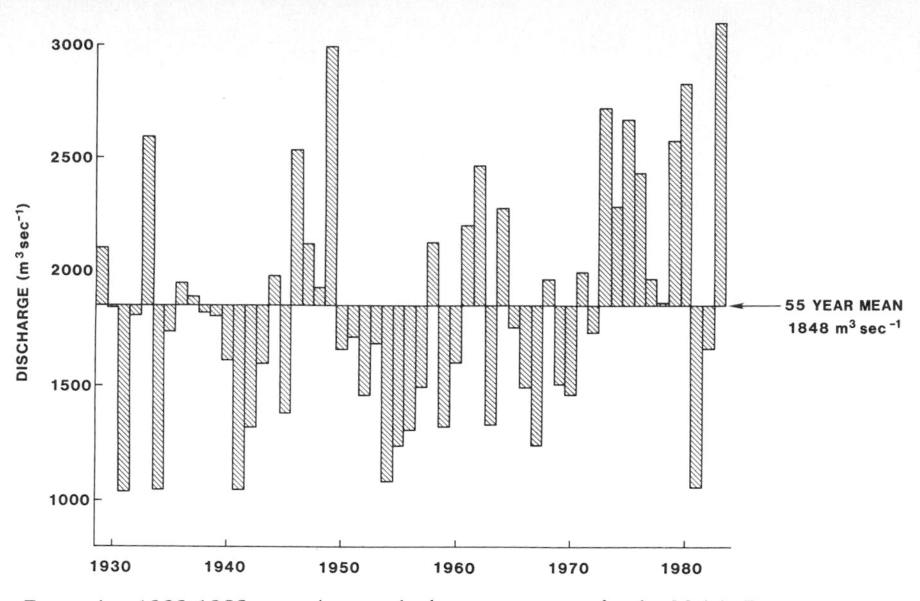

Figure 4. 1929-1983 annual mean discharge time series for the Mobile River system.

water level lagging the wind by approximately $\pi/10$ radians (18°). Both Port of Mobile and Dauphin Island water levels are coherent with east-west wind stress at frequencies lower than 0.015 cph (periods longer than 3 days) (Figs. 6C, D). In both cases the phase lag increases with frequency to $\pi/2$ at the highest frequencies. Water levels at both sites are also coherent with river discharge at very low frequencies, 0.001 to 0.0004 cph (40 to 100 days) at Dauphin Island and below 0.001 cph (periods longer than 40 days) at Port of Mobile (Fig. 6E). The phase lag corresponds to 4 to 8 days, which is in agreement with the 5 to 9 day lag period for transit time because of the distance between the gauging stations and Mobile Bay (Schroeder 1979).

Focusing attention now on the water-level slope within the Bay (root-mean-square low-passed slope $= 2 \times 10^{-6}$), the slope is coherent with (Fig. 6F) and lags the north-south wind stress across the frequency band, although the response is strongest at the higher subtidal frequencies. Slope is also coherent with river discharge at the lowest frequencies. Strong coherence is also exhibited between Port of Mobile water-level and slope, except within a narrow band centered around 0.003 cph (14 days) (Fig. 6G).

Net volume exchange with the shelf was estimated as the time derivative of the average of the water-levels at Dauphin Island and Mobile. This is justified, since water level within the Bay varies nearly in phase at subtidal frequencies. The spectrum of this volume flux (Fig. 7A) exhibits a peak near 0.01 cph (4 day periods). Flux variations within this band are coherent with the north-south wind stress and out of phase (Figs. 7C, E). They are even more coherent with the east-west wind stress with phase near $\pi/2$ (Figs. 7B, D). At frequencies between 0.005 and 0.002 cph (periods between 8 and 20 days), the weaker flux variations are

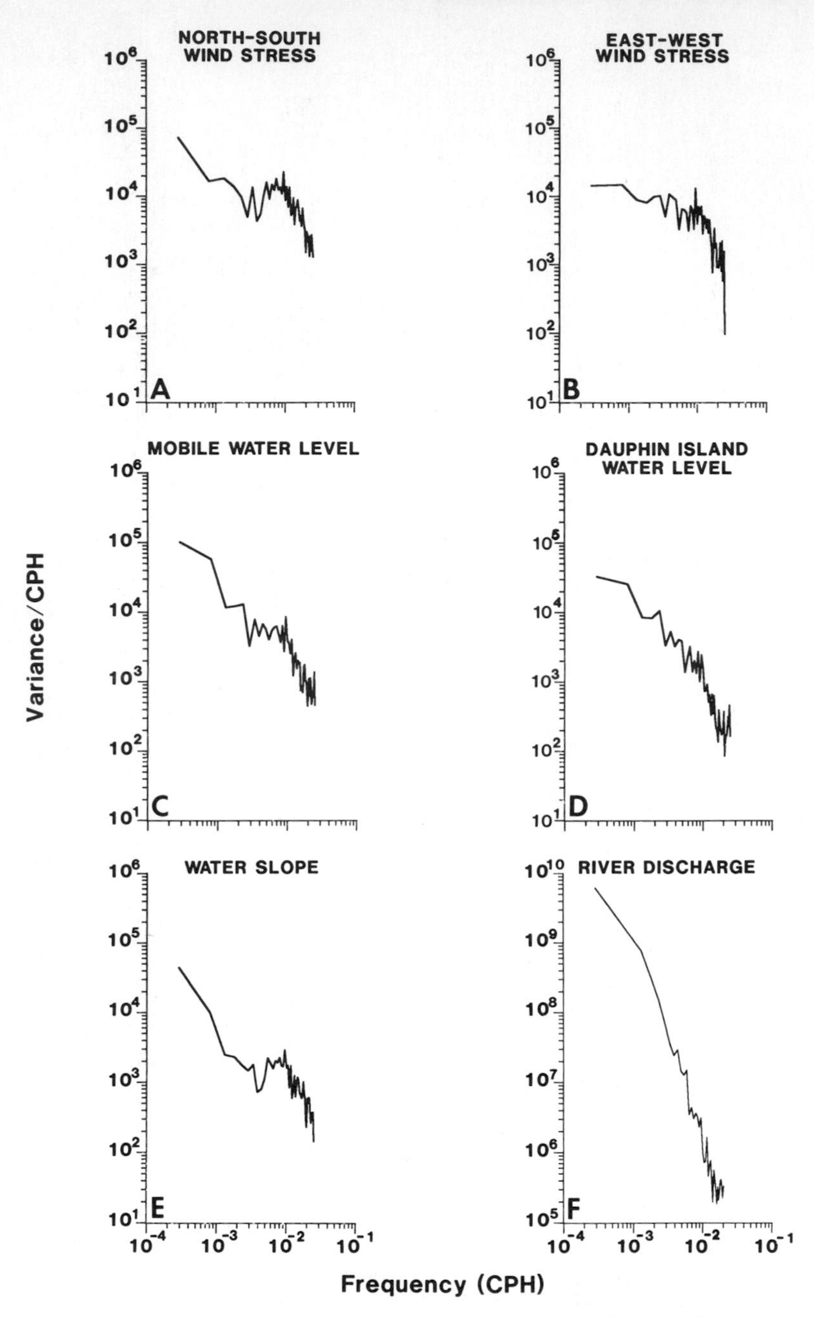

Figure 5. *Energy spectra of selected time-series. Ordinate units are cycles per hour. Abscissa units are variance units per cycle per hour. These will vary with the time-series being analyzed. A. North-south wind stress. B. East-west wind stress. C. Port of Mobile water level. D. Dauphin Island water level. E. Water slope. F. Mobile River system discharge.*

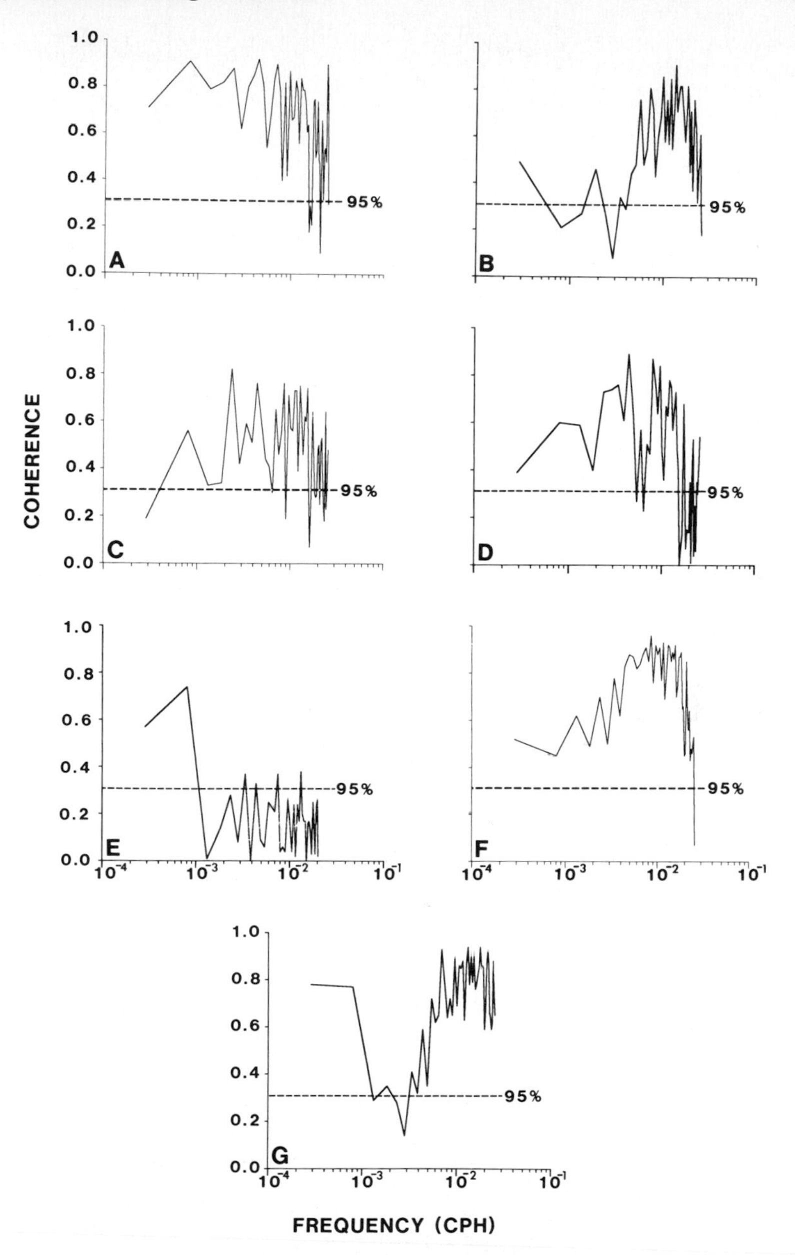

Figure 6. *Coherence estimates between selected time-series pairs. A. Port of Mobile water level versus Dauphin Island water level. B. Port of Mobile water level versus north-south wind stress. C. Port of Mobile water level versus east-west wind stress. D. Dauphin Island water level versus east-west wind stress. E. Mobile River system discharge versus Port of Mobile water level. F. Water-level slope versus north-south wind stress. G. Port of Mobile water level versus water-level slope.*

coherent with the east-west wind stress with a phase lag of slightly less than $\pi/2$ (Figs. 7B, D).

Mobile Bay is shallow and small in volume. It responds strongly to the north-south wind stress events that accompany winter frontal passages. The response is most intense in the shallow northern portion of the Bay. The energy in these fronts is greatest in the 4 to 10 day periods. Water-level slopes within the Bay develop because of the asymmetric response at the opposite ends of the Bay. The northern, upper end of the Bay is closed and shallow, while the southern, lower end of the Bay is open to the Gulf of Mexico and deeper (Fig. 2). The water-level slope is principally a local response confined to the interior of the Bay.

At periods longer than 3 days, the entire Bay responds to forcing by the east-west wind stress. The water slope, though, is not responsive to this forcing. This is a non-local Ekman effect. Ekman drift in the coastal ocean causes a sea-level rise or fall at the coast and a consequent uniform filling or draining of the Bay. At the very lowest frequencies, river discharge becomes important. Its influence is felt most strongly within the confined northern portion of the Bay because to the south the Bay widens and the runoff is transported to the Gulf of Mexico through both Main Pass and Pass-aux-Herons. Consequently, a significant barotropic pressure gradient develops within the Bay in response to the river discharge signal. Thus, the north-south wind stress drives a circulation at the higher subtidal frequencies, the east-west stress drives an intermediate-subtidal-frequency estuarine-shelf exchange, and the river discharge drives both a low-subtidal-frequency exchange with the shelf and a classical internal gravitational circulation.

Some corroboration of these interpretations is found in the hydrographic data. During 1980-81 a series of longitudinal salinity sections were collected along the eastern edge of the Mobile Bay ship channel (Marine Environmental Sciences Consortium 1980, 1981) (Fig. 8). The effects of the seasonal freshwater discharge cycle (Fig. 9) are clearly evident during cruises 3, 4, and 12 (Fig. 8). Vertical homogenization of the water column due to strong winds is often seen (Fig. 8: cruises 4, 7, 8, 9, 10).

Cruises 3 and 12 show the strong stratification and implied internal circulation resulting from spring and late winter freshets, respectively, of different strength but with weak wind conditions (Fig. 8). Cruises 6 and 11 show equally strong stratification, resulting from southward-directed wind stress events during periods of low river discharge (Fig. 8). In all these latter four cases, the stratification suggests, a strongly sheared internal circulation. Unfortunately, none of the longitudinal sections was occupied during periods of persistent moderate east-west winds.

Figure 7. Spectrum and cross-spectra of Mobile Bay volume flux with wind stress. A. Spectrum of volume flux into Mobile Bay. B. Coherence squared between east-west wind stress (positive west) and volume flux. C. Coherence squared between north-south wind stress (positive north) and volume flux. D. Phase (in radians) between east-west wind stress and volume flux. E. Phase (in radians) between north-south wind stress and volume flux.

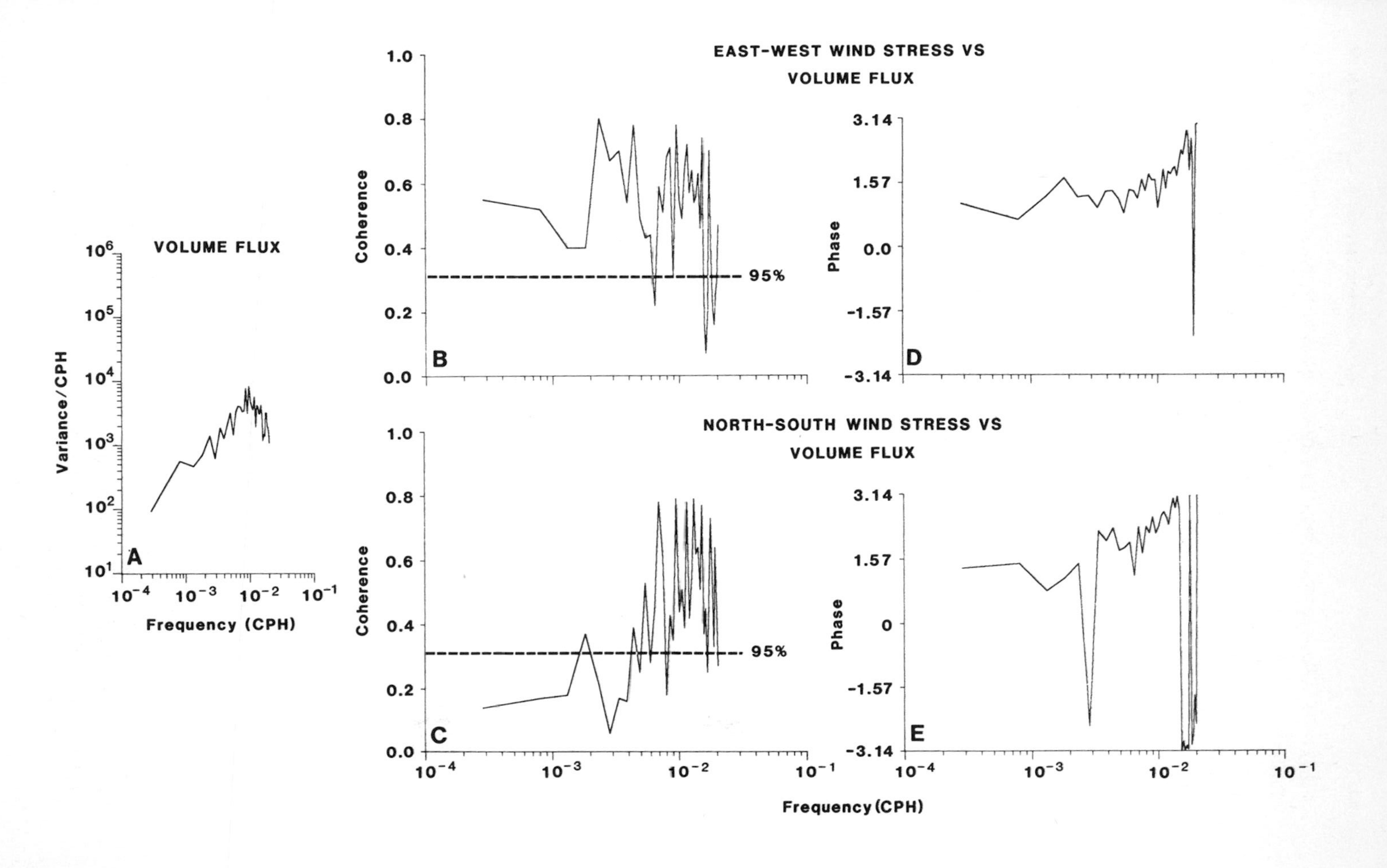
VOLUME FLUX
Variance/CPH
Frequency (CPH)
A
EAST-WEST WIND STRESS VS
VOLUME FLUX
Coherence
95%
B
Phase
D
NORTH-SOUTH WIND STRESS VS
VOLUME FLUX
Coherence
95%
C
Phase
E
Frequency (CPH)

Subtidal Exchange Processes in Other Northern Gulf of Mexico Estuaries

A number of related studies, both published and unpublished, from other estuarine areas along the northern Gulf Coast can be compared to the present results (Table 1). Coherent relationships were found between both the east-west and north-south winds and flow into Mississippi Sound at periods of 5 to 8 days (Kjerfve and Sneed 1984). These authors point out the extremely complex dynamics of this region because of the offshore topography and its location adjacent to a right-angle turn in the regional coastline. A similar conclusion with respect to Chandeleur Sound was reached by Schroeder *et al.* (1985) from inspection of LANDSAT imagery.

Chuang and Swenson (1981) have shown that water level in Lake Pontchartrain is coherent with the east-west wind at periods between 2 and 15 days. A linear model reproduced the dominant chacteristics of the observations. The authors suggested that the response is due to Ekman convergence at the coast, although other recent studies (Chuang and Wiseman 1983) implied that at the shorter periods a local frictional balance with the wind stress might be driving the flow into the estuary. Shore parallel winds were also found to control water levels within Caminada Bay at subtidal scales (Kjerfve 1975).

The bathymetry offshore of Atchafalaya Bay is anomalously shallow. Because of this, coastal water level is most responsive to cross-shore wind stress (Chuang and Wiseman 1983). These coastal water-level fluctuations drive low-frequency shelf-estuarine exchanges. Long-term current records from the mouth

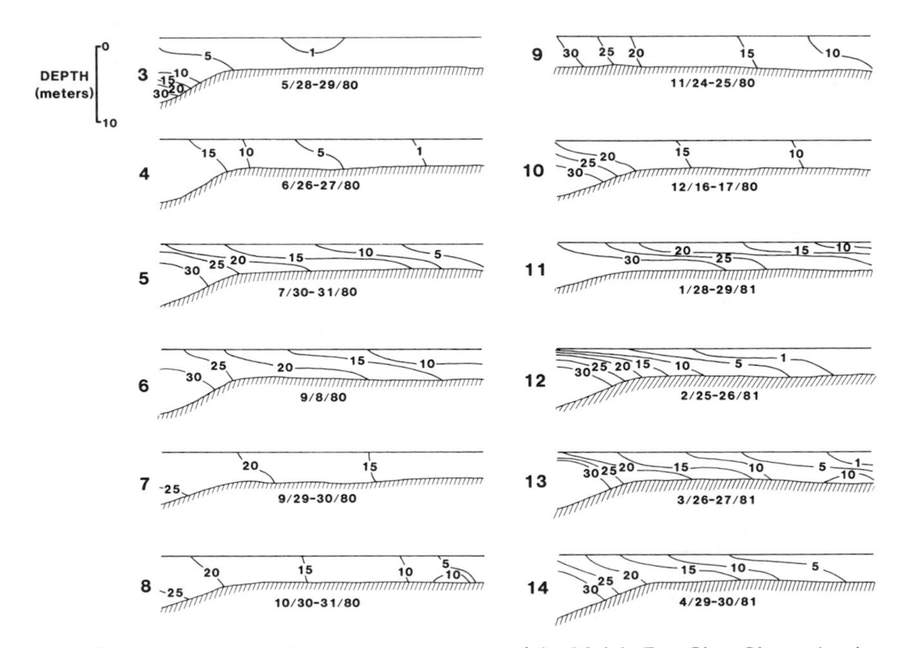

Figure 8. Longitudinal salinity sections just east of the Mobile Bay Ship Channel, taken on cruises 3-14, 1980-1981.

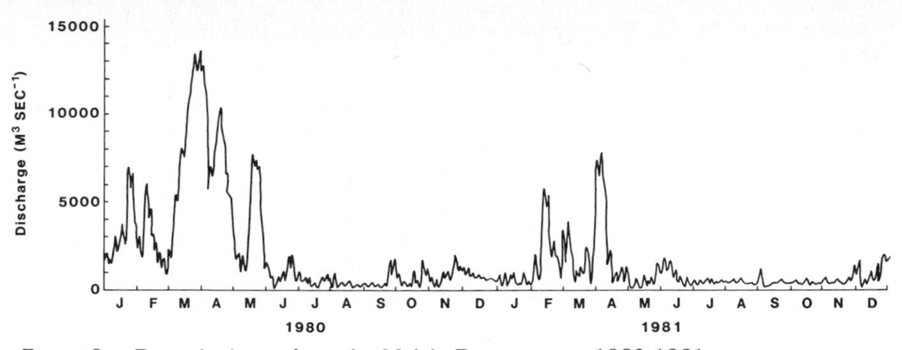

Figure 9. River discharge from the Mobile River system, 1980-1981.

of the Atchafalaya River are also strongly coherent with the cross-shore wind stress at periods from 3 to 8 days (C.Y. Wu, pers. comm.).

Farther west, at the mouth of Galveston Bay, it is the alongshore wind stress that controls coastal water level and shelf-estuarine exchange (Chuang and Wiseman 1983), while at Corpus Christi Bay, low-frequency water level fluctuations are coherent with the cross-shore wind stress at periods of 2 to 4 days and with the alongshore wind stress at somewhat longer periods (Smith 1977). Smith (1978) noted that at very low frequencies Corpus Christi Bay exchanges water with the shelf in response to a semi-annual mean water level fluctuation in the Gulf of Mexico. Because of the large ratio of bay surface area to the cross-sectional area of Aransas Pass leading into the Bay, a measurable current is produced by this exchange. The currents produced by this mechanism do not appear to be very important elsewhere in the Gulf.

Thus, across the northern Gulf of Mexico, the along-estuary wind stress drives an estuarine-shelf exchange at periods of a few days. At longer time scales, Ekman convergence/divergence driven by the alongshore wind stress controls the estuarine-shelf exchanges. Local bathymetry occasionally modifies these patterns.

Summary

Analysis of 2 years of data from Mobile Bay suggests that subtidal, nongravitational flow within the Bay and estuarine-shelf exchange are driven by three different mechanisms in adjacent frequency bands (Table 1). At the highest subtidal frequencies, cross-shore (along-estuary) winds generate a surface water-level slope within the Bay, thus driving the resultant circulation. At somewhat lower frequencies, the Bay exchanges water with the shelf as Ekman convergence/divergence at the coast is driven by alongshore winds. At seasonal scales, river discharge alters the sea-surface slope within the Bay and drives a barotropic flow as well as the more documented gravitational circulation. The suggested role of river discharge appears to be one of the few examples of the importance of this forcing mechanism. Wong and Garvine (1984) considered variations in freshwater influx during their analysis of a 40-day current record in the upper Delaware estuary. Because of the short duration of their data set, they could only conjecture on the role of riverine forcing, which is dominated by very low frequency events.

Table 1. Summary aspects of shelf-estuarine exchange: northern Gulf of Mexico

Location	Forcing	Period	Reference
Mobile Bay	Alongshore wind Cross-shore wind River discharge	>3 days 2-8 days >42 days	This study
Mississippi Sound	Alongshore wind Cross-shore wind	5-8 days 5-8 days	Kjerfve and Sneed (1984)
Lake Pontchartrain	Alongshore wind	2-15 days	Chuang and Swenson (1981)
Caminada Bay	Alongshore wind	"subtidal"	Kjerfve (1975)
Atchafalaya Bay	Cross-shore wind	>2 days 3-8 days	Chuang and Wiseman (1983) Wu, C-Y (pers. comm.)
Galveston Bay	Alongshore wind	3-30 days	Chuang and Wiseman (1983)
Corpus Christi Bay	Alongshore wind Cross-shore wind Mean sea level	>4 days 2-4 days Semi-annual	Smith (1977) Smith (1978)

Elsewhere along the northern Gulf Coast, estuarine flows also respond to subtidal wind-stress variations (Table 1). At the highest subtidal frequencies or in the shallowest waters, the cross-shore winds frictionally drive the flow. At lower subtidal frequencies or where offshore depths are great, the alongshore wind stress forces an Ekman convergence/divergence at the coast, which drives the low-frequency estuarine flows.

Acknowledgments

Funding for this work was provided by the University of Alabama Marine Sciences Program, the Dauphin Island Sea Lab, the Mississippi-Alabama Sea Grant Consortium, and the Louisiana Universities Sea Grant Program, parts of the National Oceanic and Atmospheric Administration of the U.S. Department of Commerce. The Port of Mobile tide data were provided by Mr. Geary McDonald, Sr., Mobile District, Corps of Engineers. The figures were prepared by Ms. C. Harrod. This publication is referenced as Contribution No. 86 from the Aquatic Biology Program, University of Alabama, and Contribution No. 85 from the Marine Environmental Sciences Consortium, Dauphin Island, Alabama.

References Cited

Blackman, R. B. and J. W. Tukey. 1958. *The Measurement of Power Spectra.* Dover Publ. Inc., New York, 190 pp.

Cannon, G. A. 1969. Observations of motion at intermediate and large scales in a coastal plain estuary. *Chesapeake Bay Institute Technical Report 52.* The Johns Hopkins University, Baltimore, Md., 114 pp.

Carter, H. H., T. O. Najarian, D. W. Pritchard and R. E. Wilson. 1979. The dynamics of motion in estuaries and other water bodies. *Rev. Geophys. Space Phys.* 17:1585-1590.

Chuang, W.-S. and E. M. Swenson. 1981. Subtidal water level variations in Lake Pontchartrain, Louisiana. *J. Geophys. Res.* 86:4198-4204.

Chuang, W.-S. W. J. Wiseman, Jr. 1983. Coastal sea level response to frontal passages on the Louisiana-Texas shelf. *J. Geophys. Res.* 88:2615-2620.

Elliott, A. J. and D.-P. Wang. 1978. The effect of meteorological forcing on the Chesapeake Bay: The coupling between an estuarine system and its adjacent coastal waters, pp. 127-145. *In:* J. C. J. Nihoul (ed.), *Hydrodynamics of Estuaries and Fjords.* Elsevier Scientific Publ. Co., Amsterdam.

Jones, R. H. 1965. A reappraisal of the periodogram in spectral analysis. *Technometrics* 7:531-542.

Kjerfve, B. 1975. Tide and fair-weather wind effects in a bar-built Louisiana estuary, pp. 47-62. *In:* L. E. Cronin (ed.), *Estuarine Research,* Vol. II. Academic Press, New York.

Kjerfve, B. and J. E. Sneed. 1984. Synthesis of oceanographic conditions in the Mississippi Sound offshore region. *Final Report: Volume I, Department of the Army, Mobile District, Corps of Engineers,* Mobile, Alabama, 253 pp.

Marine Environmental Sciences Consortium. 1980. Biological baseline studies of Mobile Bay: May through December, 1980. *Dauphin Island Sea Lab Interim Technical Reports III to X,* Dauphin Island, Alabama, 1480 pp.

Marine Environmental Sciences Consortium. 1981. Biological baseline studies of Mobile Bay: January through March, 1981. *Dauphin Island Sea Lab Interim Technical Reports XI to XIII,* Dauphin Island, Alabama, 572 pp.

Schroeder, W. W. 1978. Riverine influence on estuaries: A case study, pp. 347-364. *In:* M. L. Wiley (ed.), *Estuarine Interactions.* Academic Press, New York.

Schroeder, W. W. 1979. The dispersion and impact of Mobile River system waters in Mobile Bay, Alabama. *WRRI Bull. 37,* Water Resources Research Institute, Auburn University, Auburn, Alabama, 48 pp.

Schroeder, W. W., O. K. Huh, L. J. Rouse, Jr., and W. J. Wiseman, Jr. 1985. Satellite observations of the circulation east of the Mississippi Delta: Cold-air outbreak conditions. *Remote Sensing Environ.* 18:49-58.

Smith, N. P. 1977. Meteorological and tidal exchanges between Corpus Christi Bay, Texas, and the Northwestern Gulf of Mexico. *Estuar. Coastal Mar. Sci.* 5:511-520.

Smith, N. P. 1978. Long-period, estuarine-shelf exchanges in response to meteorological forcing, pp. 147-159. *In:* J. C. J. Nihoul (ed.), *Hydrodynamics of Estuaries and Fjords.* Elsevier Scientific Publ. Co., Amsterdam.

Wang, D.-P. 1979a. Subtidal sea level variations in the Chesapeake Bay and relations to atmospheric forcing. *J. Phys. Oceanogr.* 9:413-421.

Wang, D.-P. 1979b. Wind-driven circulation in the Chesapeake Bay, winter 1975. *J. Phys. Oceanogr.* 9:564-572.

Weisberg, R. H. 1976. The nontidal flow in the Providence River of Narragansett Bay: A stochastic approach to estuarine circulation. *J. Phys. Oceanogr.* 6:721-734.

Wiseman, Wm. J., Jr. In press. Estuarine-shelf interactions. *In:* C. N. K. Mooers (ed.), *Baroclinic Processes on Continental Shelves, Coastal and Estuarine Sciences, Vol. 3.* Amer. Geophys. Union, Washington, D.C.

Wong, K.-C. and R. W. Garvine. 1984. Observations of wind-induced subtidal variability in the Delaware Estuary. *J. Geophys. Res.* 89:10589-10597.

PHYTOPLANKTON SPATIAL DISTRIBUTION IN SOUTH SAN FRANCISCO BAY: MESOSCALE AND SMALL-SCALE VARIABILITY

Thomas M. Powell

Division of Environmental Studies
University of California at Davis
Davis, California

James E. Cloern

U.S. Geological Survey
Menlo Park, California

and

Roy A. Walters

U.S. Geological Survey
Tacoma, Washington

Abstract: Horizontal transects of surface salinity and *in-vivo* fluorescence indicate the existence of three distinct spatial regimes in South San Francisco Bay. A mid-Bay region of low phytoplankton biomass with little small-scale variance is bounded to the north and south by water masses having higher *in-vivo* fluorescence and enhanced small-scale variability. Autocorrelation analyses demonstrate that the length scale of phytoplankton patchiness is longest in the mid-Bay region. The persistent discontinuities of *in-vivo* fluorescence and salinity are associated with topographic features—a large shoal to the north and a constriction to the south. The three spatial regimes are consistent with measured zooplankton distributions, existing current meter data, estimated longitudinal transports, and numerical simulations of residual circulations that show one (and perhaps two) large-scale gyre(s) bounded by the northern shoal and southern constriction. Topographic features are the most important physical factors controlling mesoscale (~ 10 km) variability of phytoplankton in South San Francisco Bay. We speculate that vertical current shear and salinity stratification (and their effects upon turbulence and diffusion) control small-scale patchiness, but quantitative estimates are needed to determine the influence of large-scale (and local) phytoplankton growth and loss processes.

Introduction

San Francisco Bay has been the site of considerable basic research into estuarine processes and properties during the past decade (Conomos 1979; Cloern and Nichols 1985). One reason for the scientific interest is that the Bay comprises two different estuary types in a single body of water. The North Bay is a river-dominated estuary while the South Bay is a lagoon-type estuary with no large riverine inflow. Phytoplankton population dynamics and productivity have been studied in the South Bay nearly continuously since 1978 (Cloern *et al.* 1985). These investigations have focused on the temporal dynamics of mean properties (e.g., chlorophyll-*a* concentration or primary productivity) averaged over large geographic sub-regions. Although these studies have described variability of average properties on a crude spatial scale, the problems of (1) characterizing and (2) defining mechanisms of spatial variability have not been addressed previously. These problems motivated the study of the South Bay described here.

South San Francisco Bay is a shallow embayment with a seaward connection through a deep basin (Central Bay) inside the Golden Gate (Fig. 1). This basin is a mixing zone for waters originating from the nearshore coastal ocean, the North Bay (San Pablo and Suisun Bays, Fig. 1), and the South Bay. The southern portion of the Bay receives inputs of treated sewage effluent and freshwater runoff through small local tributaries during the winter-spring wet season (Conomos 1979). Bathymetry of the South Bay is characterized by broad lateral shoals (e.g., note the position of the 2-m isobath in Fig. 1) incised by a channel about 10-20 m deep; a transverse shoal (San Bruno shoal, Fig. 1) reduces water depth in the channel to about 7-8 m. Phytoplankton dynamics are characterized by an annual cycle with an intense spring bloom, beginning in March or April and lasting for about one month, followed by persistently low biomass the remainder of the year. During the bloom, biomass can approach 40-50 $\mu g\ l^{-1}$ chlorophyll-*a* and daily primary productivity can exceed 2 $g\ C\ m^{-2}\ d^{-1}$, in contrast to biomass and productivity of 2-5 $\mu g\ l^{-1}$ chlorophyll *a* and $< 0.5\ g\ C\ m^{-2}\ d^{-1}$ during other times of the year (Cloern *et al.* 1985). Salinity stratification plays a critical role in triggering the spring bloom (Cloern 1984); the onset coincides with a neap tide when stratification is strong. Although the spring bloom is a baywide event, it is not clear whether it begins in the channel following stratification of the deep water column there, or in the shoals. Chlorophyll concentrations in these lateral shoals, however, are always greater than or equal to those in the channel.

Physical processes in San Francisco Bay have also been investigated in recent years (Conomos 1979; Walters *et al.* 1985). The dominant flows are, of course, tidal with primarily semi-diurnal periods (Cheng and Gartner 1985). Whereas the tidal flows dominate the instantaneous current speed, low frequency variations in currents are important to the long-term distribution of solutes and suspended constituents (Cheng and Casulli 1982; Walters 1982; Walters and Gartner 1985; Walters *et al.* 1985). These latter include tidally-driven and density-driven residual currents, and wind-driven flows. For example, South Bay current records indicate eastward, tidally-driven residual currents along the northern edge of the San Bruno shoal, with northward flows along the eastern boundary of the main channel (Walters *et al.* 1985). The overall picture suggests: a clockwise gyre in the entrance to South Bay north of the San Bruno shoal and a counterclockwise rotating flow in and adjacent to the main channel south of the San Bruno shoal. Model studies (Walters and Cheng 1980; Cheng and Casulli 1982) also suggest a larger scale counterclockwise gyre in the central portion of the South Bay (between the San Bruno shoal and the San Mateo Bridge), but direct current meter evidence is not available to confirm this assertion. Residence times for the South Bay (estimated by either the replacement

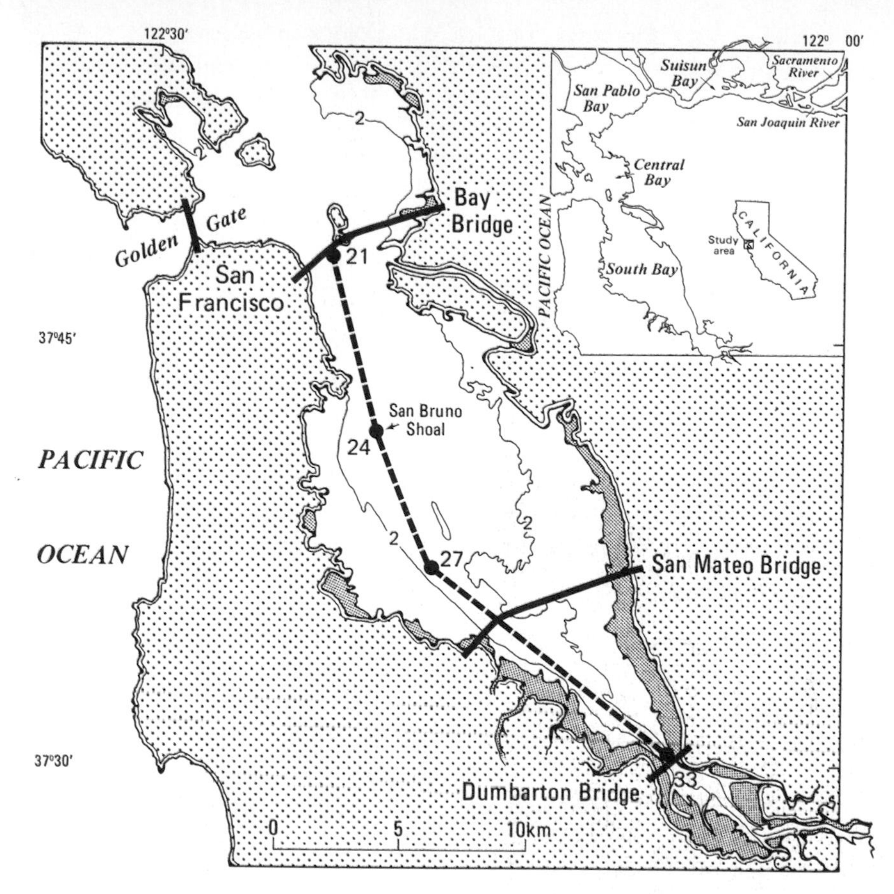

Figure 1. *South San Francisco Bay, California, showing the sampling transect between the Bay Bridge and Dumbarton Bridge. The 2-m isobath and position of the transverse San Bruno shoal are also shown.*

time for the freshwater fraction or the hydraulic flushing time) are on the order of months. This long residence time may be related to the suggested gyre circulation that could retain fluid in the central basin. During brief episodes in spring, freshwater inflow from the North Bay alters density stratification sufficiently to drive a gravitational circulation. This enhances mixing between the South Bay and the Central Bay, reducing South Bay residence times to less than a month (Walters *et al.* 1985). Local mixing in the South Bay is most intense around the San Bruno shoal.

From the above sketches of past studies we conclude that horizontal variability is an important aspect of the estuarine environment of South San Francisco Bay. For example, lateral shoal areas differ from the main channel: phytoplankton biomass, turbidity, and net primary productivity are higher over the shoals, while nutrient concentrations and tidal current speed are lower there (Cloern and Nichols 1985). Moreover, physical and biological processes in the central basin of the South Bay may differ from those north of the San Bruno shoal or south of the San Mateo Bridge, because of the enhanced residence time and relatively high local mixing there. We are thus led to ask two questions about San Francisco Bay. First, what are the characteristic spatial scales over which physical and biological processes vary? Second, are the scales of physical variability related to the scales of biological variability, and, if so, how? Here we limit ourselves to the study of the South Bay and consider only variability along the main axis of the estuary—approximately north-south along the main channel. We are particularly interested in defining and characterizing features of spatial variability that persist over the annual cycle. This precludes study of spatial changes that may occur over shorter time scales, such as day-to-day, as studied by Wilson *et al.* (1979) and Wilson and Okubo (1980) in Long Island Sound. Accordingly, we defer study of the rapid spatial changes that occur during the short period of the spring bloom.

Recent reviews (Denman and Powell 1984; Mackas *et al.* 1985) examined the coupling between biological and physical processes in a number of aquatic environments, although estuaries were not emphasized. The two works explored the idea that variability occurring at a given spatial scale in, for example, physical processes, might be expected to lead to variability in biological processes at the same scale. This rationale motivated the present study. Many investigators have looked at the general question of the scales of spatial variability in estuaries (Platt *et al.* 1970; Platt 1972; Lekan and Wilson 1978; Lewis and Platt 1982), and several have considered how physical processes acting at a particular spatial scale affect estuarine phytoplankton. For example, Bowman and Esaias (1978), Bowman *et al.* (1981), and Tyler and Seliger (1978, 1981) explored the impacts of fronts and gravitational circulation on the distribution and life-histories of phytoplankton. The present study is the first to explore the progression of spatial variability in an estuary over an annual cycle, though a similar study has been done in a large lake (Abbott *et al.* 1982).

Sampling Methods

From 27 March 1984 to 16 April 1985, 19 transects were made along the main channel in South San Francisco Bay between the Dumbarton Bridge and the Bay Bridge (Fig. 1). The sampling dates were chosen to include anticipated periods of high and low phytoplankton biomass, high and low density stratification [from an empirical model of Cloern (1984)], and high and low tidal currents (i.e., neap-spring variation), thus ensuring a broad spectrum of physical and biological conditions. Water was pumped continuously from a subsurface port at 2-m depth to a fluorometer (Turner Designs Model 10) and an induction salinometer (Schemel and Dedini 1979) while the ship was underway. Fluorescence, salinity, and water depth were measured every second and mean values recorded every five seconds on a digital-data-acquisition system. The five-second sampling frequency corresponds to a distance of approximately 18 m, based on the average speed of the vessel.

Four to eight water samples were collected during each transect. From these we measured salinity [using the method of Lewis (1980) to calibrate the salinometer], and chlorophyll-*a* [using the spectrophotometric technique of Lorenzen (1967) to calibrate the fluorometer]. The regression between fluorescence and chlorophyll had an overall r^2 of 0.84. We used separate regressions from each cruise to calculate chlorophyll-*a* from *in-vivo* fluorescence at each sample point.

Results

Spatial variability occurs on all scales in aquatic systems. The first part of our presentation focuses on the large scale, where quantities vary only over distances approximating the size of the estuary; the second concentrates on much smaller patterns.

Large-Scale Variability

Figure 2 presents a plot of the fluorescence and salinity taken during one of the transects—27 March 1984. Figures 3A-D show chlorophyll data from four representative transects in 1984 (including 27 March). The two traces of Fig. 3A, B (March and April) represent chlorophyll distributions during the modest spring bloom of 1984, and the two of Fig. 3C, D (May and October) are generally representative of traces in the remainder of the year. Qualitative examination of Fig. 2 shows two breaks—positions along the transect where the character of the record changes significantly—at about 17 and 28 km. Below 17 km, the small-scale variability in fluorescence was high; beyond 17 km the record became smoother. Below 28 km the salinity trace was very smooth, but

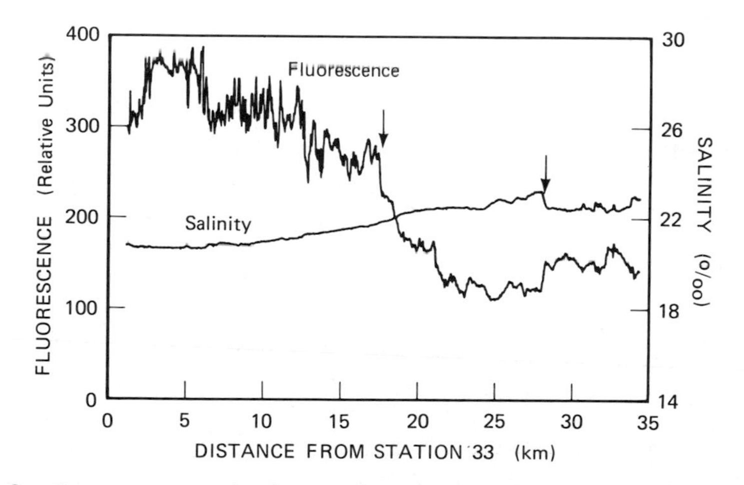

Figure 2. Continuous records of in-vivo *fluorescence (relative units) and salinity along the transect from Dumbarton Bridge (station 33; Fig. 1) to Bay Bridge, on 27 March 1974. Arrows denote breaks in the qualitative nature of the fluorescence trace (~17 km) and the discontinuity in the salinity trace (~28 km).*

beyond the prominent discontinuity at 28 km small-scale salinity features were more evident. Fluorescence exhibited a parallel discontinuity at 28 km, and salinity and fluorescence appeared to be inversely correlated beyond that discontinuity. In all records we were able to choose two similar break points: one in the south near 15 km on the basis of qualitative changes in the nature of the fluorescence record; and one in the north near 27 km from discontinuities in the salinity trace. Moreover, the segment of the fluorescence record between these breaks contained the fluorescence minimum of South Bay, except during spring bloom periods.

Chlorophyll data derived from the fluorescence records of all 19 transects exhibited the general characteristics of the fluorescence trace of Fig. 2. Four representative samples are shown in Figs. 3A-D.

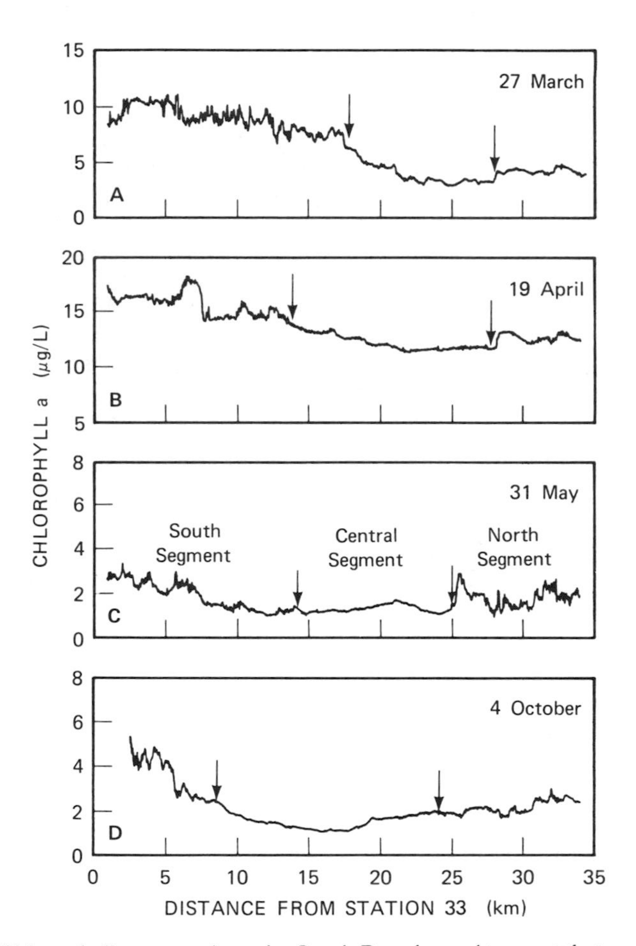

Figure 3. Chlorophyll-a traces along the South Bay channel transect that are representative of bloom (A, B) and non-bloom (C, D) periods in 1984. Arrows show the locations of break points identified from qualitative changes in the corresponding fluorescence/salinity traces.

The southern segment of the traces (approximately south of the San Mateo Bridge, at 15 km) was characterized by high variability in chlorophyll, that is the southern portions of the records were more "spiky" and patchy than the portions to the north (Fig. 3). The central segment of the traces was characterized by low variability in both chlorophyll and salinity. The northern segment (north of the San Bruno shoal at 28 km) was characterized by higher variability in salinity and chlorophyll than found in the central segment; the average levels of salinity and chlorophyll were also higher than those measured in the central segment.

Although the transects were made randomly with respect to the semidiurnal tidal cycle, and the tidal excursion along the channel axis is about 10 km (Walters *et al.* 1985), these breaks appeared consistently near 15 km and 27 km. These boundaries correspond to topographic features in the bathymetry and morphometry of the South Bay (Fig. 1): the boundary at 15 km corresponds to the narrowing seen where the San Mateo Bridge (with its closely spaced pilings, and shallow depth) crosses the bay, and the break at 27 km corresponds to the northern limit of the San Bruno shoal.

An important exception to the general picture sketched above can occur during the spring bloom period. A transect was made on 21 March 1985 (Fig. 4), at the initiation of the 1985 spring bloom. Peak values of chlorophyll (40 $\mu g\ l^{-1}$) occurred one week after that—on 29 March 1985. A chlorophyll *maximum* occurred in the central segment on this transect, not the typical minimum.

Small-Scale Variability

The difference in variability at small scales has been one of the signatures of the three larger regimes we have identified in South San Francisco Bay. For example, we noted that the southern segment had a higher level of chlorophyll variation than the central segment. We will now describe this small-scale variability focusing

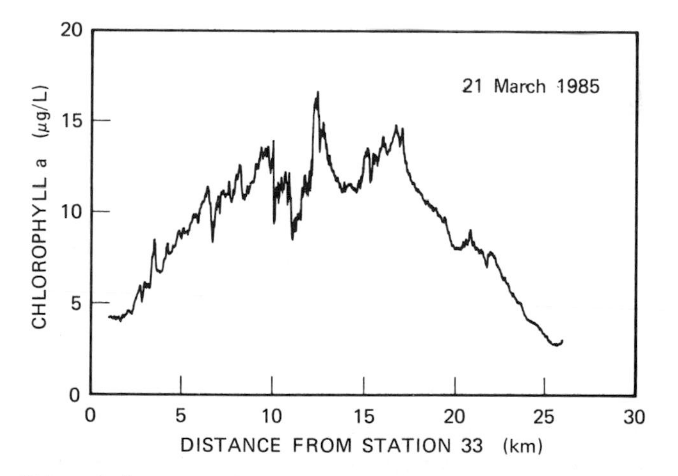

Figure 4 *Chlorophyll-*a *traces along the South Bay Channel (same as Fig. 3, but for spring bloom conditions during 1985).*

on the southern and central segments. The northern segment of South Bay was characterized by highly variable salinity, presumably resulting from the tidal advection of water into South Bay from the mixing basin north of the Bay Bridge. We believe much of the spatial variability here results from the incomplete mixing of coastal oceanic water and estuarine waters derived from both the North Bay and South Bay. Because spatial variability in this northern segment may be dominated by circulation and mixing of oceanic and estuarine waters—processes that are complex and not well understood in South San Francisco Bay—we focus our analysis of small-scale variability on the two southern segments which are more representative of resident South Bay water. Moreover, our analysis is restricted to spatial variability of fluorescence, although we recognize the critical need for parallel studies of small-scale salinity variation.

Analysis of small-scale fluctuations focuses on residuals about trends identified as large-scale variability. Figure 5A shows the variability in the fluorescence trace from one representative transect segment (in this case, the southern seg-

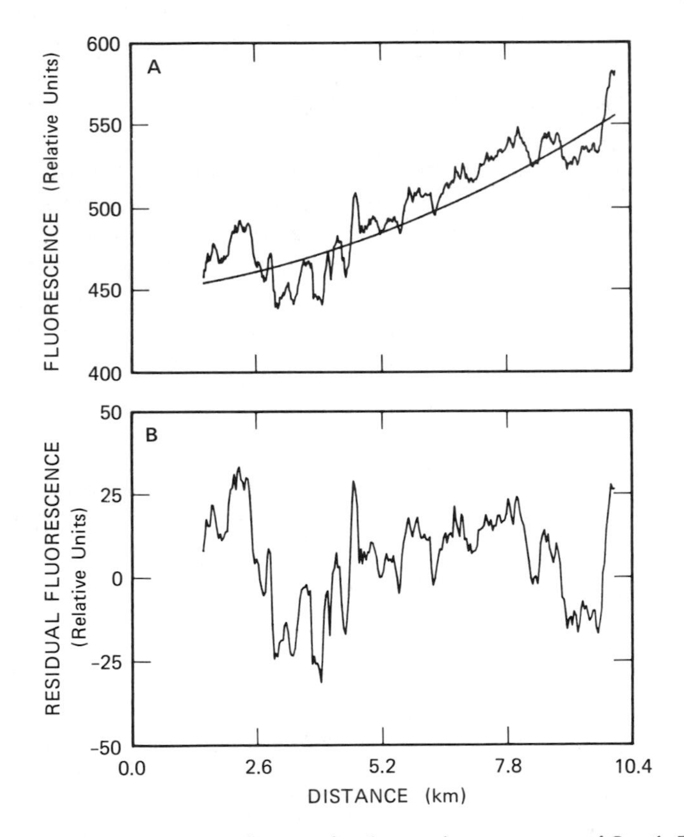

Figure 5. (A) Fluorescence vs. distance for the southern segment of South Bay on 29 March 1985 (ship direction was north to south). Also shown is the quadratic function fitted to the fluorescence. (B) Residuals obtained from the data of (A) by subtracting the fitted quadratic equation from the raw fluorescence data.

ment on 29 March 1985). Also shown is a least squares quadratic fitted through the points. For all but three sampling dates, similar quadratic curves fit the large-scale variation in both the southern and central segments (the coefficient of determination r^2, averaged over the entire set of transects was ~0.73). Residuals in fluorescence were calculated about these least square lines to examine the statistical characteristics of the small-scale fluctuations. The fitting of the quadratic was a simple way of filtering—a high pass filter in this case (Kendall 1976). Figure 5B shows the residuals for the record of Fig. 5A with the large-scale trend removed.

The standard error of the residuals in the southern segment for all 19 transects was consistently larger than the standard error in the central segment. The standard error (SE) = $[(\text{variance of the residuals in fluorescence})/(\text{mean fluorescence})^2]^{1/2}$. The ratio of standard errors, $SE_{south}/SE_{central}$ averaged over all 19 transects, was 2.2 ± 0.3 (Fig. 6). This demonstrates quantitatively that the small-scale variability of chlorophyll in the southern segment was higher than that in the central segment, as we saw earlier (Fig. 3). This robust result was violated during the 1985 spring bloom (Fig. 6), again pointing out how anomalous that period is.

The autocorrelation function (ACF) gives further quantitative information about the "patchy" structure of the small-scale variations. The autocorrelation function, the lagged correlation of the spatial record with itself, is a function of lag distance (Kendall 1976; Chatfield 1984). As an example, Fig. 7 shows the autocorrelation function of the residuals from the southern segment on 29 March 1985 (Fig. 5B). The initial exponential decay as a function of lag distance is apparent. For small lag distances, ℓ, all autocorrelation functions (from all transects in both the southern and the central segments) could be fitted by the form

$$\text{ACF}(\ell) = e^{-\ell/\lambda} \tag{1}$$

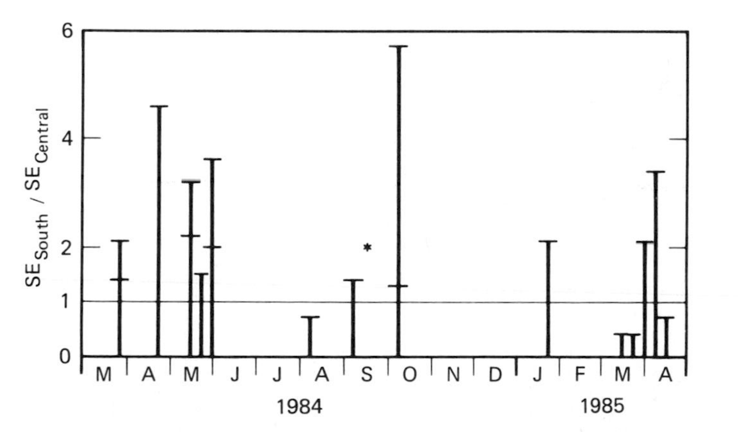

Figure 6. Ratio of the standard error of residuals (e.g., Fig. 5B) for the southern segment to that calculated for the central segment of South San Francisco Bay. Replicate transects were made on four dates.

The average r^2 for all exponential fits was 0.97. Here λ is the distance where the ACF falls to e^{-1} of its value at zero lag (λ is called the "e-folding length"). We interpret λ as the patch length in an environment in which there is variability at many overlapping (even continuously overlapping) scales. This definition of patch "size" has appeal because it encompassses our intuitive picture of discrete patches. Note that for a tow through a large patch, high phytoplankton biomass would be positively correlated with lagged, but still high, phytoplankton abundances for large distances (i.e., many lags) and therefore lead to large value of λ. The converse would be true for small patches.

The average patch size (λ) in the central segment was 30 lags (= 0.54 km), larger than the average patch size in the southern segment, 21 lags (= 0.38 km; Fig. 8). Moreover, λ was more than twice as variable in the southern segment than in the central segment: the coefficient of variation for λ in the southern segment was 50%, while in the central segment it was 24%. The impact that spring bloom conditions have upon patch size is unclear. For example, the smallest patches found for all 19 transects were obtained during the early stages of the spring bloom in both 1984 and 1985 in the southern segment—approximately 7 lags (= 0.13 km) on 27 March 1984 and 9 lags (approximately 0.17 km) on 14 and 21 March 1985. However, in the central segment during the 1985 spring bloom (Fig. 4), the patch size remained between 27 and 30 lags (= 0.49 to 0.54 km), statistically indistinguishable from the annual average of 30 lags (Fig. 8B).

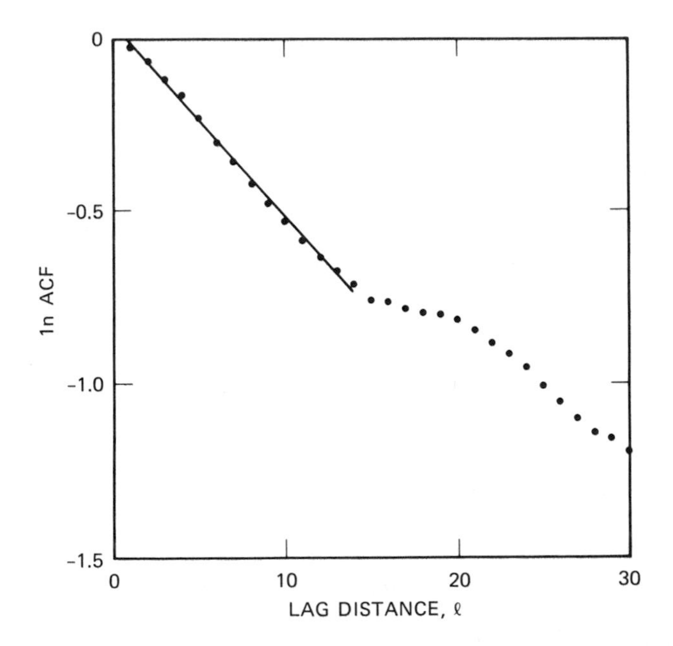

Figure 7. Autocorrelation function (ACF) of the residuals vs. lag distance, for fluorescence data collected in the southern segment on 29 March 1985 (Fig. 5B).

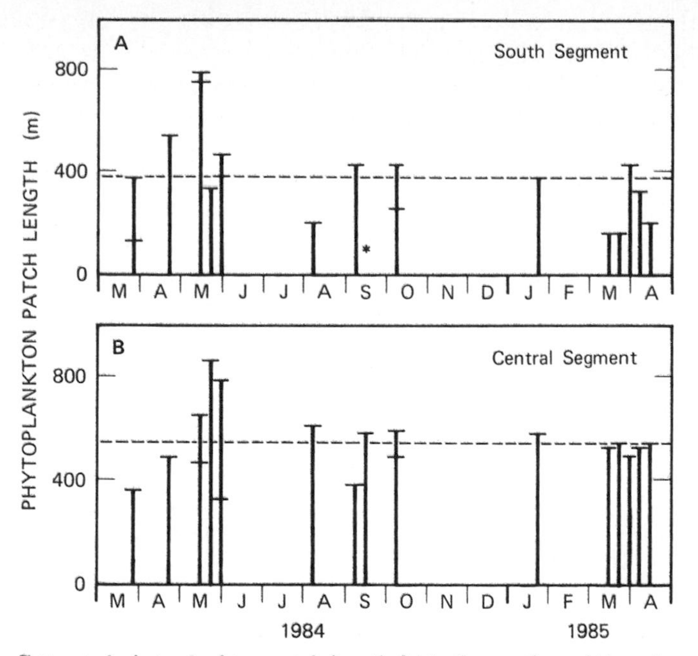

Figure 8. *Estimated phytoplankton patch length (λ) in the southern (A) and central (B) segment of South San Francisco Bay during 1984-85. The dashed line shows the mean value of λ in each segment when all transects are averaged.*

Discussion

Our analysis of chlorophyll and salinity variability has partially answered the questions posed earlier. First, we have identified two important spatial scales of horizontal variability in the South Bay channel: (1) mesoscale (~10 km) variability which suggests that the South Bay is partitioned into three distinct spatial regimes, and (2) small-scale (~0.3-0.5 km) variability that differs among these spatial regimes. Second, the role of physical processes in determining the mesoscale variability of both chlorophyll and salinity seems plausible, if not proven. In particular, the inferred boundaries between spatial regimes coincide with major topographic features (a shoal to the north and channel constriction to the south) that can induce variability in physical processes. Examples of such processes might be: (1) enhanced vertical mixing caused by the San Bruno shoal; (2) the influence of basin morphometry on residual circulations in the South Bay, including the inferred gyre(s) in the central segment; and (3) the prolonged residence time in the central segment, presumably the result of the interaction between vertical and horizontal transports there. The persistently smooth distributions of salinity and chlorophyll in the central segment could be a direct consequence of the enhanced vertical mixing generated by the San Bruno shoal.

Our picture of three large-scale regimes is supported by additional information. Ambler *et al.* (1985) examined zooplankton distributions in San Francisco

Bay and found a distinct boundary at the San Bruno shoal between a northern zooplankton community that included neritic species (e.g., *Paracalanus parvus*) and a community to the south dominated by estuarine *Acartia* species. They also showed a persistent temperature discontinuity at the San Bruno shoal with cooler temperatures to the north. These discontinuities support our conception of a northern regime with a close coupling to the coastal ocean and a southern regime with its own particular estuarine character, largely derived from the vertical mixing and residual circulations found therein.

That boundaries between large-scale regimes are associated with topographic features is a familiar result from other bodies of water. For example, a model of residual circulation in the North Sea showed several large closed eddies substantially controlled by topography (Nihoul and Ronday 1975). Nihoul (1975) further demonstrated the importance of such circulations for biological and sedimentary processes with data from a gyre off the Belgian coast that corresponded closely to an eddy in the Nihoul-Ronday model.

Topographic features may also be important to the small-scale structure observed in the South Bay, but biological processes can be of equal or greater importance. The spatial distribution of phytoplankton is controlled by the simultaneous action of two processes: those that distribute phytoplankton in any body of water, like advection or turbulent diffusion; and those that contribute to *in situ* gains or losses, like growth, death, sinking, or grazing (Kierstead and Slobodkin 1953; Denman and Platt 1976; Denman *et al.* 1977; Okubo 1980; Bennett and Denman 1985). The San Bruno shoal causes increased current speeds, leading to increased horizontal (and vertical) shear, turbulence, and diffusion (horizontal and vertical) of substances in the water column. This should increase the size of patches, and could account for the large patch dimensions in the central segment. On the other hand, the large patch sizes in mid-ay could also be explained by biological processes that result in low net growth rates. Small patches which are growing slowly will be rapidly diffused away, so diffusion will only permit large patches which are growing slowly (Okubo 1978). This, if the low chlorophyll levels in the central segment reflect low net growth rate, then the large patch dimensions there could result from low net growth rate. One might speculate, then, that patch sizes should be smallest at the time of the spring bloom when net growth rates increase. This may be true in the southern segment; but, perversely, the patch dimensions in the central segment were remarkably constant during the entire 1985 spring bloom. We note, however, that the spring bloom occurs when vertical stratification is most intense and freshwater infusions can induce an estuarine circulation in the central basin (Walters *et al.* 1985). Both effects can enhance horizontal transport and mixing, thus increasing the patch size even when growth rate increases.

This paper presents the initial descriptive stages in what we hope will become a detailed analysis of spatial variability in San Francisco Bay. Our preliminary results have defined some robust features of spatial heterogeneity in the South Bay and they have identified specific questions that now should be the focus of future work. The qualitative coupling of topographic features to such physical processes as mixing, residual circulations, and increased residence time seems straightforward, but the detailed mechanisms allowing such processes to generate the three mesoscale regimes in phytoplankton variability remain to be elucidated. Similarly, mechanisms that permit the persistent chlorophyll minimum in the central region of South Bay cannot be explained from fluorescence records alone. Understanding of this feature will require direct measurement of those biological and physical processes that jointly control phytoplankton biomass. The observation of near-constant phytoplankton patch size in mid-South bay is surprising. Although net rates of phytoplankton growth vary markedly over the annual cycle (Cloern *et al.* 1985), our one measure of spatial heterogeneity (patch size) seems remarkably constant. A related question is why patch sizes are small in the southern segment, both during bloom and non-bloom conditions. Moreover, the relationship of horizontal variability along the axis of the South Bay channel to that transverse to the channel is unexplored. The high degree of small-scale variability in the southern segment could result from the nearness of the narrow channel to the lateral shoals there (Fig. 1). Finally, anomalies associated with the spring bloom (decreased patch size in the southern segment; a chlorophyll maximum in the central basin; the breakdown of the three mesoscale regimes) suggest that the general "rules" governing spatial heterogeneity during most of the year may be changed during the spring bloom. Because the spring bloom coincides with periods of enhanced density stratification, the interactions between vertical and horizontal structure (Okubo 1971; Evans 1978; Kullenberg 1978; Wilson and Okubo 1980) should also be included in more detailed and quantitative studies of spatial variability.

Acknowledgment

We thank Sally Wienke for chlorophyll analyses; Byron Richards, and the crew of the *R.V. Polaris,* for their diligent aid during our field data collection endeavors; and Dr. R.T. Cheng for several helpful discussions. T.P. thanks Director A.O.D. Willows and the staff of the Friday Harbor Laboratories of the University of Washington for their hospitality during the writing of this paper.

References Cited

Abbott, M. R., T. M. Powell, and P. J. Richerson. 1982. The relationship of environmental variability to the spatial patterns of phytoplankton biomass in Lake Tahoe. *J. Plank. Res.* 4:927-941.

Ambler, J. W., J. E. Cloern, and A. Hutchinson. 1985. Seasonal cycles of zooplankton from San Francisco Bay. *Hydrobiologia* 129:177-198.

Bennett, A. F., and K. L. Denman. 1985. Phytoplankton patchiness: inferences from particle statistics. *J. Mar. Res.* 43:307-335.

Bowman, M. B., and W. E. Esaias (eds.). 1978. *Oceanic fronts in coastal processes.* Springer-Verlag, Berlin. 114 pp.

Bowman, M. B., W. E. Esaias, and M. B. Schnitzer. 1981. Tidal stirring and the distribution of phytoplankton in Long Island and Block Island Sounds. *J. Mar. Res.* 39:587-603.

Chatfield, C. 1984. *The analysis of time series: an introduction* (Third ed.). Chapman and Hall, London. 286 pp.

Cheng, R. T., and V. Casulli. 1982. On Lagrangian residual currents with application in South San Francisco Bay. *Water Resources Res.* 18:1652-1662.

Cheng, R. T., and J. W. Gartner. 1985. Harmonic analysis of tides and tidal currents in South San Francisco Bay, California. *Estuar. Coastal Shelf Sci.* 21:57-74.

Cloern, J. E. 1984. Temporal dynamics and ecological significance of salinity stratification in an estuary (South San Francisco Bay, U.S.A.). *Oceanol. Acta* 7:137-141.

Cloern, J. E., and F. H. Nichols (eds.). 1985. *Temporal variability in an estuary: San Francisco Bay. Developments in Hydrobiology No. 30.* Dr. W. Junk Publisher. The Hague, Neth. 237 pp.

Cloern, J. E., B. E. Cole, R. L. J. Wong, and A. E. Alpine. 1985. Temporal dynamics of estuarine phytoplankton: a case study of San Francisco Bay. *Hydrobiologia* 129:153-176.

Conomos, T. J. (ed.). 1979. *San Francisco Bay: the urbanized estuary.* Pacific Division, Amer. Assoc. Adv. Sci., San Francisco. 493 pp.

Denman, K. L., and T. Platt. 1976. The variance spectrum of phytoplankton in a turbulent ocean. *J. Mar. Res.* 34:593-601.

Denman, K. L., A. Okubo, and T. Platt. 1977. The chlorophyll fluctuation spectrum in the sea. *Limnol. Oceanogr.* 22:1033-1038.

Denman, K. L., and T. M. Powell. 1984. Effects of physical processes on planktonic ecosystems in the coastal ocean. *Oceanogr. Mar. Bio. Ann. Rev.* 22:125-168.

Evans, G. T. 1978. Biological effects of vertical-horizontal interaction. pp. 157-179. *In:* Steele, J. H., (ed.), *Spatial pattern in plankton communities.* Plenum, New York. 470 pp.

Kendall, M. G. 1976. *Time series* (2nd ed.) Griffin, London 197 pp.

Kierstead, H., and L. B. Slobodkin. 1953. The size of water masses containing plankton blooms. *J. Mar. Res.* 12:141-147.

Kullenberg, G. E. B. 1978. Vertical processes and vertical-horizontal coupling, pp. 43-71. *In:* Steele, J. H., (ed.). *Spatial pattern in plankton communities.* Plenum, New York, 470 pp.

Lekan, J. F., and R. E. Wilson. 1978. Spatial variability of phytoplankton biomass in the surface water off Long Island. *Estuar. Coastal Mar. Sci.* 6:239-250.

Lewis, E. L. 1980. The practical salinity scale 1978 and its antecedents. *I.E.E.E. J. Ocean Eng.* OE-5:3-8.

Lewis, M. R. and T. Platt. 1982. Scales of variability in estuarine ecosystems, pp. 3-20. *In:* Kennedy, V. S. (ed.). *Estuarine Comparisons.* Academic Press, New York.

Lorenzen, C. J. 1967. Determination of chlorophyll and phaeopigments: spectrophotometric equations. *Limnol. Oceanogr.* 12:343-346.

Mackas, D., K. L. Denman, and M.R. Abbott. 1986. Patchiness: biology in the physical vernacular. *Bull. Mar. Sci.* (in press).

Nihoul, J. C. 1975. Effect of the tidal stress on residual circulation and mud deposition in the Southern Bight of the North Sea. *Pure Appl. Geophys.* 113:577-581.

Nihoul, J. C., and F. C. Ronday. 1975. The influence of the "tidal stress" on the residual circulation; application to the Southern Bight of the North Sea. *Tellus.* 27:484-490.

Okubo, A. 1971. Horizontal and vertical mixing in the sea, pp. 89-168. *In:* Hood, D. W. (ed.). *Impingement of man on the Oceans.* Wiley-Interscience, New York.

Okubo, A. 1978. Horizontal dispersion and critical scales for phytoplankton patches, pp. 21-42. *In:* Steele, J. H. (ed.). *Spatial pattern in plankton communities.* Plenum, New York.

Okubo, A. 1980. *Diffusion and ecological problems: mathematical models.* vol. 10 *Biomathematics.* Spring-Verlag, Berlin. 254 pp.

Platt, T. 1972. Local phytoplankton abundance and turbulence. *Deep-Sea Res.* 19:183-187.

Platt, T., L. M. Dickie, and R. W. Trites. 1970. Spatial heterogeneity of phytoplankton in a near-shore environment. *J. Fish. Res. Bd. Canada.* 27:1453-1473.

Schemel, L. E., and L. A. Dedini. 1979. A continuous water-sampling and multiparameter system for estuaries. *U.S. Geol. Survey Open-File Rep.* 79-272.

Tyler, M. A., and H. H. Seliger. 1978. Annual subsurface transport of a red tide dinoflagellate to the bloom area: water circulation patterns and organism distributions in the Chesapeake Bay. *Limnol. Oceanogr.* 23:227-246.

Tyler, M. A., and H. H. Seliger. 1981. Selection for a red tide organism: physiological responses to the physical environment. *Limnol. Oceanogr.* 26:310-324.

Walters, R. A. 1982. Low-frequency variations in sea level and currents in South San Francisco Bay. *J. Phys. Oceanogr.* 12:658-668.

Walters, R. A., and R. T. Cheng. 1980. Calculations of estuarine residual currents using the finite element method, pp. 60-69. *In:* D. H. Norris (ed.), *Proces. Third Int. Conf. Finite Elements in Flow Problems.* University of Calgary, Calgary, Alberta, Canada.

Walters, R. A., and J. W. Gartner. 1985. Subtidal sea level and current variations in the northern reach of San Francisco Bay. *Estuar. Coastal Shelf Sci.* 21:17-32.

Walters, R. A., R. T. Cheng, and T. J. Conomos. 1985. Time scales of circulation and mixing processes of San Francisco Bay waters. *Hydrobiologia* 129:13-36.

Wilson, R. E., and A. Okubo. 1980. Effects of vertical-horizontal coupling on the horizontal distribution of chlorophyll *a*. *J. Plank. Res.* 2:33-47.

Wilson, R. E., A. Okubo, and W. E. Esaias. 1979. A note on time-dependent spectra for chlorophyll variance. *J. Mar. Res.* 37:485-491.

MODELING ESTUARINE VARIABILITY

Richard G. Wiegert, Convenor

MODELING SPATIAL AND TEMPORAL SUCCESSION IN THE ATCHAFALAYA/TERREBONNE MARSH/ESTUARINE COMPLEX IN SOUTH LOUISIANA

Robert Costanza, Fred H. Sklar and John W. Day, Jr.

Coastal Ecology Institute
Center for Wetland Resources
Louisiana State University
Baton Rouge, Louisiana

Abstract: A spatial simulation model was constructed to help understand the historical changes in the Atchafalaya/Terrebonne marsh/estuarine complex in south Louisiana and to project impacts of proposed human modifications. The model consists of almost 3,000 interconnected cells each representing 1 km^2. Each cell in the model contains a dynamic, nonlinear, simulation model. Variables include water volume and flow, sediment and salt concentrations, organic standing crop, and productivity. The balance between sediment deposition and erosion as influenced by these variables in this rapidly subsiding area is particularly critical to habitat succession and the productivity of the area. Primary input data for the model were detailed, digitized habitat maps prepared by the U.S. Fish and Wildlife Service for 1956 and 1978. In addition, long time series of field measurements are available for some variables.

The data base assembled for the model includes annual changes in environmental forcing functions, such as river discharge, and human modifications, such as canals and levees. At present, we are in the preliminary stages of running and calibrating the model. Our current results mimic the spatial seasonal patterns of sediment, salinity and water flow reasonably well. Starting with the 1956 initial conditions the model correctly predicts the habitat type in 1978 of 68% of the almost 3000 cells. Changes in salinity zones were accurately predicted but the model predicts higher land loss rates in the fresh zone than those actually observed and does not do well at predicting the growth of the Atchafalaya delta. The former problem may have to do with the prevalence of floating marsh in the fresh zone, which the model does not adequately consider at present.

This paper discusses: (1) the structure of the model; (2) the spatial data base necessary to run the model; (3) some preliminary results concerning the temporal and spatial distributions of some of the state variables; and (4) future directions for the model. In general, this approach seems to be applicable to modeling spatial ecosystem dynamics, but the size and computational complexity of the model makes calibration and verification difficult.

Introduction

The patterns of ecosystem development in time and space are the result of complex interactions of physical and biological forces. However, most ecological modeling work has focused on temporal changes with little or no spatial articulation (cf. Costanza and Sklar 1985). It is becoming clear that spatial dynamics need to be more explicitly included if ecological models are to be truly useful tools for understanding and managing real ecosystems (Risser *et al.* 1984).

Many of the decisions made by agencies charged with the management of ecological systems require information on both the spatial and temporal responses of the systems to various management scenarios. We have developed a model to project long-term, spatially articulated habitat changes ("succession") in the Atchafalaya/Terrebonne marsh/estuarine complex in south Louisiana (Fig. 1). Although our model is designed for a particular coastal ecosystem, the problem is general and our approach represents a potentially general solution. The model development was stimulated by studies and management needs in the Louisiana coastal zone where a major new Mississippi delta lobe (the Atchafalaya delta) is being formed. The introduction of large quantities of fresh water and sediments is leading to major successional shifts in habitats, both temporally and spatially. The balance among sediment deposition, erosion, and subsidence is critical to these ecological changes and it is important to be able to predict how these changes are affected by man's activities. In such systems we believe that spatial modeling can be a valuable theoretical and management tool. In this paper we describe our generalized approach (conceptual and mathematical) to the modeling of spatial successional change and apply it to a specific coastal wetland system. Our results for this application are still preliminary, and will improve as we continue the calibration and verification processes.

Overview of the Study Area

The study area, the wetlands and estuaries of western Terrebonne Parish, Louisiana, is part of the Atchafalaya delta region, which is one of the most geologically dynamic regions of North America. The Atchafalaya River is one of the two principle distributaries of the Mississippi River, carrying about 30% of the Mississippi flow to the Gulf of Mexico. During the last 50 years sediments carried by the Atchafalaya River filled in several large lakes in the Atchafalaya River basin. As increasing quantities of sediments passed through the basin, the bay area became progressively shallower, and in 1973 a new subaerial delta appeared (Roberts *et al.* 1980). This delta has since grown and now has an area of about 50 km^2. Over the past 5,000 years the Mississippi River has changed course numerous times and formed seven major deltaic lobes (Frazier 1967). The new Atchafalaya delta marks the initial stages of the next major delta lobe.

Along with the dramatic emergence of the new delta, more subtle but nonetheless profound changes have occurred in adjacent wetlands. The western Terrebonne marshes are becoming fresher while salinity is increasing in the eastern part of the study area. Analysis of vegetation maps from 1956 to 1978 (Wicker 1980) show how the boundary between fresh and brackish marshes has moved closer to the Gulf in the western marshes and further inland in the east (Fig. 1). Although the overall study region is experiencing significant wetland loss like most of the Louisiana coastal zone (Boesch 1982), land loss rates in the western Terrebonne marshes have slowed or reversed over the past two decades because of riverine deposition (Baumann and Adams 1982, Baumann *et al.* 1984). Finally, the hydrology of the area has been extensively

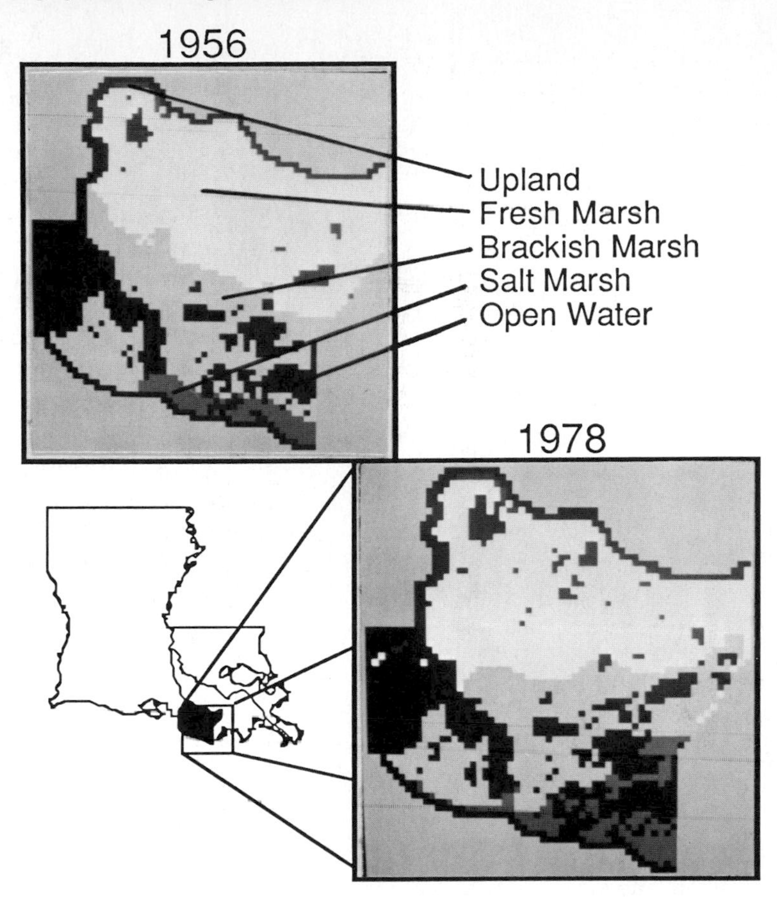

Figure 1. Habitat maps digitized into 1-km² cells for 1956 and 1978. The changes in habitat areas are indicated in Table 1.

altered as a result of the dredging of many waterways and petroleum exploration access canals (Wang 1984, Wang *et al.* 1983).

One management project currently under consideration by the U.S. Army Corps of Engineers is the extension of a levee along the east bank of the Atchafalaya River to improve navigation and control backwater flooding. This project would shunt most of the sediment-laden river water into the Gulf of Mexico and alter the spatial pattern of water and sediment inputs to the Terrebonne marshes to the east. In order to assess the impact of projects like this, it is important to understand the total sediment and water inputs to the marshes, their spatial distribution, and the projected changes in marsh habitat type and quality that result. We must also know how alternatives to the proposed levee extension would affect water, sediment, and habitat distribution.

A large base of spatial data exists for this area in the form of: (1) 1956 and 1978 habitat maps (Wicker 1980); (2) field measurements of water levels, plant productivity, salinity, nutrient concentrations, etc. taken over a number of years (Denes 1983; Caffrey 1983; Teague 1983; DeLaune *et al.* 1978; Miller 1983); and (3) results of hydrodynamic models (Wang 1984). Total marsh habitat in the Terrebonne marsh complex decreased from 1,594 km^2 in 1956 to 1,470 km^2 in 1978 (Table 1) but wetland area adjacent to the Atchafalaya River and Four League Bay increased. Open water habitats increased by 137 km^2 and upland habitats (including spoil banks and levees increased by 4 km^2.

Goals and Background of the Model

Our overall objective is to predict spatial as well as temporal changes in habitat areas (along with information on the relative health of the habitats) as a result of natural processes and various management strategies. We are using a general modeling approach that combines dynamic simulation with spatial articulation. Our model is essentially an array of interacting "cells" that represent fixed areas in a study region. Each cell contains a dynamic, nonlinear simulation model for a specific habitat that incorporates important forcing functions and processes. In the Atchafalaya/Terrebonne area these are water levels and flow, subsidence, river and tidal inputs, salinity, sedimentation, and plant production. Each cell is potentially connected to its neighbors by exchanges of water that can carry suspended sediments, dissolved salts, nutrients, and biomass between cells.

As with most large-scale modeling studies, we did not possess the resources to collect all the data necessary to build and test our models, and we relied in large part on the existing data base. This reality both constrains and guides the model construction in that we must structure the model to take best advantage of the existing data to maximize the model's effectiveness (Costanza and Sklar 1985). We first aggregated the more than 100 habitat types distinguished by Wicker (1980) to the 20 major types used in Costanza *et al.* (1983). Only 6 of these 20 occur sufficiently within the study area to be included here.

To help in developing the model, and to make understanding its structure and behavior easier, we developed a generalized version consisting of nine interacting cells under hypothetical (but not totally unrealistic) conditions that could

Table 1. Change in area of major habitat types from 1956 to 1978 in the Terrebonne marsh complex of the Atchafalaya delta, Louisiana, based on the digitized habitat maps shown in Fig. 1.

	Area (number of 1-km^2 cells)		
Habitat Type	1956	1978	Change: 1978-1956
Fresh Marsh	864	766	−98
Swamp	130	113	−17
Brackish Marsh	632	554	−78
Salt Marsh	98	150	52
Open Water	742	879	137
Upland	13	17	4
Total	2479	2479	

be considered "typical" cells in a larger model (Sklar *et al.* 1985). The current paper discusses the application of this general model to the Atchafalaya/Terrebonne marsh/estuarine complex using a 3,000-cell structure.

Model Structure

Details of the model structure are given in Sklar *et al.* (1985). Here we give only a brief sketch. We chose a square, fixed grid of equally sized cells to represent space in the model, because this arrangement is simple and does not impose any *a priori* structure on the system. An alternative approach is the finite element method used in hydrodynamic modeling (Wang *et al.* 1983) which uses a variable-sized mesh connecting nodes. The finite element method is appropriate for systems with fixed hydrologic structure or for time intervals over which the structure is not expected to change. Our concern, however, is with long-term simulations during which the hydrologic structure *is* expected to change. Thus we selected a square grid that is flexible enough to allow for changing hydrologic structure. The grid approach is used in modern global, general atmospheric circulation models with some success (Kasahara and Washington 1967; Williams *et al.* 1974). We felt atmospheric circulation modeling was akin to the fluid nature of coastal wetlands, especially when viewed in the long term.

The square cells have exchanges of water and materials across their four sides to adjacent cells (Fig. 2). We did not use a hexagonal or triangular grid or a square one that allowed exchange across the diagonals mainly because we wanted the simplest arrangement that would work reasonably well for our purposes and was easy to program.

Figure 2 shows diagramatically the water, suspended sediment (SS), and bottom sediment (BS) components of the model for a typical cell. The volume of

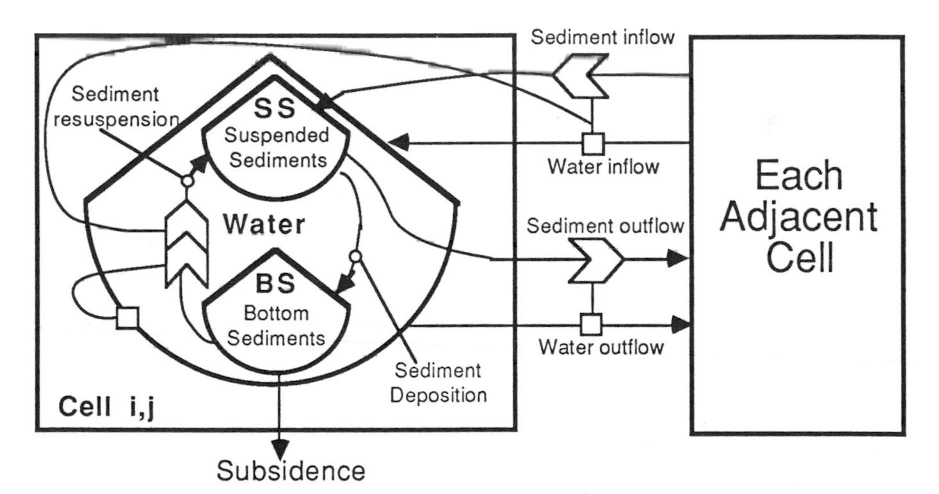

Figure 2. Storages (tank symbols) and flows (lines) of water, suspended sediments (SS), and bottom sediments (BS) for a typical cell. Inflows and outflows of suspended sediments are proportional to water flows.

water crossing from one cell to another is a function of water storage (W) and connectivity (K) such that unidirectional water flow across a single boundary is KW. Water head differential is due mainly to differences in water volumes in equally sized cells rather than elevation gradients because of the flatness of coastal marsh habitats. Connectivity is a function of (1) habitat type, (2) drainage density, (3) waterway orientation, and (4) levee height.

We used a simple Euler numerical integration technique for the differential equations initially, with the option to apply more elaborate schemes (i.e., Runge-Kutta) should the model prove unstable. For the conditions used, the Euler method proved adequate with a reasonable time step (one week) but this may not be true in general.

The change in water level in each cell is determined in the model by water exchanges in both directions across all four boundaries plus surplus rainfall (precipitation minus evapotranspiration) input as follows:

$$\begin{aligned} dW_{ij}/dt = \; & (K_{i,j+1,i,j})(W_{i,j+1}) - (K_{i,j,i,j+1})(W_{i,j}) \\ & + (K_{i,j-1,i,j})(W_{i,j-1}) - (K_{i,j,i,j-1})(W_{i,j}) \\ & + (K_{i+1,j,i,j})(W_{i+1,j}) - (K_{i,j,i+1,j})(W_{i,j}) \\ & + (K_{i-1,j,i,j})(W_{i-1,j}) - (K_{i,j,i-1,j})(W_{i,j}) \\ & + SR_{i,j} \end{aligned} \tag{1}$$

where:

$K_{i,j+1,i,j}$ = Overall water flow connectivity parameter from cell **i,j+1** to cell **i,j** (a function of habitat type, drainage density, waterway orientation, and levee height);
$W_{i,j}$ = Volume of water in cell **i,j**; and
$SR_{i,j}$ = Surplus rainfall input to cell **i,j**.

Change in abiotic material concentrations are a function of water flow between cells and concentration of materials in the cells, along with internal deposition and resuspension. For example, change in suspended sediment (SS) in a cell is given by:

$$\begin{aligned} dSS_{i,j}/dt = \; & (J_{i,j+1,i,j})(SS_{i,j+1}) - (J_{i,j,i,j+1})(SS_{i,j}) \\ & + (J_{i,j-1,i,j})(SS_{i,j-1}) - (J_{i,j,i,j-1})(SS_{i,j}) \\ & + (J_{i+1,j,i,j})(SS_{i+1,j}) - (J_{i,j,i+1,j})(SS_{i,j}) \\ & + (J_{i-1,j,i,j})(SS_{i-1,j})(J_{i,j,i-1,j})(SS_{i,j}) \\ & - (KSED)(SS_{i,j} + (TV)(JIN_{i,j})(BS_{i,j}) \end{aligned} \tag{2}$$

where:

TV = Turbulence vector parameter;
KSED = Sedimentation parameter;
$JIN_{i,j}$ = Total water flux into cell **i,j**;
$SS_{i,j}$ = Concentration of SS in cell **i,j**;
$BS_{i,j}$ = Bottom sediments (relative elevation) for cell **i,j**; and
$J_{i,j,i,j-1}$ = Water flux from cell **i,j** to cell **i,j − 1** = $(K_{i,j,i,j-1})(W_{i,j-1})$.

Other abiotic material balances (i.e., salts, nutrients) are handled in a similar fashion, as are the intercell, waterborn transport aspects of biotic components.

The removal of suspended sediments from the water column to bottom sediment storage (BS) is a function of its concentration (SS) and the sedimentation rate (KSED*SS). The sedimentation parameter (KSED) is dependent upon water depth, habitat type, and vegetation density. Bottom sediment level was modeled as:

$$dBS_{i,j}/dt = (KSED)(SS_{i,j}) - (KSUB)(BS_{i,j}) - (TV)(JIN_{i,j})(BS_{i,j}) \quad (3)$$

where:

KSED = Sedimentation coefficient;
KSUB = Subsidence coefficient;
TV = Turbulent mixing parameter;
$JIN_{i,j}$ = Total water flux into cell **i,j**;
$SS_{i,j}$ = Suspended sediments in cell **i,j**; and
$BS_{i,j}$ = Bottom sediments in cell **i,j**.

For each cell in the model the build-up of new land or the development of open water habitats depends on the balance between net inputs of suspended sediments and outputs because of subsidence. Subsidence is principally a regional process unaffected by local overburdening and was modeled as such.

Since minor changes in elevation in this flat coastal landscape have a large influence on habitat type and succession, the accuracy of the sediment/water balance component of the model is critical to the accuracy of the overall model. Therefore, we spent much of our initial effort on this part of the model.

In each wetland cell, organic matter production is a function of temperature, salinity, elevation, nutrients, and habitat type. In aquatic cells it is a function of temperature, suspended sediments, nutrients, and salinity. Gaussian distribution curves were used for each of the factors except habitat type, which determines the parameter values. This formulation generates an optimum production rate for each habitat at some particular conditions of temperature, salinity, etc., with falling production as one departs from the optimal conditions. Figure 3 illustrates this function for salinity. As salinity increases fresh marsh productivity falls while brackish marsh production increases. Habitat succession occurs in the model (after a certain time lag) when one habitat type becomes more productive than the others because of changing conditions.

Model Input Data

Some of the data necessary to run the model for the Atchafalaya/Terrebonne marsh/estuarine complex using the 1-km² grid discussed above are summarized in Fig. 1 and Figs. 4-5. Figure 1 shows habitat distribution digitized into the 1-km² grid for initial conditions (1956) and 1978. Each cell was assigned the

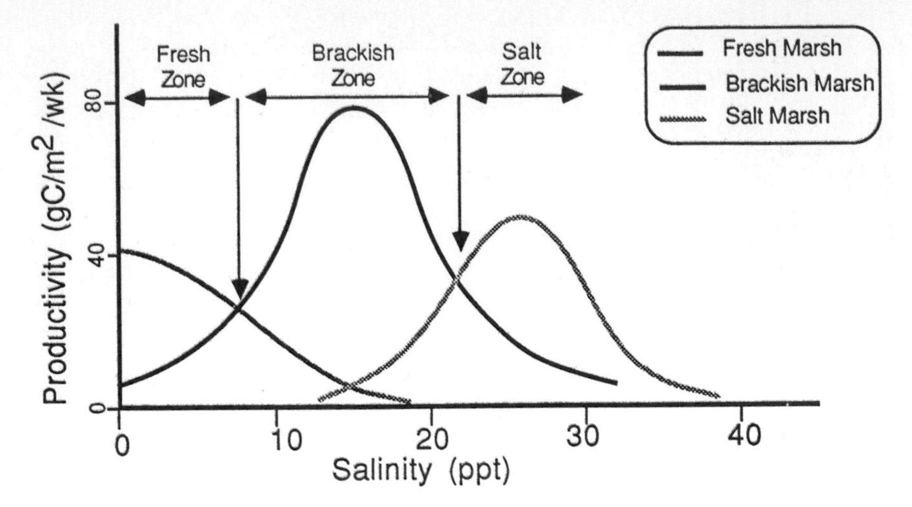

Figure 3. Plant primary production as a function of salinity and habitat type, showing points of habitat switching in the model.

habitat that covered the largest amount of the cell. While this digitization process resulted in the loss of spatial resolution, the major features and spatial changes are still clearly visible, particularly the movement of the fresh/brackish interface line as a result of increased Atchafalaya River flow. The emergence of the new delta and the breaking up of the marshes in the western part of the study area are also still clearly visible.

The 1-km² grid digitized in this way was much too coarse to allow waterways, canals, and levees to be treated as separate cells. Because these features are very important in determining water flow we added them to the model in the following way. In addition to overland flow, water can exchange with adjacent cells via canals or natural bayous, or it may be prevented from exchanging by the presence of levees. Figure 4 shows the digitized waterway location maps and the digitized levee location maps for 1956 and 1978. The overall water flow connectivity parameter ($K_{i,j,k,l}$ in equation 1) is adjusted to reflect the presence and size of waterways or levees at the cell boundaries. If a waterway is present at a cell boundary a large K value is used, increasing with the size of the waterway. If a levee is present a K value of 0 is used until water level exceeds the height of the levee. The width of the lines in Fig. 4 indicate the size of the waterways and levees. The model's canal and levee network is updated each year during a simulation run. Man-made canals and levees are added to the model's hydrologic structure at the beginning of the year they were built.

Forcing functions must also be specified as boundary conditions for the model over the simulation period (1956-1978). Figure 5 shows some of the required forcing functions in the form of time series of Atchafalaya and Mississippi river discharge, Gulf salinity, river sediments, sea level, and air movement.

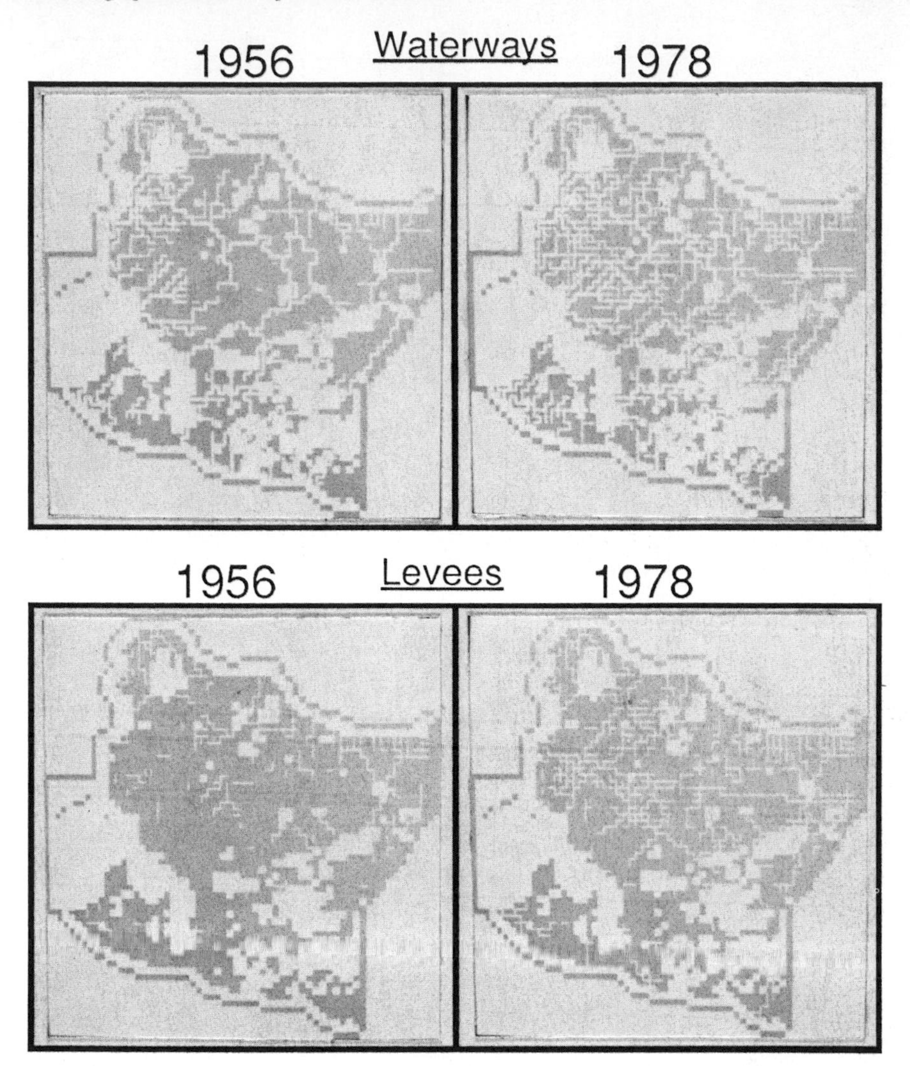

Figure 4. Waterway (upper) and levee location (lower) maps for 1956 (left) and 1978 (right). Line widths indicate the relative size of the waterway.

Preliminary Simulation Results

Our initial objective was to simulate the historical temporal dynamics of spatial patterns in the region. This required adjusting the model's parameters so that starting with the 1956 initial conditions, the model reproduces the 1978 conditions as closely as possible.

The model produces a huge amount of output, essentially a map of all eight state variables plus habitat types for each week of the simulation run. For our initial calibration we ran the model for 24 years (1248 weekly itterations)

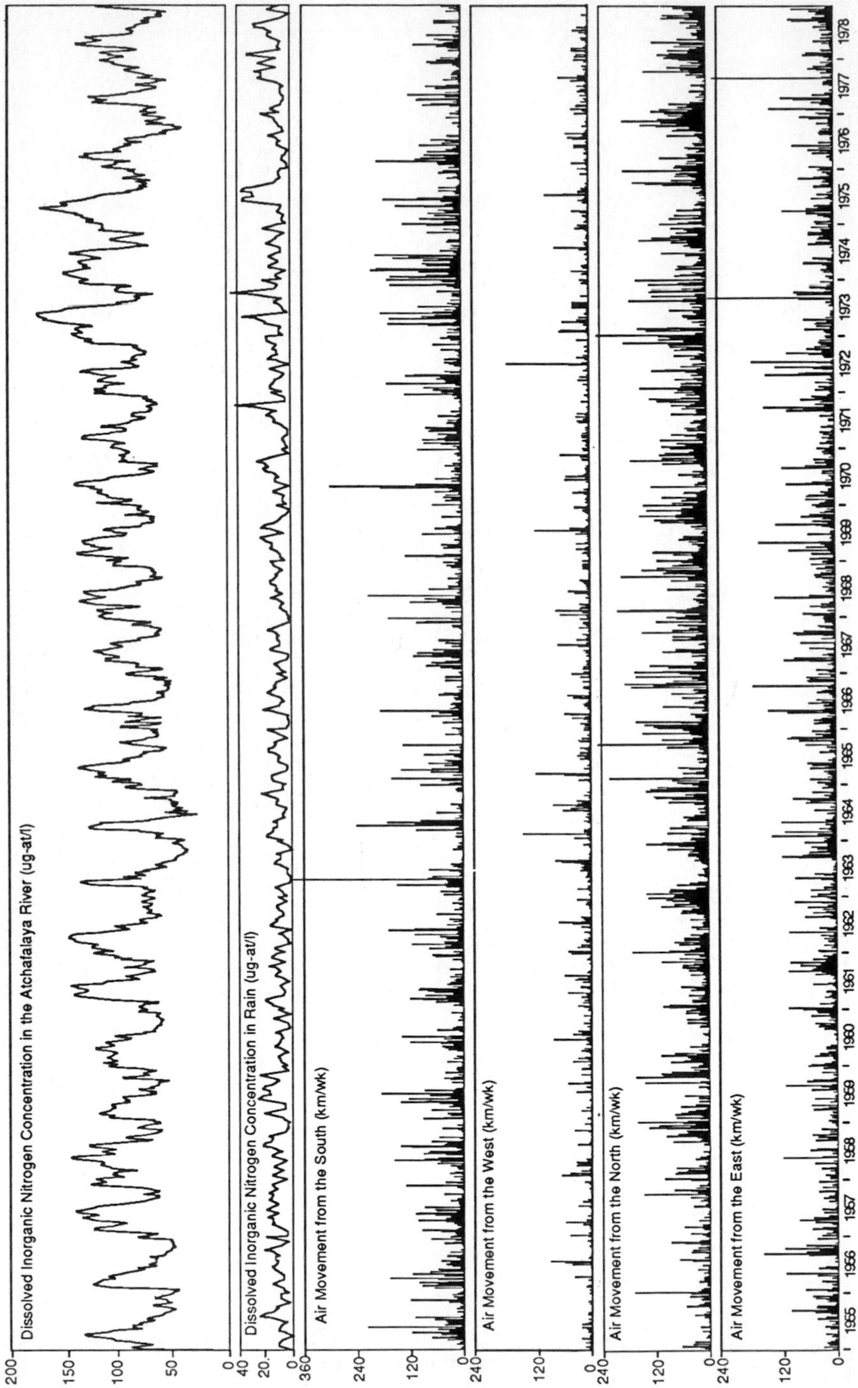
Dissolved Inorganic Nitrogen Concentration in the Atchafalaya River (ug-at/l)
Dissolved Inorganic Nitrogen Concentration in Rain (ug-at/l)
Air Movement from the South (km/wk)
Air Movement from the West (km/wk)
Air Movement from the North (km/wk)
Air Movement from the East (km/wk)
200
150
100
50
0
40
20.
0
360
240
120
0
240
120
0
240
120
0
240
120
0
1955
1956
1957
1958
1959
1960
1961
1962
1963
1964
1965
1966
1967
1968
1969
1970
1971
1972
1973
1974
1975
1976
1977
1978

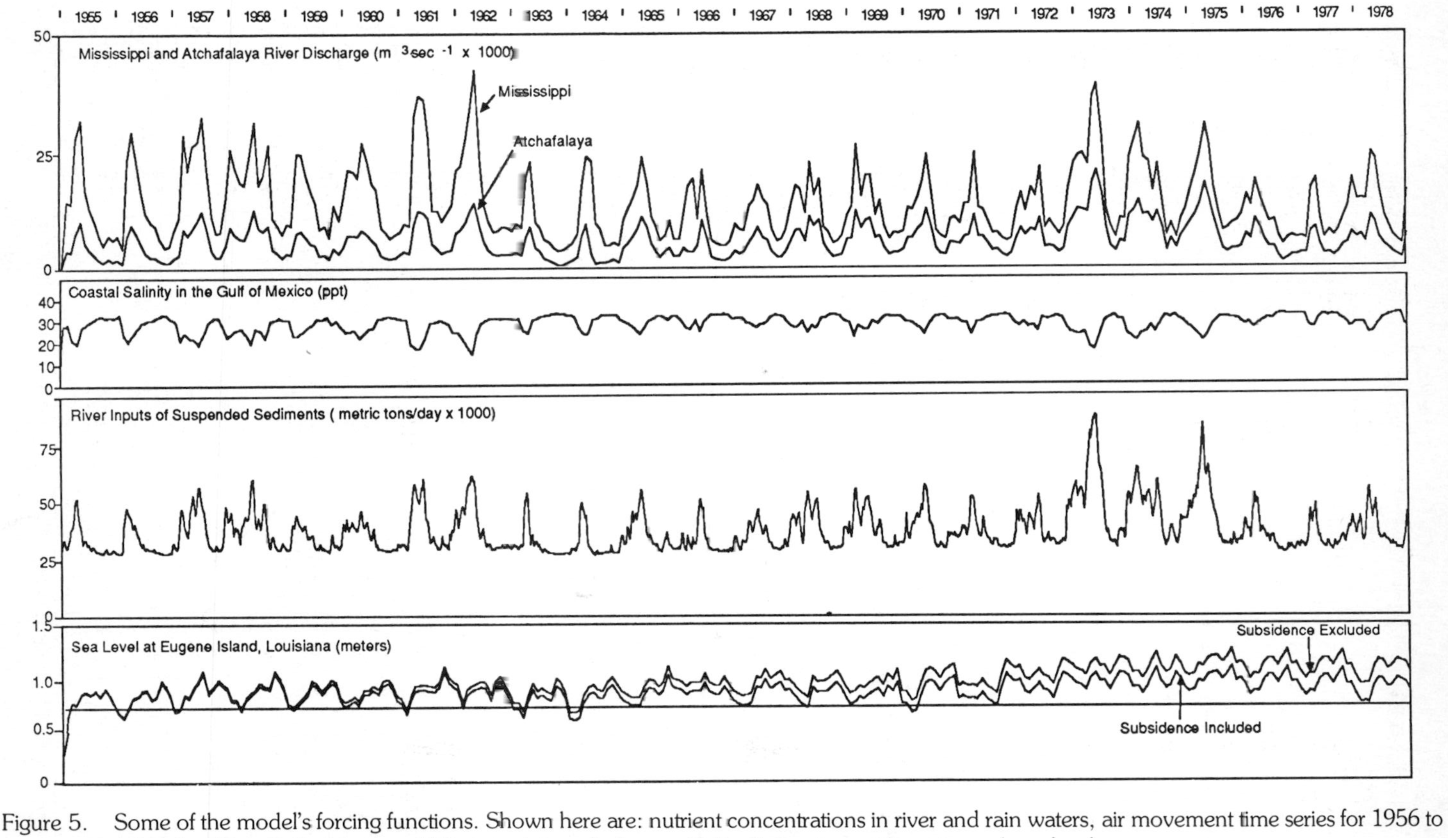

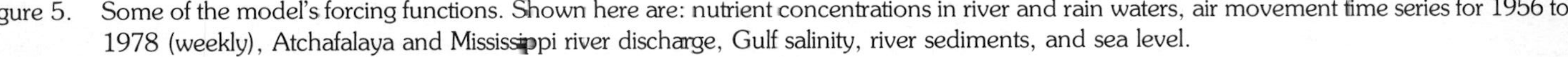
Figure 5. Some of the model's forcing functions. Shown here are: nutrient concentrations in river and rain waters, air movement time series for 1956 to 1978 (weekly), Atchafalaya and Mississippi river discharge, Gulf salinity, river sediments, and sea level.

starting in 1956 and ending in 1978. These runs used the 1956 initial condition habitat distribution, waterway, levee location, and forcing function data discussed above.

The best way to get a feel for the model's dynamic spatial behavior is to use a video display of the mapped variables in a time-series display loop. In this paper we can show only a few example snapshots. Examples of the spring (week 12), summer (week 24), fall (week 36), and winter (week 52) simulated patterns are shown for nutrients (Fig. 6) and suspended sediments (Fig. 7) for

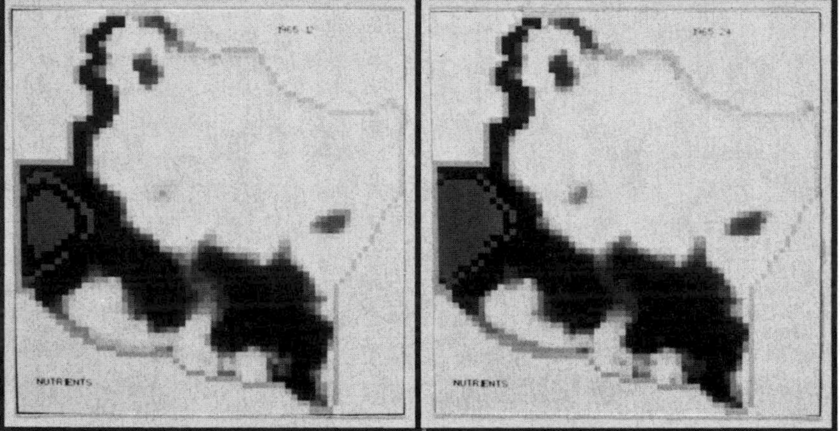

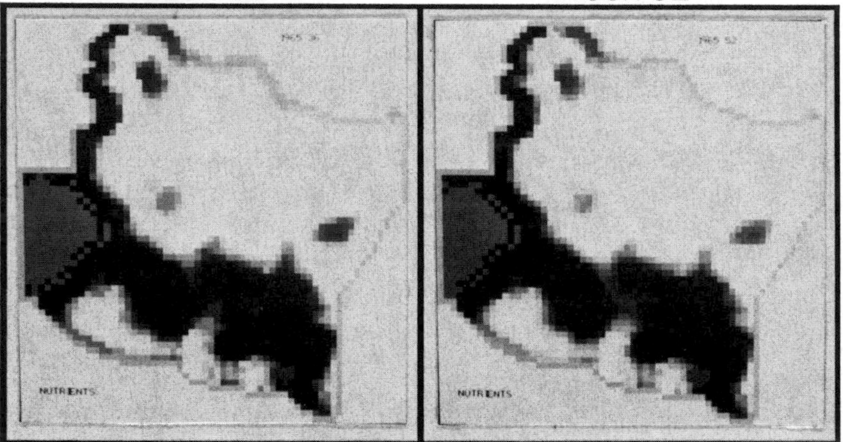

Figure 6. Seasonal pattern of simulated nutrient changes for 1965. Shown are week 12 (upper left), week 24 (upper right), week 36 (lower left) and week 52 (lower right). Shading reflects concentration of total nitrogen, ranging from <20 mg l^{-1} for unshaded areas to 120 mg l^{-1} for the darkest areas.

1965. Fresh water, sediment, and nutrient inputs via the Atchafalaya River all increased in the spring in the northwestern part of the study area. The series shows the gradual intrusion of suspended sediments and nutrients into the system over the course of the year, mainly through the lower brackish and saline zones.

Figure 8 shows a long-term series of simulated maps of suspended sediments for week 52 of 1960, 1965, 1970, and 1975, respectively. This suspended sediment series mimics the gradual increase of sediment inputs to the system that has occurred over the period because of increased Atchafalaya River

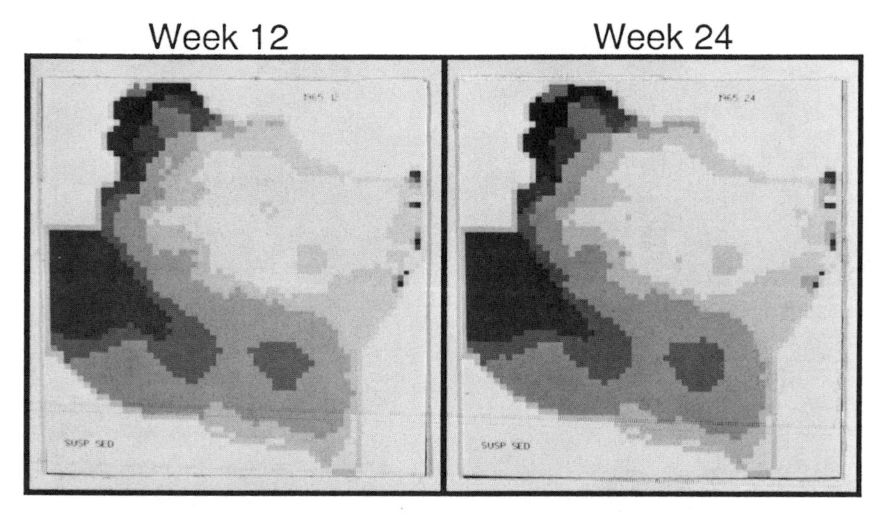

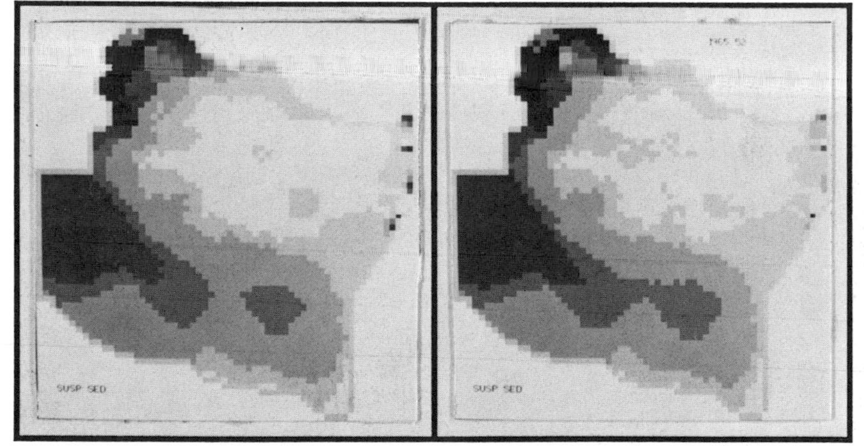

Figure 7. Seasonal pattern of simulated suspended sediment changes for 1965. Shown are week 12 (upper left), week 24 (upper right), week 36 (lower left) and week 52 (lower right). Unshaded areas have <60 g l^{-1} suspended sediments, while darkest areas are >220 g l^{-1}.

discharge. Water volume and salinity (not illustrated) also behave in a manner generally consistent with what we know about the historical behavior of system.

Figure 9 is a comparison of our simulated 1978 habitat map with the actual 1978 habitat map. The degree of fit between these two maps was calculated as the percentage of corresponding cells in the two maps with the same habitat type. The fit measured this way is 68%. We stress that these are preliminary results and the fit should improve dramatically as we continue the calibration process. Changes in salinity zones were accurately predicted but the model

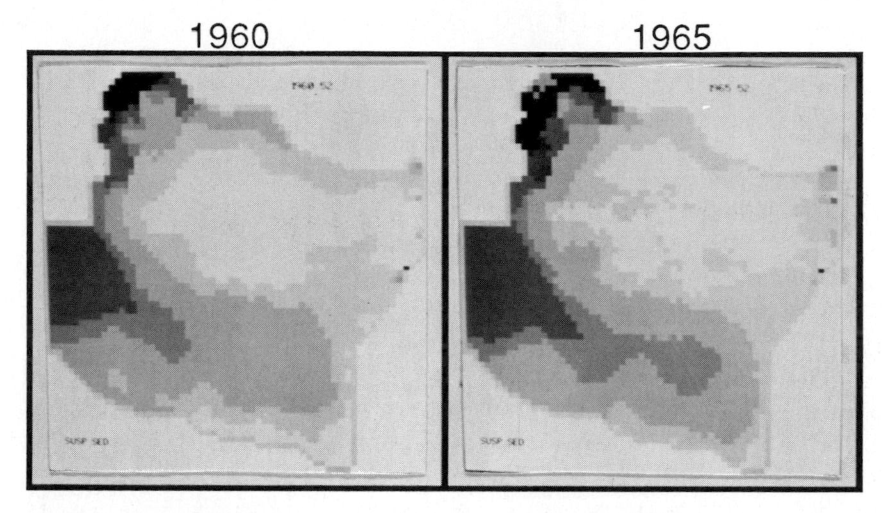

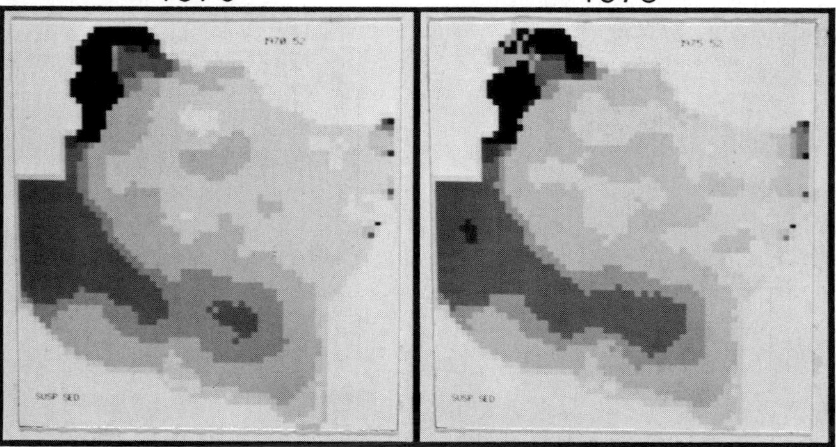

Figure 8. Simulated long-term changes in suspended sediment. Shown are week 52 of 1960 (upper left), 1965 (upper right), 1970 (lower left) and 1975 (lower right). Unshaded areas have <60 g l^{-1} suspended sediments, while darkest areas are >220 g l^{-1}.

predicts higher land loss rates in the fresh zone than those actually observed and does not do well at predicting the growth of the Atchafalaya delta. The former problem may have to do with the prevalence of floating marsh in the fresh zone, which the model does not adequately consider at present. In fact, there is very little data on the distribution of floating vs. non-floating fresh marsh in the study area, since it is almost impossible to differentiate these habitats from the air, especially from high-altitude platforms. The current algorithm in the model evaluates water levels and replaces fresh marsh with open water when the water levels exceed a set value. In fact the marsh may simply have detached and floated on the surface rather than becoming open water as water levels increased. We need to allow for this possibility in the model. The model's difficulty with reproducing the Atchafalaya delta formation probably has to do with improperly specified initial (1956) conditions of bottom elevations in Atchafalaya Bay. We are currently gathering better data on this variable that should improve the model's performance.

Discussion and Conclusions

The preliminary runs on our 3,000-cell dynamic spatial model were intended to simulate long-term behavior in the Atchafalaya/Terrebonne marsh/estuarine complex. These preliminary results are encouraging but they are only the beginning. The next (and most difficult) step is to calibrate the model to reproduce as closely as possible the historical sequence of as many variables as possible. In addition to the 1978 habitat map, we will use actual data from the Corps of Engineers on salinity, water volume, and suspended sediment patterns for various times in the area for calibration purposes. A habitat distribution map for 1984 has also been prepared, and will provide a further basis for verifying the model's predictions. After calibrating with the 1978 data we will run the model to 1984 and compare the results with the 1984 data. The degree of fit between these two maps will provide an unbiased estimate of the model's predictive accuracy.

Our calibration process will involve two steps: (1) manual adjustment of the parameters, as described in this paper, to bring the model "in the ballpark" relative to actual patterns, and (2) application of a computer parameter optimization algorithm to "fine tune" the model. This algorithm iteratively reduces the sum of the squared error between the model's output and the observed salinity, water, sediment, and 1978 habitat data. To do this it must run the model several hundred times and look at the way the error changes with small changes in critical parameters. At the conclusion of this process, we will have a model that reproduces the historical patterns as closely as possible for our particular model structure. More importantly, we will know exactly how well the model fits the historical data (in terms of an overall degree of fit measure). Then we can verify the model using the 1984 habitat distribution data (which will not be used as part of the calibration process) as a basis for judging its predictive success.

The most significant general problem with this approach to spatial modeling is the large size and complexity of the models required to achieve a reasonably high degree of spatial resolution. To avoid making the task insurmountable,

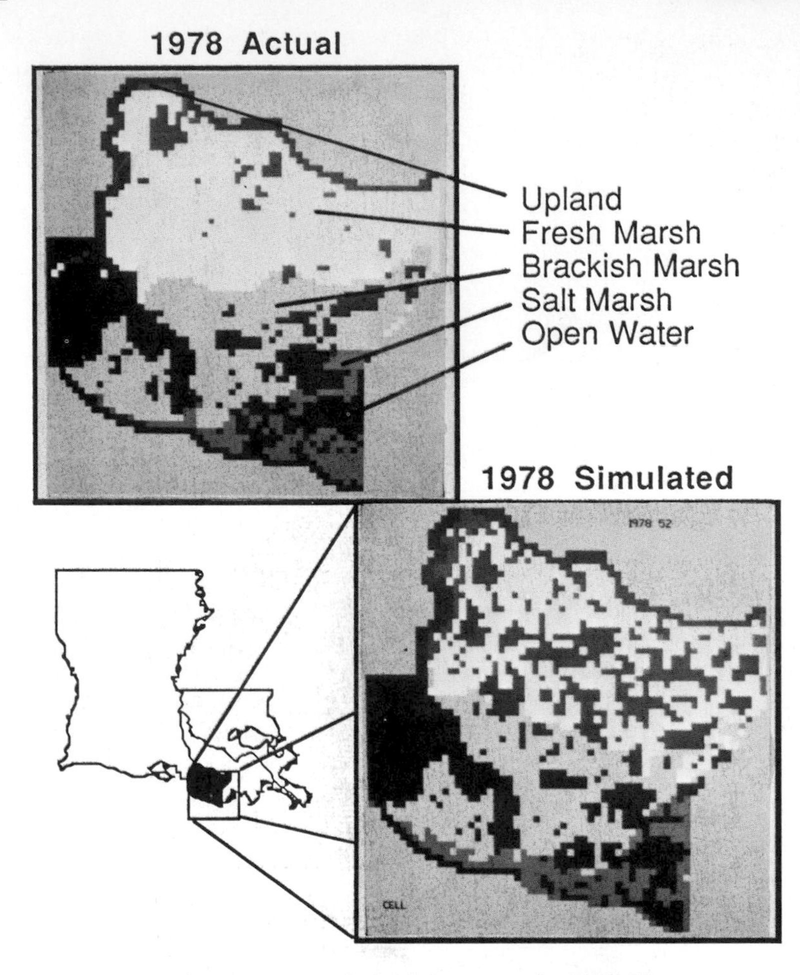

Figure 9. Actual (left) vs. simulated (right) habitat maps for 1978. The percentage of the corresponding cells that are the same habitat is 68%.

the model must be kept conceptually simple by limiting the number of state variables and their methods of interaction. The model described in this paper is computationally very large and costly (one full-scale run from 1956 to 1978 currently takes about 1 hr of CPU time and costs about $400 on an IBM 3081 mainframe) but it is conceptually relatively simple, being a relatively straightforward mass balance model with eight state variables duplicated 3,000 times. There is, of course, a price to be paid for this conceptual simplicity—the underlying physical and ecological processes must be somewhat crude approximations. Striking the proper balance between spatial and conceptual resolution is part of the art of model building, and is dependent on the specific questions being asked, the available data, and the computational facilities available. Computational costs have been a major factor limiting spatial ecosystem modeling in the past. The level of spatial resolution we chose for this model was not computationally affordable until very recently. Our present analysis highlights the difficulties and limitations of implementing and parameterizing a model of this size and complexity with existing, readily accessible computational facilities. As the cost of computation continues to fall over the next decade, and as supercomputers and array processors become more available, spatial dynamic ecological models of the type outlined in this paper will become more practical tools for understanding and managing natural systems.

Acknowledgments

This research was supported in part by grants from the U.S. Fish and Wildlife Service, U.S. Department of the Interior; the U.S. Army Corps of Engineers; and the Louisiana Sea Grant College Program, a part of the National Sea Grant Program maintained by the National Oceanic and Atmospheric Administration, U.S. Department of Commerce. This is contribution No. LSU-CEI-85-14, Center for Wetland Resources, Louisiana State University, Baton Rouge, LA 70803-7503.

References Cited

Baumann, R. H. and R. D. Adams. 1982. The creation and restoration of wetlands by natural processes in the lower Atchafalaya River System: possible conflicts with navigation and flood control objectives. pp. 1-24. *In:* R. H. Stovall (ed.), *Proceedings of the eighth conference on wetlands restoration and creation.* Hillsborough Community College, Tampa, Florida.

Baumann, R. H., J. W. Day, Jr. and C. A. Miller. 1984. Mississippi deltaic wetland survival: sedimentation versus coastal submergence. *Science.* 224:1093-1094.

Boesch, D. F. (ed.). 1982. *Proceedings of the conference on coastal erosion and wetland modification in Louisiana: causes, consequences, and options.* FWS/OBS-82-59. U.S. Fish and Wildlife Service, Division of Biological Services, Washington, D.C. 256 pp.

Caffrey, J. M. 1983. The influence of physical factors on water column nutrients and sediments in Fourleague Bay, Louisiana. M.S. Thesis, Louisiana State University, Baton Rouge. 85 pp.

Costanza, R. and F. H. Sklar. 1985. Articulation, accuracy, and effectiveness of mathematical models: a review of freshwater wetland applications. *Ecol. Mod.* 27:45-68.

Costanza, R., C. Neill, S. G. Leibowitz, J. R. Fruci, L. M. Bahr and J. W. Day, Jr. 1983. *Ecological models of the Mississippi Deltaic Plain Region: data collection and presentation.* FWS/OBS-82/68. U.S. Fish and Wildlife Service, Division of Biological Services, Washington, D.C. 342 pp.

DeLaune, R. D., W. H. Patrick and R. J. Buresh. 1978. Sedimentation rates determined by ^{137}Cs dating in a rapidly accreting salt marsh. *Nature.* 275:532-533.

Denes, T. A. 1983. Seasonal transports and circulation of Fourleague Bay, Louisiana. M.S. Thesis. Louisiana State University, Baton Rouge. 112 pp.

Frazier, D. E. 1967. Recent deltaic deposits of the Mississippi River: their development and chronology. *Trans. Gulf Coast Assoc. Geol. Soc.* 17:287-315.

Kasahara, A. and W. M. Washington. 1967. NCAR global general circulation model of the atmosphere. *Monthly Weather Review.* 95:389-402.

Miller, C. A. 1983. Sediment and nutrient inputs to the marshes surrounding Fourleague Bay, Louisiana. M.S. Thesis, Louisiana State University, Baton Rouge. 68 pp.

Risser, P. G., J. R. Karr, and R. T. T. Forman. 1984. Landscape ecology: directions and approaches. Special Publication Number 2. Illinois Natural History Survey, Champaign. 18 pp.

Roberts, H. H., R. D. Adams and R. W. Cunningham. 1980. Evolution of sand-dominated subaerial phase, Atchafalaya delta, Louisiana. *Am. Assoc. Pet. Geol. Bull.* 64:264-278.

Sklar, F. H., R. Costanza and J. W. Day, Jr. 1985. Dynamic, spatial simulation modeling of coastal wetland habitat succession. *Ecol. Mod.* 29:261-281.

Teague, K. G. 1983. Benthic oxygen uptake and net sediment-water nutrient fluxes in a river-dominated estuary. M.S. Thesis, Louisiana State University, Baton Rouge. 72 pp.

Wang, F. C. 1984. Simulation of levee extension and marsh flooding. *Hydraulic Eng.* (in press).

Wang, F. C., C. A. Moncrieff and J. A. Amft. 1983. Wetland hydrology and hydrodynamics. Sea Grant Publication No. LSU-T-83-001, Center for Wetland Resources, Louisiana State University, Baton Rouge. 152 pp.

Wicker, K. M. 1980. The Mississippi deltaic plain region habitat mapping study. FWS/OBS-79/07. U.S. Fish and Wildlife Service, Division of Biological Services, Washington, D.C. 464 maps.

Williams, J., R. G. Barry and W. M. Washington. 1974. Simulation of the atmospheric circulation using the NCAR global circulation model with ice age boundary conditions. *J. Appl. Meteor.* 13:305-317.

MODELING SPATIAL AND TEMPORAL VARIABILITY IN A SALT MARSH: SENSITIVITY TO RATES OF PRIMARY PRODUCTION, TIDAL MIGRATION AND MICROBIAL DEGRADATION

Richard G. Wiegert

Department of Zoology
University of Georgia
Athens, Georgia

Abstract: A 23-compartment model of a coastal Georgia salt marsh was used to simulate perturbations of three different processes that affect spatial variability: 1) tidal migrations of consumers; 2) primary production of creekbank and highmarsh cordgrass (*Spartina*); and 3) anaerobic and aerobic microbial degradation.

Increasing the migrant consumers depressed several compartments and lowered the amount of carbon exported tidally from the system. In general, removing migrant consumers reversed these effects. Standing stocks of aerobic microbes, benthic infauna and meiofauna were depressed, for differing reasons, whether migrant consumers were increased or removed.

Increased net primary production by cordgrass caused substantially greater accretion of carbon in the anaerobic sediments and somewhat greater rates of aerobic degradation. Increasing or decreasing the rate of anaerobic degradation also directly affected the accretion of carbon, but had little affect on other processes, at least over the two-year duration of the simulations.

Small increases in the permissible rates of aerobic degradation produced a model with two quite different locally-stable steady states: a carbon-fixing, carbon-exporting marsh versus a carbon-degrading, carbon-importing marsh. Variations in the standing stock of aerobic microbes determined which of the two states was obtained in the model.

Introduction

The coastal salt marshes of Georgia comprise one of the largest relatively undisturbed tidal ecosystems in the eastern United States. These marshes have been the focus of ecological study at the University of Georgia Marine Institute since the mid-1950s. Until 1972 this research was conducted by individuals or small groups—often without an explicit ecosystem focus. In that year a long-term, coordinated study of the marsh ecosystem was begun, with the overall objective of studying the dynamics of carbon transfer and transformation in the marsh to determine whether the system acted as a source or a sink for fixed carbon.

The coastal tidal salt marshes differ from terrestrial and lotic ecosystems by the daily bidirectional movement of tidal water. This transport mechanism, plus the finding that net primary production exceeded the heterotrophic demand on the marsh, led to the "outwelling hypothesis". According to this idea the twice daily inundation of the marsh provided a mechanism for removing large quantities of organic carbon, especially particulate organic carbon, from the marsh into the estuary, and eventually into the nearshore coastal region (Nixon 1980; Pomeroy and Wiegert 1981). In 1972 virtually no data existed to support or refute this hypothesis, and the University of Georgia group made it the focus of their research. Ecosystem modeling was incorporated as an integral part of the field and laboratory measurements and experiments in this program.

The first model, constructed in 1973 largely from published data (Wiegert *et al.* 1975), suggested useful measurements and experiments that resulted in refinements and a major revision of the model in 1977 (Wiegert and Wetzel 1979). At this point the model predicted a large surplus of fixed carbon that could be exported, but it included no explicit mechanisms for simulating the dynamics of physical transport nor did the necessary data exist to construct such mechanisms. Studies in the Duplin River provided data on standing stocks and on diffusive and advective transport of carbon (Imberger *et al.* 1983). Incorporating this new information into the model provided new predictions, that led to a two-year "flume" study of carbon and nitrogen transport in the marsh and in the Duplin River (Chalmers *et al.* 1985). This research suggested that most of the excess carbon produced in the marsh could not, by any reasonable manipulation of the hydrological data, be physically transported into the estuary, thereby refuting the "outwelling hypothesis" in its simplest form.

The current hypothesis (Chalmers *et al.* 1985) is that the twice-daily tides, instead of eroding material from the marsh surface on each ebb, actually deposit a thin film of particulate organic carbon on the marsh at slack high tide. This layer accumulates until it is removed by the runoff from rain on the exposed marsh (light rains of a few millimeters suffice) or until it is utilized by microorganisms in the aerobic surface layer. The latter process transforms the carbon into microbial biomass that enters the food web and supports the larger migrant populations that use the marsh as a foraging area during high tide. Thus loss of carbon from the marsh may occur 1) through *in situ* metabolic transformation into carbon dioxide; 2) by physical transport in the guts and tissues of migrants that use the marsh and move into the estuary; or 3) by physical diffusion or convection augmented by the aperiodic flushing effect of large amounts of fresh water from severe storms (Imberger *et al.* 1983). The current hypothesis further assumes that the daily tidal deposition of carbon onto the marsh, followed by rain-induced erosion back into the upper parts of the tidal creeks, augments the *in situ* metabolic loss by maintaining a thin aerobic layer of particulate organic carbon (POC).

To test this expanded, current hypothesis beyond the purely descriptive measurements of carbon deposition and standing stocks required detailed knowledge of the populations and movements of the major groups involved in these food chains, as well as major changes in the salt marsh model. Studies of specific migrant groups were begun in the early 1980s with penaeid shrimp (Vetter 1983) and are continuing with the blue crab (*Callinectes*). These results plus more detailed information on the interactions between soil water and the productivity of *Spartina alterniflora,* have resulted in a major expansion of the number of compartments in the salt marsh model, and refinement of the biological control incorporated in the equations (Wiegert, Wetzel and Vetter, ms. in prep.). I have used the model to examine the nature of temporal interactions between 1) the marsh proper, with its relatively sedentary populations (rooted plants, benthic algae, burrowing invertebrates, etc.) and 2) the water-borne system that moves onto the marsh with each high tide and is resident in the estuary and large tidal creeks at low tide. This paper describes results of modeling experiments on the interactions between these two spatial divisions that are primarily responsible for variability in the tidal marsh ecosystem.

A simulation model of this type incorporates hypotheses explaining how the system operates, and manipulation of the model is equivalent to manipulation of the hypotheses in order to make predictions that can be tested by further observation or experiment. While scientific hypotheses can sometimes be made to produce predictions by simple logical deduction, the logical deduction may be so complicated that experiments on the natural system are required. A simulation model, viewed in this light, is simply a tool for organizing and conducting some of the preliminary experimentation, it is not a way to avoid the observation and experiment necessary for testing the conclusions.

The Model

The revised model has 23 compartments, 15 biotic and 8 abiotic (Fig. 1). The major expansion from the original 14-compartment versions of the model was to split compartment 7 (heterotrophs in the water) into seven compartments, two migratory with the tides, four resident and one, zooplankton (X23), that remains with the water and is thus only passively migratory. Migrants are *particle feeders* (X27) (shrimp, small crabs, killifish, etc.) and *top carnivores* (X36) (large crabs, sea trout, drum etc.). Residents are *filter feeders* (X28) (mussel and oysters), *benthic infauna* (X34) (polychaete worms), *particle feeders* (X35) (fiddler crabs) and *meiofauna* (X39) (protozoa and nematodes). The algal compartment was separated into benthic algae (X21) and phytoplankton (X22). A new detritus compartment (X33 and associated aerobic microorganisms X32) was added in this version of the model to represent the thin aerobic flocculent layer deposited by successive tidal inundations. This is the transition layer between the water-borne POC (X9) and the anaerobic soil organic carbon (X13 and associated microorganisms X8) compartments.

As in the original versions, the modeled system can be visualized best as compartmentalized into soil, water and a quasi-terrestrial component represented by the aerial shoots of *Spartina* and their consumers (X4). These consumers might have a large impact on the salt marsh if they strongly influenced the survival and growth of *Spartina*, but there is no evidence for this at present and this group is relatively unimportant in the transport and transformation of carbon.

The soil-plant association in the high marsh is markedly different from that in the creek banks or levees. The high marsh is less productive, the production is generally limited by factors other than light (Giurgevich and Dunn 1982;

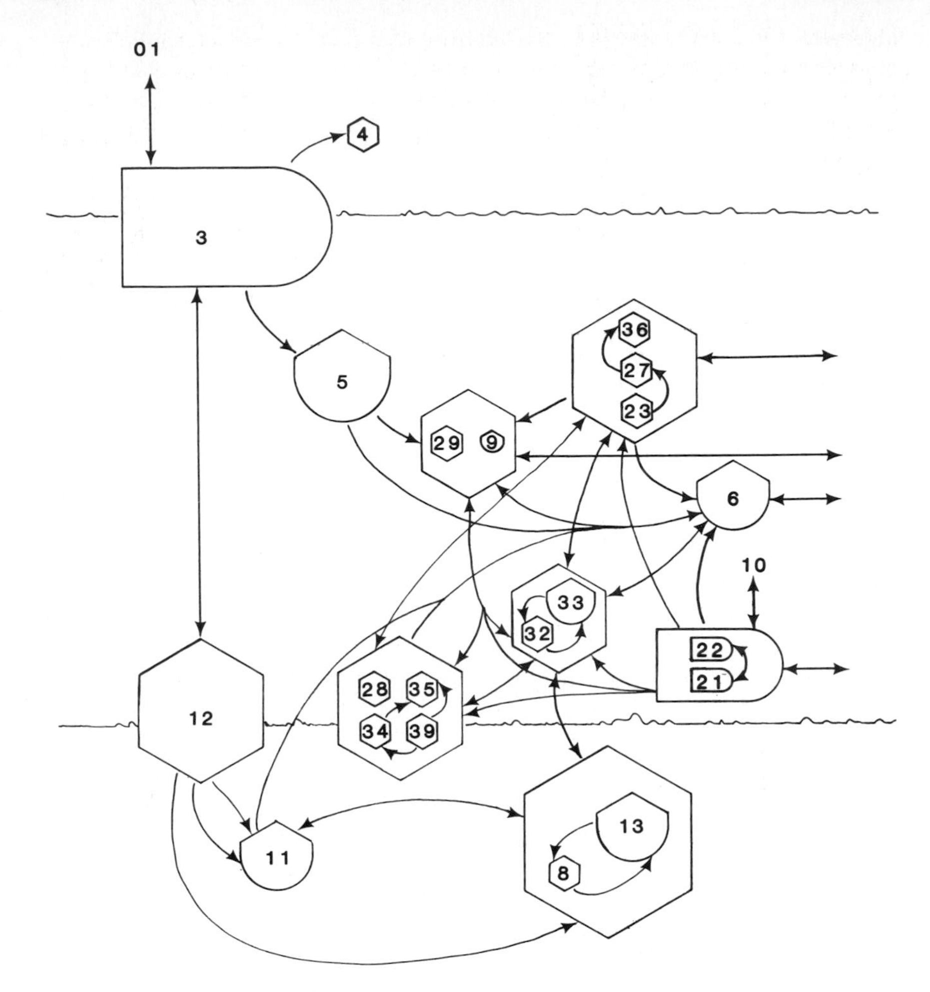

Figure 1. *Diagram of the salt marsh model showing the major flux pathways for carbon transfer. The state variables are: (X1) CO_2 in the air (not shown); (X3) Shoots of* Spartina alterniflora; *(X4) Consumers of* Spartina *shoots; (X5) Standing dead plant material; (X6) Dissolved organic carbon (DOC) in the water; (X8) Anaerobic microbes in the sediment; (X9) Particulate organic carbon (POC) in the water; (X10) CO_2 in the water (shown only as a number); (X11) POC in the sediment interstitial water (anaerobic); (X12) Roots and rhizomes of* Spartina; *(X13) POC in the sediment (anaerobic); (X21) Benthic algae; (X22) Phytoplankton; (X23) Zooplankton; (X27) Migrant particle feeders (shrimp, small crabs, killifish); (X28) Filter feeders (oysters, mussels); (X29) Aerobic microbes in the water; (X32) Aerobic microbes in the flocculent layer; (X33) POC in the flocculent layer; (X34) Benthic infauna (polychaete worms); (X35) Resident particle feeders (fiddler crabs, snails); (X36) Migrant top consumers (larger fish, crabs); (X39) Meiofauna.*

Chalmers 1982) and there is little movement of soil water. The creekbank cordgrass is commonly two meters or more in height, annually produces about 2500 g C m^{-2} net and productivity is usually light-limited. Creekbank type marsh occupies approximately 8 percent of the total marsh surface (Pomeroy and Wiegert 1981).

The controls used in the model follow the ecologically realistic functions described for earlier versions (Wiegert *et al.* 1975; Wiegert and Wetzel 1979 and Wiegert *et al.* 1981).

The Experiments

Three kinds of experiments were undertaken with the expanded model (designated MARSH8.V1): 1) those in which the importance of the migrants in the spatially-distinct tidal water was evaluated; 2) simulated comparison of the spatially distinct creekbank versus the high marsh; and 3) evaluation of the effects of changing the rates of and controls on microbial degradation of carbon within the marsh proper and upper tidal creek. All experiments are discussed with reference to a "nominal" run that incorporates the best information on parameter values and functional control forms. For those compartments retaining the identities of the previous versions of the model, the seasonal dynamics and magnitude of the standing stocks of the nominal simulations were virtually identical to those produced earlier. Thus the following description focuses on the new capabilities for discrimination permitted by the compartmental subdivision.

The Nominal Simulation

Migrant populations comprising small fish, shrimp and crabs (X27) and larger predatory fish (X36) normally move onto the marsh during flood tide and depart with the ebb. To simulate the tidal behavior of these migrants, each tidal cycle was divided into 4 periods, beginning at slack low tide. During the first period (20% of the cycle) all migrants enter the marsh with the flood. During the second period (40% of the cycle) they remain on the marsh and feed, subject to whatever controls due to scarcity of food or crowding are operating on each population. During the third period (20% of the cycle) one half of the migrants leave the marsh. This simulates the behavior of those species, such as penaeid shrimp, that leave the marsh as soon as the ebb current begins. During the final 20% of the cycle, the remaining migrants leave the marsh. This routine requires an iteration interval of 0.1 day in order to simulate two tides per day. A typical diel immigration/emigration of the migrant particle feeders and top predators is shown in Fig. 2 simulating two tides per day with an iteration interval of 0.1 day. However, the nominal *seasonal* behavior of the system components simulated with an iteration interval of 0.2 day (one tidal cycle per day with twice the duration) was virtually indistinguishable from the nominal run using an interval of 0.1 day. Thus the experiments reported here were all run with an iteration interval of 0.2 day.

Simulated seasonal standing stocks of the compartments representing *Spartina alterniflora* are shown in Fig. 3. The majority of the live biomass is

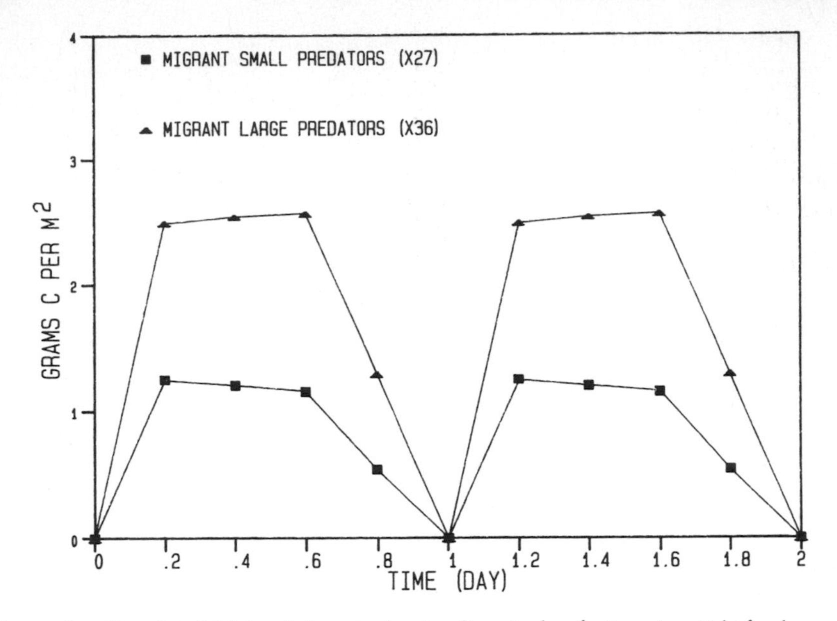

Figure 2. Simulated tidal variations in the standing stocks of migrant particle-feeders and top predators.

found in the root/rhizome compartment (reflecting the dominant influence of the short *Spartina* characteristic of the high marsh). A summer peak in both roots and shoots is followed by the fall die-back. The "standing dead" remains in place and undergoes degradation throughout the winter. The rate of degradation (Newell *et al.* 1985) became known too late for inclusion in these model studies. The error incurred in total rate and amount of degradation by omitting the rate of decomposition of standing dead was minimized by assuming that the material quickly enters the POC compartments due to the action of physical and biotic processes.

Aerobic degradation in the present model occurs only after the standing dead material is immersed in the water and becomes part of the POC, either in the water (suspended, X9) or as part of the erodable, sediment flocculent layer (X33), that is washed from the surface of the marsh with each light rainfall at low tide and then redeposited by subsequent successive tides (Chalmers *et al.* 1985).

The seasonal behavior of the aerobic microbial standing stocks is given in Fig. 4. Because of the much lower standing stock of suspended POC compared to the flocculent layer, the suspended microbial standing stock is much lower throughout the year. In the simulations, the aerobic microbial compartments show a rich dynamical behavior, due only partly to the seasonality of parameter values. A large part of the seasonal variation in standing stock of aerobic microbes results from changes in the consumers of microbes (which also vary

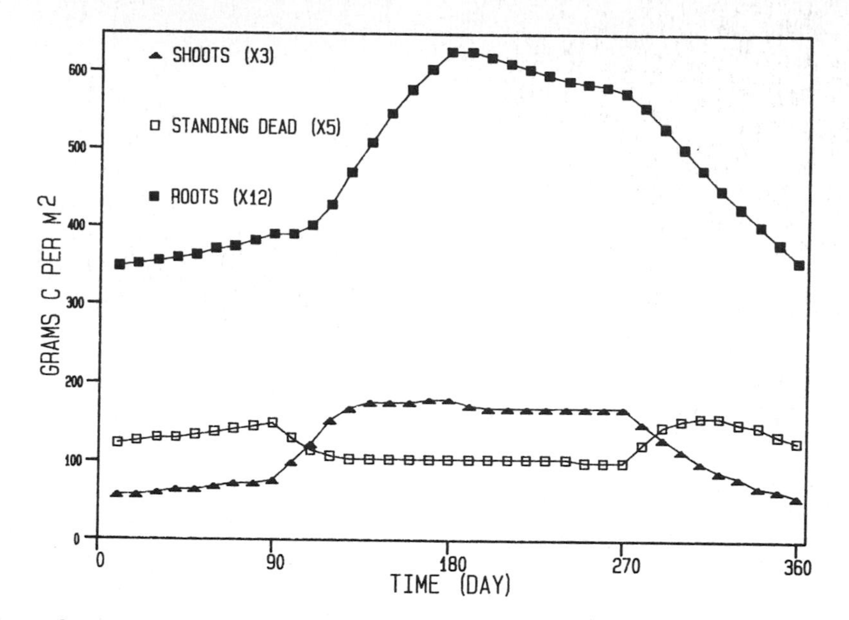

Figure 3. *Simulated seasonal variation in the standing stocks of* Spartina alterniflora *shoots, standing dead and roots/rhizomes, for the nominal case.*

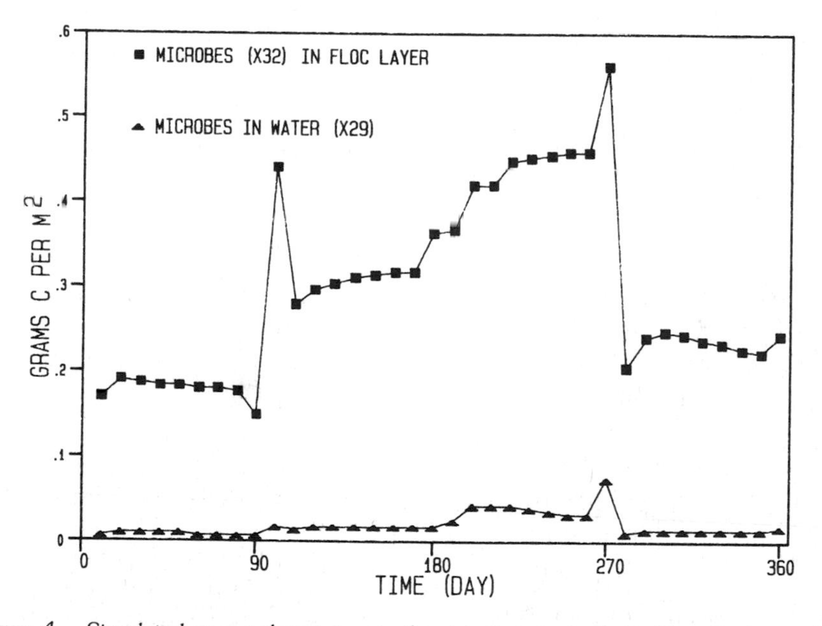

Figure 4. *Simulated seasonal variation in the standing stock of aerobic microbes in the water and in the flocculent layer of the Duplin R. marshes (nominal case).*

seasonally). For example, the aerobic microbes in the water (X29) by definition use the suspended POC (X9) for a physical substrate as well as for organic resources, and a close correlation between the seasonal variation in the suspended POC and the standing stock of microbes was expected (Fig. 5). During the summer, however, the dynamics of the microbe population depart from this close correlation. The microbial population declines during midsummer and increases sharply during late summer, just prior to the seasonal parameter changes for autumn. The reasons for this departure from the expected simulation are coupled to the zooplankton (X23) and the migrant particle feeders (X27) (Fig. 6). The zooplankton, a major consumer of the aerobic microbes (and particles) suspended in the water, are held in check in early summer by the migrant small consumers such as killifish (*Fundulus*). The onset of the summer blooms of phytoplankton, a preferred zooplankton resource, allow the zooplankton to escape from control by their predators, however, and deplete the aerobic microbes. The sharp decline in zooplankton standing stock during the last week of the summer period is due to the occurrence of a simulated storm on model day 262, eight days before the onset of autumn. Such storms may remove 95 percent of the zooplankton and phytoplankton standing stocks, flushing them into the estuary (Wiegert *et al.* 1981). During the ensuing 10 days the phytoplankton, with a great potential rate of increase, had recovered suffi-

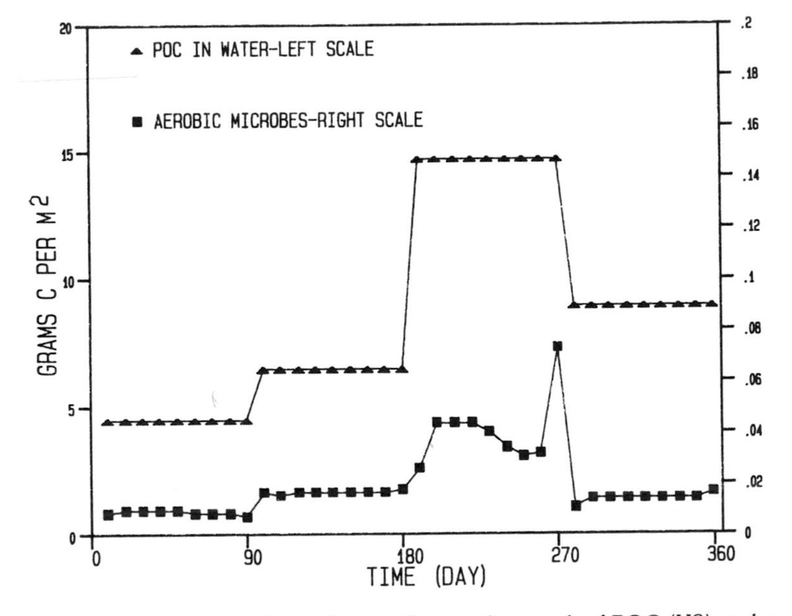

Figure 5. *Simulated seasonal variations in the standing stock of POC (X9) and aerobic microbe population (X29) in the Duplin R. marshes.*

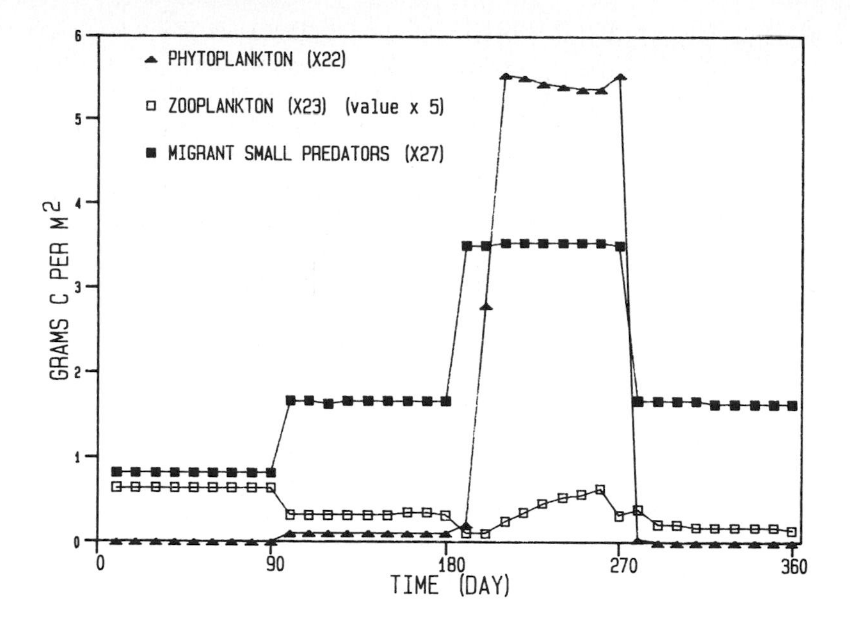

Figure 6. Nominal simulated seasonal variation in the standing stock of phytoplankton, zooplankton and migrant particle feeders in the Duplin R. marshes.

ciently such that only a small decrease in the standing stock was apparent (Fig 6). The zooplankton had recovered to less than half of the standing stock present before the storm (< 2 g C m^{-2}), permitting the sharp increase in microbes (Fig. 5). The zooplankton continued to increase during the first week of autumn, but the autumn decline of the phytoplankton eventually caused a crash in their consumers.

The revised model, as reflected by the nominal simulation, supports the amended outwelling hypothesis discussed previously. The fluxes summarized for the nominal simulation (Table 1) show an export (at least to the upper portion of the Duplin estuary) of 598 g C m^{-2} y^{-1} and an import (fish, shrimp and zooplankton) of 160 g C m^{-2} y^{-1} for net export of 438 g C m^{-2} y^{-1}. The remainder of the surplus carbon is lost as respiration by the migrants and the consumers of lower tropic levels. The net increase of carbon storage in the sediments (20 g C m^{-2} y^{-1}) is consistent with geological estimates of the mean annual rise in sea level and marsh elevation. For example, a 2-mm y^{-1} rise (Hicks *et al.* 1983; Stevenson *et al.*, this volume), coupled with measured values for carbon content of the sediments (Letzsch and Frey 1980; Christian *et al.* 1974), gives a calculated deposition of 48-66 g C m^{-2} y^{-1}. Whether the predicted *distribution* of this export is correct, however, and how much of it finally reaches the nearshore coastal zone is a matter for future research. Experimental manipulation of the revised model generates predictions of spatial and

Table 1. Cumulative fluxes (g C m^{-2} y^{-1}) from eleven different simulation runs with model MARSH8.V1, an expanded 23-compartment model of the coastal Georgia salt marsh. All values are from the second year of a two year simulation. The system was in steady state except for the anaerobic sediment and microbes, thus the total system change in carbon is given by the change in X13 + X8. The runs are: 0) The nominal simulation. 1) Doubled daily input of migrant fish and shrimp (X27) and migrant top predators (X36). 2) Zero migrant compartments. 3) Increased production of *Spartina* to equal that of the creek banks. 4) Decreased (50%) threshold controls for microbes (X8) degrading POC in the anaerobic sediments (X13). 5) Increased (200%) thresholds for X8. 6) Decreased (50%) thresholds for microbes (X29 and X32) degrading POC (X9 and X33) in aerobic zones (water and floc on the surface of the marsh). 7) Increased (200%) thresholds for microbes (X29 and X32) degrading POC (X9 and X29) in the aerobic zones (in the water and in the floc on the surface of the marsh). Nominal initial conditions. 8) Increased (1000%) initial conditions of aerobic microbes X29 and X32. 9) Increased (200%) maximum assimilation rates of X29 and X32 with nominal initial conditions. 10) Increased (25%) thresholds for microbes (X29 and X32) degrading POC (X9 and X33) in aerobic zones (water and floc on the marsh surface), plus high initial conditions for X29 and X32 (1000%). 11) Increased (200%) maximum assimilation rates of X29 and X32 with high initial conditions (1000%) for X29 and X32.

Flux Categories	Experiment Number											
	0	1	2	3	4	5	6	7	8	9	10	11
Net export by part, feeders (X27)	−52	−221	0	−49	−52	−52	−53	−52	−52	−42	−1	−1
Net export by top predators (X36)	62	163	0	60	62	62	62	61	62	57	57	57
Net Export as POC (X9 + X29)	207	206	205	207	207	207	207	207	207	207	167	216
Export as DOC (X6)	146	139	123	309	146	146	151	144	146	147	−244	−169
Export as phytoplankton (X22)	183	0	0	165	183	183	185	182	183	178	0	0
Export as zooplankton (X23)	−108	−148	−46	−108	−109	−109	−112	−108	−109	−89	−29	−28
Net POC (X9 + X29) to Floc (X33 + X32)	640	701	620	1583	640	640	637	641	640	638	218	165
Change in anaerobic sediment (X13 + X8)	20	72	−27	897	579	−905	23	19	21	−59	−284	−385
Net prod.: benthic algae (X21)	27	36	18	24	27	27	28	27	27	23	23	21
phytoplankton (X22)	243	15	11	221	243	243	242	243	243	254	16	16
Spartina (X3 + X12)	1574	1574	1574	2751	1574	1574	1574	1574	1574	1574	1579	1574

temporal variability in the marsh, and can help identify productive avenues for this future research.

Changes in Migration On and Off the Marsh

The current hypothesis about carbon transport of the marsh assigns considerable importance to transport by migrants as well as the secondary effects they may have as consumers while on the marsh. Thus the effect of doubling and zeroing the input of migrant consumers in the two compartments X27 and X36 was examined.

Doubling the net daily immigration rate of consumers onto the marsh depressed several compartments. The zooplankton decreased as a result of increased consumption by the mobile small consumers (Fig. 7). The combined predation by both groups of migrant consumers reduced the resident particle feeders, the benthic infauna and the meiofauna (Fig. 8). The secondary effects of this simulated increase in predation included stimulation of benthic algae (by release from grazing pressure by residents) and depression of the summer phytoplankton peak.

Important fluxes changed in the system (Table 1, Expt. 1). The nominal slight positive import of organic carbon via migration became a net export of 58 g C m^{-2} y^{-1}. Net export of phytoplankton decreased to zero and the net import of zooplankton increased by 50%. Net production by benthic algae and

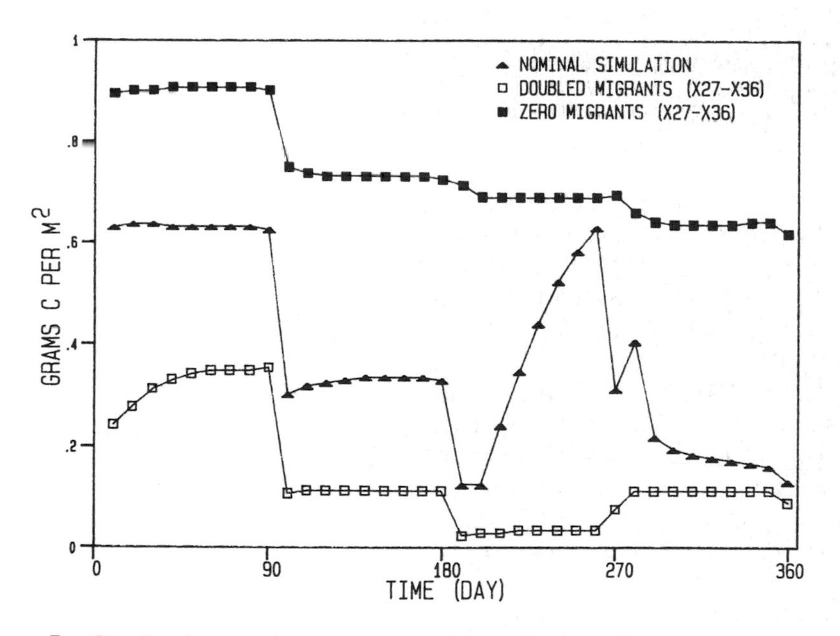

Figure 7. Simulated seasonal variation in standing stock of zooplankton (X23) for three experiments: 1) nominal, 2) doubled migrants and 3) zero migrants.

phytoplankton decreased from 270 to 51 g C m^{-2} y^{-1} with the result that total net system export of organic carbon (sum of "exports" in Table 1) decreased from the nominal 436 to 139 g C m^{-2} y^{-1}.

Removing (zeroing) the movement of migrants onto the marsh in general reversed the effects of doubling. That is, phytoplankton, zooplankton (Fig. 7) and benthic particle feeders increased whereas benthic algae decreased. Both doubling and zeroing however, reduced aerobic bacteria, benthic infauna (Fig. 8) and meiofauna! This similarity in effects of two opposite manipulations is explained by examining the interactions between the migrants and resident consumers. For example, the benthic infauna are consumed by both resident and migrant particle feeders. Doubling the latter directly affected the benthic infauna enough to offset the positive effects of the migrant's simultaneous reduction of the resident particle feeders. On the other hand, removing all of the migrant predators and particle feeders resulted in an increase of resident particle feeders thereby reducing the major resources of the benthic infauna [benthic algae and aerobic bacteria in the flocculent layer (X32 and X33)] and precipitating a decline in standing stock. Hence the opposing manipulations produced similar changes (Fig. 8), although different in magnitude and for different reasons!

The overall system effects of this manipulation (Table 1, Expt. 2) were to reduce the net import of zooplankton to 46 g C m^{-2} y^{-1} and cause a net loss of carbon to the anaerobic sediments. The total net system export of organic carbon rose to 282 g C m^{-2} y^{-1}, 154 less than the export predicted by the nominal model.

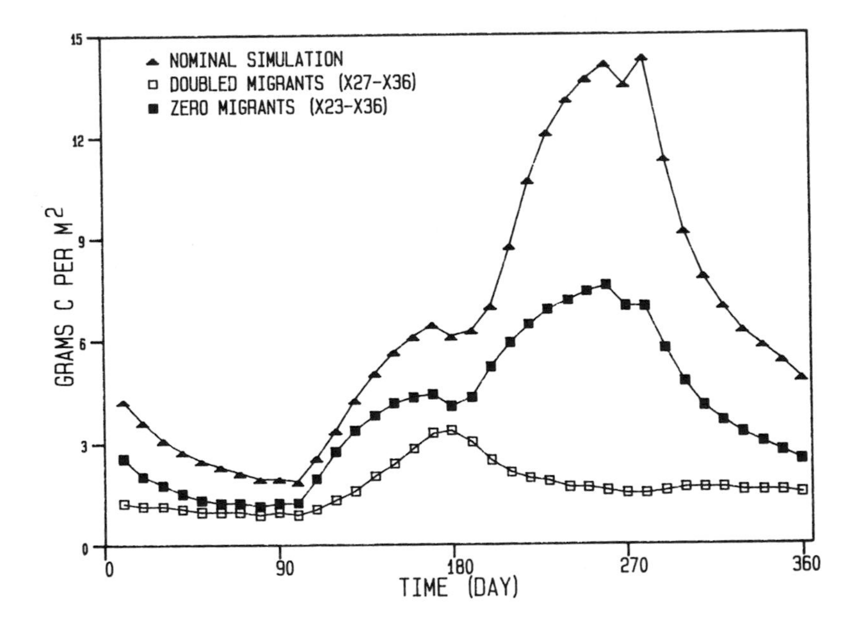

Figure 8. Simulated seasonal variation in standing stock of benthic infauna (X34) for three experiments: 1) nominal 2) doubled migrants 3) zero migrants.

Increased Net Production by *Spartina*

The levees and creek banks of Sapelo area marshes are significantly more productive than the high marsh areas. Studies on the Sapelo marshes estimate the total net production in the creek bank zone at about 2500 g C m^{-2} y^{-1} (Pomeroy and Wiegert 1981). However, creek banks and levees occupy only a small fraction of the total marsh. Ideally, simulation of such large-scale spatial variability would be done with two different models run simultaneously since the position of the two community variations remains fixed. As a first approximation to this ideal, I increased the shoot carrying capacity of the overall marsh to approximate that of creek bank plants. Additional changes were made in other parameters necessary to more closely approximate the physiological characteristics of the creek bank and levee vegetation.

The shoot standing crop increased and the root/rhizome standing crop decreased (Fig. 9), in agreement with the field situation (Gallagher and Plumley 1979; Gallagher *et al.* 1980). Net production by shoots, roots and rhizomes increased to 2571 g C m^{-2} y^{-1}(Table 1, Expt. 3). This greatly increased production in turn, increased the accretion to the sediments (X13) and the standing stocks of anaerobic microbes (X8). The increased production by the dominant vascular plants caused only minor changes in the standing stocks of the other components of the system. The major ecosystem effects of the change in parameters controlling plant production (Table 1, Expt. 3) were an increase in the ex-

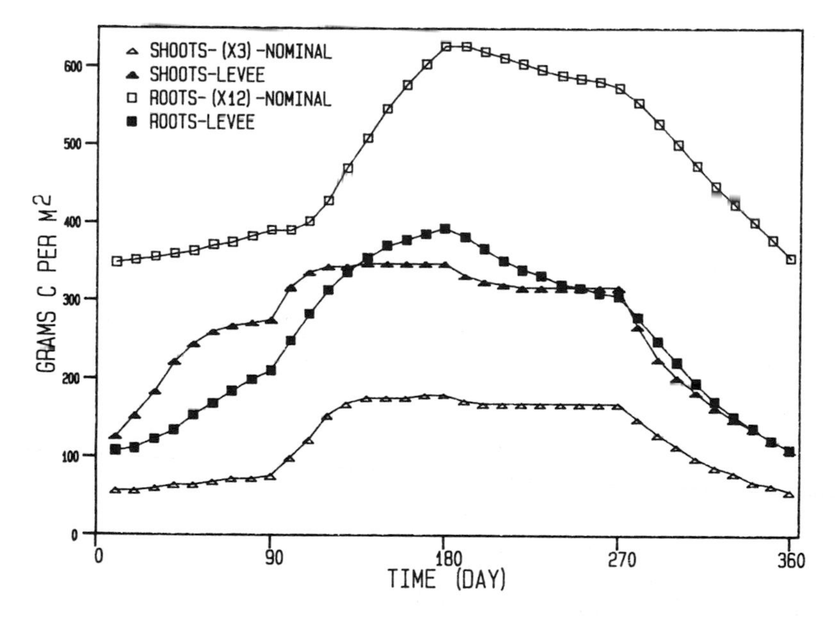

Figure 9. Simulated seasonal variation in standing stocks of Spartina *shoots and roots/ rhizomes for nominal productivity versus the increased rates characteristic of levees and creek banks.*

port of DOC from 146 to 309 g C m^{-2} y^{-1} and a large increase in the deposition of POC (plus microbes) from the water to the flocculent layer (640 g C to 1583 g C). Much of this flux eventually contributed to the large increase in the net accretion of carbon in the anaerobic sediment which increased from 20 g C m^{-2} y^{-1} in the nominal case to 897 g C m^{-2} y^{-1}. The total net system export to the estuary increased from 435 to 583 g C m^{-2} y^{-1}.

Changed Microbial Degradation

The final series of simulation experiments (4-11) concerned the effects of manipulating the thresholds and rates of degradation by both anaerobic and aerobic microorganisms on the salt marsh ecosystem. These last experiments were suggested by the discovery that rather small changes in the potential microbial rates of degradation could lead to two locally-stable steady states with greatly different character with respect to the transformation and transport of carbon. Thus the potential exists for spatial variability on a much smaller scale than that of creek bank versus high marsh.

The feedback functions used in the model to control the rates of microbial degradation employ thresholds which are not absolute amounts of carbon but are instead ratios of microbial carbon to available substrates (Wiegert *et al.* 1975; Christian and Wetzel 1978). Above the lower or response threshold ratio, microbial rates begin to be limited. The upper threshold is the carrying capacity.

In Expt. 4 the threshold limits for the ratio of anaerobes (X8) to POC (X13) were decreased, thus reducing the carrying capacity of the POC for microbes. This change decreased the standing stock of the microbes and increased the accretion of carbon in the anaerobic sediment. By the second year of simulation (Table 1) the annual rate of accretion (579 g C m^{-2} y^{-1}) to the anaerobic sediment was far higher than any reasonable estimate based on the geological data. The standing stocks of the aerobic parts of the marsh system were affected very little. In the short run (2 years) the system affects of this manipulation were virtually unchanged from the nominal run expect for the greatly increased accretion of carbon in the sediments.

The opposite manipulation, i.e., increasing the threshold ratios controlling anaerobes (Expt. 5), increased only the standing stock of anaerobic microbes, causing an annual *loss* from the sediments of 905 g C m^{-2}. Because these two simulations affected only the anaerobic system processes in the salt marsh, I conclude they are weakly coupled to the remainder of the system, at least over the short term. Similarly, Christian *et al.* (1978) found little change in the sediment anaerobic system when it was experimentally divested of organic matter input from plant shoots, roots and rhizomes for more than one year. The simulation experiments suggest also that the reverse coupling is weak over at least a period of one year or more. That is, any increase in organic matter (and associated microbes) in the anaerobic zone is not reflected by change in the aerobic processes in the system.

Changing the parameters governing the aerobic microbial processes was expected to cause significant changes in the ecosystem function, however, because of the direct connection between microbial grazers and the important consumer compartments in the water. Thus it was surprising to find little change in system behavior when the threshold ratios governing the degradation rate of aerobic microbes (X29 and X33) were varied (Table 1, Expts. 6, 7). In Expt. 6 the threshold ratios governing the degradation rates of aerobic microbes in the flocculent layer (X32) and in the water were decreased by one half. In Expt. 7 these same threshold ratios were doubled. In each case the important cumulative system fluxes and the seasonally varying standing stocks remained essentially unchanged from the nominal simulation. A further effort to perturb the system (Table 1, Expt. 8) was made by increasing the initial standing stocks of aerobic microbes (X29 and X32) by tenfold, again with no apparent effect; the minor perturbation caused by the initial condition change had virtually disappeared by the second year.

These changes failed to influence overall system behavior because, in the nominal simulation, both microbe compartments, X29 and X32, are controlled by grazing to levels far below the densities where limitation by either scarcity of resource or by crowding on the particles would be felt. Thus, no increases or moderate decreases of these limits produced any effect, a phonomenon also found in the model used by Christian and Wetzel (1978). Increasing the initial standing stocks temporarily increased the resource available to the microbial grazers but caused no lasting change in system behavior.

Doubling the maximum assimilation rates of the two microbial compartments (Expt. 9) produced a modest 27 g C m^{-2} increase in the annual export by the system (mostly by decreasing the net import of migrant particle feeders and zooplankton). Sediments exhibited an annual loss of 59 g C m^{-2} in contrast to the 20 g C m^{-2} y^{-1} accretion in the nominal case. Benthic algae and phytoplankton standing stocks decreased on average about 20% and 15% respectively and zooplankton, resident particle feeders and meiofauna increased between 5 and 15% because of the increased productivity and availability of aerobic microbes.

Increases in either the threshold ratios or the maximum potential assimilation rates produced an important change in the model: it now has two locally-stable steady states, completely different from each other insofar as the transformation and transport of carbon are concerned. Which of these states is realized now depends solely on the perturbations applied to the standing stock of aerobic microbes.

In Expt. 10, thresholds controlling aerobic microbes were increased only 25% (not doubled as in Expt. 7) and the system was perturbed by initializing it with a tenfold grater standing stock of microbes. As a result the microbes escaped control by grazers. The standing stock of DOC in the water was reduced greatly (Fig. 10); movement of DOC from the sediment to the water was thus accelerated, as was movement of POC out of the anaerobic sediment and into the flocculent layer, where it was subjected to increased degradation associated with the greatly increased standing stock of the aerobic microbes in the floc. All groups feeding on bacteria increased, but they were no longer effective in holding the aerobic microbes to low levels. Microbes increased quickly to the point where the major controlling factor was crowding.

Relative to the nominal simulation benthic algae increased in spring, decreased in summer and increased again in the autumn (Fig. 11). Phytoplankton increased slightly above the levels in the nominal simulation only in the winter and autumn. This group was severely depressed in spring and the large summer bloom was almost completely suppressed. The benthic algae and phytoplankton do not directly utilize the aerobic microbes or their products (except CO_2 which was never a limiting factor in the simulations). Thus the response of the algae to the increase in the aerobic microbes is an indirect effect of changes in the algal consumers, specifically seasonal switching by omnivorous consumers between algae and the aerobic microbes.

The results of Expt. 10 at the system level (Table 1) were to increase the net export from the combined migrants to 58 g C, change the export of DOC to a net import of 244 g C, reduce the export via phytoplankton to less than 1 g C, reduce the import of zooplankton to 28 g C and reduce the net movement of POC/microbes from the water to the flocculent layer to 218 g C. The net production by benthic algae was reduced only slightly, but because of the suppression of the large summer blooms, the production by phytoplankton was reduced by more than 90% to 16 g C. Net system exchange of carbon with the estuary was changed from an export of 435 g C to an import of 49 g C m^{-2} y^{-1}. This, together with the change to a net annual decrease in the carbon in the anaerobic sediments amounting to 351 g C m^{-2} y^{-1} shifted the salt marsh

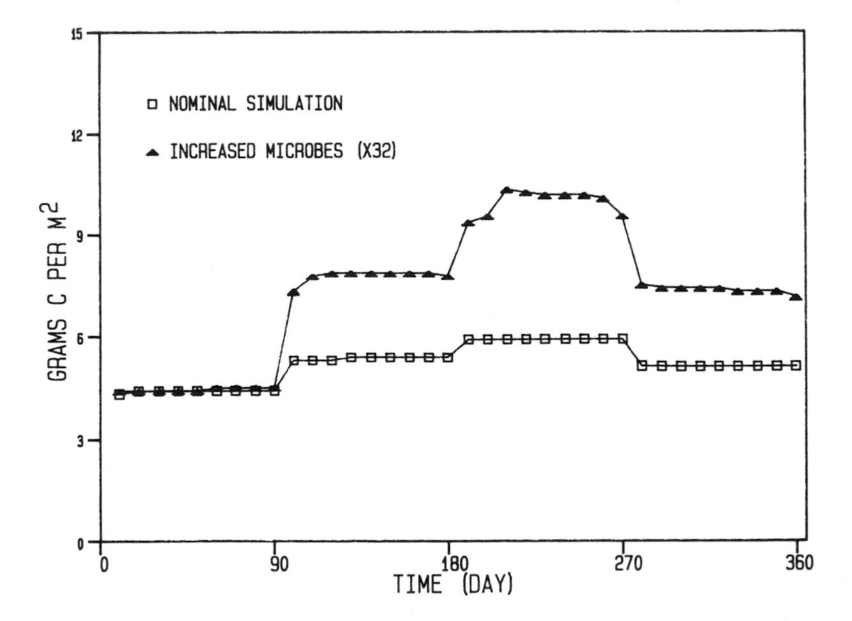

Figure 10. Simulated seasonal variation in the standing stock of DOC (X6) for nominal versus increased microbial threshold rates.

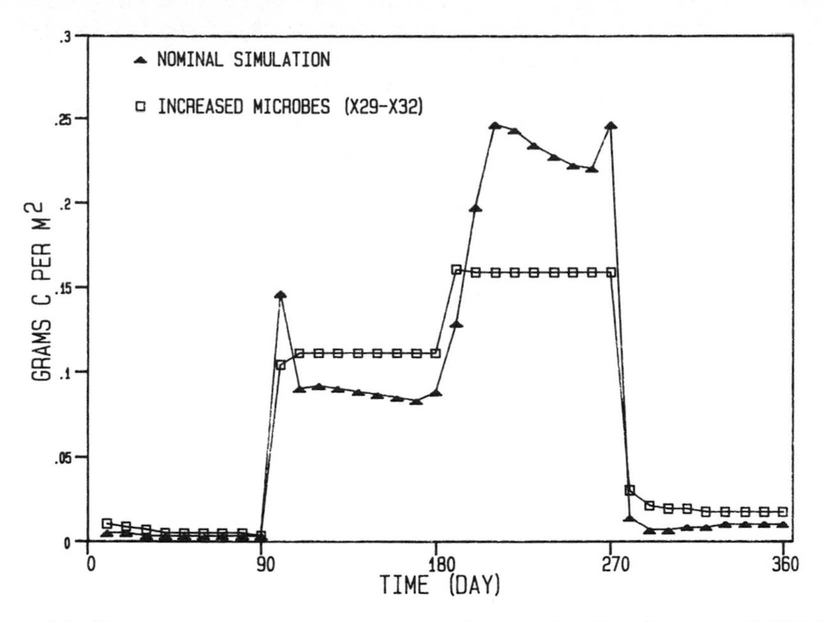

Figure 11. Simulated seasonal variation in standing stock of benthic algae (X21) for nominal versus increased microbial threshold ratios.

ecosystem from a system accreting and exporting organic carbon to one importing and degrading organic carbon.

In Expt. 11 the doubling of the maximum potential rates of assimilation by aerobic microbes, when coupled with a tenfold increase in initial values of the microbes resulted in changes similar to those produced in Expt. 10, namely, the microbes escaped from control by grazing. The magnitude of change in standing stocks, however, were on average somewhat lower than those of Expt. 10. Figures 12 and 13 contrast the standing stocks of the two aerobic microbe compartments, X29 and X32, in the nominal case and in the two simulations where these compartments escaped from control by grazers. In each of the latter cases the microbe populations have increased to the point of severe space limitation. The seasonal changes are caused by changes in the availability of POC in the water (Fig. 12) and changes in the abundance of grazers on the bacteria in the flocculent layer (Fig. 13).

Experiment 11 produced overall changes in annual system fluxes (Table 1) that were almost identical with those produced by the manipulation of thresholds in Expt. 10. Even though the standing stocks of microbes were not as great as they were in Expt. 10, the net loss of carbon from the anaerobic sediment was even greater, 385 g C m^{-2} y^{-1}. The salt marsh system was still exporting 124 g C m^{-2} y^{-1}, but in view of the large annual loss from the sediments, this can be regarded only as a temporary condition. Overall, this manipulation also

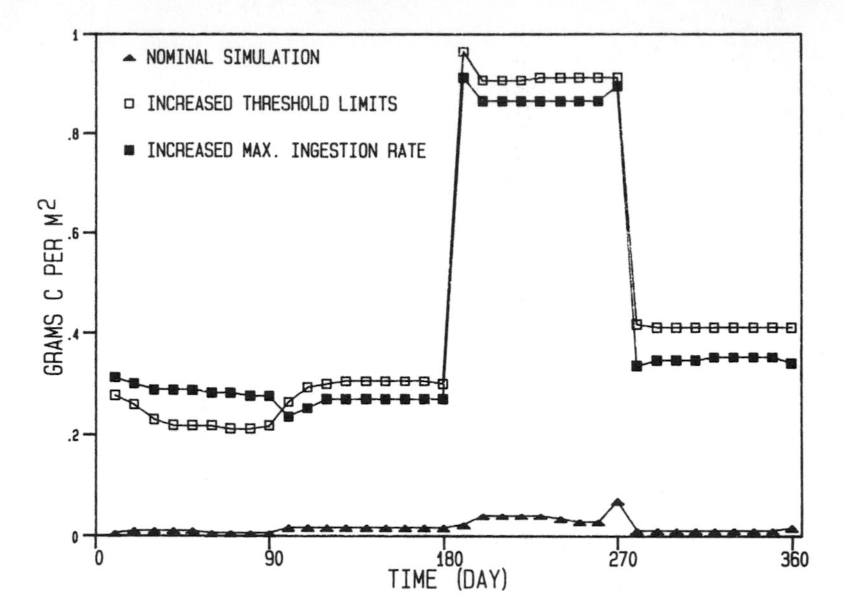

Figure 12. Simulated seasonal variation in standing stock of aerobic microbes in water (X29) for 3 experiments: 1) nominal; 2) increased microbial threshold ratios; and 3) increased maximum assimilation rates.

transformed the system from a carbon fixing, "source" ecosystem, to a carbon degrading, "sink" ecosystem.

The manipulations of aerobic microbial degradation (Expts. 4-11) showed the nominal system is in a stable steady state with respect to perturbations of either the initial standing stocks of aerobic decomposers or the parameters controlling their rates of degradation of POC. Relatively modest changes in the control parameters can, however, change the model to one that is sensitive to changes in the initial standing stocks of microbes. The system shifts to a different locally-stable steady state upon sufficiently large perturbation of the standing stock of aerobic microbes. The two possible steady states are very different: in one the system is accreting and exporting carbon whereas in the other the system becomes a degrader of organic carbon.

Discussion and Conclusions

Experimental manipulation of the newest version of the salt marsh model has underscored the value of simulation models in collating data and synthesizing the consequences of hypotheses about how components of ecosystems interact. Although it may be relatively easy to trace out the causal mechanism once a simulation result is at hand, an *a priori* prediction of the simulation result, let alone an explanation, would be difficult even in such simple cases as those

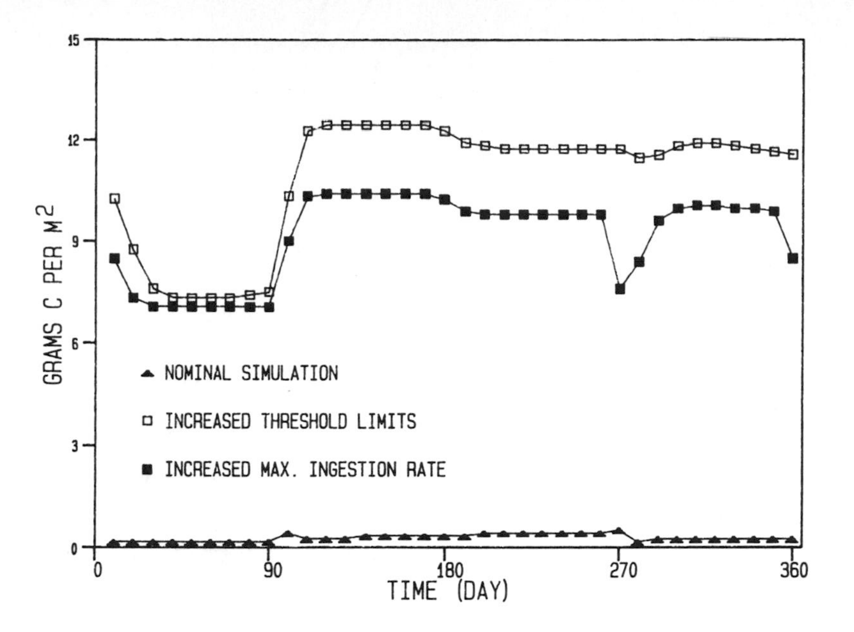

Figure 13. Simulated seasonal variation in standing stock of aerobic microbes in the flocculent detrital layer (X32) for 3 experiments: 1) nominal; 2) increased microbial threshold ratios; and 3) increased maximum assimilation rates.

described in this paper. With more complex series of interactions it would be virtually impossible. Our experiments showed that certain components and processes in the marsh model are strongly coupled whereas others appear only weakly connected, at least in the short term. Our results also suggest some new avenues for research designed to increase our understanding of this complex ecosystem.

Manipulating the migrant particle feeders and top predators showed that these two groups can have an important influence on the standing stocks and the fluxes of carbon in the marsh system. In particular, they can act not only as a mechanism for transporting carbon onto and off of the marsh, but can secondarily influence the behavior of the resident compartments. The finding that increasing and decreasing the migrants both served to decrease the standing stocks and carbon fluxes associated with the benthic infauna and meiofauna is counterintuitive and needs to be investigated further. Are the controls used for these groups in the model accurate? How sensitive is the result to other types of manipulations of parameters, initial conditions and seasonal variation? These are the kinds of questions suggested by these first approximations. We know very little about the ecology of most of the migrant populations moving onto and off the coastal salt marshes. Our model results suggest that we need to know more before an accurate picture of carbon exchanges can be constructed.

The manipulation of the productivity of *Spartina* underscored the fact that this is the dominant component insofar as the transformation of carbon is concerned. We have quite good data on the annual fixation of carbon, at least for the marshes around Sapelo Island. However, the model is still weak in the level of realism incorporated in the factors controlling this production. Current work on the effects of subsurface water movement on the productivity of *Spartina* will alleviate this problem somewhat. But further research on the high productivity creek bank/levee subsystem is needed.

One of the major sources of spatial variation in the Georgia salt marshes is the discontinuity between the lushly vegetated creek banks and levees and the more stressed, lower-productivity high marsh. The causal factor largely responsible for the difference in productivity between these two areas is the amount of subsurface water movement (Wiegert *et al.* 1983). But the effect of this spatial variability in *Spartina* production on the rest of the system has been little studied. Comparison of system responses in the nominal simulation and Experiment 3 suggests the potential range of variability in salt marsh characteristics and processes.

Manipulations of the factors controlling microbial transformations of carbon in the marsh produced the most interesting results. The demonstration of the weak coupling between the anaerobic processes and the other subsystems in the marsh, at least over the short run, suggests that experimental manipulations of the anaerobic systems in the laboratory might not sacrifice as much reality as might have been suspected.

Understanding the rates and controls of the aerobic microbial parts of the system, however, is vital. If, for given combinations of threshold ratios or of maximum rates, the system indeed has more than one locally stable steady state, then we must face the possibility that some relatively minor perturbation could drastically alter what appears to be a temporally very stable system. Wiegert (1979) described the rate and control function conditions necessary for more than one locally-stable steady state to be possible in a system model. These conditions were met in the present model and the two states were accidentally discovered as a result of the manipulations of simulation Expts. 6-11. The difficulty is that almost nothing is known about the possibility of error nor about the range of rates and thresholds actually found in the natural salt marsh. For instance, can the aerobic microbes be given sufficient initial advantage in standing stock to increase to the levels predicted? Is it possible that the system is heterogeneous enough so that each of these steady states can exist simultaneously in different parts of the marsh? Furthermore, the thresholds or maximum rates of aerobic degradation in the flocculent layer may vary not only seasonally but over periods longer than annual cycles. If the behavior of the system with respect to the transformation and transport of carbon is highly sensitive to relatively small changes in many of these parameters associated with aerobic microbes, then a number of significant research questions can be formulated along the lines outlined above.

The degradation of organic carbon by aerobic microbial processes, however, is very closely coupled to the operation of the rest of the system, as we have seen in earlier versions and manipulations of the model. Indeed, the hypothesis put forward by Chalmers, *et al.* (1985) invokes the periodic deposition and resuspension of the flocculent detritus layer, and its aerobic degradation and assimilation by organisms in the water, as the mechanism responsible for transforming, (to carbon dioxide), the excess carbon reported in previous investigations, (Pomeroy and Wiegert 1981). Unfortunately, we know relatively little about many of the processes important in determining and controlling the rates of aerobic degradation of organic carbon in salt marshes—a process that appears from these model studies to be one of the most vital links in the chain of our understanding of the spatial and temporal variability in salt marsh ecosystems.

Acknowledgments

This research was supported by the NSF through Grant #28-21-RR099-057 and The Georgia Sea Grant Program through Grant #10-21-R100-148. I thank Drs. Elizabeth Vetter, Richard Wetzel, Robert Christian, Alice Chalmers and Ron Kneib for assistance with the parameter estimates. Ms. Sara Koenig helped prepare the illustrations.

This is Contribution #549 from the University of Georgia Marine Institute.

References Cited

Chalmers, A. G., R. G. Wiegert, and P. Wolf. 1985. Carbon balance in a salt marsh: interactions of diffusive export, tidal deposition and rainfall-caused erosion. *Estuar. Coastal Shelf Sci.* 21:757-771.

Chalmers, A. G. 1982. Soil dynamics and the productivity of *Spartina alterniflora.* pp. 231-243. *In:* V. S. Kennedy (ed.), *Estuarine Comparisons.* Academic Press, New York.

Christian, R. R. and R. L. Wetzel. 1978. Interactions between substrate microbes and consumers of *Spartina* "detritus" in estuaries. pp. 93-114. *In:* M. Wiley (ed.), *Estuarine Interactions.* Academic Press, New York.

Christian, R. R., K. Bancroft and W. J. Wiebe. 1978. Resistence of the microbial community within salt marsh soils to selected perturbations. *Ecology* 59(6):1200-1210.

Gallagher, J. L. and F. G. Plumley. 1979. Underground biomass profiles and dynamics in Atlantic coastal marshes. *Am. J. Bot.* 66:156-161.

Gallagher, J. L., R. J. Reimold, R. A. Linthurst and W. J. Pfeiffer. 1980. Aerial production, mortality and mineral accumulation dynamics in *Spartina alterniflora* and *Juncus roemerianus* in a Georgia salt marsh. *Ecology* 61(2):303-312.

Giurgevich, J. R. and E. L. Dunn. 1982. Seasonal patterns of daily net photosynthesis, transpiration and net primary productivity of *Juncus roemerianus* and *Spartina alterniflora* in a Georgia salt marsh. *Oecologia* 52:404-410.

Hicks, S. D., H. A. Debaugh Jr., and L. E. Hickman. 1983. *Sea level Variations for the United States 1855-1980.* U.S. Dept. Commerce, National Oceanic and Atmospheric Administration, Rockville, Maryland. 170 pp.

Imberger, J., T. Berman, R. R. Christian, E. B. Sherr, D. E. Whitney, L. R. Pomeroy and W. J. Wiebe. 1983. The influence of water motion on the distribution and transport of materials in a salt marsh estuary. *Limnol. Oceanogr.* 28(2):201-214.

Letzsch, W. S. and R. W. Frey. 1980. Organic carbon in a Holocene salt marsh, Sapelo Island, Georgia. *Georgia J. Sci.* 39:15-23.

Newell, S. Y., R. D. Fallon, R. M. Cal Rodriguez, and L. C. Groene. 1985. Influence of rain, tidal wetting and relative humidity on release of carbon dioxide by standing-dead salt-marsh plants. *Oecologia* 68(1):73-80.

Nixon, S. W. 1980. Between coastal marshes and coastal waters—A review of twenty years of speculation and research on the role of salt marshes in estuarine productivity and water chemistry. pp. 437-525. *In:* P. Hamilton and K. MacDonald (eds.), *Estuarine and Wetlands Process.* Plenum Press, New York.

Pomeroy, L. R. and R. G. Wiegert (eds.) 1981. *The Ecology of a Salt Marsh.* Springer-Verlag, New York. 271 pp.

Vetter, E. F. 1983. The ecology of (*Penaeus setiferus*): habitat selection, carbon and nitrogen metabolism, and simulation modeling. Ph.D. Dissertation. The University of Georgia, Athens. 151 pp.

Wiegert, R. G. 1979. Population models: experimental tools for the analysis of ecosystems. pp. 233-279. *In:* D. Hain, G. Stairs, and R. Mitchell (eds.), *Ecosystem Analysis.* Ohio State University Press, Columbus.

Wiegert, R. G. and R. L. Wetzel. 1979. Simulation experiments with a fourteen-compartment model of a *Spartina* salt marsh. *In:* R. F. Dame (ed.), *Marsh-Estuarine Systems Simulation.* University of South Carolina Press, Columbia, pp. 7-39.

Wiegert, R. G., R. R. Christian, J. L. Gallagher, J. R. Hall, R. D. H. Jones and R. L. Wetzel. 1975. A preliminary ecosystem model of coastal Georgia *Spartina* marsh. pp. 583-601. *In:* L. E. Cronin (ed.), *Estuarine Research,* Vol. 1. Academic Press, New York.

Wiegert, R. G., R. R. Christian and R. L. Wetzel. 1981. A model view of the marsh. pp. 183-218. *In:* L. R. Pomeroy and R. G. Wiegert (eds.), *The Ecology of a Salt Marsh.* Springer-Verlag, New York.

Wiegert, R. R., A. G. Chalmers and P. F. Randerson. 1983. Productivity gradients in salt marshes: the response of *Spartina alterniflora* to experimentally manipulated soil water movement. *Oikos* 41:1-6.

A MODEL OF CARBON FLOW IN THE *SPARTINA ANGLICA* MARSHES OF THE SEVERN ESTUARY, U.K.

P. F. Randerson

Department of Applied Biology,
University of Wales Institute of Science and Technology
Cardiff, Wales, United Kingdom

Abstract: A model of carbon flow in the *Spartina anglica* marshes of the Severn Estuary, U.K. is described. Export of organic carbon to the tidal water was estimated at 460 g C m^{-2} y^{-1} by the steady state model and over 600 g C m^{-2} y^{-1} when seasonal variation was included. Values of the standing stocks and export of carbon were particularly sensitive to factors relating to *Spartina* production. Degradation of detritus and dead root material in the marsh soil appears to be more important than direct loss from standing dead for the passage of carbon to tidal water, because of the large underground production by *Spartina*. Also, in less exposed locations, a greater proportion of standing dead *Spartina* is probably degraded by soil organisms than is lost directly to the tidal water. Long-term temporal variation reflecting successional changes requires a different modeling approach from that suited to short-term annual fluctuations. Spatial variation may arise from edaphic, physiographic or climatic differences between locations, which emphasizes the difficulty of extrapolating from one site to another. Aspects of variation within and among *Spartina* marshes are discussed and the value of adapting a model to reflect such differing situations is considered.

Introduction

As part of the feasibility studies for the proposed Severn Tidal Barrage, the salt marshes and mud flats fringing the Severn Estuary, U.K. were studied to assess the relative importance of these areas as suppliers of organic carbon to the estuary. Benthic invertebrates inhabiting intertidal mud and sand flats of the estuary support large populations of wading birds. The international importance of these populations has led to the formal recognition of the Severn Estuary as a site of special importance to nature conservation.

The study took the form of a literature review and modeling exercise, supported by some field observations. Data from the relevant literature were used as the basis for a model of the dynamics of carbon in the salt marsh and mud flat ecosystems. The literature review and carbon flow model for the salt marsh described here revealed the considerable variability encountered within and among different salt marsh locations. This paper describes the analysis of carbon flow in the Severn Estuary and efforts to assess the model's ability to encompass the range of variation encountered in the field.

The Severn Estuary

The Severn Estuary is a large funnel-shaped inlet in the south-west of the U.K. (Fig. 1). It has the third-highest normal maximum tidal range in the world

(Shaw 1977). The mean range at Avonmouth is 6.5 m, while the mean spring range is 12.3 m and the extreme spring range is 14.5 m. The normal excursion of spring tides extends some 70 km from the Holm Islands to Maisemore Weir (Winters 1973).

The intertidal margin of the Severn Estuary has three distinct zones. Extensive tidal mudflats, estimated at 11,000 ha in area (14,000 ha including Bridgwater Bay) provide a substrate for micro-algal colonization as in other large estuaries such as the Wash (Natural Environment Research Council 1976; Coles 1979) and the Emms-Dollard (van Es 1977). The lower salt marsh zone, totaling 440 ha (770 ha including Bridgwater Bay) is dominated by *Spartina anglica* (Hubbard), which has a seaward limit just above the level of neap high

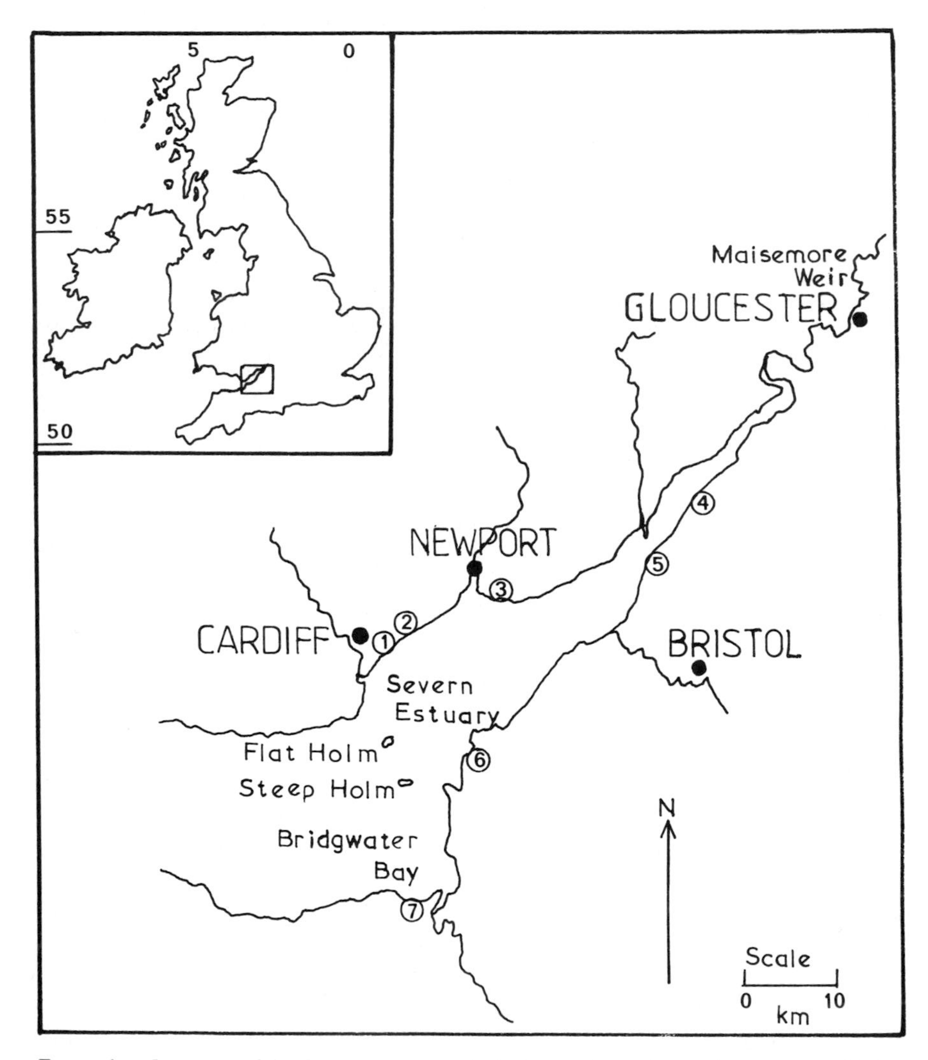

Figure 1. Location of the Severn Estuary. Study Sites 1-7 are those of Allan et al. *(1980).*

tides (Teverson 1980). This area will be referred to throughout this paper as the *Spartina* marsh. An upper marsh zone dominated by *Puccinellia maritima* Hudson occurs intermittently landward of the *Spartina* marsh and is often separated from it by an earth cliff, indicating a history of erosion and recolonization.

Organic carbon enters the Severn Estuary from several sources, including allochthonous inputs, *Spartina* marsh, upper marsh, mudflat, phytoplankton and rocky shore systems. Allochthonous sources, arising mainly from sewage inputs, may be as high as 130,000 metric tons C y^{-1} (T. D. Ruxton *pers. comm.*). Because of the high suspended sediment levels in the Severn Estuary (Kirby and Parker 1977), phytoplankton production within the estuary is relatively low. At an estuarine site on the River Usk (a tributary of the Severn), phytoplankton production during August 1978 averaged only 2.2 g C m^{-2} (Karim 1978). In similar high turbidity situations, Cadee and Hegeman (1974) and Moll (1977) found only 13 and 11.7 g C m^{-2} y^{-1} for phytoplankton production in the Wadden Sea and Long Island, New York, respectively. Although the productivity of macro-algae on rocky shores can be as high as 5000 g C m^{-2} y^{-1} (Sieburth and Jensen 1970), the Severn Estuary contains very little of such habitat. Only 4.17 km^2 of the 123.17 km^2 of the outer estuary intertidal area consists of rocky shore, and not all of this is suitable for macro-algal colonization (I. R. Joint, *pers. comm.*). The upper marsh system, which is covered only occasionally by high spring tides, is extensively grazed by domestic stock during the summer (Smith 1979) so that a large proportion of the annual production is lost to the estuary through consumption. It is likely that the *Spartina* marsh and mudflat systems contribute in excess of 90% of autochthonous carbon inputs to the estuary. Only the *Spartina* marsh is considered in further detail here: a carbon flow model for the mudflat system has been developed by Allan *et al.* (1980).

A Carbon Flow Model of the Spartina Marsh

An initial model was constructed to simulate the dynamics of organic carbon in the *Spartina* marsh in a steady state condition, *i.e.*, where all reservoirs of carbon are in dynamic equilibrium and rates of throughput of carbon represent peak conditions. Seasonal variation in selected parameters related to productive and degradative processes was subsequently introduced, following the approach of Wiegert and Wetzel (1979) in their carbon flow model for the Sapelo Island marshes of Georgia (USA).

Steady state values for the compartments and for the fluxes of carbon among them have been derived from field data or from the literature relating as closely as possible to the Severn Estuary. For simplicity, transfers between the *Spartina* marsh and other intertidal systems are ignored. Inputs of carbon are assumed to arise only through photosynthesis, with outputs only to respiration and transfer to the estuarine water. The major reservoirs of carbon in the model and transfers among them are shown in Tables 1 and 2. Data on which the model was calibrated are given in Allan *et al.* (1980), and discussed below. Although Wiegert and Wetzel (1979) made a clear distinction between the pathways of dissolved organic carbon and particulate organic carbon in their model

Table 1. State variables for the carbon flow model of the Severn Estuary.

Symbol	State Variable	Standing Stock at 1 January (g org. C m^{-2})
X00	Carbon in Water	0.0
X01	Spartina Shoots	60.0
X02	Spartina Roots	190.0
X03	Spartina Dead	215.0
X04	Spartina Grazers (insects)	0.11
X05	Carbon in Marsh Soil	25.0
X06	Heterotrophs in Soil	2.5

Table 2. Rate parameters for the carbon flow model of the Severn Estuary. Each unidirectional rate parameter is identified by its departure and destination code, *e.g.*, R0102; transfer from compartment X01 to X02.

Symbol	Rate Parameter	Rate Type Recipient (R) or Donor (D)	Seasonal (S) or Non-Seasonal (N)
RPP01	Spartina Photosynthesis	R	S
R01RR	Spartina Respiration	D	S
R0102	Transfer to Roots	D	S
R0103	Spartina Shoot Death	D	S
R0104	Spartina Grazing	R	N
R0100	Spartina Exudation	D	S
R02RR	Spartina Root Respiration	D	S
R0205	Spartina Root Death	D	S
R0305	Dead Shoots to Marsh Soil	D	S
R0300	Dead Shoots to Water	D	S
R04RR	Spartina Grazers Respiration	D	N
R0405	Spartina Grazers Death/Excretion	D	S
R0506	Soil Heterotrophic Consumption	R	N
R0500	Marsh Soil to Water	D	N
R06RR	Soil Heterotrophic Respiration	D	N
R0605	Soil Heterotrophic Death/Excretion	D	S

of carbon flow in a Georgia salt marsh, this distinction has not been attempted in the present model. Similarly, heterotrophic consumers of dead organic carbon are considered as a single compartment. Simulated compartments representing *S. anglica* are referred to below as *Spartina*.

The model consists of a set of differential equations representing the rate of change of each state variable calculated as the net of its inputs and ouputs, using the mass-balance approach of Kelly (1976). Each pathway between state variables is identified as either donor-determined or recipient-determined (Wiegert *et al.* 1975). Discontinuities are introduced at defined thresholds of the controlling (donor or recipient) variables (Wiegert and Wetzel 1979, Table 2).

Simulation of Carbon Dynamics in the *Spartina* Marsh

The interactive computer simulation language ISIS (Simulation Sciences 1980), was used to run the model with respect to time. Initial conditions for the state variables are shown in Table 1. A 5th-order variable-step integration method was used with a relative error condition of 0.1 and an initial iteration interval of one day. Sine wave functions with a 365-day period and differing phases were generated to introduce seasonal changes in selected rate parameters (Wiegert and Wetzel 1979). Parameter values were switched between low, medium and high states at appropriate times during the year.

In the initial steady-state model, parameters defining specific rates of transfer per unit of standing stock (g org. C m^{-2} d^{-1} per g org. C) were calculated from each defined average standing stock and daily flux of carbon (Fig. 2). The total annual export of organic carbon to the estuarine water was calculated as 461.2 g C m^{-2} y^{-1}. Over the total area of *S. anglica* marsh in the Severn (including Bridgwater Bay) this amounts to about 3550 metric tons C y^{-1}. In contrast the total input of carbon to the Severn Estuary from allochthonous sources, mainly sewage discharges, was estimated at 138,000 metric tons C y^{-1} (Allan *et al.* 1980). The modeled estimate is subject to error because of the known spatial and temporal variability of the model components. The paucity of data relating specifically to the Severn marshes precluded reliable calculation of values for seasonally variable rate parameters. Instead, the model was calibrated with respect to two criteria: (1) the annual peak value of each state variable should approximate the available data from the field or from the literature; and (2) the state variables should show cyclic stability, *i.e.*, return to the same value on January 1 each year.

A simulation run of the model generates a time-course of the dynamic changes in each of the seven state variables (Fig. 3). Seasonal changes in standing stocks of *Spartina* shoots, roots and dead material are apparent, as are the effects of the discontinuities imposed by the seasonal switching in the magnitude of rate parameters (Fig. 3A). For example, carbon in live *Spartina* shoots (X01) rises rapidly in the early summer (from about day 170) as photosynthesis achieves its maximum rate. This growth is curtailed in the late summer (about day 260) when the rate of shoot death increases. A peak standing crop of approximately 130 g C m^{-2} is maintained during September (until about day 290) after which the rate of photosynthesis is reduced by continued rapid shoot death, resulting in rapid decline in the standing stock. Ratios of root:shoot carbon values (X01:X02) vary from 1.3 in September to 3.2 in January. It remains to be determined how accurately this portrays the variation in the field, and whether we have timed the seasonal switches appropriately for the transfer of carbon from shoots to roots. Dead *Spartina* (X03) accumulates at the end of the growing season and is progressively lost by transfer to the soil and estuarine water compartments during the winter and spring. The small population of *Spartina* insect grazers (X04) grows to a peak in autumn, and mortality increases as *Spartina* declines (Fig. 3B). Soil carbon (X05) shows a general seasonal trend of depletion in summer and accumulation in winter, together with the effects of

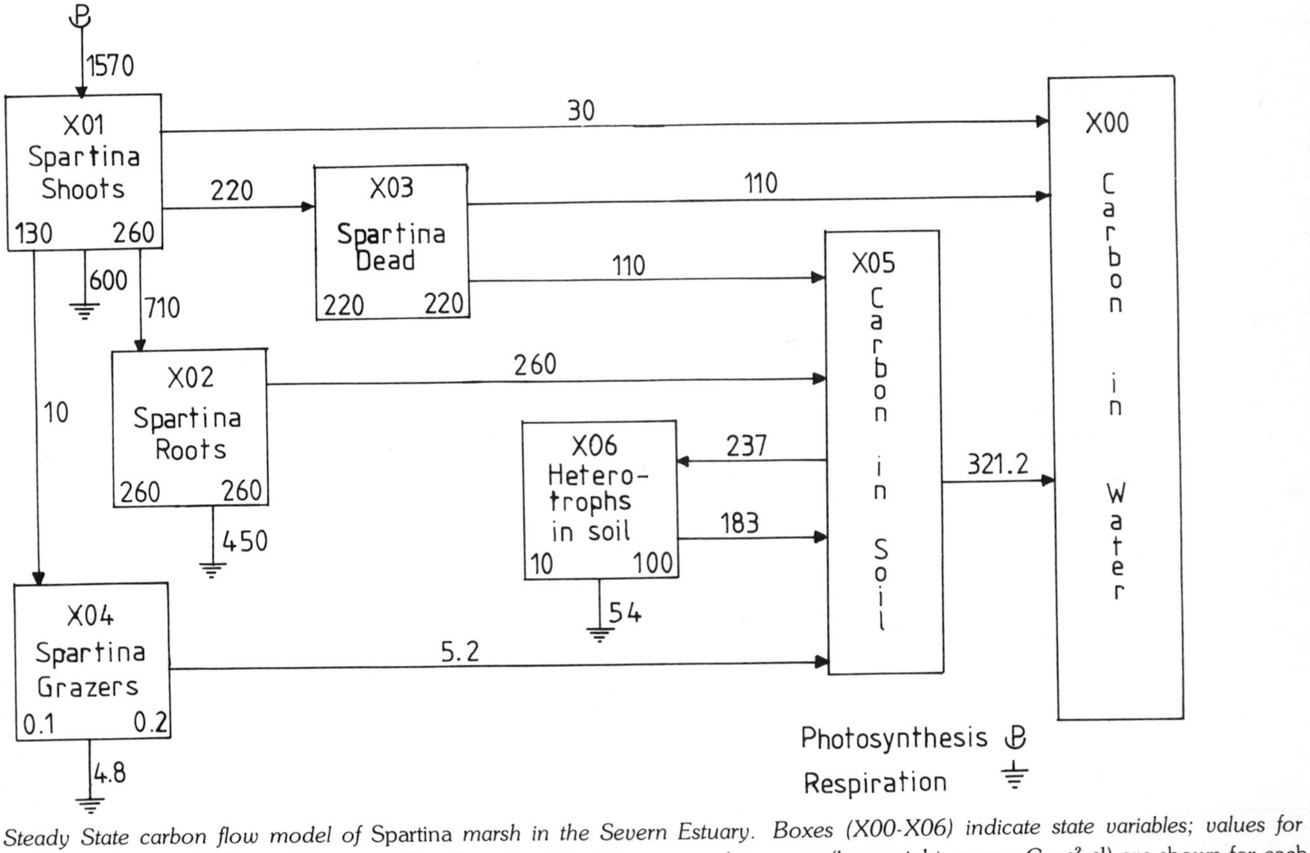

Figure 2. *Steady State carbon flow model of* Spartina *marsh in the* Severn Estuary. *Boxes (X00-X06) indicate state variables; values for peak standing stocks (lower left, g org. C m^{-2}) and production or annual turnover (lower right, g. org. C $m^{-2}y^{-1}$)* are shown for each. Arrows represent annual carbon fluxes (g org. C $m^{-2}y^{-1}$).

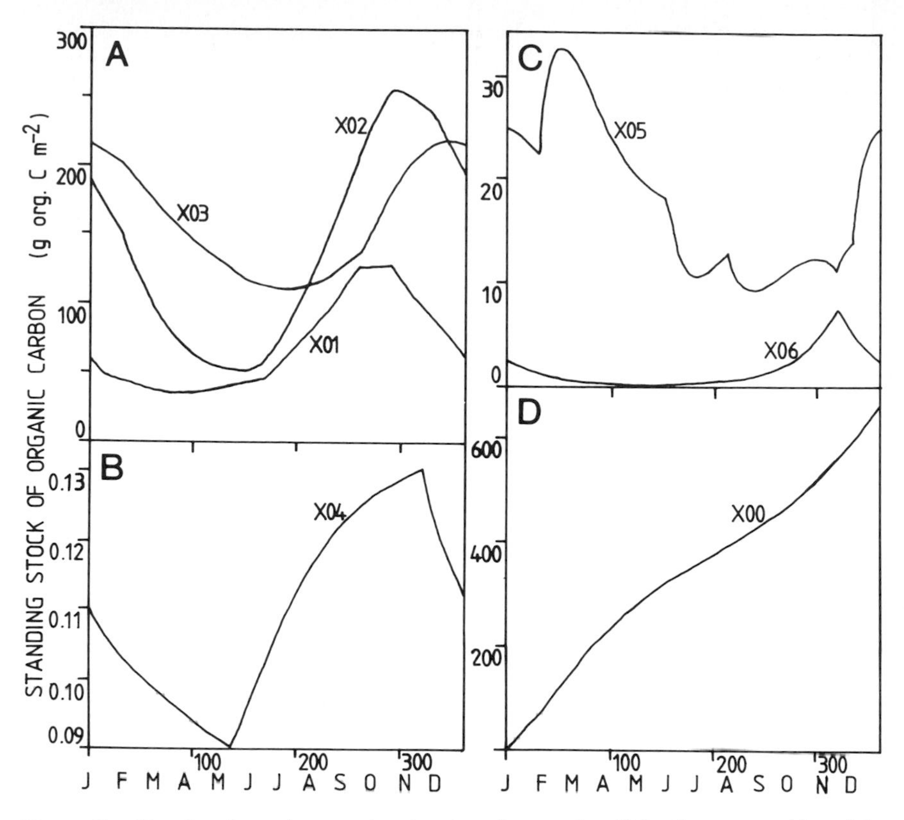

Figure 3. Simulated standing stocks of carbon (g org. C m^{-2}*) for 7 state variables of the seasonal model during the course of one year: (A)* Spartina *stocks X01 (shoots), X02 (roots), X03 (dead); (B) invertebrate grazers X04; (C) soil carbon and soil heterotrophs X05, X06; (D) carbon in water X00. Note the different scales for organic carbon in different compartments. Initial conditions as in Table 1.*

consumption and release by soil heterotrophs (X06), which increase in activity during the summer (Fig. 3C). Unlike other compartments, carbon in the estuarine water (X00) represents an external sink for organic carbon exported from the salt marsh. Starting from an initial condition of zero, the rate of accumulation in this compartment during the course of one year (over 600 g C m^{-2}), provides a measure of carbon throughout in the system (Fig. 3D). The assumption is made that, having once entered the estuarine water, carbon is not returned to the *Spartina* marsh surface. In view of the rapid movement of tidal water on and off the marsh surface and the extent of mixing of the estuarine water in the Severn, this may be appropriate for the majority of the *Spartina* marsh area. Mats of *Spartina* detritus typically accumulate along the spring tide strand line and on the upper saltings in the late autumn but no estimate of these localized deposits is available.

The sensitivity of the model to parameter variation can be assessed experimentally in relation to effects on the values of all state variables and in particular on the overall export of carbon to the estuarine water.

Sensitivity of Model Parameters

In successive simulations, various parameter values were changed, and the behavior of the model was noted. Sample simulations are discussed here in relation to the model's ability to incorporate variability in the *Spartina* marsh system.

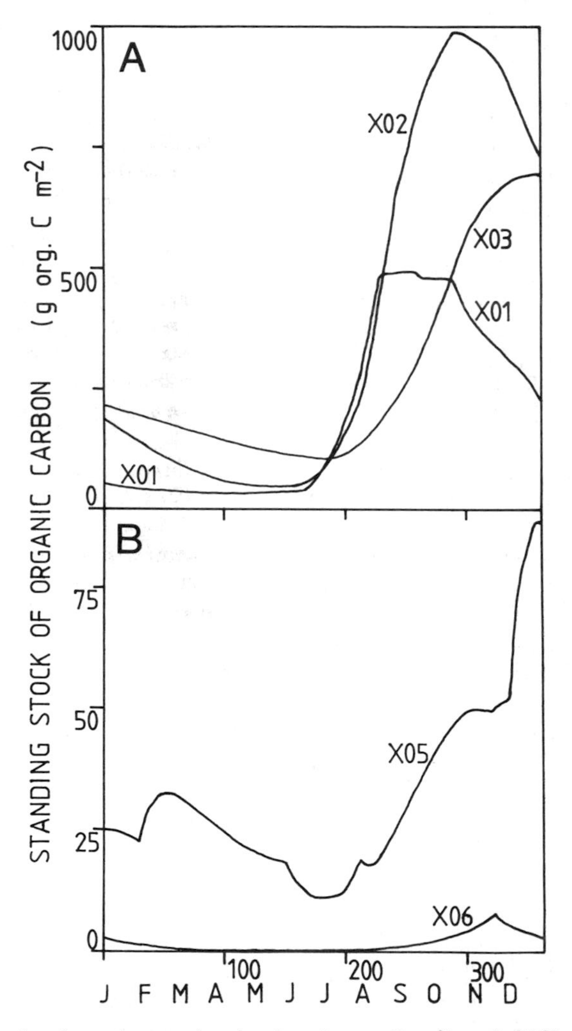

Figure 4. *Simulated standing stocks of carbon (g org. C m^{-2}), with 20% increase in summer rate of photosynthesis (R01PP); (A)* Spartina *stocks X01 (shoots), X02 (roots), X03 (dead); (B) soil carbon and soil heterotrophs X05, X06.*

The model showed greatest sensitivity to changes in parameters associated with *Spartina* production. For example, increasing the summer rate of photosynthesis by only 20% results in a greatly enhanced *Spartina* standing stock (X01) and increase in the dead *Spartina* (X03) with a consequent increase towards the end of the year in the soil carbon (X05) (Fig. 4). Export of carbon to estuarine water is almost doubled. Growth of *Spartina* is held in check in the model by feedback thresholds representing self-limitation of production at high population levels. Photosynthetic influx was progressively reduced above 450 g C m^{-2}, reaching zero at a value of 500 g C m^{-2} (Wiegert and Wetzel 1979).

In other simulations, the growth limitation thresholds were reduced to 20% and 10% of their original levels, to represent growth of *Spartina* in poor environments. As a result of the reduced production of dead plant material, carbon export was reduced to 90% and 80% respectively. *Spartina* standing stocks were similarly sensitive to increases in respiration rate. Enhancing summer respiration of *Spartina* shoots prevented the onset of the growth phase and caused a decline in standing stocks, unless a corresponding enhancement of photosynthesis was also provided. The model is clearly capable of representing a range of productivities of *Spartina* but concurrent adjustments in photosynthesis and respiration must be made with care if given levels of standing stocks are to be maintained.

Increasing the thresholds of carrying capacity for *Spartina* grazers (X04), so that the population is not constrained by its own density, allowed an expansion of 40% in its peak standing stock to occur before the onset of winter mortality, but this had a negligible effect on *Spartina*. In contrast, increasing by a factor of 10 the initial levels of both heterotrophic components (X04, X06) had a severe effect on *Spartina* in the subsequent growing season, and reduced the amount of carbon exported. Unless adjustments were made to growth rate or consumption parameters, such a response by the model was short-lived (because the standing stocks of heterotrophs were depleted through lack of a food source), and *Spartina* standing stocks returned to normal in subsequent years.

Doubling the initial value for dead *Spartina* (X03), representing perhaps an influx to the marsh of tidal litter or other organic matter, produced a small increase in soil organic carbon and hence a 17% increase in soil heterotrophs (X06). This illustrates the dependence of these consumers in the model on the availability of soil carbon as a food source. Consumption is modeled as a flux of carbon determined by both donor and recipient feedback (Wiegert and Wetzel 1979).

The basic model was also extended by inclusion of a flux of carbon out of the *Spartina* shoots compartment (X01) to represent potential consumption by domestic stock. This flux was made subject to donor feedback control such that *Spartina* shoots were heavily cropped (40% removal of standing stock per day) whenever the standing stock exceeded 100 g C m^{-2}, with the grazing effect progressively declining to zero as the stock decreased to 80 g C m^{-2}. In the simulation, *Spartina* growth was truncated to a level of 80 g C m^{-2} during the summer period but export of carbon to the estuary was only reduced by about

10%, indicating the substantial contribution of carbon exported *via Spartina* roots and soil carbon in the model.

Stochastic effects can readily be introduced into the model to represent periodic events or irregular changes in rate parameters. Losses of dead material (X03) to estuarine water were subjected to random tenfold increases in daily rate with a probability of 0.7, with a view to simulating the removal of dead material by storms. This resulted in an irregular reduction in dead standing stock (X03) towards the end of the year, and an enhanced export of organic carbon (Fig. 5). The rate of photosynthesis of *Spartina* was also varied stochastically in an attempt to simulate day-to-day fluctuations, but this produced little change in carbon export.

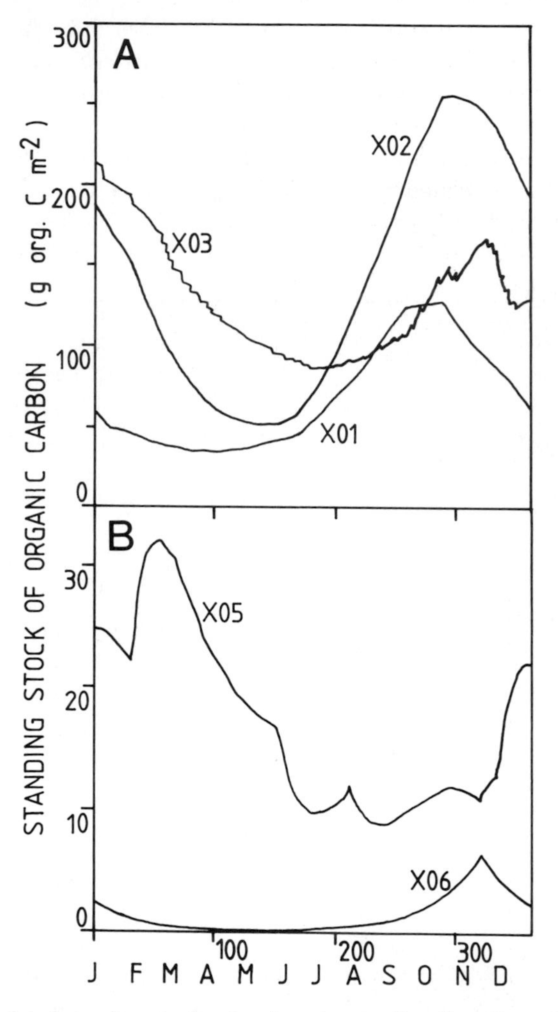

Figure 5. Simulated standing stocks of carbon (g org. C m^{-2}*), with random increases in losses of dead* Spartina*; (A)* Spartina *stocks X01 (shoots), X02 (roots), X03 (dead); (B) soil carbon and soil heterotrophs X05, X06.*

Discussion

The following sections consider the potential of the relatively simple carbon-flow model to simulate selected aspects of the variation reported in the extensive literature on salt marshes.

Temporal Variability: Annual Fluctuations

Seasonal variation of parameter values controlling carbon fluxes can be modeled conveniently by switching at appropriate times of the simulated year, as described above. Greater precision in simulating changes in standing stocks could be achieved by switching parameters more frequently over the annual cycle. Alternatively, realistic values of a forcing variable such as temperature, or stochastic variations of it, could be incorporated. Even with infrequent discontinuous switches, however, the model generated adequately smooth responses (Fig. 3A) that effectively represented field observations, such as in the *Spartina* marsh at Blakeney Point, Norfolk, U.K. (Fig. 6).

Introducing stochastic elements into the model effectively simulates events that are better defined by a probability of occurrence than by a continuous rate. For example, in their earlier model of carbon dynamics on a *S. alterniflora* marsh in Georgia, Wiegert and Wetzel (1979) found that rates of removal of

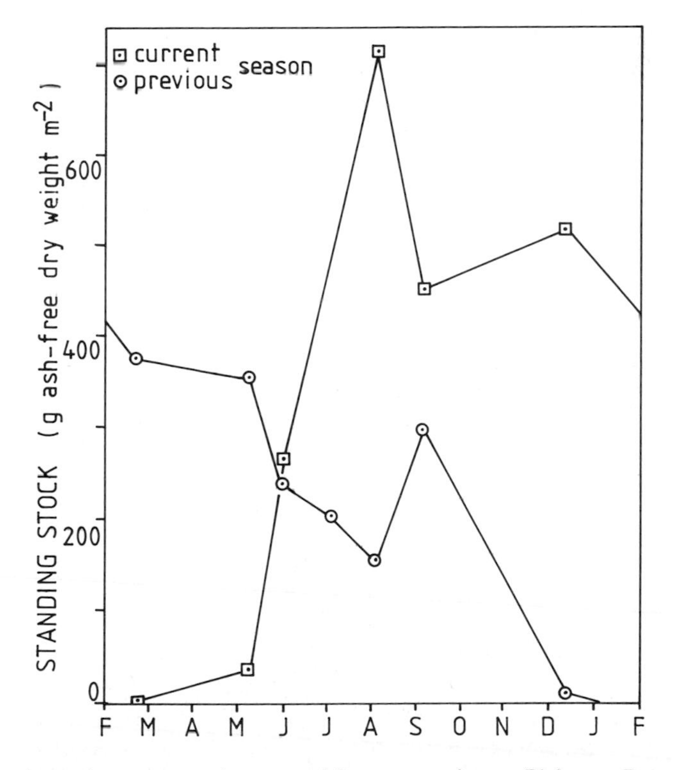

Figure 6. Annual variation in biomass of Spartina anglica *at Blakeney Point, Norfolk, U.K. (Randerson 1975a).*

dead material measured in the field were not sufficient to prevent accumulation of dead organic carbon. Further field investigations showed that storms were effective in washing out organic detritus from the marshes and tidal creeks to the estuarine water (Imberger *et al.* 1983). To improve the realism of the model, therefore, and avoid accumulation of dead material, periodic brief episodes of depletion of dead organic carbon were included in the model as random events.

Temporal Variation: Long-Term

The current model was designed to simulate annual fluctuations and to remain stable over successive years, with no allowance for longer term changes. In many cases, however, successional changes occur, notably where significant accretion of tide-borne or wind-borne sediment occurs, as in the Wash (Randerson 1975b) and the Severn Estuary (Ranwell 1964). These encompass biotic factors such as the species composition and productivity of the vegetation, edaphic factors such as aeration of sediment, and hydrologic factors such as the frequency and period of tidal inundation. These additional factors must be incorporated if longer term changes are to be simulated. Such a model, involving interactions between plant populations and the accretion of sediment, was developed previously to show successional changes on the salt marshes of the Wash (Randerson 1978). In this model, sediment accretion caused the gradual elevation of the marsh, with consequent progressive reduction in submergence by the tide. This affected the onset and rates of growth of different salt marsh species, and the presence of vegetation in turn enhanced the rate of accretion of sediment. Studies in progress on the Wash are attempting to verify rates of sediment accretion on the salt marsh surface and the resultant long-term effects of elevation change on the species composition of the plant community.

Spatial Variation

Salt marshes show a wide range of variation, both within a single marsh area and among sites, with respect to species composition and productivity, as well as edaphic and hydrologic factors. Although every site could be regarded as unique, such aspects of spatial variation can be accommodated within a general model structure, given suitable calibration of parameter values or the inclusion of specific components.

The range of variation in *Spartina* biomass (Table 3) and its productivity (Table 4) are well known, and the present model could be calibrated readily to accommodate any desired value. Although variation in productivity is derived to some extent from the methods of assessment (Long and Mason 1983), there is a well-established trend in productivity with respect to latitude for *Spartina alterniflora* along the east coast of N. America (Turner 1976).

Wide variations of biomass and productivity within given sites may be superimposed on such geographical variation. These differences may reflect various stages of long-term succession as described above. In some instances, edaphic factors dependent on sediment particle size may differ significantly, with resulting variation in plant biomass. In the Wash (Randerson 1978), for example,

Table 3. Measurements of above ground biomass (g org. C m^{-2}) for various species of *Spartina*. Dry weight data have been converted using Westlake's (1963) estimate of 38% organic carbon.

Location	Biomass	Reference
Spartina alterniflora		
Delaware	157	Morgan 1961
New Jersey	114	Good 1965
N. Carolina	95 - 789	Williams and Murdoch 1966
N. Carolina	98 - 500	Stroud and Cooper 1969
Virginia	506	Wass and Wright 1969
N. Carolina	207	Williams and Murdoch 1969
New Jersey	146 - 198	Nadeau 1972
Louisiana	387	Kirby and Gosselink 1976
Louisiana	276	Goodd and Walker 1978
Spartina anglica		
Essex	300	Long and Woolhouse 1979
Norfolk	270	Randerson 1975a
Wash	255 - 374	Randerson 1975b
Severn	130	Allan *et al.* 1980
Suffolk	150	Jackson *et al.* 1986
Spartina patens		
Texas	418 - 684	Borey *et al.* 1983
Spartina maritima		
South Africa	199 - 258	Pierce 1983

sites with a high proportion of fine particles (<0.02 mm) to tend to support pioneer colonizing vegetation with relatively high biomass (largely *Spartina anglica*), whereas sites with predominantly sandy (>0.02 mm) sediment derived by wind accretion from offshore intertidal sandbanks, supported sparser pioneer vegetation with about one-third of the biomass (mainly *Puccinellia maritima* and *Salicornia fragilis* P. W. Ball & Tutin). At the same sites in the Wash, elevation of the pioneer zone with respect to tidal level, increases with increasing sand content of the sediment. Thus establishment of pioneer vegetation may be inhibited in less cohesive sediments.

At six sites in the marshes of the Severn Estuary, measurements of - *Spartina anglica* biomass made in March-April 1980 showed landward-to-seaward variation (Table 5). Means of the landward (1, 2) and seaward (3, 4, 5) stations were significantly different (*a priori* t-test, $P<0.001$). Variances for the five stations were homogeneous (Bartlett's test, $P>0.05$), and the six sites were not significantly different (two-way analysis of variance, $P>0.1$). The higher biomass on landward areas of the *Spartina* marsh reflects the longer establishment of vegetation and possibly the more frequent and intense exposure of the seaward zone to wave action (Allan *et al.* 1980). In contrast, the salt marshes of

Table 4. Measurements of primary production (g org. C m^{-2} y^{-1}) for different species of *Spartina*. Dry weight data have been converted using Westlake's (1963) estimate of 38% organic carbon.

Location	Production	Reference
Spartina alterniflora		
Georgia	370	Smalley 1959
Georgia	760 - 1254	Odum 1961, 1971
Delaware	169	Morgan 1961
N. Carolina	380	Williams and Murdoch 1966
N. Carolina	125 - 492	Stroud and Cooper 1969
N. Carolina	247	Williams and Murdoch 1969
N. Carolina	156 - 551	Marshal 1970
Louisiana	558 - 1086	Kirby 1971
Maryland	179	Cahoon 1975
Nova Scotia	200 - 300	Mann 1975
Georgia	118 - 494	Reimold *et al.* 1975
Louisiana	503 - 1055	Kirby and Gosselink 1976
Spartina anglica		
Somerset	365	Ranwell 1961
Norfolk	372	Jefferies 1972
Essex	267	Dunn 1981
Suffolk	220 - 280	Jackson *et al.* 1986
Spartina maritima		
South Africa	181 - 235	Pierce 1983

Table 5. *Spartina anglica* shoot biomass (g org. C m^{-2}), March-April 1980 at six sites (Fig. 1) in the Severn Estuary (Allan *et al.* 1980). Stations were numbered from landward (1) to seaward (5) at each site. Carbon biomass was derived from dry weight data using Westlake's (1963) estimate of 38% organic carbon. Site 5 was omitted because of intense grazing by sheep.

	Stations				
Site	*Spartina anglica* zone		Seaward erosional zone		
	1	2	3	4	5
1	128	96	61	61	50
2	123	—	63	58	65
3	156	255	99	58	65
4	133	—	40	73	17
6	78	—	19	20	—
7	183	151	19	34	24
station means	133.5	167.3	50.2	50.7	44.2
mean for zone	144.8			48.6	
standard error	17.3			5.7	

the southeastern USA exhibit well-documented variation in productivity and biomass of *S. alterniflora* that is associated with the tall form of creek edges as opposed to the short form of the major areas of flat marsh. Studies involving experimental drainage of areas of intermediate height *S. alterniflora* (Wiegert *et al.* 1983) strongly suggest that increased interstitial water exchange with its concomitant reduced waterlogging (and not simply reduced salinity stress) is responsible for higher productivity under drained conditions.

With the variety of possible causal mechanisms for spatial variation in salt marsh vegetation, the modeler is faced with the choice of adopting an empirical coefficient to define a site-specific rate of productivity, decomposition, *etc.* or alternatively attempting to include the relevant causal factors as explicitly modeled elements.

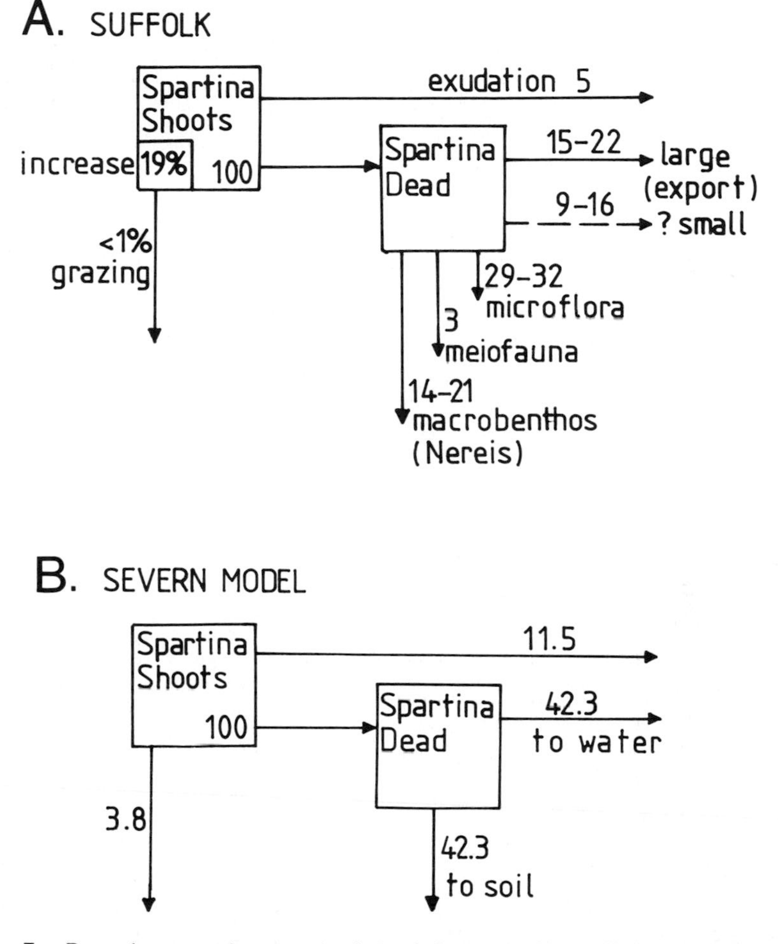

Figure 7. Degradation and removal of dead Spartina. *(A) Suffolk, U.K. data, (B) Severn Estuary steady state model. Proportions of the annual net shoot production are shown for each route.*

Recent studies of a salt marsh in Suffolk, U.K. (Jackson 1984; Jackson *et al.* 1986) exemplify the differences between salt marsh sites that may arise from differing environmental conditions. Rates of productivity and degradation of *S. anglica* were measured and the export of dead organic matter was determined using cages to trap the detritus released from standing dead material and surface litter. Proportions of shoot production assigned to various routes of degradation are contrasted (Fig. 7) with those derived from the Severn Estuary in the steady state model described in this paper. In the Suffolk marsh the proportion of the annual net *Spartina* production passing *via* litter to the heterotrophs in the soil was almost twice that tidally exported from the marsh. This marsh had an unusually high population of macro-invertebrates, largely the polychaete *Nereis diversicolor,* whose diet consisted largely (>85%) of *S. anglica* (Jackson 1984). Together with meiofaunal and microbial degradation, this consumption accounted for up to 56% of total losses of *Spartina.* Export of tidal litter was estimated to be only 15-22% although the losses unaccounted for (9-15%) probably represented small particles of dead material which were not retained by the litter traps. Live biomass of *Spartina* increased by 19% in both years of measurement. Consumption by sub-aerial invertebrates (mainly the leafhopper *Phileanus spumaris)* accounted for an insignificant proportion of the *Spartina* production, whereas losses of soluble carbon by exudation were assumed to be about 5%. In contrast, for the Severn model it has been assumed, following Day *et al.* (1973) and van Es (1977), that 50% of dead organic matter passes directly to the estuarine water.

The two sites differ considerably in their exposure to wave action during high tide. The Suffolk marsh is enclosed by shingle banks, with tidal access restricted to a single creek inlet, while the Severn marshes are situated on open coast. The greater shelter at the Suffolk site may explain its higher invertebrate populations. On the exposed open coast, wave action is likely to move *Spartina* litter more readily from the marsh and may be responsible for a greater degree of plant damage *in situ* leading to higher leaf mortality and leaching losses.

Conclusions

Aspects of the variability inherent in *Spartina anglica* marshes can be investigated with the aid of a relatively simple model representing throughput of carbon. The model provides a framework against which the relative importance of different processes and routes of carbon transfer in different situations can be assessed.

The present model was calibrated with respect to the Severn Estuary, and the major standing stocks of carbon and the major flows involved the three *Spartina* compartments. Of these, both roots and dead *Spartina* exceeded the live shoots, although all compartments underwent large seasonal variation. The passage of carbon from *Spartina via* the marsh soil to the tidal water was more than twice the magnitude of losses directly from the shoots and standing dead material. Insufficient data were available to validate the simulated variation in root:shoot ratio or to confirm the importance of the soil reservoir in degrading *Spartina* detritus originating from shoot and root production.

Adjustment of rate parameters and threshold values can easily be made to accommodate the range of variation reported at different locations arising from differing environmental conditions. In *Spartina* marshes of the U.K., landward-to-seaward variation in biomass appears to be related to exposure to wave action and, in some situations, to differences in particle-size composition of the sediment. In contrast, *Spartina* marshes of the eastern USA typically exhibit large differences in production between well drained creek sides and extensive inter-creek areas where interstitial water movement is restricted. Differences in the output of carbon to tidal water at different locations probably depend on the relative exposure to tidal water flow and wave action. Manipulation of model parameters may provide a simulated fit to a given situation but does little to illuminate mechanisms underlying such variation. In the same way, inclusion of random factors has no direct homology with the fluctuations they are intended to represent. Different modeling approaches would be appropriate to represent long-term successional trends (in which progressive changes in the local environment are important), as opposed to short-term temporal variation related to season fluctuations.

A model has important roles, however, in summarizing the state of knowledge of a system in a holistic way, in representing its dynamics and in communicating this information to other workers. It is therefore of value in a comparative study of related marsh systems with differing species and environmental factors, as for example those of the Georgia Coast and the Severn Estuary. A model inevitably exposes the need for further site-specific data, and may provoke investigation of alternative hypotheses when its predictions fail to match observations, leading to refinement or elaboration of the model in parallel with field studies.

Acknowledgments

I am grateful to the Estuarine Research Federation for providing funds to cover costs to travel to this conference. I wish to thank Prof. R. G. Wiegert and Prof. R. W. Edwards for financial assistance with the visit. I acknowledge the work of J. Allan and D. R. Blatchford in reviewing the literature and developing the original computer model using C.S.M.P. (Continuous System Modelling Program). The work was carried out under a grant from the Energy Technology Support Unit, Dept. of Energy as part of the Severn Tidal Power Pre-Feasibility Study (Contract No. E/5A/CON/1618/51/055).

References Cited

Allan, J. G., D. R. Blatchford and P. F. Randerson. 1980. A carbon flow model of the intertidal areas of the Severn Estuary. *Severn Tidal Power Pre-feasibility Study*. Report to Dept. of Energy, U.K. Contract No. E/5A/CON/1618/51/055. Dept. of Appl. Biol., University of Wales Institute of Science and Technology, Cardiff, U.K. 112 pp.

Borey, R. B., P. A. Harcombe and F. M. Fisher. 1983. Water and organic carbon fluxes from an irregularly flooded brackish marsh on the upper Texas coast. USA. *Estuar. Coastal Shelf Sci.* 16:379-402.

Cadee, G. C. and J. Hegeman. 1974. Primary production of the benthic microflora living on tidal flats in the Dutch Wadden Sea. *Neth. J. Sea Res.* 8:260-291.

Cahoon, D. R. 1975. Net productivity of emergent vegetation at Horn Point salt marsh. M.S. Thesis. University of Maryland, College Park. 94 pp.

Coles, S. M. 1979. Benthic microalgal populations on intertidal sediments and their role as precursors to salt marsh development pp. 25-42. *In:* R. L. Jefferies and A. J. Davy (eds.), *Ecological Processes in Coastal Environments.* Blackwell, London.

Day, J. W., W. G. Smith, P. R. Wagner and W. C. Stowe. 1973. Community structure and carbon budget of a salt marsh and shallow bay estuarine system in Louisiana. Pub. LSU-SG-72-04. Louisiana State University Center for Wetland Resources, Louisiana State Univ., Baton Rouge. 80 pp.

Dunn, R. 1981. Effects of temperature on the photosynthesis, growth and productivity of *Spartina townsendii (sensu lato)* in controlled and natural environments. Ph.D. Thesis, University of Essex, Colchester, U.K. 218 pp.

Good, R. E. 1965. Salt marsh vegetation. Cape May, New Jersey. *Bull. N. J. Acad. Sci.* 10:1-11.

Good, R. E. and R. Walker. 1978. Cited in: Whigham, D.F. *et al.* 1978. Biomass and primary production in freshwater tidal wetlands of the middle Atlantic coast. pp. 3-20. *In:* R. E. Good, D. F. Whigham and R. L. Simpson (eds.). *Freshwater Wetlands Ecological Processes and Management Potential.* Academic Press, New York.

Imberger, J., T. Berman, R. R. Christian, E. B. Sherr, D. E. Whitney, L. R. Pomeroy, R. G. Wiegert and W. J. Wiebe. 1983. The influence of motion on the distribution and transport of materials in a salt marsh estuary. *Limnol. Oceanogr.* 28(2):201-214.

Jackson, D. 1984. Salt marsh populations and the fate of organic matter produced by *Spartina anglica. British Ecological Soc. Bull.* 15:192-196.

Jackson, D., S. P. Long and C. F. Mason. 1986. Net primary production decomposition and export of *Spartina anglica* on a Suffolk salt marsh. *J. Ecol.* (in press).

Jefferies, R. L. 1972. Aspects of salt-marsh ecology with particular reference to inorganic plant nutrition. pp. 61-85. *In:* R. S. K. Barnes and J. Green (eds.), *The Estuarine Environment.* Applied Science Publishers, London.

Karim, R. 1978. Photosynthetic activities of the phytoplankton in the River Usk Estuary. M.Sc. Thesis. University of Wales Institute of Science and Technology, Cardiff, U.K. 53 pp.

Kelly, R. A. 1976. Conceptual ecological model of the Delaware Estuary. pp. 3-45. *In:* B. C. Patten (ed.), *Systems Analysis and Simulation in Ecology.* Vol. IV. Academic Press, London.

Kirby, C. J. 1971. The annual net primary production and decomposition of the salt marsh grass *Spartina alterniflora* in the Barataria Bay estuary of Louisiana. Ph.D. Thesis. Louisiana State University, Baton Rouge. 73 pp.

Kirby, C. J. and J. G. Gosselink. 1976. Primary production in a Louisiana Gulf Coast *Spartina alterniflora* marsh. *Ecology* 57:1052-1059.

Kirby, R. and W. R. Parker. 1977. Sediment dynamics in the Severn Estuary. pp. 41-52. *In:* T. L. Shaw (ed.), *An Environmental Appraisal of the Severn Barrage.* Dept. Civ. Engrg., Univ. of Bristol, U.K.

Long, S. P. and C. F. Mason. 1983. *Saltmarsh Ecology.* Blackie, London. 160 pp.

Long, S. P. and H. W. Woolhouse. 1979. Primary production in *Spartina* marshes. pp. 333-352. *In:* R. L. Jefferies and A. J. Davy (eds.), *Ecological Processes in Coastal Environments.* Blackwell, London.

Mann, K. H. 1975. Relationship between morphometry and biological functioning in three coastal inlets of Nova Scotia. pp. 634-644. *In:* L. E. Cronin (ed.), *Estuarine Research.* Vol. 1. Academic Press, New York.

Marshall, D. E. 1970. Characteristics of *Spartina* marsh which is receiving treated sewage wastes. pp. 317-359. *In:* H. T. Odum and A. F. Chestnut (eds.), *Studies of Marine Estuarine Ecosystems Developing with Treated Sewage Wastes.* Inst. Mar. Sci. Ann. Rept. 1969-70. Univ. North Carolina, Wilmington.

Moll, R. A. 1977. Phytoplankton in a temperate-zone salt marsh: Net production and exchange with coastal waters. *Mar. Biol.* 42:109-118.

Morgan, M. H. 1961. Annual angiosperm production on a salt marsh. M.S. Thesis. University of Delaware, Newark. 34 pp.

Nadeau, R. J. 1972. Primary production and export of plant materials in the salt marsh ecosystem. Ph.D. Thesis. Rutgers Univ., New Brunswick, New Jersey. 175 pp.

Natural Environment Research Council 1976. *The Wash Water Storage Scheme Feasibility Study: A Report on the Ecological Studies.* Nat. Environ. Res. Counc. Publ. Ser. C. No. 15, London. 36 pp.

Odum, E. P. 1971. *Fundamentals of Ecology.* 3rd Edition. W. B. Saunders, Philadelphia. 574 pp.

Odum, E. P. 1961. The role of tidal marshes in estuarine production. *The Conservationist* (N.Y. State Cons. Dept.) 15:12-15.

Pierce, S. M. 1983. Estimation of the non-seasonal production of *Spartina maritima* (Curtis) Fernald in a South African estuary. *Estuar. Coastal Shelf Sci.* 16:241-254.

Randerson, P. F. 1975a. An ecological model of succession on a Norfolk salt marsh. Ph.D. Thesis. University of London. 313 pp.

Randerson, P. F. 1975b. The salt marshes of the Wash. *The Wash Water Storage Scheme Feasibility Study. Ecological Report, D.* Nat. Environ. Res. Counc., London. 22 pp.

Randerson, P. F. 1978. A simulation model of the salt marsh ecosystem. pp. 48-67. *In:* B. K. Knights and J. L. Phillips (eds.), *Coastal and Estuarine Land Reclamation.* Saxon House, London.

Ranwell, D. S. 1961. *Spartina* salt marshes in southern England. I. The effects of sheep grazing at the upper limits of *Spartina* marsh in Bridgwater Bay. *J. Ecol.* 49:325-340.

Ranwell, D. S. 1964. *Spartina* salt marshes in southern England. II. Rate and seasonal pattern of sediment accretion. *J. Ecol.* 52:79-95.

Riemold, R. J., J. L. Gallagher, R. A. Linthurst and W. J. Pfeiffer. 1975. Detritus production in coastal Georgia salt marshes. pp. 217-228. *In:* L. E. Cronin (ed.), *Estuarine Research,* Vol. 1. Academic Press, New York.

Shaw, T. L. 1977. Tides, currents and waves. pp. 1-34. *In:* T. L. Shaw (ed.), *An Environmental Appraisal of the Severn Barrage.* Dept. Civ. Engrg., Univ. of Bristol, U.K.

Sieburth, J. McN. and A. Jensen. 1970. Production and transformation of extracellular organic matter from littoral marine algae: a resume. pp. 203-223. *In:* D. W. Hood (ed.), *Organic Matter in Natural Waters.* Inst. Mar. Sci. Occasional Publ. No. 1, Univ. Alaska, Fairbanks.

Simulation Sciences. 1980. *ISIS Users Manual.* Simulation Sciences, Yatton, Avon, U.K. 107 pp.

Smalley, A. E. 1959. The growth cycle of *Spartina* and its relation to the insect populations in the marsh. pp. 96-100. *In: Proc. Salt Marsh Conf.* Mar. Inst. Univ. of Georgia, Sapelo Island.

Smith, L. 1979. *A Survey of the Salt Marshes in the Severn Estuary.* Nature Conservancy Council, London. 100 pp.

Stroud, L. M. and A. W. Cooper. 1969. Color-infrared aerial photographic interpretation and net primary productivity of a regularly-flooded North Carolina salt marsh. *Water Resources Res. Inst. Rept.* No. 14. Univ. of North Carolina, Wilmington. 86 pp.

Teverson, R. 1980. Salt marsh ecology in the Severn Estuary. *Severn Tidal Power Pre-feasibility Study.* Report to Dept. of Energy, U.K. Contract No. E/5A/CON/1619/51/054. Dept. Zool., Univ. of Bristol, U.K. 109 pp.

Turner, R. E. 1976. Geographic variation in salt marsh macrophyte production: a review. *Contr. Mar. Sci.* 20:47-68.

van Es, F. B. 1977. A preliminary carbon budget for a part of the Emms Estuary: The Dollard. *Helgolander wiss. Meeresunters* 30:283-294.

Wass, M. L. and T. D. Wright. 1969. Coastal wetlands of Virginia. Interim Rept. to the Governor and General Assembly. *Virginia Inst. of Mar. Sci. Spec. Rept. in Applied Mar. Sci. and Ocean Eng.* No. 10. Gloucester Point, Virginia. 154 pp.

Westlake, D. F. 1963. Comparisons of plant productivity. *Biol. Rev.* 38:385-425.

Wiegert, R. G. and R. L. Wetzel. 1979. Simulation experiments with a fourteen-compartment model of a *Spartina* salt marsh. pp. 7-39. *In:* R. F. Dame (ed.), *Estuarine Systems Simulations.* Belle Baruch Libaray in Marine Science, No. 8. Univ. South Carolina Press, Columbia, South Carolina.

Wiegert, R. G., R. R. Christian, J. L. Gallagher, J. R. Hall, R. D. H. Jones and R. L. Wetzel. 1975. A preliminary ecosystem model of coastal Georgia *Spartina* marsh. pp. 583-601. *In:* L. E. Cronin (ed.), *Estuarine Research.* Vol. 1. Academic Press, New York.

Wiegert, R. R., A. G. Chalmers and P. F. Randerson. 1983. Productivity gradients in salt marshes: the response of *Spartina alterniflora* to experimentally manipulated soil water movement. *Oikos* 41:1-6.

Williams, R. B. and M. B. Murdoch. 1966. Annual production of *Spartina alterniflora* and *Juncus roemerianus* in salt marshes near Beaufort, North Carolina. *Assoc. SE Biologists Bull.* 13:49.

Williams, R. B. and M. B. Murdoch. 1969. The potential importance of *Spartina alterniflora* in conveying zinc, manganese and iron into estuarine food chains. pp. 431-439. *In:* D. J. Nelson and F. C. Evans (eds.), *Proc. 2nd Nat. Symp. Radioecology.* CONF-670503. U.S. Atomic Energy Commission TID 4500, Washington, D.C.

Winters, A. 1973. A desk study of the Severn Estuary. pp. 106-112. *In:* A. L. H. Gameson (ed.), *Mathematical and Hydraulic Modelling of Estuarine Pollution.* Tech. Paper No. 13. Water Pollution Research Laboratory, Dept. of Environment, Stevenage, U.K.

CHESAPEAKE BAY PHYSICAL MODEL INVESTIGATIONS OF SALINITY RESPONSE TO NEAP-SPRING TIDAL DYNAMICS: A DESCRIPTIVE EXAMINATION

Mitchell A. Granat and David R. Richards

Hydraulics Laboratory
U.S. Army Engineer Waterways Experiment Station
Vicksburg, Mississippi

Abstract: Automation and computer control capabilities have advanced state-of-the-art physical modeling practice, enabling dynamic testing with computer-generated variable source tides and computer-controlled freshwater discharges. Several large-scale dynamic tests conducted at the Chesapeake Bay hydraulic model illustrate distinctive salinity responses to neap-spring tidal dynamics. Oscillations from well-mixed spring tide conditions to partially-stratified neap tide conditions are indicated. These responses vary by location within the main estuary and its many subestuaries. Freshwater discharge variations can modify these responses, though they are not totally obscured. The hydraulic model is well suited for dynamic neap-to-spring investigations since boundary conditions can be carefully controlled (no meteorological disturbances to contaminate the records) and the entire estuarine system can be easily sampled at concurrent times.

Introduction

Recent field investigations indicate that tidal dynamics play a major role in mixing and transport processes within certain estuarine environments (Haas 1977; Allen *et al.* 1980; Boersma and Terwindt 1981; Haas *et al.* 1981). It has long been a practice in physical estuarine modeling to examine spring, mean, and neap tidal conditions, but under "steady-state" repetitive diurnal or semidiurnal tide conditions. Several ranges of tides have been used separately to investigate the full range of potential impacts associated with engineering works or for planning purposes. Recent improvements in computer control technology have advanced physical modeling capabilities to allow dynamic testing with computer-generated variable source or forcing tides and computer-controlled freshwater discharges.

This paper illustrates results from some large-scale studies conducted under such dynamic testing conditions at the Chesapeake Bay hydraulic model, which was operational from May 1976 to August 1984. A series of sensitivity studies were conducted to document more thoroughly the day-to-day salinity response to the neap-spring tidal cycle. Freshwater discharge variation is shown to modify this response, although it generally does not totally obscure it.

The Prototype Bay

Chesapeake Bay (Fig. 1) located on the east coast of the United States, is one of the largest, most productive, and most diversely used estuaries in the world. The main bay extends approximately 305 km north from the ocean entrance to the mouth of the Susquehanna River. In geologic terms, Chesapeake Bay is a submerged river valley, a dynamic remnant of the ancestral Susquehanna River. The average depth of the bay is about 8.5 m although a natural channel deeper than 15 m traverses the bay for more than 60 percent of its length. The maximum depth of 53 m is located in the upper bay near Bloody Point, Kent Island, Maryland.

Like many coastal plain estuaries, the bay is irregular in shape varying in width from 6.5 km, between Annapolis and Kent Island, to 48 km in the middle bay off the Potomac River. Runoff enters the bay from more than 165,000 km^2 of drainage area through more than 50 different tributary systems. Five major western shore rivers (Susquehanna, Potomac, James, York, and Rappahannock) provide approximately 90 percent of the 2,000 $m^3 s^{-1}$ annual freshwater discharge. The Susquehanna River at the head of the bay contributes approximately one-half of the total freshwater inflow to the bay.

Hydrodynamically, the bay is sufficiently long to accommodate one complete tidal wave at all times. Tides are mixed semidiurnal showing progressive wave characteristics at the ocean entrance and standing wave characteristics at the heads of the bay and tributaries. Tidal amplitudes and associated tidal currents are relatively low, generally under 0.5 m and below 1 m s^{-1} respectively. The interchange and mixing of fresh riverine water and salty ocean water within the estuary help maintain the Chesapeake Bay system as a productive natural resource.

The Chesapeake Bay Physical Model

The Chesapeake Bay physical model, located adjacent to the prototype bay on Kent Island, was the largest and one of the most sophisticated estuarine models of its kind. It was a 3.5-ha concrete fixed-bed model, housed in a 5.5-ha shelter approximately 300 m long and 180 m wide. The molded area of the model (Fig. 1) extended from approximately 48 km offshore in the Atlantic Ocean to the head of tide for all tributaries emptying into Chesapeake Bay.

The hydraulic model was based on the equality of model and prototype Froude numbers:

$$\mathbb{F}_n = \frac{\text{inertial force}}{\text{gravitational force}} = \frac{\text{velocity}}{\sqrt{\text{gravity/length}}}$$

reflecting similitude of gravitational effects. Economics (small enough to be manageable and minimize costs) and hydrodynamics (large enough for proper physics and measurement) were considered in choosing the horizontal 1:1,000 and vertical 1:100 model-to-prototype scales; this reflects a distortion ratio of 10:1. The vertical dimension is the characteristic length for distorted-scale models. These scales and Froudian model laws defined the following model-to-prototype ratios: time 1:100; velocity 1:10; volume 1:10^8; and discharge 1:10^6. The general practice of maintaining salinity in a 1:1 ratio was followed.

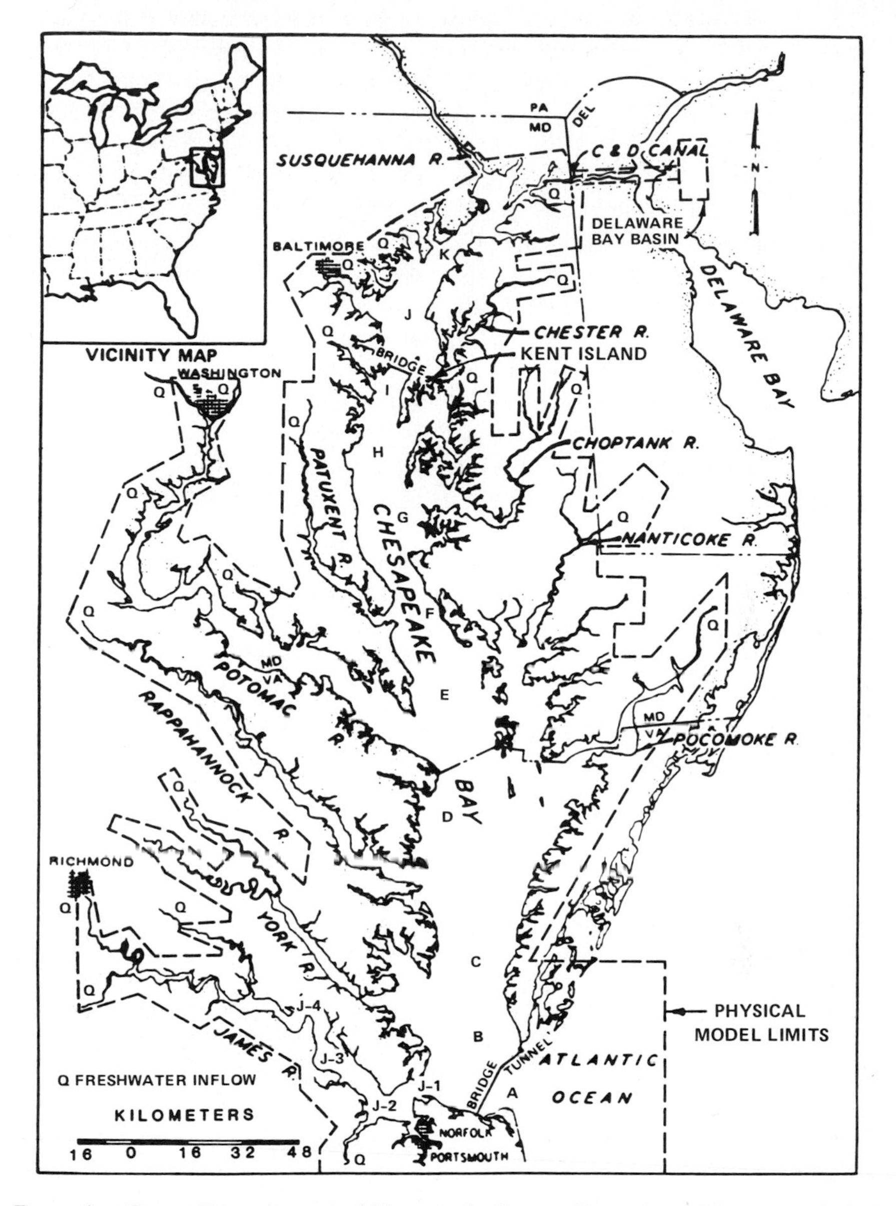

Figure 1. General location map of Chesapeake Bay and boundary of the area included in the physical model. Numbered locations are hydraulic model sampling locations.

As in other distorted-scale models, additional roughness was used to reproduce boundary friction and to ensure that the flow regime remained turbulent so that proper hydrodynamics and reproduction of tidal heights, tidal velocities, and salinity distributions could be achieved. In shallow-water areas (generally less than 2 m prototype) additional roughness was achieved by crosshatching the concrete surface during model construction. In deeper areas, 1.25-cm-wide stainless steel resistance strips were embedded in the model bottom and allowed to project to the water's surface. Over 700,000 such strips were placed in the model during construction. These resistance or roughness strips were used to adjust longitudinal and lateral flow patterns. Final roughness distribution was determined during calibration by systematically bending up or bending down these strips until proper amplitude and phasing of tidal heights and velocities were obtained through trial and error. An induced-mixing bubbler system was used to enhance vertical mixing.

The model was designed to include all necessary appurtenances for the reproduction of selected prototype boundary conditions and the measurement of the model responses to these conditions. A detailed description of these features may be found in Granat *et al.* (1985).

Minicomputers, including a PDP 11/44, performed a variety of tasks ranging from model control and data logging to analysis and graphical display of model test data. Source or forcing tides could be computer-controlled to generate any desired tidal sequence. Fig. 2 illustrates the 12-tidal constituent, 28-lunar-day, 56-cycle tidal sequence generated during the various studies reported in this paper. This sequence was derived from a harmonic analysis of NOAA tide records (July 1971–June 1972) from Old Point Comfort at the entrance to the James River.

Freshwater inflow into the model was controlled through 21 strategically located discharge points (Fig. 1) chosen to best represent the combined total freshwater discharge into Chesapeake Bay. Inflow at each location was controlled by a digital valve system consisting of 12 solenoid-actuated discharge ports. These valves were capable of either manual or computerized operation enabling the accurate reproduction of variable freshwater discharge hydrographs. Inflow was monitored by various meters that covered the full range of desired discharge. A computerized closed-loop feedback system was used to compare actual flows with desired flows and automatically make digital valve adjustments to ensure that the desired flow was achieved.

Salinity throughout the model was determined using Beckman RA-5 salinity meters (Beckman Industries, Essex Co, New Jersey). Water samples were generally collected in vials during slack-before-flood or slack-before-ebb tidal conditions using a series of vacuum pumps attached to stationary probes. The main bay and tributaries were divided into 39 separate cotidal areas such that sampling time at any station was never more than 30 min (prototype) from the desired slack-water sampling time. An automated data logging system was used to improve the accuracy of salinity measurements and data analysis.

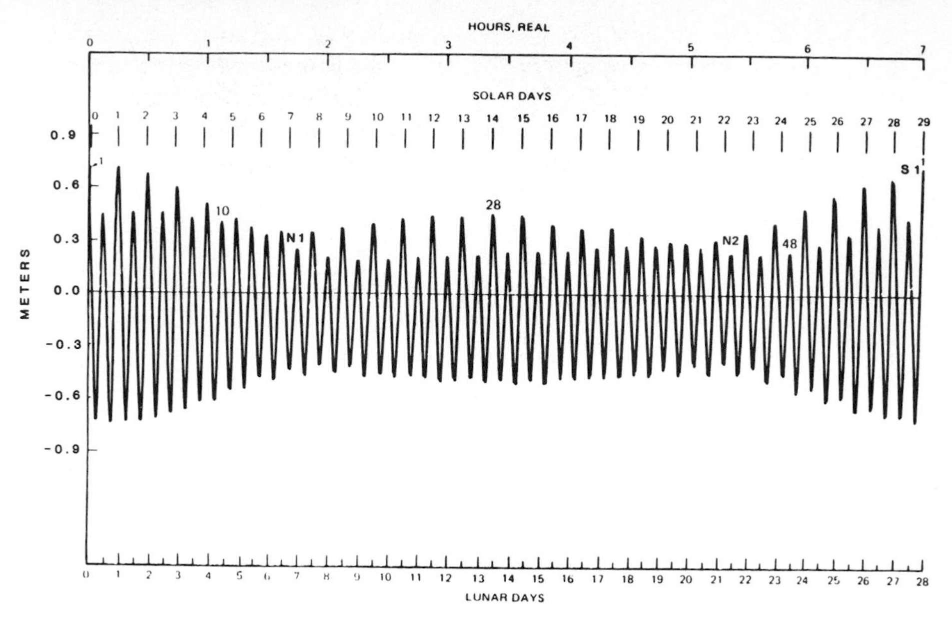

Figure 2. *Atlantic Ocean 12-tidal-constituent, 28-lunar-day tide used as the forcing tide for all model studies presented in this report.*

Physical Model Salinity Investigations

Salinity sampling on the Chesapeake Bay physical model was conducted under variable tide conditions associated with the generation of a 12-tidal-constituent, 28-lunar-day source tide (Fig. 2). Samples were generally collected four times during each 28-lunar-day cycle at times corresponding to spring, mean, and neap tidal conditions at either slack-before-flood (SBF) or slack-before-ebb (SBE) conditions.

Figure 3 presents a typical SBE salinity plot from the Baltimore Harbor and Channels Deepening Study (Granat and Gulbrandsen 1982) for a station in the entrance to the James River (near Station J-1). The sawtooth pattern in salinity variation, characteristic of this sampling mode, is associated with neap-to-spring tidal variations (Fig. 3). Additional tidal energy associated with the larger range spring tide resulted in a better mixed water column. This phenomenon of increased mixing with spring tides has been noted in field studies of lower bay subestuaries (Haas 1977).

Salinity and freshwater discharge characteristically exhibit an inverse relationship (Fig. 3). Surface-to-bottom salinity gradients are larger during the higher discharge periods, with the surface waters generally undergoing a larger salinity reduction than the bottom waters. During the higher discharge periods, the neap-to-spring salinity variations (a mixing and dispersion phenomenon) are therefore amplified and are easier to detect.

A computer failure caused by a power outage shortly after lunar day 700 of this testing program provided an interesting comparison to the steady-state

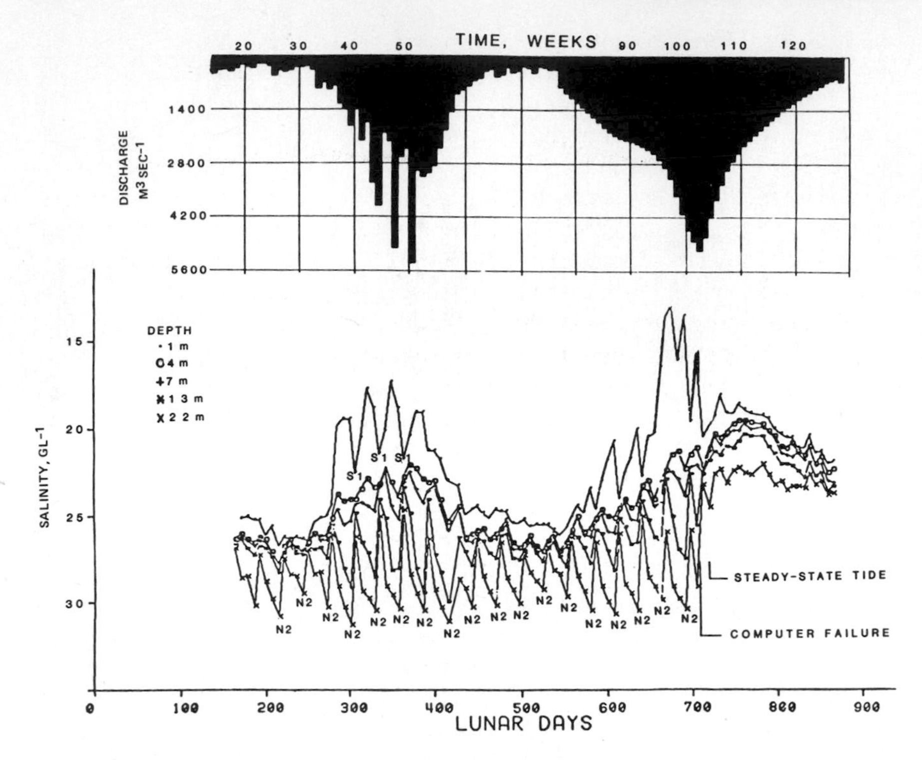

Figure 3. *Time-history of slack-before-ebb salinity at different depths of the James River entrance in the physical model, and freshwater discharge for the same period (Granat and Gulbrandsen 1982). The sawtooth pattern in salinity variations through lunar day 700 is associated with the computer-generated 28-lunar-day variable tide. S1 corresponds to the large spring tide sample, sampled every tide 1, and N2 corresponds to the preceding neap tide sample, sampled every tide 44 (Fig. 2). The change in salinity response after lunar day 700 was caused by loss of power and computer-control, which required establishment of a repetitive steady-state tide as the forcing tide condition.*

testing technique (Fig. 3). Following the computer failure, a repetitive cosine tide of spring tide range was generated until the conclusion of the test. The loss of the dynamic nature of salinity response to the variable tide is clearly illustrated. The subsequent paper in this volume (Richards and Granat 1986) addresses the importance of studies conducted using the dynamic approach.

Figure 4 illustrates the time-histories of model surface and bottom SBF salinity values collected during the Oct 1969-Oct 1973 simulation period used for verification (Granat *et al.* 1985), for four stations progressing up the main axis of Chesapeake Bay. Dramatic neap-to-spring salinity variations are illustrated in the lower main bay (Station B) and again in the upper main bay (Station K). In the middle main bay area (Stations E through I), this salinity response is greatly reduced or does not exist. A complete time-history data set for the 11 main bay stations (Stations A through K) is given in Granat *et al.* (1985).

Variations in salinity response to tidal dynamics are further documented in time-history salinity plots from the Baltimore Harbor and Channels Deepening Study (Granat and Gulbrandsen 1982) and the Norfolk Harbor and Channels Deepening Study (Richards and Morton 1983). These plots indicate that salinity response to the neap-spring tidal cycle is present to varying degrees throughout the main bay and its subestuaries. The strongest neap-to-spring variations are located in the lower main bay and lower bay subestuaries.

Figure 5 presents results from a specifically conducted neap-to-spring salinity sensitivity study. These model data were collected at each SBF throughout the 28-lunar-day tidal cycle during a constant 2,000 $m^3\ s^{-1}$ total bay discharge condition. Increased mixing at Stations A, B, and K corresponds to

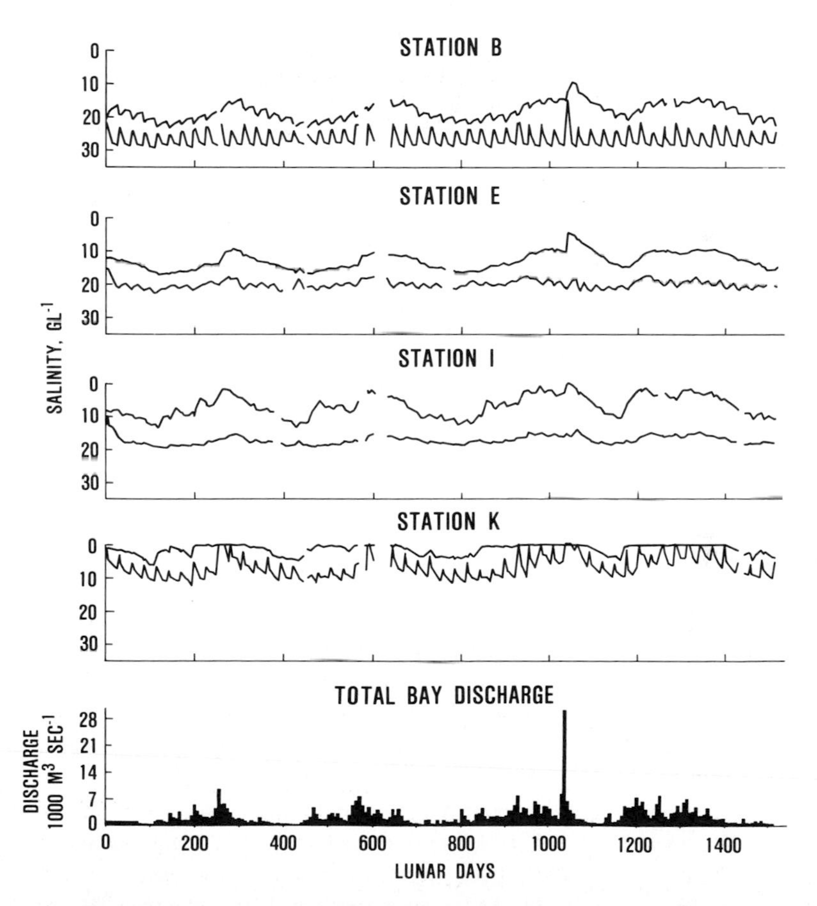

Figure 4. Time-histories of model slack-before-flood salinity at four stations during the October 1969-October 1973 verification simulation period. Tides 1, 10, 28, and 48 (Fig. 2) were sampled, and station locations are shown in Fig. 1.

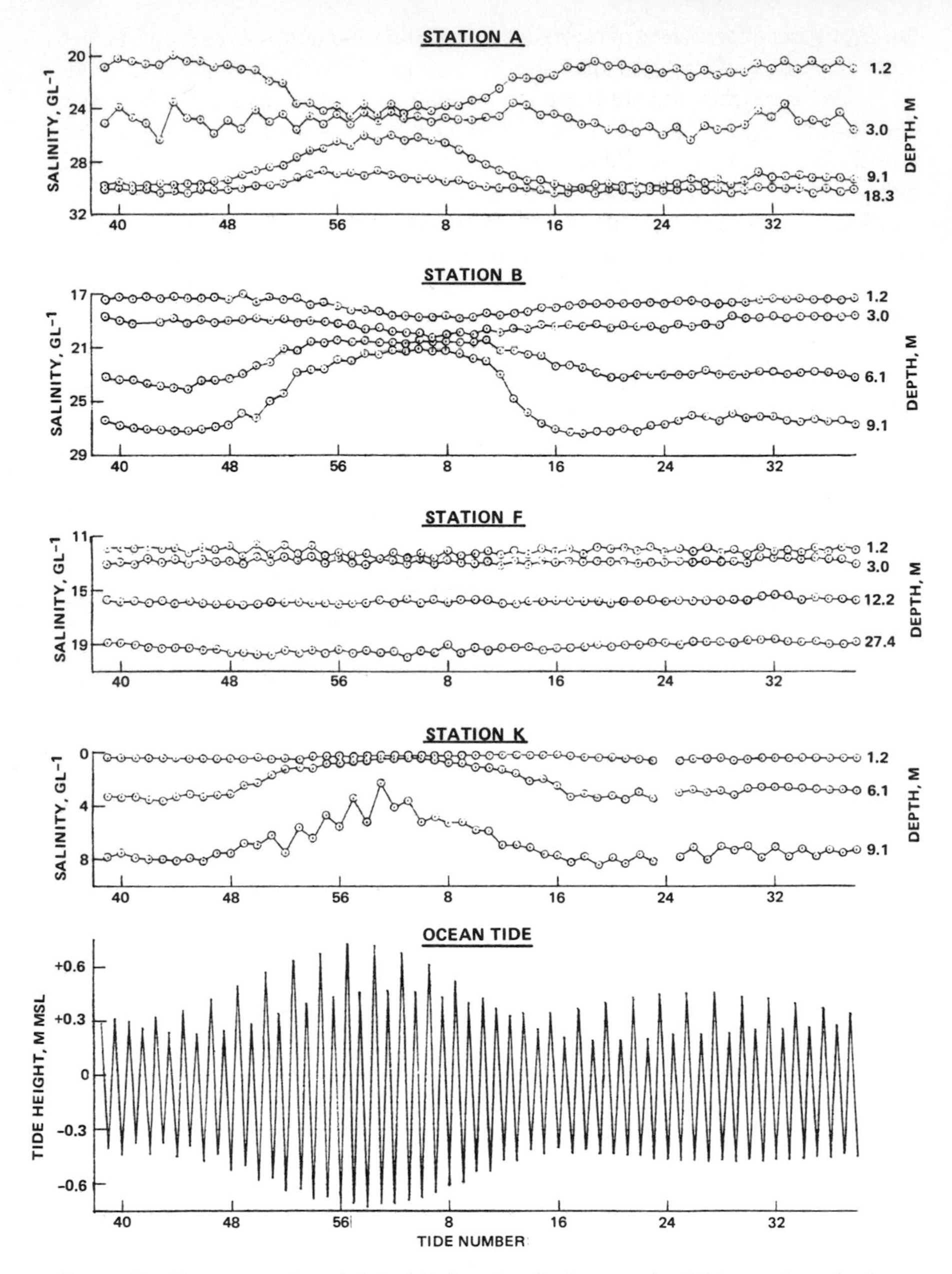

Figure 5. *Response of model slack-before-flood salinity to the 28-lunar-day tide during constant 2,000 $m^3 s^{-1}$ total bay freshwater discharge. Enhanced mixing is indicated at Stations A, B, and K (Fig. 1) during the higher range spring tide period (around tide 1). Little salinity change is indicated at Station F, in the middle bay.*

the high spring tide period. The reduced tidal response described for middle bay stations is clearly shown at Station F.

The distinctive salinity response to neap-spring tidal conditions was also observed during a similar 85 $m^3 s^{-1}$ discharge salinity-sensitivity test for main bay Station B and the James River entrance Station J-1 (Fig. 6). This figure illustrates that both vertical and longitudinal differences may occur. A vertical redistribution of salt is apparent at Station B while reduced salinity intrusion (a longitudinal response) into the James River (Station J-1) occurred during the high spring tide period compared with the neap tide period for this fixed discharge condition.

Varying freshwater discharge, meteorology, and sampling interval (approximately monthly) limit the amount of field data available to verify the existence of this salinity response to neap-spring tidal dynamics in the main bay. Field data from four SBF periods that illustrate this neap-to-spring salinity response during the verification period at main bay Station B are shown in Fig. 7. Stratified conditions existed during these neap tide periods while well-mixed conditions existed during the next spring tide periods. The bottom two examples (Fig. 7) demonstrate that these structural responses are not due to discharge variations. The July-October 1972 time frame was a period of continuously decreasing freshwater discharge following Tropical Storm Agnes (Agnes hit Chesapeake Bay on 23 June 1972). Restratification in the 19 September data

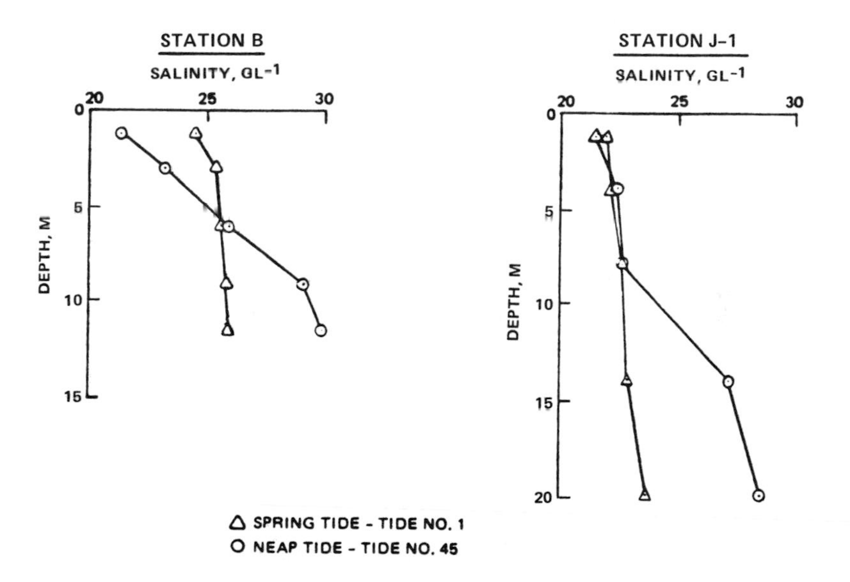

Figure 6. *Comparison of model slack-before-ebb salinity response at neap (tide 45) and spring (tide 1) tide conditions during constant 850 $m^3 s^{-1}$ total bay freshwater discharge. During the spring tide, Station B, in the main bay, indicates increased vertical mixing, while Station J-1, in the entrance to the James River, indicates reduced salinity intrusion.*

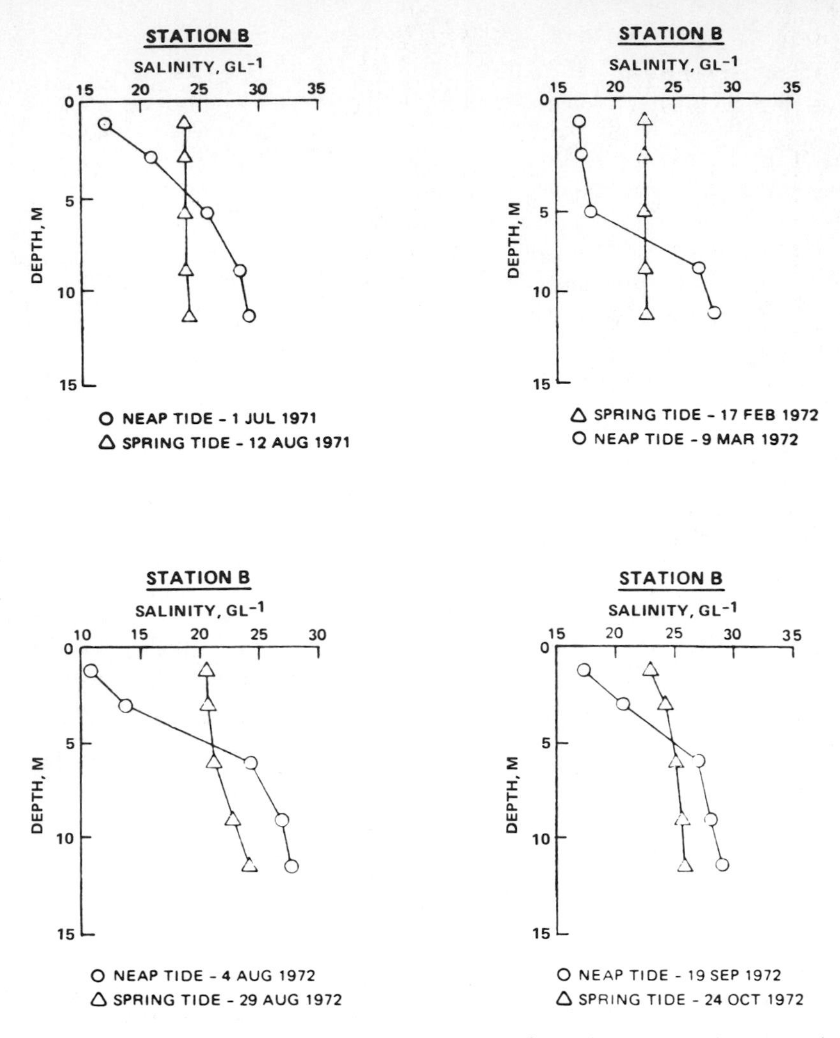

Figure 7. Field data from main bay Station B verifying the neap and spring salinity response observed (Fig. 6) in the physical model.

(neap tide) compared with the 29 August data (spring tide) indicates this response to be tide-related rather than discharge-related. Figure 8 demonstrates the agreement between the model and the prototype neap-to-spring salinity for the July-August 1971 and February-March 1972 sampling periods. This agreement supports the use of the hydraulic model in obtaining a better understanding of neap-to-spring salinity responses.

Figures 9 and 10, respectively, provide an intensive look at 28-lunar-day SBF and SBE salinity values for main bay Station B and the four James River stations during a salinity-sensitivity investigation at 875 $m^3 s^{-1}$ constant total bay

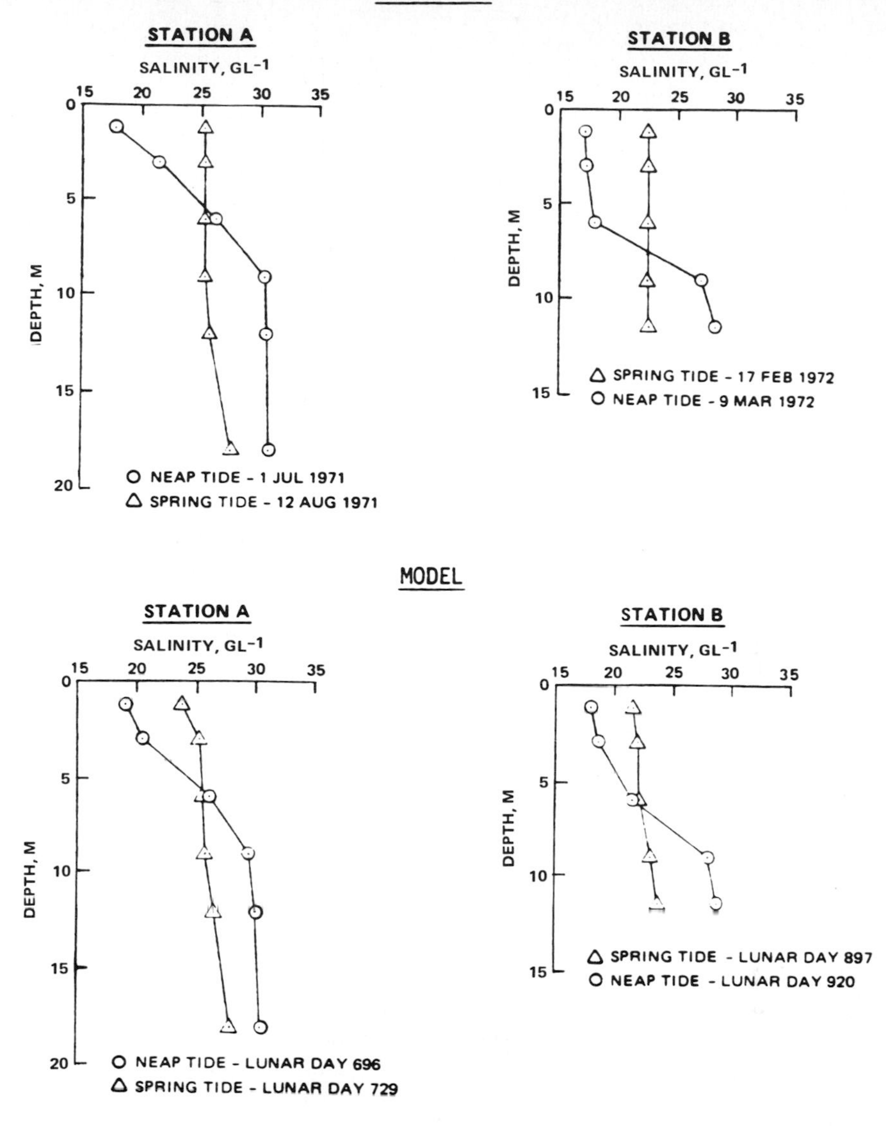

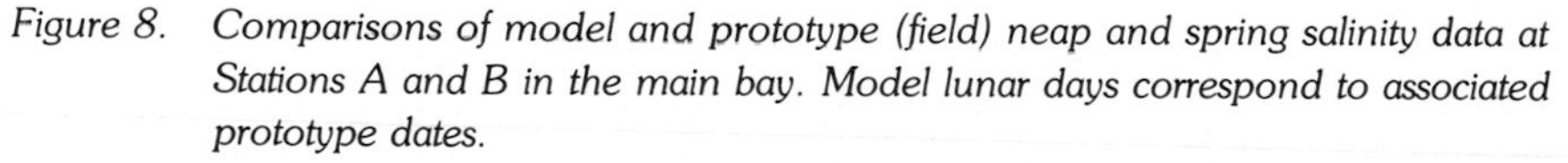

Figure 8. Comparisons of model and prototype (field) neap and spring salinity data at Stations A and B in the main bay. Model lunar days correspond to associated prototype dates.

discharge. Reduced surface-to-bottom salinity differences (generally interpreted as increased mixing) were associated with the high spring tide period at both tide stages. This reduction generally began a few tidal cycles before the highest tide (tide 1) and lasted a few tidal cycles after it. The most stratified conditions generally were associated with the tidal cycles around the neap tide periods of

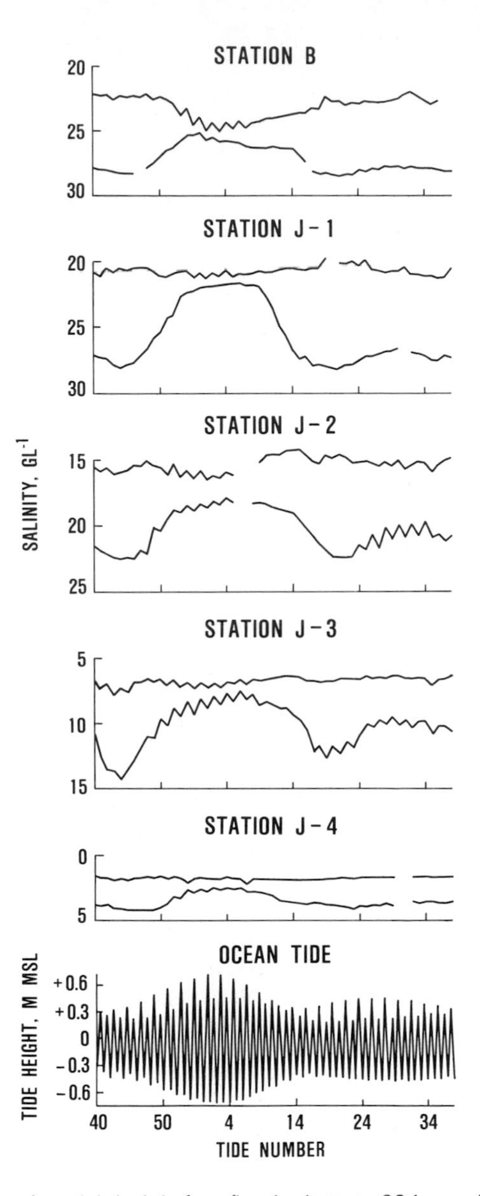

Figure 9. *Response of model slack-before-flood salinity to 28-lunar-day tide during constant 875 $m^3 s^{-1}$ freshwater discharge condition. Station locations are shown in Fig. 1.*

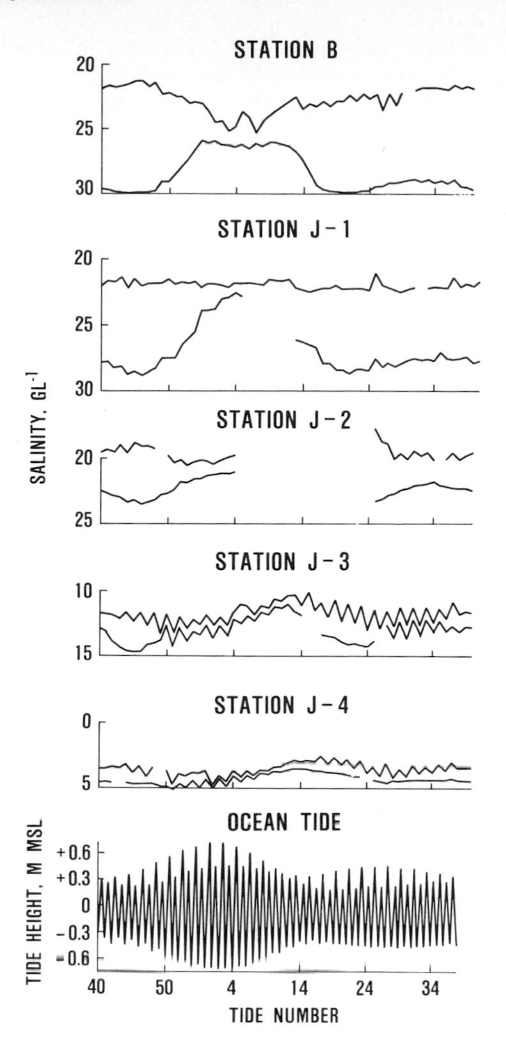

Figure 10. Response of model slack-before-ebb salinity to 28-lunar-day tide during constant 875 $m^3 s^{-1}$ freshwater discharge condition. Station locations are shown in Fig. 1.

tides 18 and 44. The lower spring period, around tide 28, also enhanced mixing conditions, but much less than the higher range spring tide period.

Discussion

Tidal dynamics and estuarine transport, mixing, and dispersion processes are important concepts in understanding estuarine salinity variability. The mechanics directly associated with salinity response to neap-spring tidal dynamics are not suitably described in the literature nor are they sufficiently understood. An intensive, elaborate, and costly sampling effort designed especially for neap-spring investigations would be required to examine the extent, and understand the mechanics, of this dynamic process from field observations. Discharge variations, wind effects, setups, setdowns, and other variable external forces would complicate or obscure the neap-to-spring salinity response.

Physical models have been used during the past 40 years to investigate three-dimensional characteristics of estuarine variability caused by natural and/or man-induced changes to the environment. They have also been used to develop, test, and expand our understanding of physical estuarine processes and responses. The hydraulic model is an ideally suited tool because boundary conditions can be carefully controlled, thereby avoiding undesirable interaction with other external processes. Also, the entire estuarine system may be easily sampled. Automation and computer-control capabilities continue to advance physical modeling practice enabling dynamic testing associated with computer-generated variable source or forcing tides and computer-controlled freshwater discharges. As shown in this paper, these new capabilities provide the opportunity to expand understanding of the complex three-dimensional aspects of neap-spring tidal dynamics and estuarine mixing and transport processes.

Stratification is reduced during periods of increased tide range associated with high spring tide periods (tide ranges of around 1 m or greater) relative to neap tide periods. This response varies throughout the main bay as well as its subestuaries. The neap-to-spring variation is more pronounced during high freshwater discharge periods (Fig. 3) when surface-to-bottom salinity gradients are amplified (i.e., increased stratification periods as a result of surface layers demonstrating larger salinity reductions than bottom layers). The strongest neap-to-spring salinity variations are located in the lower main bay and lower bay subestuaries. Reduced neap-to-spring variation in the middle bay may be associated with reduced tidal range and tidal energy.

During spring tide periods, tidal range, tidal excursion, and associated tidal currents are at their highest magnitudes; one would therefore expect highest salinity values under SBE conditions as water is transported farthest into the estuary. During SBF conditions when water is transported or extruded farthest out of the estuary, lowest salinity values would be expected. During neap tide periods, tidal range, tidal excursion, and tidal velocity are at their lowest magnitudes, and intermediate salinity values would be expected. The salinity variations presented in this paper are different from the above expectations.

Table 1 provides SBF and SBE summary values for the 875 $m^3\ s^{-1}$ discharge condition (Figs. 9, 10) for a representative spring tide (tide 4) and a representative neap tide (tide 45). These results provide additional insight into the nature of the salinity response to neap-spring tidal dynamics. At each station, bottom-to-surface salinity differences (Dif, a stratification index value) illustrate increased stratification (larger differences) during the neap tide period relative to the spring tide period for both SBE and SBF conditions. The same depth-averaged salinity is maintained at main bay Station B for neap and spring conditions, indicating a true vertical mixing phenomenon. The depth-averaged

Table 1. Neap (tide 45) and spring (tide 4) salinity summary during constant 875 m^3 s^{-1} freshwater discharge condition at main bay Station B and James River Stations J-1—J-4. Range = surface-to-bottom salinity, Dif = bottom-to-surface salinity difference, $\overline{X}$ = depth-averaged salinity.

Station	Tide No.	Slack Before Ebb			Slack Before Flood		
		Range	Dif	$\overline{X}$	Range	Dif	$\overline{X}$
B	45	21.3-29.9	8.6	25.7	22.4-28.3	5.9	24.9
	4	24.8-26.4	1.6	25.7	24.3-25.7	1.4	24.8
J-1	45	21.5-28.5	7.0	24.7	20.7-27.9	7.2	24.4
	4	21.8-22.5	0.7	22.2	20.7-21.7	1.0	21.4
J-2	45	18.9-23.3	4.4	21.4	15.8-22.4	6.6	19.6
	4	19.8-21.0	1.2	20.5	15.9-17.9	2.0	17.0
J-3	45	12.4-14.7	2.3	13.4	7.7-13.6	5.9	10.8
	4	11.6-12.2	0.6	12.0	6.9- 7.7	0.8	7.2
J-4	45	3.8- 4.7	0.9	4.4	2.0- 4.2	2.2	2.8
	4	3.7- 4.0	0.3	3.9	1.7- 2.5	0.8	2.0

salinity increased under both SBE and SBF conditions at all four James River stations during the neap tide period compared with the spring tide period. A reduced intrusion of salt, resulting from increased mixing in the main lower bay, is responsible for the apparent better-mixed spring condition in the James River. During neap tide periods, when main bay stratification is again established, vertical density gradients transport deep, higher salinity bay water into the river at the deeper depths. Field data (Haas *et al.* 1981) illustrate a similar response of reduced salt intrusion during spring tide periods for the York River.

The salinity response to the approaching high spring tide varies in magnitude but generally occurs about the same time throughout the James River and at main bay Station B (Figs. 9, 10). This response begins around tide 46, following the lowest range neap tide (tide 44), and generally reaches its most mixed conditions between tides 3 and 9. Restratification generally begins around tide 10 and reaches the most stratified conditions by tide 20. The response to the low spring tide condition (tide 28) is much reduced compared to the high spring condition. As indicated, the variations from the most stratified neap conditions to the better-mixed spring conditions follow the 28-lunar-day tidal cycle in a relatively smooth and steady fashion. As tidal range and associated tidal energy increase, the degree of destratification also increases; as the tidal range decreases from spring tides, stratification increases.

Conclusions

Computer-control technology has advanced physical modeling capabilities by allowing dynamic testing associated with computer-generated forcing tides and computer-controlled variable freshwater discharge hydrographs. Dynamic testing results presented here indicate that salinity conditions in Chesapeake Bay show a systematic variation with the neap-spring tidal cycle. These salinity responses vary by location in the main bay and its many subestuaries. The most dramatic neap-to-spring variations occur in the lower main bay and the lower bay subestuaries. The middle bay and its subestuaries exhibit reduced salinity variation, and the neap-spring response redevelops in the upper bay.

Intensive SBF and SBE salinity sampling in the lower main bay and the James River during 28-lunar-day constant freshwater discharge conditions provides additional insight into the complex three-dimensional aspects of neap-to-spring tidal dynamics and estuarine mixing and transport processes. Increased surface-to-bottom salinity gradients exist during neap tide periods relative to better-mixed spring tide periods. In the main bay, true vertical mixing occurs during the higher range tides, maintaining the neap tide average water column salinity value. In the James River, however, reduced salt intrusion occurs during the spring tide period, resulting in a reduced water column salinity. During neap tide periods, vertical density gradients in the main bay are reestablished, resulting in additional salt intrusion into the James River.

Specifically designed neap-spring studies are required to investigate in detail the processes and mechanisms associated with salinity responses to tidal dynamics. Physical models are excellent tools for beginning such studies.

Acknowledgments

The efforts of the entire Chesapeake Bay Model staff, U.S. Army Engineer Waterways Experiment Station (WES) and the modeling staff of Acres American (1979-1981) and Tetra Tech (1981-1983), contractors for model operation and maintenance, are acknowledged. The enthusiasm and energy of all the dedicated employees greatly contributed to the success of the model and its studies. The US Army Engineer Districts, Baltimore and Norfolk, provided funds for the referenced model studies, and the Office, Chief of Engineers, has given permission to publish this report.

References Cited

Allen, G. P., J. C. Salomon, P. Bassoullet, Y. Du Penhoat and C. De Grandpre. 1980. Effects of tides on mixing and suspended sediment transport in macrotidal estuaries. *Sed. Geol.* 26:69-90.

Boersma, J. R. and J. H. J. Terwindt. 1981. Neap-spring tide sequences of intertidal shoal deposits in a mesotidal estuary. *Sedimentology* 28:151-170.

Granat, M. A. and L. F. Gulbrandsen. 1982. Baltimore Harbor and channels deepening study: Chesapeake Bay hydraulic model investigation. Tech Report HL-82-5. U.S. Army Engineer Waterways Experiment Station, Vicksburg, Miss., 174 pp.

Granat, M. A., L. F. Gulbrandsen and V. R. Pankow. 1985. Reverification of the Chesapeake Bay model; Chesapeake Bay hydraulic model investigation. Tech Report HL-85-3. U.S. Army Engineer Waterways Experiment Station, Vicksburg, Miss., 297 pp.

Haas, L. W. 1977. The effect of the spring-neap tidal cycle on the vertical salinity structure of the James, York, and Rappahannock Rivers, Virginia, U.S.A. *Estuar. Coastal Mar. Sci.* 5:485-496.

Haas, L. W., F. J. Holden and C. S. Welch. 1981. Short term changes in the veritical salinity distribution of the York River estuary associated with the neap-spring tidal cycle, pp. 585-596. *In:* B. J. Neilson and L. E. Cronin (ed.), *Estuaries and Nutrients.* Humana Press, New Jersey.

Richards, D. R. and M. R. Morton. 1983. Norfolk Harbor and channels deepening study; Chesapeake Bay hydraulic model investigation. Tech Report HL-83-13, Report 1, physical model results. U.S. Army Engineering Waterways Experiment Station, Vicksburg, Miss., 371 pp.

Richards, D. R. and M. A. Granat. 1986. Salinity redistributions in deepened estuaries, pp. 463-482. *In:* D. A. Wolfe (ed.), *Estuarine Variability.* Academic Press, New York.

SALINITY REDISTRIBUTIONS IN DEEPENED ESTUARIES

David R. Richards and Mitchell A. Granat

Hydraulics Laboratory
U.S. Army Engineer Waterways Experiment Station
Vicksburg, Mississippi

Abstract: Physical model tests were conducted in a distorted-scale model of Chesapeake Bay to determine the effect of deepening the Norfolk Harbor approach channels on salinity distributions in the lower bay. This model study used a repetitive, 28-lunar day, 56-cycle source tide that was based on 12 harmonic constituents, and a weekly-stepped, 2.5-year historical hydrograph for freshwater input. Sufficient numbers and locations of salinity stations were used to predict lateral as well as longitudinal salinity redistributions. Channel deepening resulted in increased salinity intrusion and vertical stratification. The increased intrusion was confined largely to the deepened channel areas. Shallow waters adjacent to the deepened channels experienced negligible salinity increases and in some cases a freshening. Changes in the shape of the neap-spring salinity response were detected in the form of damped variations in the bottom salinity record, primarily in the deepened channel areas.

Introduction

With today's trend toward deeper draft vessels and the competition for a limited amount of commerce among a substantial number of ports, channel deepening is a common activity. In the United States, the Corps of Engineers has the responsibility for maintaining navigable waterways at their authorized depths and dimensions as well as performing the design and environmental impact analyses for channel deepening projects. The Corps or its contractors have modeled, by physical or numerical means, most of the deepening projects in recent history.

The state-of-the-art of modeling has advanced considerably since the Corps began deepening harbors. Initial model studies were conducted in distorted-scale physical models using rather simplified boundary conditions. Usually this meant running a series of base and plan tests with constant amplitude tides and a constant freshwater discharge. Dynamic variations in the prototype tide record and freshwater hydrograph were often modeled by running a series of high and low tide ranges and freshwater discharges. More recently, computer control of physical models has allowed for highly variable tide and freshwater inputs without prohibitive costs and manpower requirements. The most recent model studies, including the one presented herein, can simulate a complete range of tidal and hydrographic variability.

This paper describes the results of studies performed at the Chesapeake Bay hydraulic model on how channel deepening affects salinity redistributions in estuaries. The model was used in studies of deepened channels approaching the ports of Baltimore (Granat and Gulbrandsen 1982) and Norfolk (Richards and Morton 1983). The design of the Norfolk deepening study allowed a more thorough examination of salinity redistributions, however, because more sampling stations were used in the project area and these stations were located so that lateral redistributions could be observed. As a result, this paper draws primarily on the Norfolk test data set for its conclusions.

The Norfolk Harbor approach channels scheduled for deepening are shown in Fig. 1 and are described as follows:

a. Atlantic Ocean Channel—A new channel to be dredged 17.4 m below mean low water.
b. Thimble Shoal Channel—Deepened from 13.7 to 16.8 m below mean low water.
c. Newport News Channel—Deepened from 13.7 to 16.8 m below mean low water.
d. Norfolk Harbor Channel—Deepened from 13.7 to 16.8 m below mean low water.
e. Elizabeth River and Southern Branch of the Elizabeth—Deepened from 12.1 to 13.7 m below low water between Lamberts Point and the N&W R.R. bridge.
f. Southern Branch of the Elizabeth—Deepened from 10.7 to 12.2 m below mean low water between the N&W R.R. bridge and U.S. Route 460 Highway crossing.

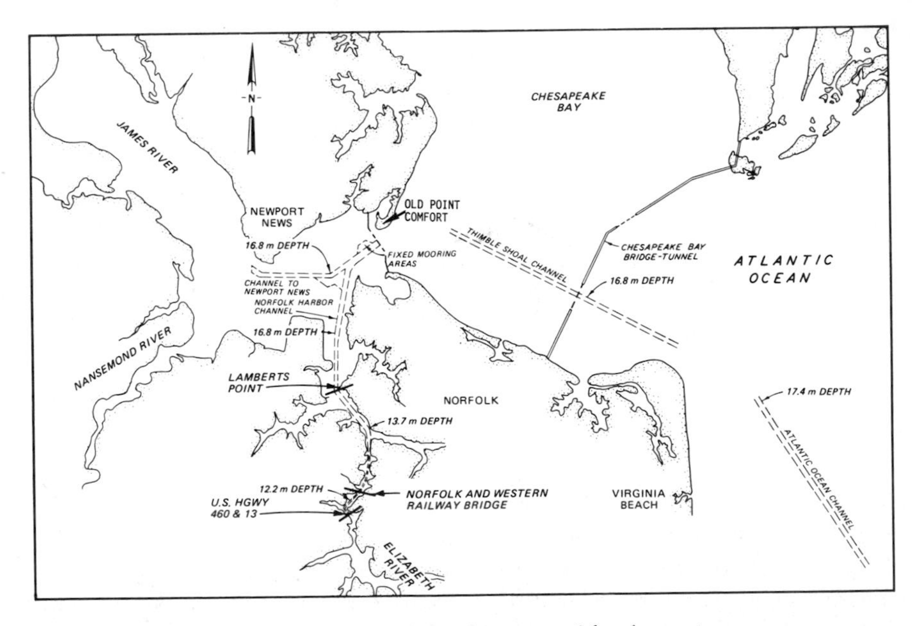

Figure 1. Norfolk Harbor entrance channels proposed for deepening.

Model Description

The Chesapeake Bay hydraulic model was located on Kent Island in Stevensville, Maryland. It was a 3.5-hectare concrete, distorted-scale, fixed-bed model. The molded area of the model represented the area from 48 km offshore in the Atlantic Ocean to the heads of tide for all tributaries emptying into Chesapeake Bay (Fig. 2). Freshwater inflows entered the model at 21 separate inflow points that represent significant tributaries. Source tides and salinity were controlled at the Atlantic Ocean boundary. Both source tides and freshwater inflows were programmed by model control computers that also record model control data. A more complete description of the model and its capabilities can be found in Granat *et al.* (1985). The model was well verified and could be used to assess base-to-plan condition changes to within ± 0.5 ppt salinity. The variable boundary conditions used and their impact on modeling deepened estuaries follows.

Tides

The source (ocean) tides used in the Norfolk Harbor study consisted of a repetitive, 28-lunar-day, 56-cycle tide sequence (Fig. 3). The 12 harmonic constituents used to construct the tide were derived from a harmonic analysis of National Oceanic and Atmospheric Administration (NOAA) tide records of July 1971 to June 1972 obtained at Old Point Comfort, Virginia, near the entrance of the James River. Once the desired tide for Old Point Comfort had been determined, it was necessary to adjust the amplitudes and phases to obtain values for the ocean source tide generated 30 miles offshore. The ocean tide is based on the equation:

$$h(t) = A_o + \sum_{i=1}^{12} \left[a_i \cos\left(\frac{2\pi}{T_i} t - \phi_i \right) \right]$$

where

$h(t)$ = tide height, m
A_o = mean water level, m
a_i = constituent amplitude, m
t = time, hr
T_i = constituent period, hr
ϕ_i = constituent phase, radians

The constituent amplitudes, periods, and phases can be found in Richards and Morton (1983). Tide ranges for the source tide vary from 1.46 m for tide 1, the largest spring tide, to 0.79 m for tide 48, a smaller neap tide (Fig. 3). Salinity was measured at each multidepth station throughout the model at the local slack before ebb on tides 1, 10, 28, and 48. These tides represent the variable tide

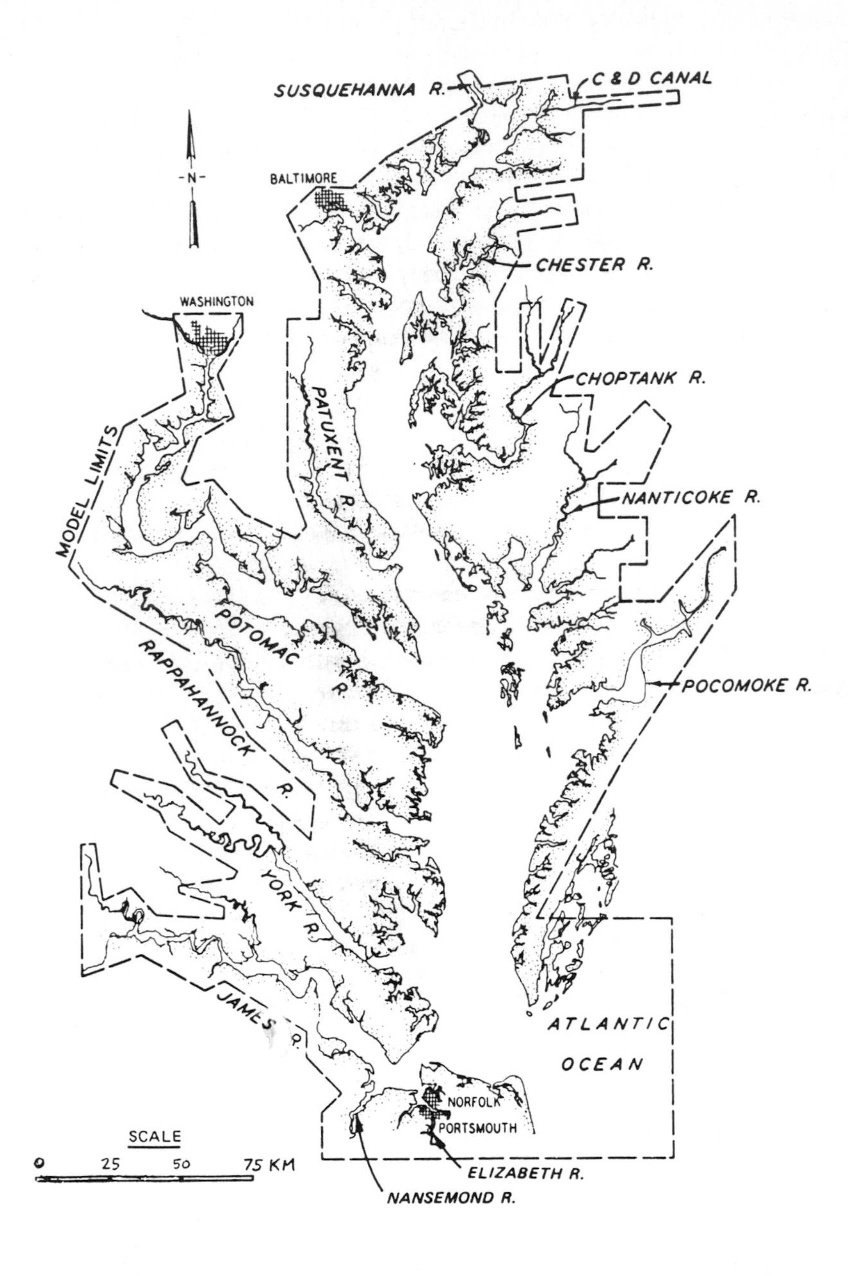

Figure 2. *Area of Chesapeake Bay included in the physical model used in the salinity redistribution study.*

conditions associated with high-spring, mean, low-spring, and low-neap range tides, respectively. Using this sampling procedure, we could observe a wide range of tide conditions that simulated true prototype conditions.

Freshwater Inflow

The Norfolk test hydrograph consisted of two parts (Fig. 4). First, there was a lead-in period (weeks 1-15) of constant discharge that reflected an average total bay discharge. The second period (weeks 16-136) depicted weekly averaged prototype conditions between 24 May 1963 and 17 August 1965.

The 15-week constant discharge period was used to observe stability prior to starting the historical hydrograph and salinity sampling. The ensuing historical hydrograph was used because it was an available data set and had been used on other Chesapeake Bay model tests. As a whole, it reflects a period of relatively low discharge to drought conditions on the bay. The hydrograph does, however, provide sufficient seasonal variations in freshwater inflow to study salinity variability. Source salinity was held at a constant 32.5 ppt throughout the test.

Existing Salinity Variability

The lower Chesapeake Bay and the western shore Virginia rivers exhibit some of the most dramatic structural variations in salinity in the entire Chesapeake Bay system. These variations are largely induced by natural fluctuations in the freshwater inflow and tidal conditions. Early attempts at classifying estuaries into mixed, partially stratified, and stratified groups were based largely on the role that varying freshwater inflows played in mixing processes. While that classification has not changed substantially, some estuaries, and most particularly the lower Chesapeake Bay, have their mixing processes controlled as much by variations in the monthly tide record. Haas (1977) first published Chesapeake Bay data that suggested the neap-spring tide variation induced fortnightly variations in salinity structure. Studies conducted soon thereafter at the Chesapeake Bay model confirmed the influence of the neap-spring tide cycle and documented its strength and presence throughout the Chesapeake Bay model (Granat and Richards, this volume). Prior to discussing changes in salinity variability brought on by channel deepening, we will describe the natural variations that occur in the James River and the lower Chesapeake Bay. Fig. 5 shows the locations (within the model) of the salinity sampling stations used in this study.

Discharge-Induced Salinity Variations

Salinity structure in the James River and Thimble Shoal Channel is strongly influenced by freshwater inflows. The James River is one of the larger inputs of fresh water to Chesapeake Bay so it is predictable that seasonal variations in the freshwater hydrograph would have a sizable effect on salinity structure. Periods of high freshwater discharge during the winter and spring seasons generally freshen the entire water column with an increased vertical salinity stratification. With the approach of the summer and early autumn low-flow periods, the salinity in the entire column increases with a net decrease in stratification.

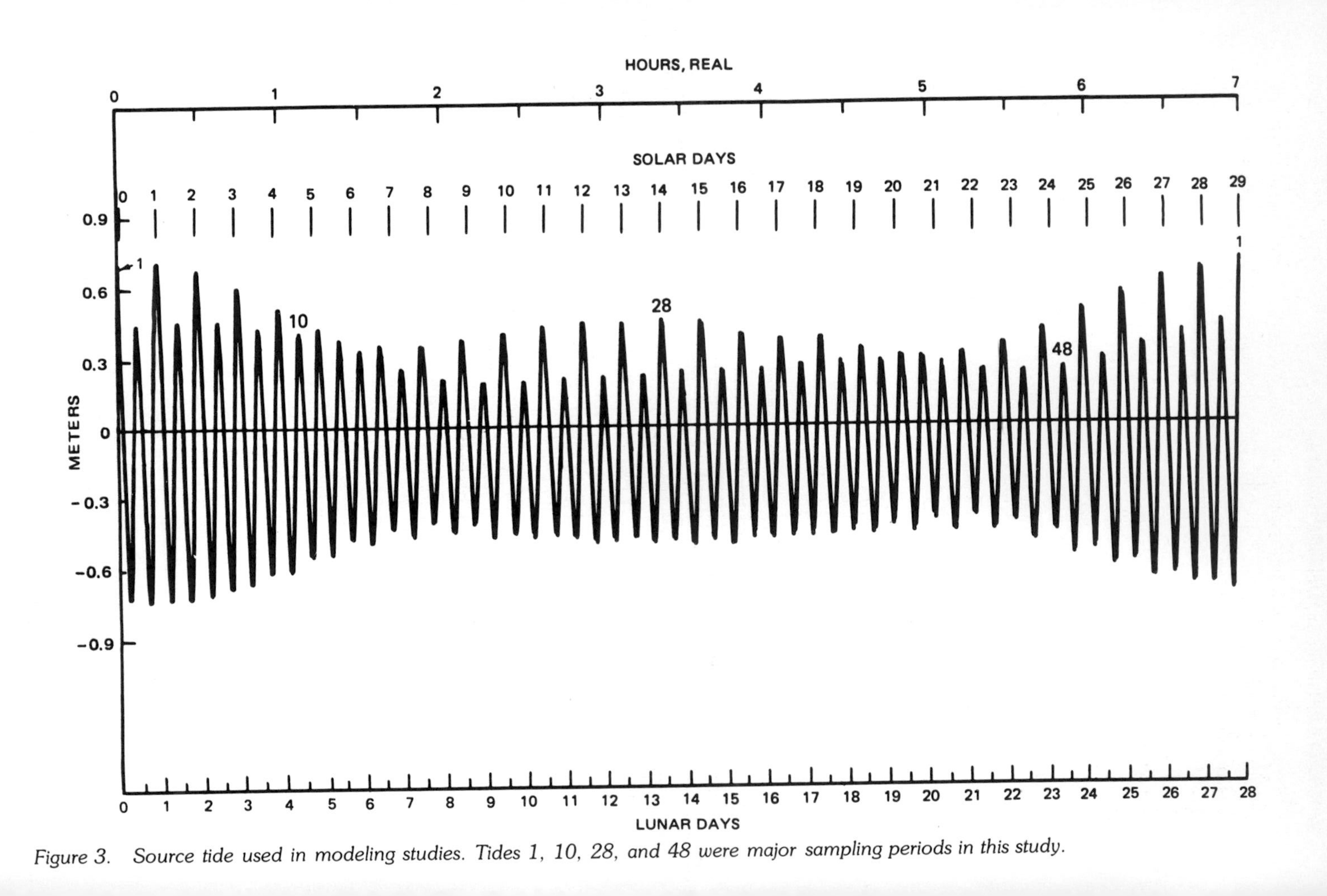

Figure 3. *Source tide used in modeling studies. Tides 1, 10, 28, and 48 were major sampling periods in this study.*

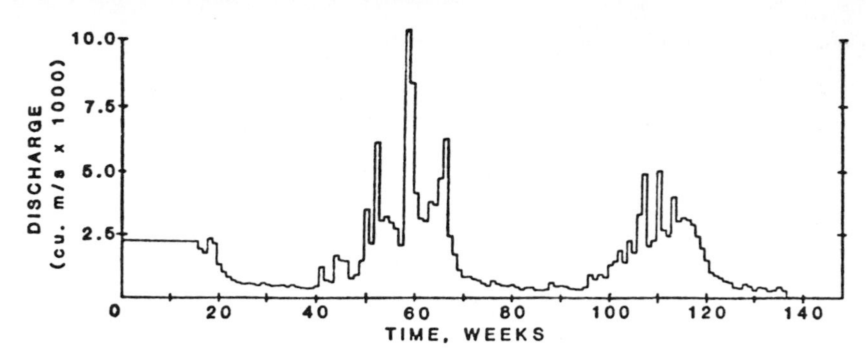

Figure 4. Hydrograph of freshwater inflow (total bay discharge) used in modeling the salinity redistribution in lower Chesapeake Bay.

The flushing effects of a high discharge rate can be observed in the model from the longitudinal isohaline profiles of the James River over a range of coincident tide conditions (Fig. 6). In general, over these tide conditions, the 1-ppt isohaline is located approximately 38 km farther downstream during the high-flow period than during the low-flow period. Similarly, the 15-ppt isohaline is about 15 km farther downstream during the high-flow conditions.

The freshwater influence of the James River and the upper bay combine to form large freshwater-induced salinity variations in the Thimble Shoal and Atlantic Channels as well (Fig. 7). High flow conditions cause increased stratification and the formation of a much stronger vertical density gradient. A similar response to increased discharge occurs in the Elizabeth River although it results largely from its entrance conditions on the James River. These dramatic, natural variations in salinity structure occur throughout the project area prior to channel deepening. They have a seasonal frequency that is associated with freshwater discharge conditions. However, other natural variations of salinity in the project area have a much shorter frequency.

Tide-Induced Salinity Variations

The lower Chesapeake Bay, along with the James, York, and Rappahannock Rivers, experiences tidally induced salinity variations that have been noticed in the prototype and confirmed in the Chesapeake Bay model (Granat and Richards, this volume). The neap-spring variations, as they are often called, result when lower amplitude tides (tides 28 and 48 in Fig. 3) allow stratification in the water column followed by larger range tides (tides 1 and 10) that mix the column again. These variations present themselves as sawtoothed patterns in the discretely sampled salinity time-histories. This response is quite strong from Station TS 0004 in the Thimble Shoal Channel upstream to Station JG 0302 near the tip of the salinity wedge in the James River (Fig. 8). Intertidal salinity changes of as much as 5 to 8 ppt are common at a single depth along with depth-averaged salt variations of 3 to 5 ppt. Extreme variations, such as at Station JG

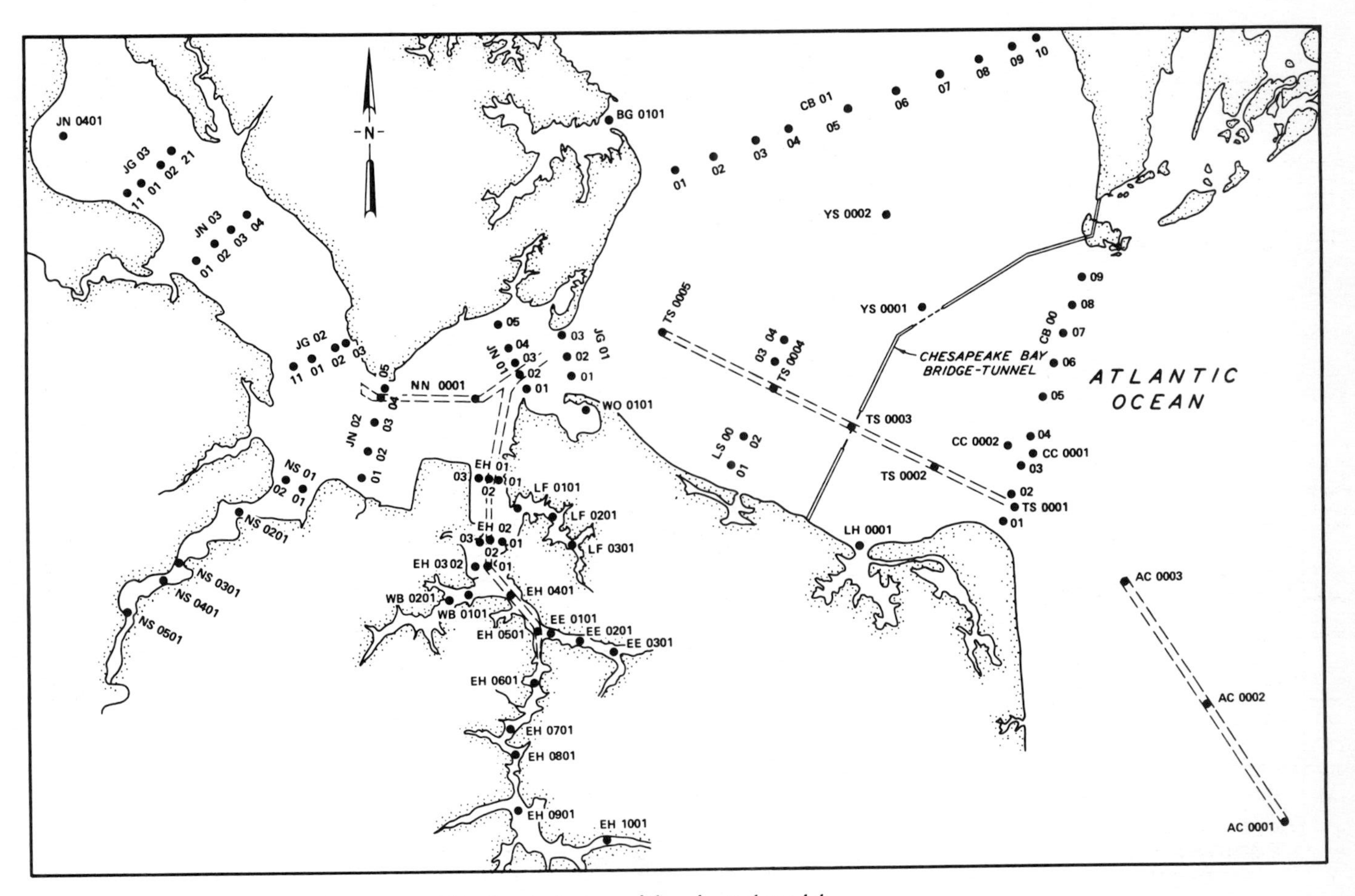

Figure 5. *Sampling stations for salinity within the study area of the physical model.*

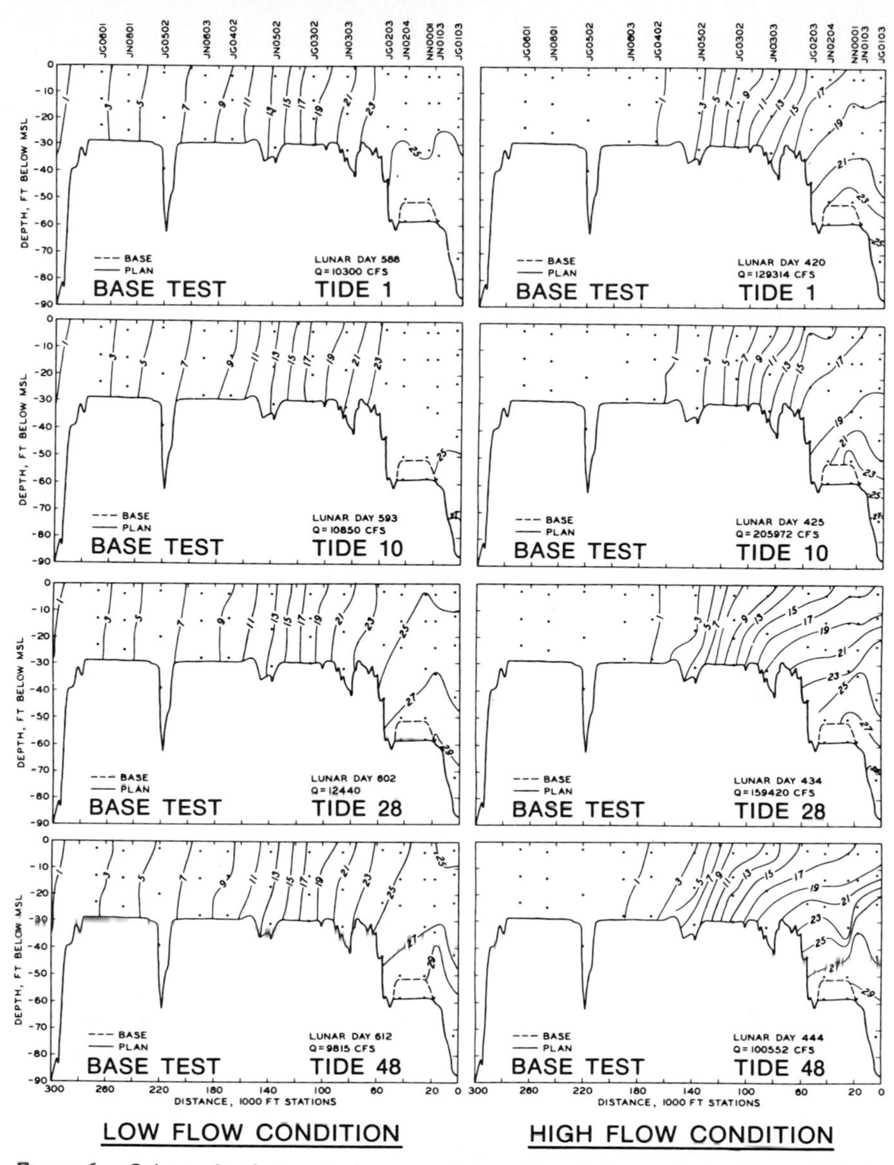

Figure 6. Salinity distribution in the lower James River during periods of low and high flow. Station locations (numbers at top) are indicated in Figure 5.

0302, can result in variations of depth-averaged and bottom salinity changes of 9 and 14 ppt, respectively. Neap-spring changes in the project area appear to be greatest during periods of high discharge when the vertical density gradient is the strongest. A longitudinal view of salinity variations resulting from the neap-spring tide cycle can be seen in the isohaline plots (Figs. 6, 7). Again, these tide-induced variations are present in the project area prior to channel deepening.

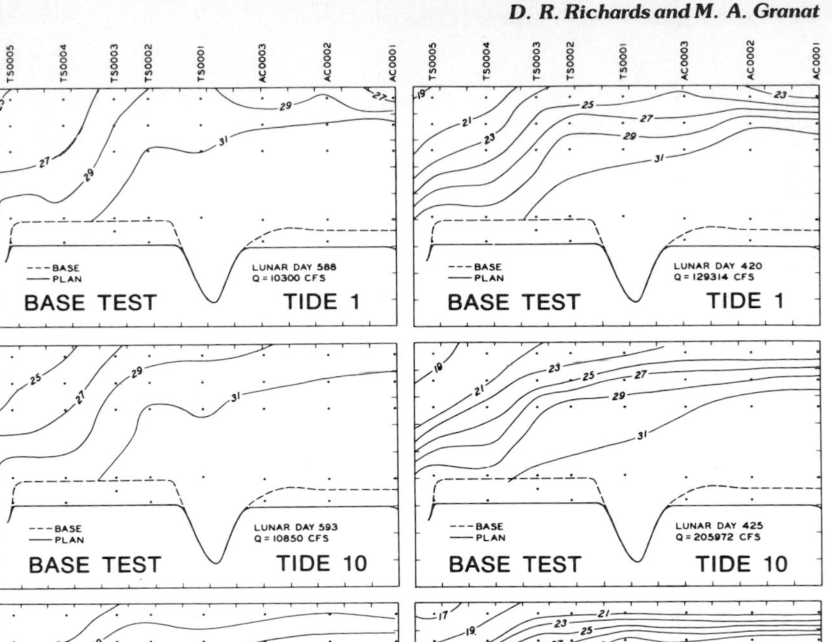

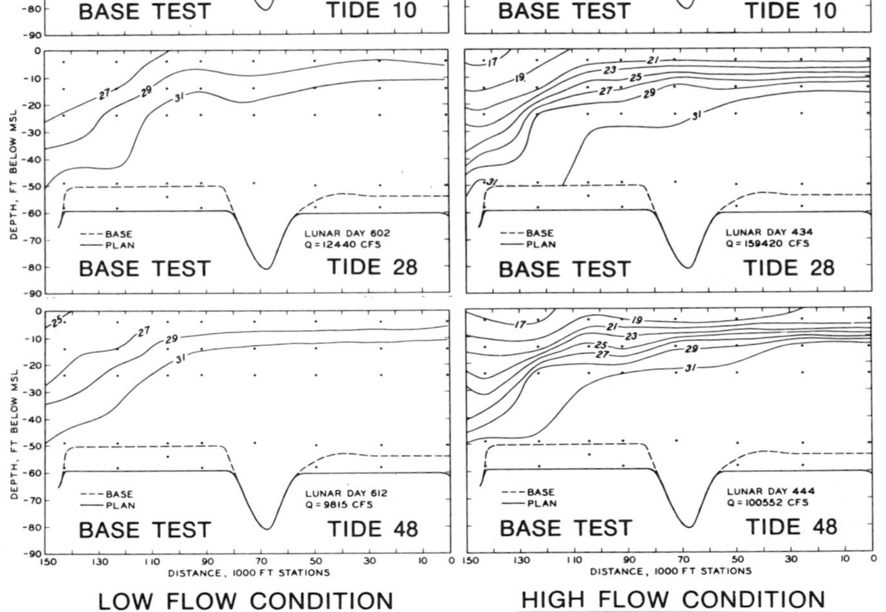

Figure 7. *Salinity distributions during low flow and high flow in the Atlantic Ocean Channel and Thimble Shoal Channel. Distance is marked from the seaward end of the Ocean Channel, and station locations are indicated in Figure 5.*

The Effects of Channel Deepening on Salinity Structure

Any time an estuary is deepened, there is increased potential for salinity intrusion. This potential is greater in stratified and partially stratified estuaries like Chesapeake Bay than in vertically mixed estuaries with smaller freshwater influences. Deepening provides easier access for more saline waters on the bottom

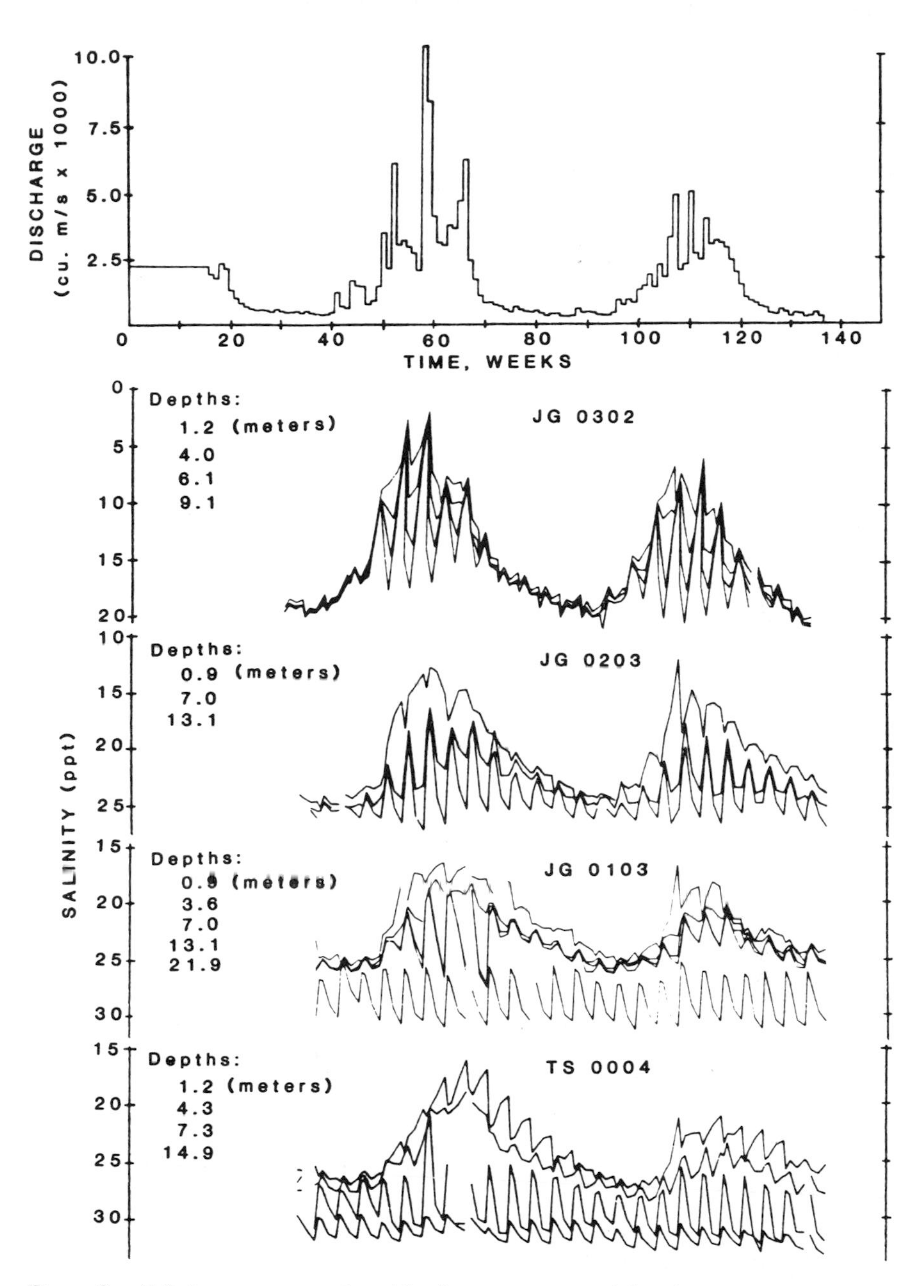

Figure 8. *Salinity responses induced by the neap-spring tidal cycle at various stations in the vicinity of Norfolk Harbor, prior to channel deepening.*

to intrude into the upper estuary. The major force resisting salinity intrusion, namely the opposing freshwater inflow, remains constant or is reduced, while the forces causing upstream density transport of the more saline waters increase.

While it is easy to predict that channel deepening will increase salinity intrusion, it is much more difficult to describe the magnitude of the increase, and more difficult still to characterize the nature of the changes with respect to salinity redistributions. In the early stages of the Norfolk deepening study design, it seemed plausible that redistributions would occur as a result of the deepening project. First, there was the possibility that deepening the approach channels into the James River would divert salt water from the Atlantic Ocean and lower Chesapeake Bay into the James River, thereby resulting in a freshened upper bay. As a result, stations were located throughout the model to detect such large-scale redistributions (Fig. 5). There was also the possibility that redistributions could occur between the extensive shallow water areas on either side of the channels and the deepened channels themselves, so samples were taken transversely across the deepened channels as well as longitudinally. The last area of concern was that deepening could cause structural changes in the neap-spring salinity response. To observe this possible phenomenon, all salinity stations were sampled at multiple depths to allow inspection of the vertical salinity gradient. In addition, accurate depth-averaged calculations of salinity transport were made to allow for further understanding of how the neap-spring tide cycle affects salinity transport. In the deepened plan test, the depths sampled were the same as in the base test, with the addition of a new bottom depth near the plan channel bottom. In each of the above cases, the model was able to answer the questions of interest.

Norfolk Test Results

Salinity redistributions caused by the proposed channel deepening were examined in each of three ways: (a) plan-minus-base differences in depth-averaged salinity to determine differences in the mass movement of salt, (b) plan-minus-base differences in salinity at each depth but most importantly the bottom depth were the biota are most sensitive and immobile, and (c) plan-minus-base variations in the structure of the salinity response to the neap-spring tide cycle.

Salinity data from 193 sampling stations throughout the model were analyzed and form the basis for the reported observations. The broad sampling indicated that there were no large-scale redistributions of salt away from the upper Chesapeake Bay to the deepened lower bay and James River areas. Increased salinities in the deepened project areas did not result in a fresher upper bay. Total salt in the entire Chesapeake Bay system seems to have been increased slightly but was confined to the project area.

Within the project area, salinity intrusion was predictably increased in the deepened condition. Plan-minus-base differences in depth-averaged salinity indicate that channel stations in the Elizabeth and lower James Rivers along with the upper end of Thimble Shoal Channel show noticeable average net salinity

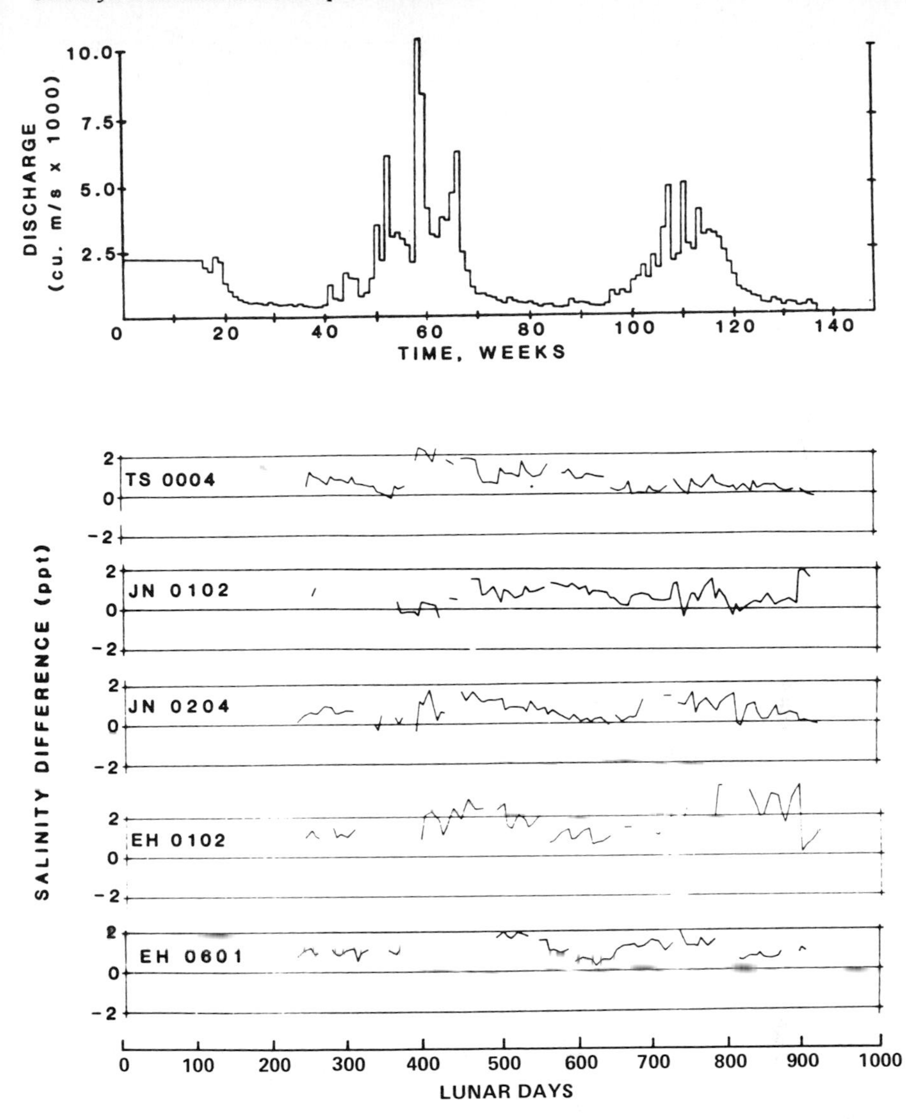

Figure 9. *Changes in depth-averaged salinity at five model stations (See Figure 5) near Norfolk Harbor, as a result of channel deepening. Positive numbers indicate salinity increases.*

increases ranging from 0.5 to 1.7 ppt (Fig. 9). The stations shown in Fig. 9 are all deepened channel stations, but this trend also occurred (to a lesser degree) in the undeepened channels farther upstream in the James River. These increases were confined largely to the channel areas.

Shallow portions of the estuary showed no noticeable increase in depth-averaged salinity. In fact, it was noticed that some shallow waters adjacent to the deepened channels experienced a freshening as a result of channel deepening

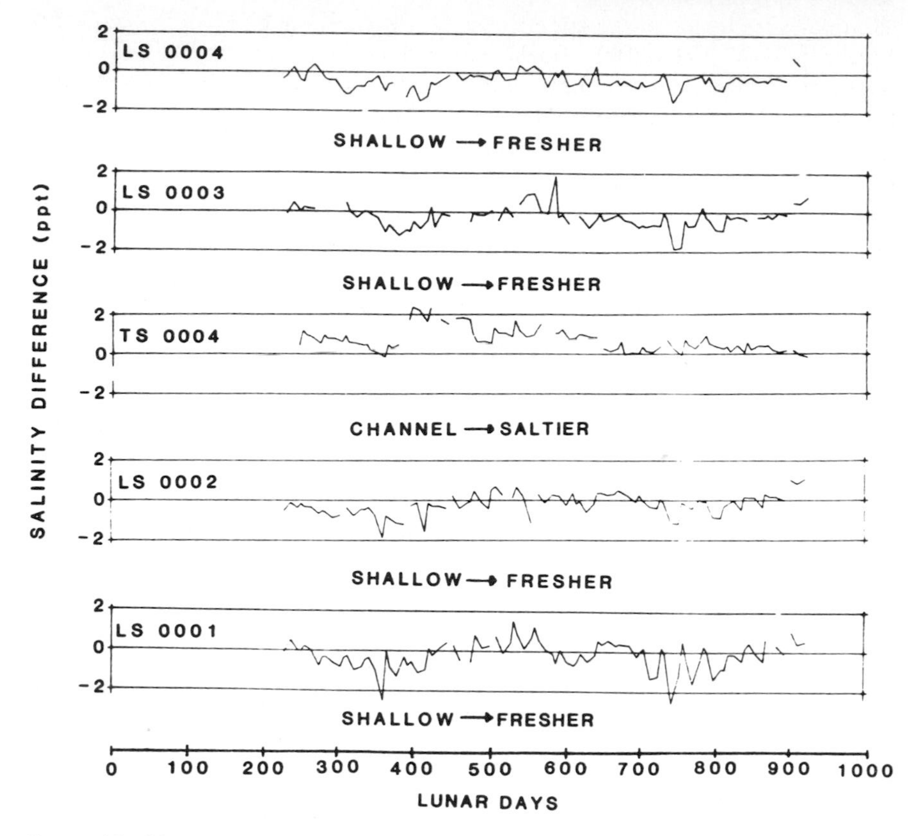

Figure 10. Changes in depth-averaged salinity at five cross-channel model stations in lower Chesapeake Bay. As a result of channel deepening, the salinity increased in the center of the channel (Station TS0004), but decreased at the shallow stations on either side (Stations LS0001-0004).

(Fig. 10). The shallow portions of the estuary, because of their immobile shellfish populations, have been identified as the areas most sensitive to increased salinity. Our test results indicated that such increases would not occur.

Since effects on shellfish are normally the primary concern in deepening projects, special attention was given to the bottom salinities throughout the lower bay. Plan-minus-base differences in bottom salinity were calculated for a variety of tidal and hydrograph conditions, and the results are summarized in Figs. 11-13. Figures 11 and 12 give results for high and low discharge periods, respectively, averaged over the tidal record, while Fig. 13 summarizes the results averaged over all discharge and flow conditions. The results confirm earlier suggestions that increased intrusion was confined largely to deepened areas with lesser impacts in shallow waters.

Changes in the shape of the neap-spring salinity response were detected in the deepened condition. This occurred in the form of damped variations in the bottom salinity record, primarily in the deepened channel areas. Fig. 14

illustrates this change at the entrance to the James River (JG 0103) and in the Elizabeth River (EH 0202). Tidally induced variations in the bottom salinity record were approximately 5 ppt at Station JG0103 in the base condition and 2.5 ppt in the deepened plan condition. Bottom salinity variation at station EH0202 was damped from 3 to 1.5 ppt. This damping was confined largely to the deepened Elizabeth River.

The biological ramifications of a damped neap-spring response are worthy of consideration. A variety of organisms can tolerate modest increases or decreases in salt concentration provided they are of limited duration. The damped neap-spring response along with the slight increase of bottom salinities could place a number of species in a stressful position. Biologists would have to determine the tolerance levels for each species versus the changes predicted in this report. The predicted changes, in either an absolute or time-varying neap-spring form, inevitably occur, however, in the less biologically active areas in the deep channels. Shallow-water areas where most shellfish thrive were relatively unaffected by this deepening project especially when compared with the magnitude of natural salinity variations that already occur in the base condition.

Conclusions

Physical model tests at the Chesapeake Bay hydraulic model indicate that channel deepening can cause several forms of salinity redistributions within the estuary. The redistributions detected in the Norfolk deepening study represented very minor salinity variations relative to the naturally-induced variability present in the project area.

Salinity redistributions were observed in three dimensions. Longitudinal redistributions were indicated by increased depth-averaged salinity in the deepened channel areas. The longitudinal axis of the Elizabeth River showed greater increased salinity intrusion than did the main axis of the Thimble Shoal Channel and James River. Increased depth-averaged salinity along the longitudinal axes resulted from increased bottom layer salinity and thus a vertical redistribution. The channel areas were generally more stratified in the deepened condition.

Lateral redistributions were minor, as well. Concurrent with the net salinity increases in the deepened channels, there were either no changes or detectable freshening of adjacent shallow-water areas. This freshening phenomenon was most pronounced when the difference in channel and shallow-water depths was greatest.

Perhaps the most unexpected effect of channel deepening on salinity was the damped response of salinity to the neap-spring tide cycle. This was prevalent in the bottom waters of the deepened channels and was not evident in any of the shallow-water areas. This observation could not have been made in numerical or physical models that use steady-state boundary conditions or short-term simulations.

For all of the redistributions that resulted from deepening Norfolk approach channels, it is important to note that plan-minus-base salinity variations rarely exceeded 2 ppt. Chesapeake Bay, in the vicinity of the project area, is a

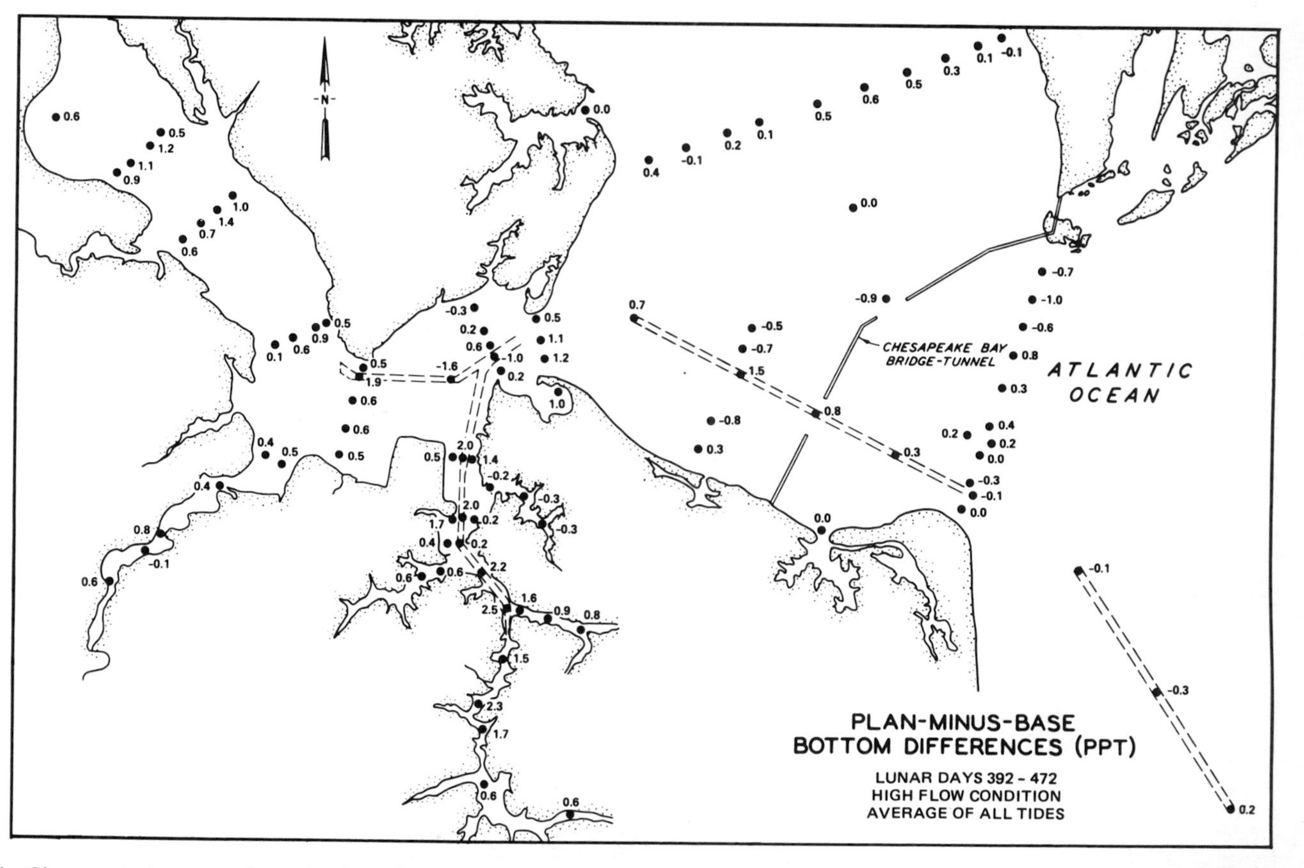

Figure 11. *Changes in bottom salinity (ppt) in the physical model due to the channel deepening project (plan-minus-base). High-flow conditions.*

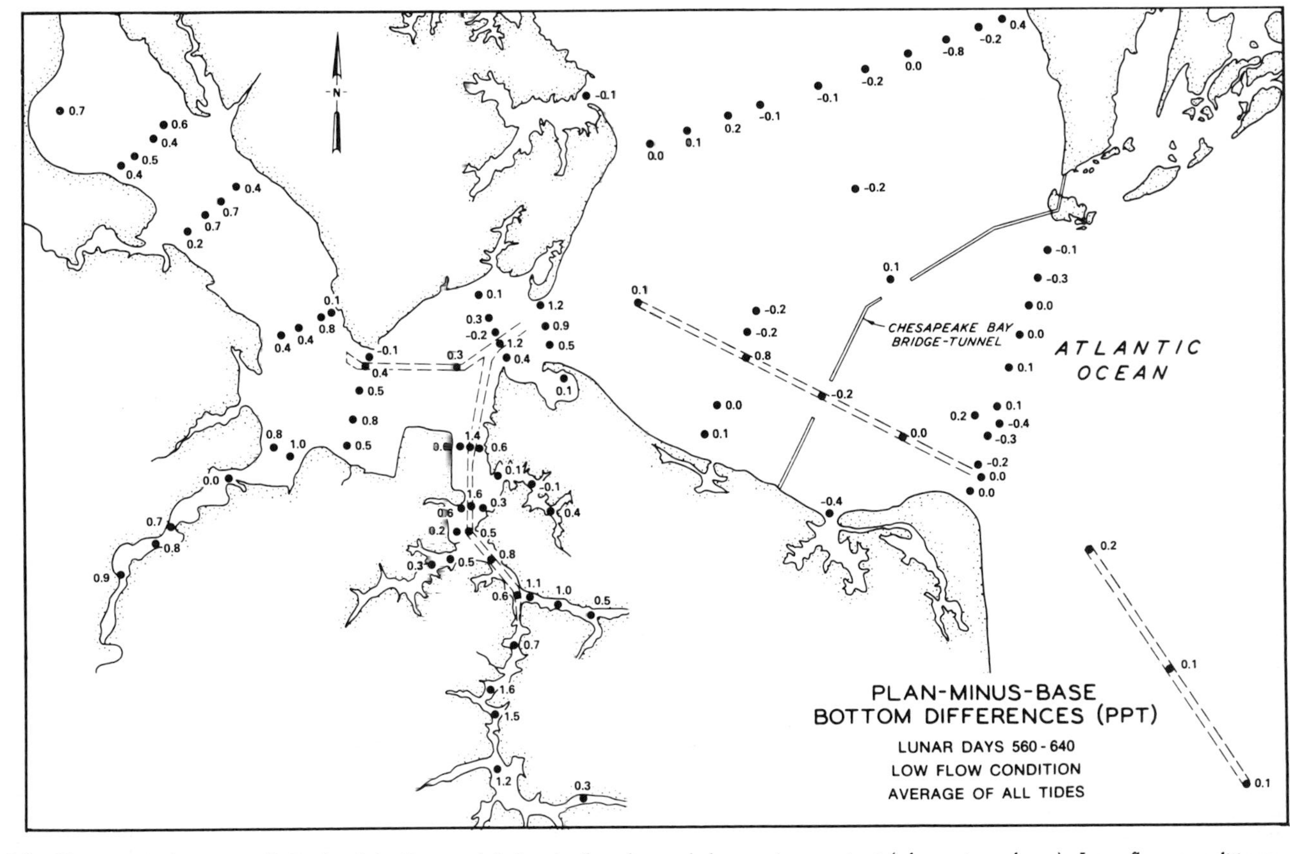

Figure 12. Changes in bottom salinity (ppt) in the model due to the channel deepening project (plan-minus-base). Low flow conditions.

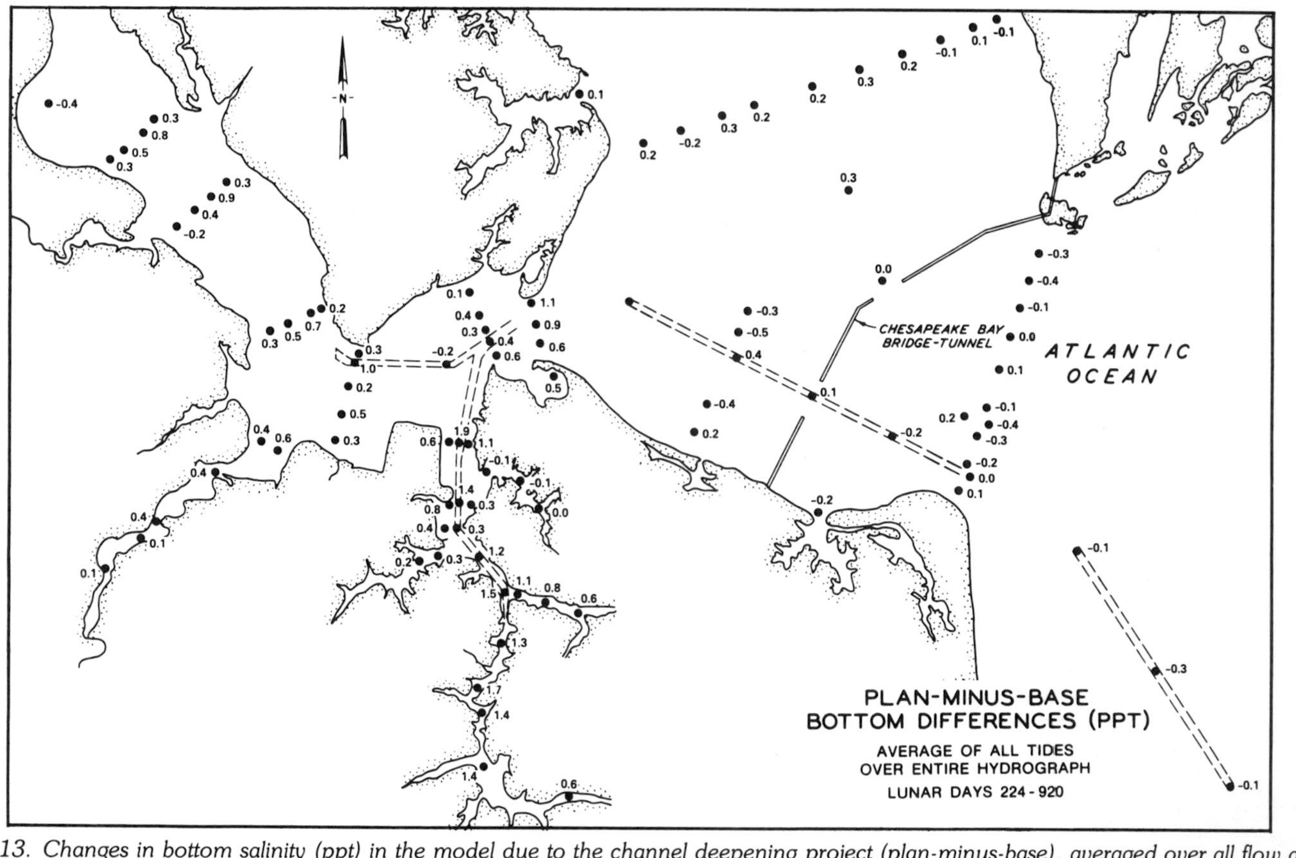

Figure 13. *Changes in bottom salinity (ppt) in the model due to the channel deepening project (plan-minus-base), averaged over all flow and tide conditions.*

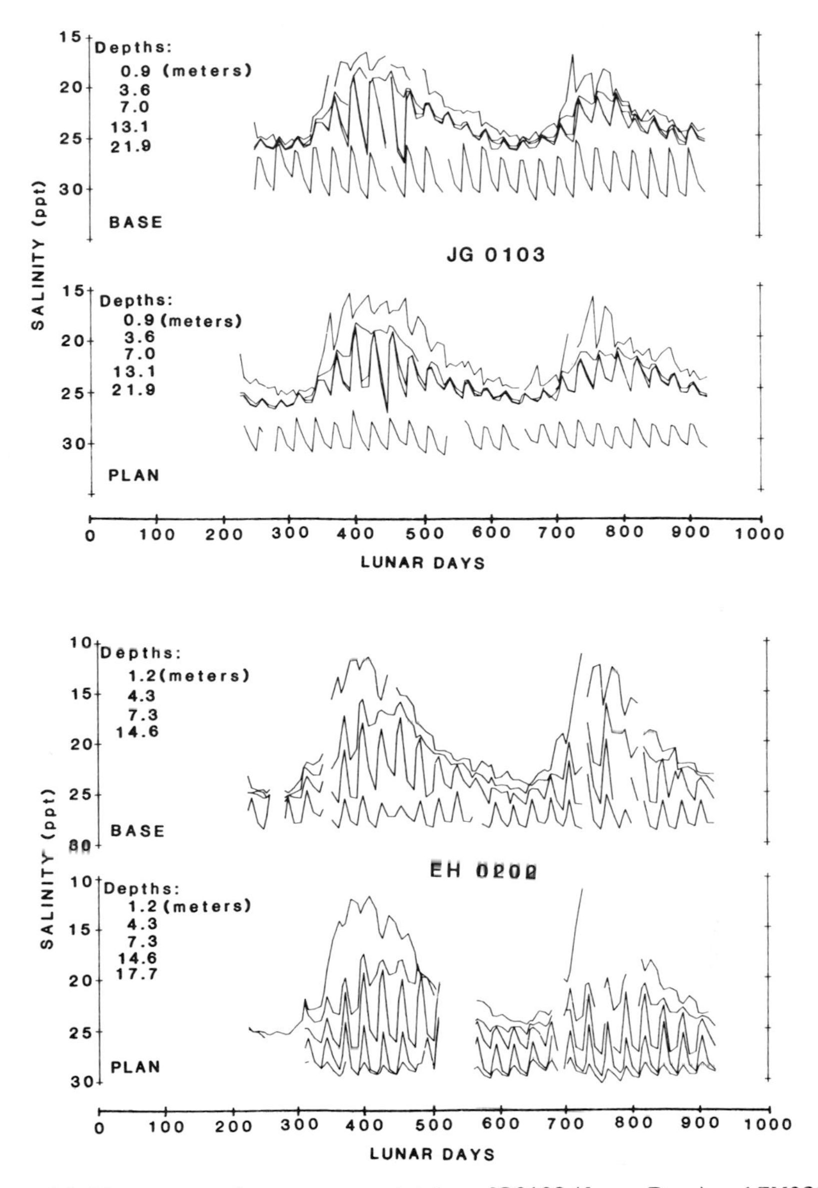

Figure 14. Neap-spring salinity response at stations JG0103 (James River) and EH0202 (Elizabeth River) for the base and plan channel depths. The amplitude of variation in bottom salinity is markedly damped in the deepened (plan) condition.

highly dynamic estuary where natural salinity fluctuations resulting from variable tide and freshwater conditions are many times greater than this value. In fact, because salinity redistributions were well defined, it was determined that the most sensitive biological areas, namely shallow-water shellfish beds, experienced negligible salinity changes.

Although Chesapeake Bay is the only estuary to be studied to date using the variable boundary conditions described herein, the observations are probably not specific only to Chesapeake Bay. The findings could apply generically to many deepened estuaries; but until other estuaries are modeled with variable boundary conditions, this cannot be established.

Acknowledgments

Special thanks go to the U.S. Army Engineer District, Norfolk, for their full cooperation throughout the conduct of the Norfolk Harbor and Channels Deepening Study. Without their cooperation and considerable funding, this study would not have been possible. The dedicated staffs of the U.S. Army Engineer Waterways Experiment Station, Hydraulics Laboratory, and their contractors, Acres American, Inc. and Tetra Tech, Inc., are also acknowledged. The Office, Chief of Engineers, has given permission to publish this paper.

References Cited

Granat, M. A. and L. F. Gulbrandsen. 1982. Baltimore Harbor and channels deepening study: Chesapeake Bay hydraulic model investigation. Technical Report HL-82-5. U.S. Army Engineer Waterways Experiment Station, Vicksburg, Miss. 174 pp.

Granat, M. A. and D. R. Richards. 1986. Chesapeake Bay physical model investigations of salinity response to neap-spring tidal dynamics: A descriptive examination. pp. 447-462. *In:* D. A. Wolfe (ed.), *Estuarine Variability*. Academic Press, New York.

Granat, M. A., L. F. Gulbrandsen, and V. R. Pankow. 1985. Reverification of the Chesapeake Bay model; Chesapeake Bay hydraulic model investigation. Technical Report HL-85-3. U.S. Army Engineer Waterways Experiment Station, Vicksburg, Miss. 297 pp.

Haas, L. W. 1977. The effect of the spring-neap tidal cycle on the vertical salinity structure of the James, York, and Rappahannock Rivers, Virginia, U.S.A. *Estuar. Coastal Mar. Sci.* 5:485-496.

Richards, D. R. and M. R. Morton. 1983. Norfolk Harbor and channels deepening study: Chesapeake Bay hydraulic model investigation. Technical Report HL-83-13, Report 1, physical model results. U.S. Army Engineer Waterways Experiment Station, Vicksburg, Miss. 371 pp.

A SCREENING MODEL FRAMEWORK FOR ESTUARINE ASSESSMENT

C. John Klein III

Ocean Assessments Division
Office of Oceanography and Marine Assessment
National Oceanic and Atmospheric Administration
Rockville, Maryland

and

J. A. Galt

Ocean Assessments Division
Office of Oceanography and Marine Assessment
National Oceanic and Atmospheric Administration
Seattle, Washington

Abstract: This screening model framework is intended to provide a preliminary assessment of hydrodynamic and/or water quality behavior within an estuary. Since the model is diagnostically designed, it can readily incorporate available information; its generic structure allows application to any system. Another characteristic, the kinematic formulation of current fields, provides an alternative to the formidable task of solving a full set of geophysical, fluid-dynamics equations. Hydrodynamic features, either known or postulated, can thus be incorporated through appropriate boundary conditions. The model is steady-state, two-dimensional, depth-averaged and uses a finite element solution technique. The model was applied to Breton Sound (Louisiana) to investigate the significance of gross circulation features and the possible effects of a proposed fresh water diversion on salinity distributions. The results contributed to the formulation of a field monitoring program to support the development of an eventual real-time modeling application that will be used in determining an operating schedule for the diversion structure. The results, as presented, demonstrate the utility of the modeling framework for identifying processes, controlling circulation, future research needs, and appropriate follow-up modeling approaches.

Introduction

This paper describes the formulation of NOAA's Water Quality Screening Model Framework and its application to a study of the effects of a proposed freshwater flow diversion from the Mississippi River, in the vicinity of New Orleans, on salinities within Breton Sound. The work was conducted for the U.S. Army Corps of Engineers, New Orleans District. The model was used to investigate the significance of pre-construction circulation features on salinities and to aid in the development of a field monitoring program to support a real-time modeling effort. The latter is to be used in directing the operation of the diversion structure after construction.

The screening model framework is part of a series of efforts by NOAA's Ocean Assessments Division to develop a capability to conduct comprehensive assessments of estuaries throughout the USA. As a screening tool, the model provides assessments of system hydrodynamics and order-of-magnitude estimates of water quality behavior. It also serves as a framework for organizing existing information into a modeling context. This latter feature takes advantage of previous work and identifies where research should be directed to refine future modeling efforts. The model is not designed to replicate near-field or short-term processes. It is intended, however, to reproduce the long-term advective and diffusive processes that drive residual circulation patterns and consequently affect the retention time of contaminants within the system.

Calculation of the flow-field component of the modeling framework differs from usual hydrodynamic models in that the residual flow patterns are formulated according to kinematic constraints. Normally such models require solution of geophysical, fluid-dynamic equations. This model assumes an incompressible, irrotational flow, and a flow field that is determined by the shape of the estuarine basin. These features are especially pertinent in estuaries where complex bathymetry exerts a greater influence on controlling currents than that exerted by local dynamics. In cases where dynamics are important or postulated to exist, they can be explicitly incorporated into the model framework by configuring the appropriate internal boundary conditions to reproduce these flow features.

With this framework, water quality profiles are inferred from computed salinity distributions. Salinities are derived by using the standard equation for conservation of mass. The diffusion processes parameterize the effects of wind, tide, and bottom friction. The model is calibrated and verified to accurately reproduce historical salinities, given the appropriate meteorological and hydrological boundary conditions. It is assumed that the same advective and diffusive properties will determine the distribution of contaminants within the estuarine system. Table 1 summarizes the general characteristics of the modeling framework.

The modeling procedure has been standardized for a "generic" estuary and automated for rapid application to individual estuaries, making it easy to implement and cost-effective. It should be stressed that this modeling approach does not replace the need for more sophisticated modeling germane to site-specific issues. It is also not suited to issues needing temporal resolution in time periods of less than a month. However, it can help in assessing the relative merits of alternative estuarine management strategies and provide guidance in the development of the more sophisticated modeling approaches. Plans are to apply the screening model framework to a subset of the 92 estuaries included in NOAA's National Estuarine Inventory (U.S. Department of Commerce 1985).

Model Formulation

The water quality component of the model framework is based upon the usual conservation-of-mass equation. Salinity is the parameter of interest and is represented as the conservative constituent. Assuming that freshwater and salt-

Table 1. Characteristics of the estuarine screening model framework.

Physical Processes:	Advection Diffusion
Water Quality Processes:	Conservative Constituents (Conservation of Mass) Non-conservative Constitutents (First Order Decay)
Model Inputs:	Hydrodynamic (Tide, Freshwater Inflow, Wind) Water Quality (Pollutant Concentration at Boundary Point)
Spatial Domain:	500 Points 2-D in X, Y Plane
Model Formulation:	Incompressible, Irrotational Flow Classical Channel Flow Theory
Mathematical Properties:	Finite Element Technique Implicit Solution
Computer Hardware:	Pre-processing Model Inputs—Apple IIe Model Execution—IBM 9000
Temporal Domain:	Steady State Period of Application Specified According to Flow Period

water are introduced only at the model boundaries, we begin with the usual distribution of variables equation:

$$\frac{\partial S}{\partial t} + \nabla(SV_r) = \nabla \bullet (K\nabla S) \tag{1}$$

If we furthermore assume that none of the variables are functions of the vertical coordinate, equation 1 can be written as

$$D\frac{\partial S}{\partial t} + \nabla(DSV_r) = \nabla \bullet (KD\nabla S) \tag{2}$$

where:

t = Time
D = $D(x,y)$ Depth
V_r = $V(x,y)$ Residual Velocity
S = $S(x,y)$ Salinity
K = $K(x,y)$ Diffusion Coefficient

and

$$x = x \text{ direction}$$
$$y = y \text{ direction}$$

Equation 2 assumes that there are no significant vertical gradients and that currents can be represented as barotropic. Therefore, since equation 2 is a function only of the horizontal coordinates, we may expand it into its component form as:

$$D\frac{\partial S}{\partial t} + \frac{\partial}{\partial x}(DSu_r) + \frac{\partial}{\partial y}(DS\, v_r) = \frac{\partial}{\partial x}\left(KD\frac{\partial S}{\partial x}\right) + \frac{\partial}{\partial y}\left(KD\frac{\partial S}{\partial y}\right) \tag{3}$$

where:

$$u_r = \text{residual velocity in the x direction}$$
$$v_r = \text{residual velocity in the y direction}$$

We now consider how to apply this equation to a typical estuary with river inflow at one end and the open ocean at the other.

The residual flow or the advective component (u_r, v_r) can be approximated kinematically using a stream function representation (Batchelor 1981)

$$Du_r = \frac{\partial \psi}{\partial y} \tag{4}$$

$$Dv_r = \frac{-\partial \psi}{\partial x} \tag{5}$$

Assuming both irrotational and incompressible flow, we have

$$\frac{\partial}{\partial x}\left(\frac{1}{D}\frac{\partial \psi}{\partial x}\right) + \frac{\partial}{\partial y}\left(\frac{1}{D}\frac{\partial \psi}{\partial y}\right) = 0 \tag{6}$$

We can now solve equation 6 with $\psi = 1$ along the right bank of the estuary and $\psi = 0$ along the left bank, yielding a mass-conserving flow pattern which can be easily scaled to river flow velocity (Galt and Payton 1981).
In particular, if ψ and x,y are non-dimensionalized, we can replace equations 4 and 5 with:

$$u_r = V_R\left(\frac{1}{D}\frac{\partial \psi}{\partial y}\right)_r' = V_R u_r' \tag{7}$$

$$v_r = V_R\left(\frac{1}{D}\frac{\partial \psi}{\partial y}\right)_r' = V_R v_r' \tag{8}$$

where:

$$V_R = \text{constant}$$

Since the model is linear, systems with multiple freshwater inflows can be handled by treating each inflow separately. The individual flow patterns are then scaled and added together for the composite field.

Now substituting equations 7 and 8 back into equation 3:

$$\frac{\partial}{\partial x}\left(KD\frac{\partial S}{\partial x}\right) + \frac{\partial}{\partial y}\left(KD\frac{\partial S}{\partial y}\right) = D\frac{\partial S}{\partial t} + DV_R\left(u_r'\frac{\partial S}{\partial x} + v_r'\frac{\partial S}{\partial y}\right) \quad (9)$$

The next problem in the application of equation 9 is how to estimate the diffusion coefficient. Mixing was approximated using a spatially variable eddy coefficient with parameters for the maximum flood tide, residual flow, and wind, and formulated according to equation 10.

$$K_{x,y} = C_t\theta V_t^2 + C_r\theta V_r^2 + C_w\theta^2\frac{W^3}{D} \quad (10)$$

where:

C_t, C_r, C_w = Calibration Coefficients
θ = Tide Period Frequency
W = Wind Speed
V_t = Flood Tide Velocity
V_r = Residual Flow
D = Depth

This approach follows the logic presented by Ketchum (1951) and expanded upon by Arons and Stommel (1951). In Ketchum's tidal prism approach, the mixing is proportional to the ratio of tidal prism volume to total volume. Arons and Stommel assumed that the diffusion coefficient was proportional to the maximum tidal current in the tidal excursion. Following this latter approach we assume:

$$K_t \propto LV_t = C_t\theta|V_t|^2 \quad (11)$$

If V_t is given in m s^{-1} and θ (the tidal period) in seconds, then a value of proportionality of $C_t\theta = 2 \times 10^{-3}$ will give a characteristic eddy coefficient of approximately 10 m^2 s^{-1} (Officer 1983).

Streamline analysis can be used in a similar fashion to calculate the maximum tidal currents as follows:

$$u_t = V_T\left(\frac{A}{D}\frac{\partial\psi}{\partial y}\right)_t' = V_T\mu_t' \quad (12)$$

$$v_t = -V_T\left(\frac{A}{D}\frac{\partial\psi}{\partial x}\right)_t' = V_T v_t' \quad (13)$$

Then,

$$|V_y|^2 = V_T^2\left(\frac{A^2}{D^2}\left(\frac{\partial\psi}{\partial x}\right)^2 + \left(\frac{\partial\psi}{\partial y}\right)^2\right)^2 = V_T^2(u_t'^2 + v_t'^2) \tag{14}$$

where:

$V_T =$ constant
$A =$ amplitude factor

The amplitude factor is used to allow for divergence associated with the spatial attenuation of the tidal prism volume. It is prescribed in such a manner as to allow a smooth gradient to represent the fractional divergence associated with the tidal wave. To satisfy these conditions, the amplitude factor is assumed to satisfy Laplace's equation with Dirichlet conditions—setting the value to zero at the head of the bay and unity at the open ocean. In areas of complex geometry, different amplitude factors can be assigned to numerous internal points to proportion the tidal prism volume by system segment. Substituting equation 14 into equation 11 yields:

$$K_t = C_t\theta V_T^2\,(u_t'^2 + v_t'^2) \tag{15}$$

In a similar fashion, the second component of the diffusion equation relates the square of the residual flow to diffusion. However, its significance is usually limited to the immediate area of the freshwater inflow where velocities are higher due to the constricted cross-sectional areas.

To extend the usefulness of our analysis, we now consider what would happen in the case where the characteristic tidal velocity becomes small, that is, $V_t < 1$ m s^{-1}. For this case, the diffusion must depend on other physical processes, and we would like the formulation to default to an alternate case. For an unstratified, non-tidal lagoon system, we would expect that wind waves would provide the major turbulent energy. Therefore, we expect the diffusion coefficient to be proportional to the kinetic energy generated by the wind-induced shear (Kraus 1977).

$$K_w \propto \frac{W^3}{D} \tag{16}$$

To convert equation 16 into dimensional form, we use a time scale (i.e., θ^2) and assume that the actual dissipation rate is constant:

$$K_w = C_w\theta^2\left(\frac{W^3}{D}\right) \tag{17}$$

Assuming a characteristic eddy coefficient of 0.1 m^2 s^{-1} for a 10 m s^{-1} wind and 10 m depth, we get an initial estimate of $C_w\theta^2 = 5 \times 10^{-5}$ s^2. Thus, equation 10 can be expanded as follows:

$$K_{x,y} = C_t\theta V_t^2 (u_t'^2 + v_t'^2) + C_r\theta V_R^2 (u_r'^2 + v_r'^2) + C_w\theta^2 \frac{W^3}{D} \quad (18)$$

We are now ready to solve equation 9. We initially consider the steady-state case where a single river and tidal currents are represented using streamline analysis. For this case V(x,y) and A(x,y) are given, and we know from equations 7 and 8:

$$u_r' = \left(\frac{1}{D}\frac{\partial\psi}{\partial y}\right)' \quad (19)$$

$$v_r' = \left(\frac{1}{D}\frac{\partial\psi}{\partial x}\right)' \quad (20)$$

and from equation 14:

$$(u_t'^2 + v_t'^2) = \frac{A^2}{D^2}\left(\left(\frac{\partial\psi}{\partial x}\right)^2 + \left(\frac{\partial\psi}{\partial y}\right)^2\right) \quad (21)$$

At steady state, equation 9 becomes:

$$DV_R\left(u_r'\frac{\partial S}{\partial x} + v_r'\frac{\partial S}{\partial y}\right) - \frac{\partial}{\partial x}\left(KD\frac{\partial S}{\partial x}\right) - \frac{\partial}{\partial y}\left(KD\frac{\partial S}{\partial y}\right) = 0 \quad (22)$$

Since equation 22 is homogenous in S, we may scale the problem such that river salinities are always zero and the open ocean salinities are always one. To ensure that freshwater inflows are not affecting the open ocean boundary, the latter must be located sufficiently seaward. This gives the boundary conditions on equation 22 as

$$S(x,y)\text{river} = 0$$
$$S(x,y)\text{ocean} = 1$$

In addition, the x and y coordinates can be scaled such that:

$$x = Lx'$$
$$y = Ly'$$

where L is the length scale of the map used in the streamline analysis. Thus, equation 22 becomes

$$DV_RL\left(u_r'\frac{\partial S'}{\partial x'} + v_r'\frac{\partial S'}{\partial y'}\right) - \frac{\partial}{\partial x'}\left(KD\frac{\partial S'}{\partial x'}\right) - \frac{\partial}{\partial y'}\left(KD\frac{\partial S'}{\partial y'}\right) = 0 \quad (23)$$

where all primed quantities are non-dimensional. We can solve the above using finite element techniques employing first-order triangular topology to describe both dependent and independent variables (Zienkiewicz 1971). The model is then calibrated to reproduce a known salinity field by adjusting the coefficients in equation 10.

The modeling software consists of 15 sub-programs covering all phases of application from digitization of system geometry to water quality predictions. The programs are written in Fortran 77, the largest being the matrix solver that requires 75K memory. The framework is set upon an IBM 9000 micro-computer. Data entry is performed using a digitizing pad interfaced with an Apple IIe micro-computer. The latter is hard-wired to the IBM for data uploading.

Application to Breton Sound

Breton Sound, near the mouth of the Mississippi River (Fig. 1), has been an important source of oyster, shrimp, and other commercially valuable seafood for several decades. However, its productivity has been threatened by a number of conditions: confinement of the lower Mississippi River by levees, dredging, subsidence of alluvial sediments, and shoreline erosion. Each of these factors

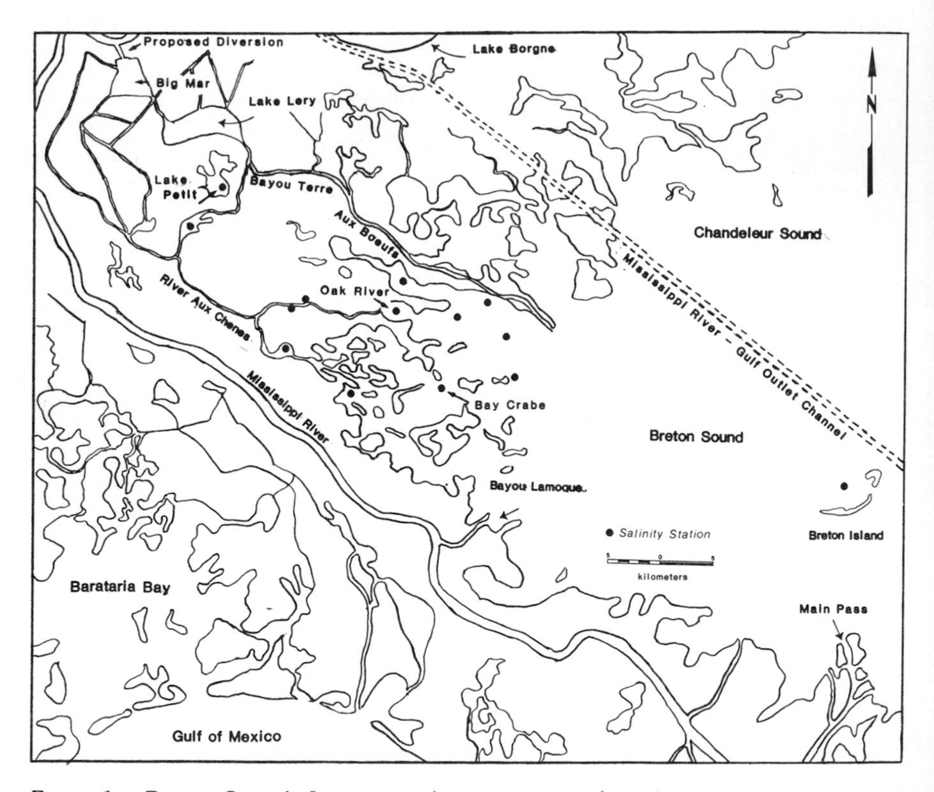

Figure 1. Breton Sound, Louisiana, showing stations for salinity measurements and location of proposed freshwater diversion (upper left).

contributes significantly to land loss, and also promotes salt water intrusion. Increased salinity levels have resulted in the gradual conversion of fresh, intermediate, and brackish wetlands and marsh to more saline types, causing loss, alteration, and relocation of habitats.

The Louisiana Department of Wildlife and Fisheries has determined desirable salinity levels for each month based upon historical oyster productivity (Chatry and Dugas 1983). The U.S. Army Corps of Engineers (1985), in designing the freshwater diversion, is evaluating how best to operate the structure to achieve these salinities. The modeling framework was used to identify the dominant circulation processes affecting the distribution of freshwater within the system. This was accomplished by addressing the sensitivity of various transport processes on existing salinity distributions. Throughout the modeling process, the Corps provided critical comment and review to help assess the significance of previous studies and determine how best to reflect them in the modeling framework.

Boundary Conditions

The model boundary (Fig. 2) extends 76 km from the marsh headwaters to a seaward boundary that lies on a transect extending from approximately the Mississippi River Gulf Outlet to Main Pass. A total of 199 points are reflected in

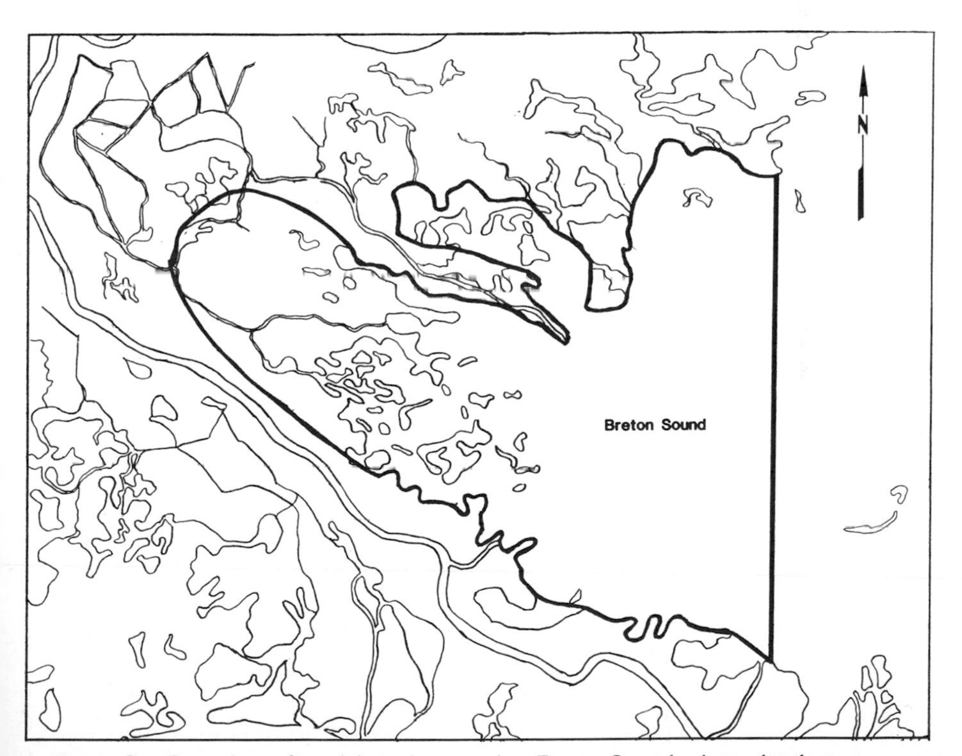

Figure 2. Boundary of model used to simulate Breton Sound salinity distributions.

the model grid (Fig. 3). They were located to capture pertinent bathymetric features; greater intensity was needed in the marsh areas to depict the numerous interconnecting canals.

The model was calibrated using monthly freshwater inflow data for 1980, and by comparing the modeling prediction with historical salinities for this same period, taken at 13 stations (Fig. 1) established by the State of Louisiana (van Beek *et al.* 1984). Table 2 shows the salinity, discharge, and prevailing wind conditions for a subset of these stations. The 1980 time period was selected because its wide range of freshwater inflow conditions allowed a study of the sensitivity of various factors on salinity.

The model was applied using monthly residual flows. The upper basin discharge was based on water budget calculations using observed precipitation and temperatures to produce average monthly rates of water surplus or deficit. Bayou Lamoque discharge was estimated from river gages upstream and downstream of the diversion and was adjusted to reflect freshwater run-off from the lower portion of the system. The actual freshwater inflow used for a particular month was lagged to reflect the influence of previous inflow in addition to the prevailing inflow during that month (U.S. Army Corps of Engineers 1985).

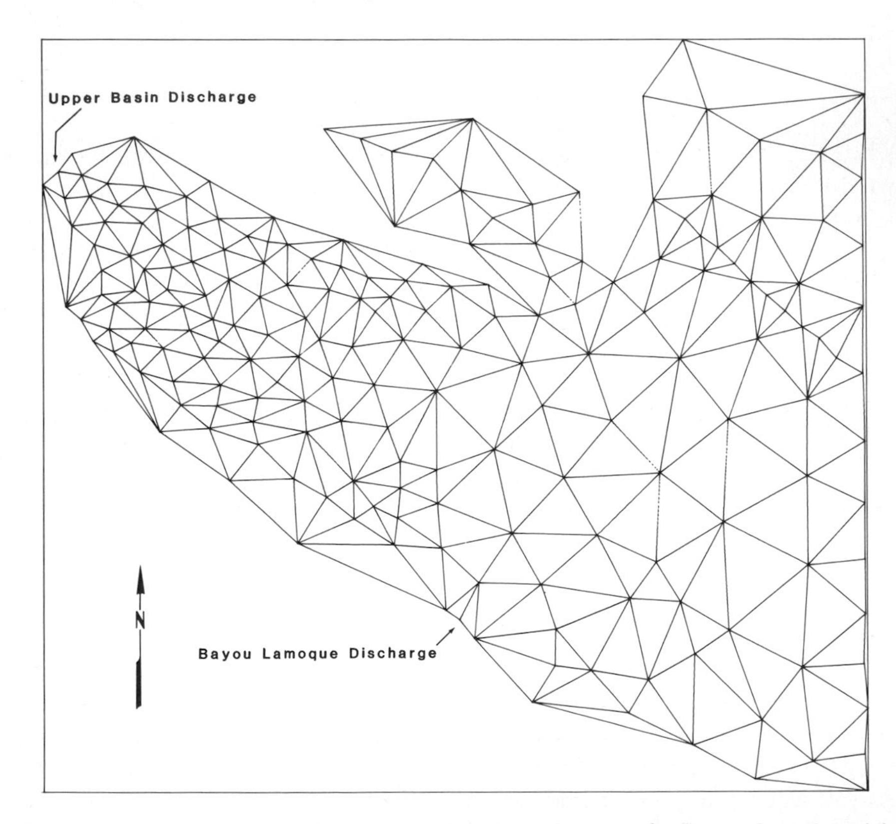

Figure 3. Finite element grid and principal discharge locations for Breton Sound model.

Table 2. Monthly freshwater discharge, salinity, and wind characteristics for Breton Sound, Louisiana in 1980. Wind data are from National Weather Service. Other data are from van Beek *et al.* (1984).

	Discharge (m^3 s^{-1})		Salinity (‰)			Wind	
	Upper Basin	Bayou Lamoque	Lake Petit	Mouth of Oak River	Bay Crabe	Direction*	Speed (m s^{-1})
Jan	30.99	244.91	6.7	9.2	14.8	60	3.6
Feb	22.27	223.85	5.3	10.2	15.2	60	3.8
Mar	28.89	199.52	7.4	12.2	17.6	90	4.7
Apr	67.44	281.64	3.7	6.8	10.3	200	4.2
May	98.43	333.71	2.3	4.2	9.8	150	3.6
Jun	55.67	252.18	3.0	7.6	13.5	220	3.5
Jul	−0.62	143.62	6.3	15.1	21.1	270	2.7
Aug	−22.27	148.29	10.1	17.1	26.0	130	2.7
Sep	−1.39	140.42	12.0	18.6	24.8	90	2.5
Oct	14.86	126.33	10.5	19.4	22.0	40	2.8
Nov	23.66	81.28	8.5	15.6	18.8	20	3.0
Dec	10.41	54.56	9.5	13.5	16.3	—	—

*Degrees from true North.

Since the model is linear, the composite pattern can be represented as a composite of individual flow scenes. This allows for the easy incorporation of additional hydrodynamic features if warranted. In the application of the modeling framework to Breton Sound, three individual flow scenes were used to create the composite residual pattern. Two of the flow patterns were reflective of discharges from the upper basin and Bayou Lamoque diversion. Their discharge points into the model grid are shown in Fig. 3, and their flow patterns are shown in Fig. 4 and 5, respectively.

Hart (1976) applied a tidal model to Breton/Chandeleur Sound. He concluded that tidal energy and related volume flow entered Breton/Chandeleur Sound at both its northern and southern inlets, but exited primarily through the southern inlet. This action produced a tide-induced residual flow pattern from north to south and resulted in the advection of higher-salinity Chandeleur water into Breton Sound. The water is then topographically guided along the Delta to the south.

In more recent work, Schroeder *et al.* (1985) investigated the influences of recurrent cold air outbreaks on circulation patterns east of the Mississippi River Delta using 12 years of Landsat multispectral scanner images. Such low-frequency meterological events were associated with water being forced south through the Breton/Chandeleur Sound, drawing water from the shelf region south of the Mississippi-Alabama barrier islands. This latter phenomenon may be enhancing the astronomically-induced residual during fall, winter, and spring, the time periods associated with cold air outbreaks.

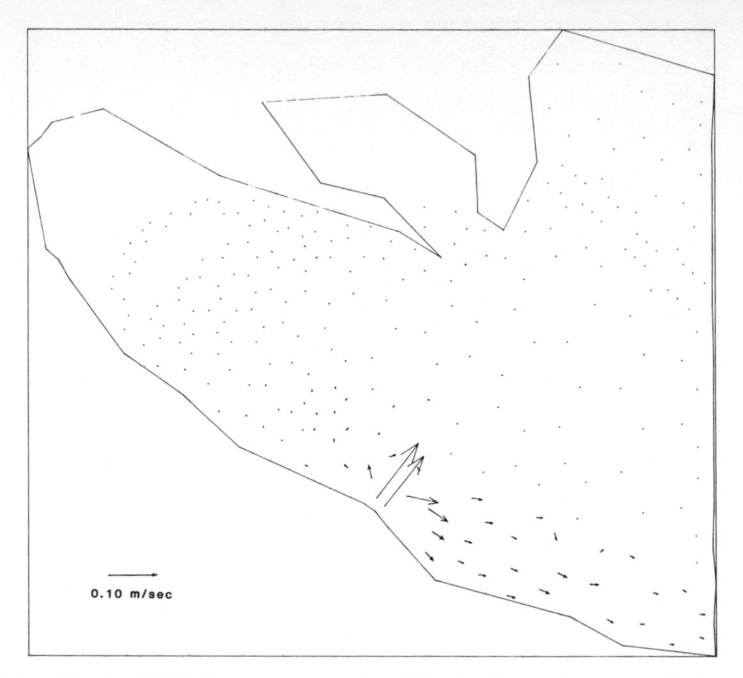

Figure 4. *October 1980 flow pattern predicted for the Bayou Lamoque discharge component of total flow.*

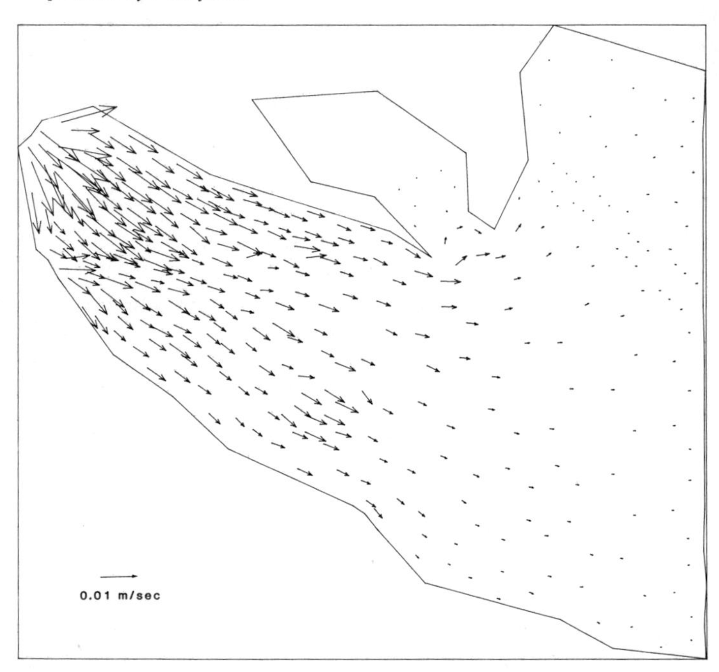

Figure 5. *October 1980 flow pattern predicted for the Upper Basin discharge component of total flow.*

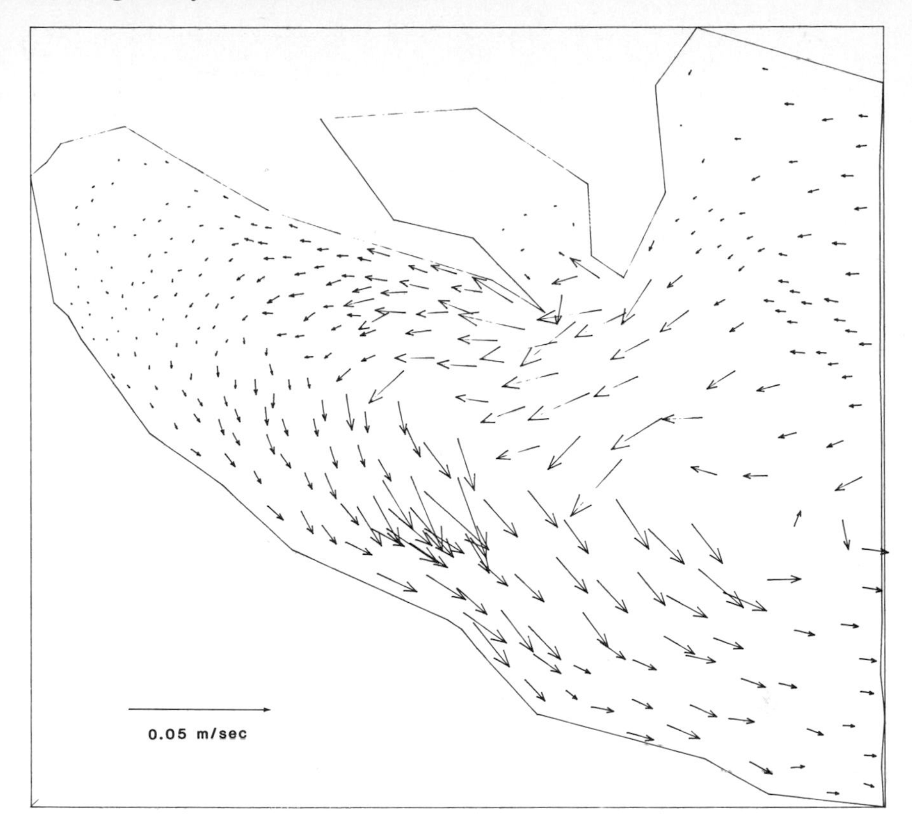

Figure 6. Tide-induced residual flow component for Breton Sound, independent of discharge-induced flows.

Based on the work of Hart (1975) and Schroeder *et al.* (1985), a third flow pattern was constructed depicting a residual flow from north to south (Fig. 6). The volume transport associated with this flow was based upon the portion of the tidal prism volume being advected through the system as determined by the phase lag between the northern and southern inlets. Standard NOAA tide tables were used to calculate the phase lag.

The tidal residual flow pattern was assumed to remain the same for each month. Based on 1980 monthly flow values, however, the discharge profiles varied by month. A typical composite residual flow pattern is shown in Fig. 7 for October of 1980.

The diffusion component, as formulated according to equation 10, is proportional to the squares of the composite residual flow and the maximum flood tide velocities, and to the cube of the average wind speed. The maximum flood tide profile (Fig. 8) is based on displacing the tidal prism volume throughout the system according to system geometry and spatial fluctuations in tide range. The approach is to segment the system based on co-phase lines and determine its volume based on tidal fluctuations.

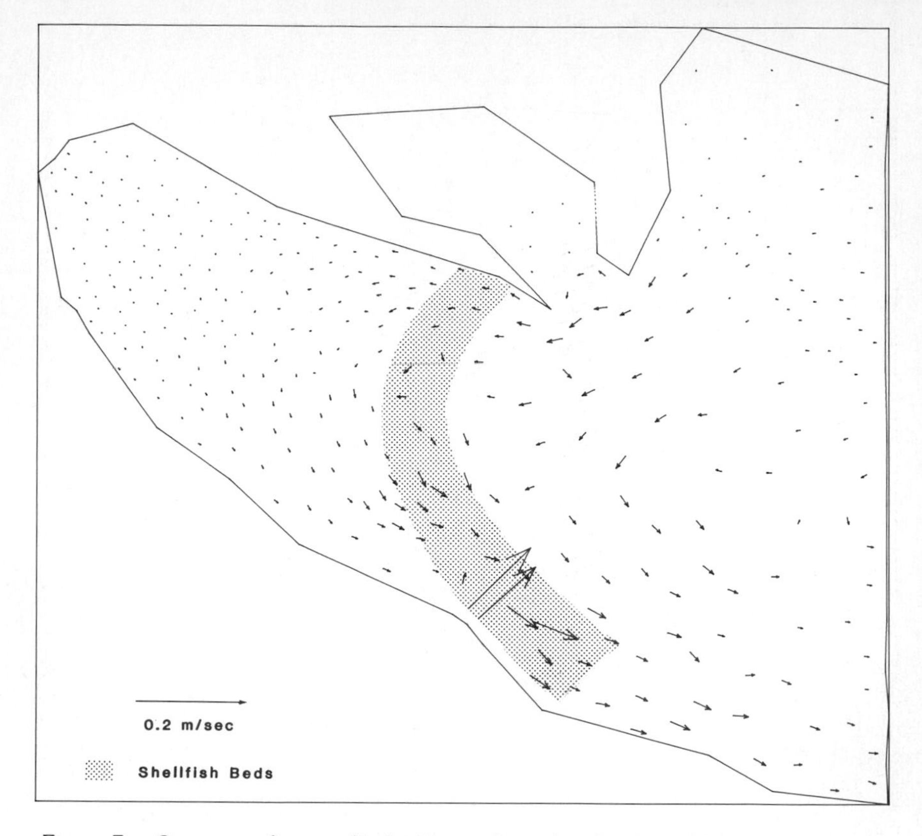

Figure 7. Composite flow profile for Breton Sound in October 1980.

Calibration

Using the advection and diffusion components as formulated above, the model was calibrated/verified to reproduce monthly salinities for 1980, resulting in the following diffusion coefficients:

Tidal Diffusion = $0.859 \times 10^2\ m^2\ s^{-1}$
Residual Flow Diffusion = $0.254 \times 10^2\ m^2\ s^{-1}$
Wind Diffusion = $1.02 \times 10^2\ m^2\ s^{-1}$

Figures 9 and 10 show comparisons between model-predicted salinities and historical data for April and August of 1980, which typify high and low flow periods respectively.

In general, modeled predictions reproduce both the structure and absolute value of salinities. However, the model consistently underestimated salinities at Bay Crabe (Fig. 1) and, during periods of low flow, near the upper basin discharge.

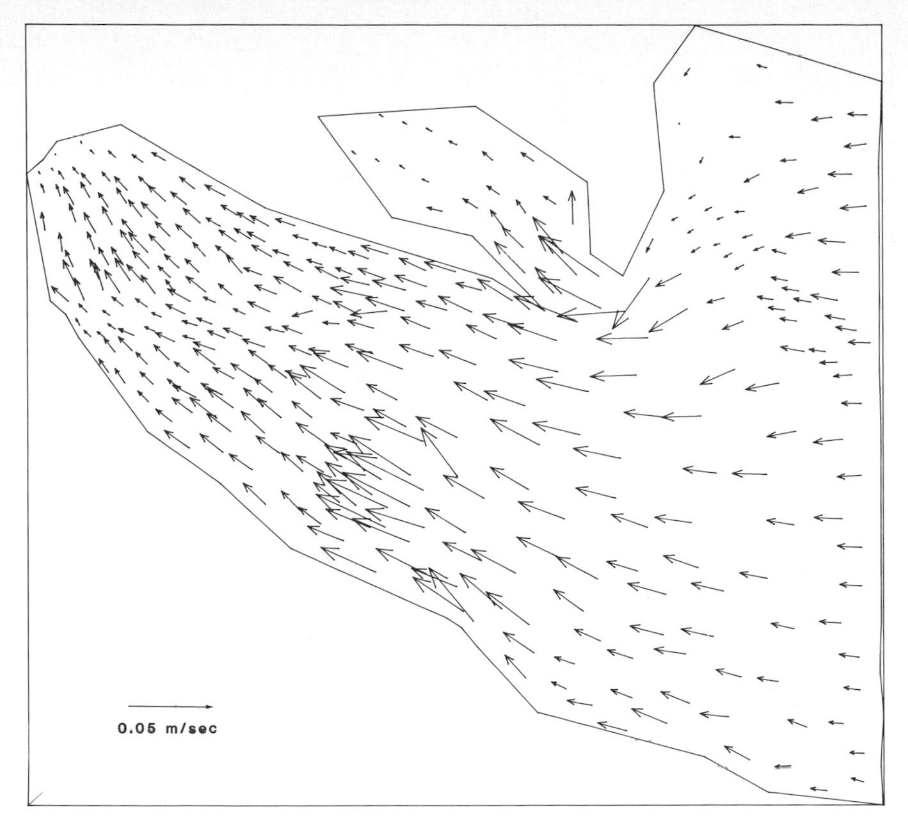

Figure 8. Flow profile for Breton Sound on flood tide.

Closer inspection of the bathymetry in the immediate area surrounding Bay Crabe indicated the presence of certain detailed bathymetric features not captured in the model. This brackish water area tended to have limited access to adjacent open waters, which the model did not resolve. As a result, the model induced high flushing rates that resulted in decreased salinities. An additional factor, possibly enhancing the difference, is the differential effect of evapotranspiration on sluggish water masses.

The salinity discrepancy in the vicinity of the upper basin during low flow is attributable to the fact that the model assigns zero salinity to all discharge points. During low-flow the salinity probably intruded further up-basin, beyond the present model boundary for freshwater inflow.

The model boundary was, however, considered reasonable since the upper basin run-off was concentrated at this discharge point. The only effect was to sacrifice near-field salinity resolution at the point of discharge. This was not observed during high flow since the freshwater inflow extended beyond the discharge point. These discrepancies between observed and predicted salinities were localized and did not affect the overall ability of the model to reproduce the temporal and spatial nature of the salinity regime.

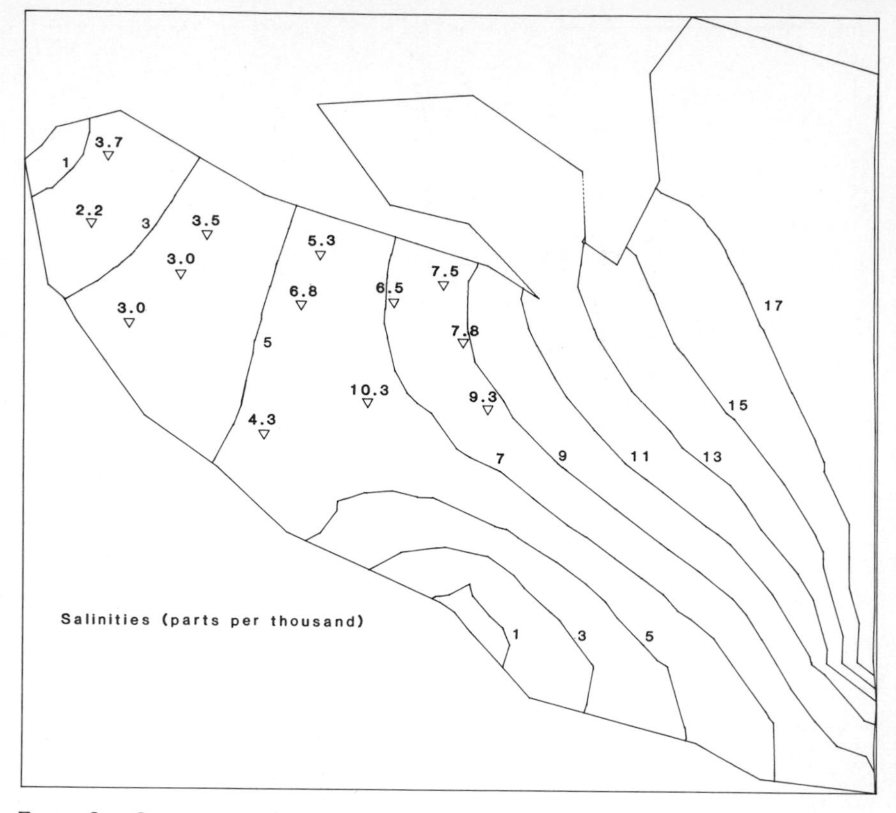

Figure 9. *Comparison of salinity isopleths predicted by the model for Breton Sound and measured salinities (triangles) for high-flow period (April 1980).*

Results and Discussion

The modeling application was used in Breton Sound to identify the dominant circulation features affecting salinity distribution, particularly within the historical oyster reef area (Fig. 7). The purpose was to design a monitoring program to support a real-time modeling effort to be used in directing operations for the proposed flow diversion.

The model calibration/verification presented in the previous section, suggests that the advective and diffusive processes, as formulated, adequately resolve the spatial and temporal variability of the salinity regime. Figure 7 is characteristic of the individual monthly composite residual flow patterns. Scaling of the velocities may differ from month to month, but the relative contributions from the individual component flow patterns remain virtually the same. Resolution of the portion of the current pattern within the oyster reef area is accomplished by reference to the individual component flow patterns (Figs. 4-6). It is important to realize that the magnitude of the vectors in Figs. 4-6 represent velocities and not mass flux. To estimate flux, the velocities must be reconciled with the bathymetry.

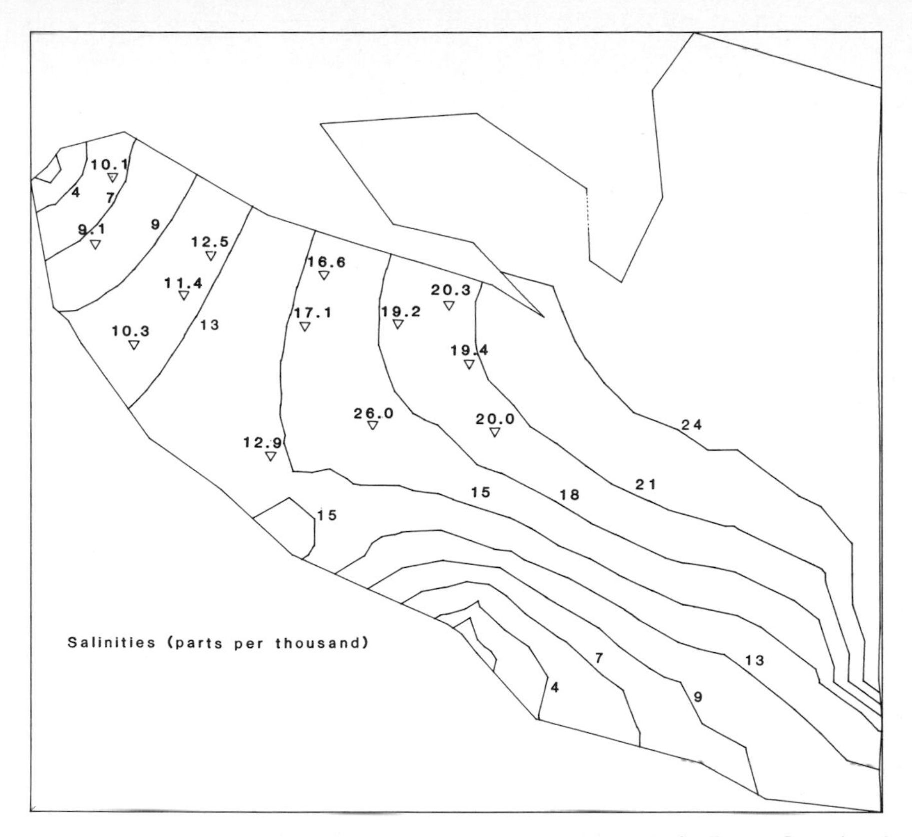

Figure 10. Comparison of salinity isopleths predicted by the model for Breton Sound and measured salinities for low-flow period (August 1980).

The northern portion of the flow pattern within the shaded area (Fig. 7) is a result of the tidally-induced residual flow pattern (Fig. 6) and its configuration is the result of bathymetric guidance. Maintenance of the desired salinity ranges in Breton Sound must therefore account for the influence of water for Chandeleur Sound and the factors affecting its salinities. During calibration/verification, the boundary salinities associated with Chandeleur Sound were set by field data; however, future modeling efforts should be expanded to incorporate the entire Chandeleur Sound.

The mid-zone of the shaded area (Fig. 7) exhibits a transition as Chandeleur water continues to rotate counter-clockwise in a more southerly direction and begins to combine with upper basin and Bayou Lamoque discharges. This interaction causes a highly complex flow pattern, exhibiting eddy formations. The latter may cause higher retention times and salinities as manifested in the observed Bay Crabe salinities.

The circulation in the southern portion is the result of waters piling up and being forced along the Mississippi Delta as Chandeleur, upper basin, and Bayou Lamoque waters combine. The two dominant vectors in this portion (Fig. 7) reflect the near-field Bayou Lamoque discharges and are not representative of overall southward and eastward transport in the area.

Conclusions

The circulation of Breton Sound is not dominated by any one transport process. Many transport processes interact to form unique zones within the area. These interactions and the ways they can affect salinity should be the focus of future monitoring programs. The regulated release of upper basin discharge may, at times, cause greater entrainment of Chandeleur water, inducing a greater flux of salt into the northern zone, and producing an effect opposite to the desired one. Such scenarios should be the basis of design for a monitoring program.

In summary, this modeling framework has demonstrated that it can readly synthesize available information and, upon calibration/verification, be used to investigate the sensitivity of certain hydrodynamic features on the distribution of salinity in estuaries. In its application of Breton Sound, the model helped to identify and resolve the influence of various transport features on salinity distributions. The results suggested that the field monitoring program should study in detail the temporal and spatial characteristics of entrainment of water from Chandeleur Sound and the factors influencing this entrainment.

Acknowledgments

The modeling application was funded by the U.S. Army Corps of Engineers, New Orleans District. We are particularly grateful to Tony Drake, the Corps Project Officer, for his help and insights. We also extend our thanks to Ned Burger and Paul Orlando, both of NOAA's Ocean Assessments Division, for their contribution to the quality of this project.

References Cited

Arons, A. B. and H. Stommel. 1951. A mixing length theory of tidal flushing. *Trans. Am. Geophys. Union* 32:419-421.

Batchelor, G. K. 1981. *An Introduction to Fluid Dynamics.* Cambridge University Press, New York. 615 pp.

Chatry, M. and R. J. Dugas. 1983. Optimum salinity regime for oyster productivity on Louisiana's state seed grounds, pp. 81-94. *In: Contributions to Marine Science, Volume 26.* University of Texas, Port Aransas Marine Laboratory.

Galt, J. A. and D. L. Payton. 1981. Finite element routines for the analysis and simulation of near shore circulation. pp. 121-132. *In:* Proceedings of the International Symposium on the Mechanics of Oil Slicks. September 2-5, 1981. Editions Anciens ENPC, Paris, France.

Hart, W. E. 1976. A numerical study of currents, circulation, and surface elevations in Chandeleur-Breton Sound, Louisiana. Ph.D. dissertation. Louisiana State University, Baton Rouge. 154 pp.

Ketchum, B. H. 1951. The exchange of fresh and salt water in tidal estuaries. *J. Mar. Res.*, 10:18-38.

Kraus, E. B. 1977. *Modeling and Prediction of the Upper Layers of the Ocean.* Pergamon Press, Oxford. 325 pp.

Officer, C. B. 1983. Physics of Estuarine Circulation, pp. 13-39. *In:* B. H. Ketchum (ed.), *Ecosystems of the World, Estuaries and Enclosed Seas.* Elsevier Scientific Publishing Company, New York.

Schroeder, W. W., O. K. Huh, L. J. Rouse Jr. and W. J. Wiseman Jr. 1985. Satellite observations of the circulation east of the Mississippi Delta: cold-air outbreak conditions. *Remote Sensing of Environment.* 18:49-58.

U.S. Army Corps of Engineers. 1985. Hydrology and Hydraulics. (Appendix A). *Mississippi Delta Region, Salinity Control Structure, Design Memorandum No. 1, Caernarvon Freshwater Diversion Structure.* New Orleans Distrit. Multiple pagination.

U.S. Department of Commerce. 1985. *National Estuarine Inventory. Data Atlas Vol. 1 Physical and Hydrologic Characteristics.* Ocean Assessments Division, National Oceanic and Atmospheric Administration. Washington, D.C. Multiple paginatic.ı.

van Beek, J. L., D. Roberts and T. Duenkel. 1984. *A Management Plan for Freshwater Diversion at Caernarvon, Louisiana.* Plaquemines Parish Commission Council, Louisiana. 77 pp.

Zienkiewicz, O. C. 1971. *The finite element method in Engineering Science.* McGraw-Hill Publishing Co., New York. 521 pp.

INDEX